
For further volumes:
http://www.springer.com/series/5584

Bernhard Kadenbach
Editor

Mitochondrial Oxidative Phosphorylation

Nuclear-Encoded Genes, Enzyme Regulation, and Pathophysiology

Springer

Editor
Bernhard Kadenbach
Philipps-Universitaet Marburg
Marburg, Germany

ISBN 978-1-4614-3572-3 ISBN 978-1-4614-3573-0 (eBook)
DOI 10.1007/978-1-4614-3573-0
Springer New York Heidelberg Dordrecht London

Library of Congress Control Number: 2012938858

Printed on acid-free paper

Springer is part of Springer Science+Business Media (www.springer.com)

Preface

ATP, the universal energy source of life, is in large part synthesized in the mitochondria by the oxidative phosphorylation (OxPhos) process. After identification of the mitochondrial metabolic pathways and the complexes of OxPhos, i.e., respiratory chain complexes I–IV and ATP synthase (complex V) in the second half of the last century, bioenergetic research concentrated on bacterial OxPhos enzymes, which contain fewer protein subunits but fulfill the same catalytic activity. Mitochondria came again into focus after identification of new mitochondrial functions like apoptosis, heat generation via uncoupling proteins and passive proton leak, formation and degradation of reactive oxygen species (ROS), and their involvement in numerous "mitochondrial diseases." This book focuses on new developments in mitochondrial morphology, biogenesis, and evolution, as well as in structure and regulatory functions of the enzyme complexes of OxPhos by experts in their fields. Particular emphasis is given to the oxygen consuming enzyme of the respiratory chain, cytochrome *c* oxidase, representing a controlling step of OxPhos. The book also opens up a new field of research, the physiological regulation of mitochondrial energy synthesis. In addition, many newly identified mitochondrial diseases based on nuclear-encoded genes and proteins are reviewed.

The editor owes thanks to the authors who contributed to *Mitochondrial Oxidative Phosphorylation: Nuclear-Encoded Genes, Enzyme Regulation, and Pathophysiology* and a great debt of gratitude for their patience with the editor. Maik Hüttemann is gratefully acknowledged for various helpful suggestions. Many thanks also to Rabia Ramzan who was of invaluable help in preparing the book and to my colleague and collaborator Sebastian Vogt who always spread good cheer, inspired our research, and supported the writing of this book.

Marburg, Germany Bernhard Kadenbach

Contents

Contributors

Susanne Arnold Institute for Neuroanatomy, Faculty of Medicine, RWTH Aachen University, Aachen, Germany

Ulrich Brandt Molecular Bioenergetics Group, Medical School, Cluster of Excellence Frankfurt "Macromolecular Complexes", Center for Membrane Proteomics, Johann Wolfgang Goethe-Universität, Frankfurt am Main, Germany

Matthieu Depuydt Zellbiologie, Universität Kaiserslautern, Kaiserslautern, Germany

Salvatore DiMauro Department of Neurology, Columbia University Medical Center, College of Physicians & Surgeons, New York, NY 10032, USA

Jeffrey W. Doan Center for Molecular Medicine and Genetics, Wayne State University School of Medicine, Detroit, MI, USA

Stefan Dröse Molecular Bioenergetics Group, Medical School, Cluster of Excellence Frankfurt "Macromolecular Complexes", Center for Membrane Proteomics, Johann Wolfgang Goethe-Universität, Frankfurt am Main, Germany

Andrew Ferree Department of Pharmacology, and Department of Neuroscience, Boston University School of Medicine, Boston, MA, USA

Maria Luisa Genova Department of Biochemistry "G. Moruzzi", University of Bologna, Bologna, Italy

Daniele Ghezzi Unit of Molecular Neurogenetics, The Foundation "Carlo Besta" Institute of Neurology, Milan, Italy

Lawrence I. Grossman Center for Molecular Medicine and Genetics, and Cardiovascular Research Institute, Wayne State University School of Medicine, Detroit, MI, USA

Karmanos Cancer Institute, Detroit, MI, USA

Johannes M. Herrmann Zellbiologie, Universität Kaiserslautern, Kaiserslautern, Germany

Maik Hüttemann Center for Molecular Medicine and Genetics, and Cardiovascular Research Institute, Wayne State University School of Medicine, Detroit, MI, USA

Karmanos Cancer Institute, Detroit, MI, USA

Bernhard Kadenbach Fachbereich Chemie, Philipps-University Marburg, Marburg, Germany

Icksoo Lee Center for Molecular Medicine and Genetics, Wayne State University School of Medicine, Detroit, MI, USA

Giorgio Lenaz Department of Biochemistry "G. Moruzzi", University of Bologna, Bologna, Italy

Thierry Letellier Laboratoire de Physiopathologie Mitochondriale, INSERM, Université Victor Segalen Bordeaux 2, Bordeaux, France

Sebastian Longen Zellbiologie, Universität Kaiserslautern, Kaiserslautern, Germany

Kazumasa Muramoto Department of Life Science, University of Hyogo, Hyogo, Japan

Denis Pierron Center for Molecular Medicine and Genetics, Wayne State University School of Medicine, Detroit, MI, USA

Laboratoire de Physiopathologie Mitochondriale, INSERM, Université Victor Segalen Bordeaux 2, Bordeaux, France

Richard K. Porter School of Biochemistry and Immunology, Trinity College Dublin, Dublin 2, Ireland

Rabia Ramzan Biomedical Research Center, Cardiovascular Laboratory, Philipps-University, Marburg, Germany

Thomas H. Sanderson Cardiovascular Research Institute, Wayne State University School of Medicine, Detroit, MI, USA

Department of Emergency Medicine, Wayne State University School of Medicine, Detroit, MI, USA

Eric A. Schon Department of Neurology, and Department of Genetics and Development, Columbia University Medical Center, New York, NY, USA

Kyoko Shinzawa-Itoh Department of Life Science, University of Hyogo, Hyogo, Japan

Orian Shirihai Department of Medicine, Boston University School of Medicine, Boston, MA, USA

Kurenai Tanji Department of Neurology, and Department of Pathology and Cell Biology, Columbia University Medical Center, New York, NY, USA

Sebastian Vogt Biomedical Research Center, Cardiovascular Laboratory, Philipps-University, Marburg, Germany

Petra Weber Biomedical Research Center, Cardiovascular Laboratory, Philipps-University, Marburg, Germany

Daniel Weckbecker Zellbiologie, Universität Kaiserslautern, Kaiserslautern, Germany

Derek E. Wildman Center for Molecular Medicine and Genetics, Wayne State University School of Medicine, Detroit, MI, USA

Margaret T.T. Wong-Riley Department of Cell Biology, Neurobiology and Anatomy, Medical College of Wisconsin, Milwaukee, USA

Shinya Yoshikawa Department of Life Science, University of Hyogo, Hyogo, Japan

Massimo Zeviani Unit of Molecular Neurogenetics, The Foundation "Carlo Besta" Institute of Neurology, Milan, Italy

Chapter 1
Introduction to Mitochondrial Oxidative Phosphorylation

Bernhard Kadenbach

Abstract The basic mechanism of ATP synthesis in the mitochondria by oxidative phosphorylation (OxPhos) was revealed in the second half of the twentieth century. The OxPhos complexes I–V have been analyzed concerning their subunit composition, genes, and X-ray structures. This book presents new developments regarding the morphology, biogenesis, gene evolution, heat, and reactive oxygen species (ROS) generation in mitochondria, as well as the structure and supercomplex formation of OxPhos complexes. In addition, multiple mitochondrial diseases based on mutations of nuclear-encoded genes have been identified. Little is known, however, of the regulation of OxPhos according to the variable cellular demands of ATP. In particular, the functions of the supernumerary (nuclear-encoded) subunits of mitochondrial OxPhos complexes, which are mostly absent in bacteria, remain largely unknown, although the corresponding and conserved core subunits exhibit the same catalytic activity. Identification of regulatory pathways modulating OxPhos activity, by subunit isoform expression, by allosteric interaction with ATP/ADP, by reversible phosphorylation of protein subunits, or by supercomplex formation, will help to understand the role of mitochondria in the many degenerative diseases, mostly based on ROS formation in mitochondria and/or insufficient energy production.

1.1 Mitochondrial Oxidative Phosphorylation

Life represents an exergonic process requiring energy for all of its manifestations. Aerobic organisms, including plants, fungi, and animals, use the high energy yield of the oxyhydrogen gas reaction ($2\ H_2 + O_2 \rightarrow 2\ H_2O$; $\Delta G^{\circ\prime} = -193$ kJ/mol under physiological conditions) to drive the endergonic synthesis of ATP from ADP and

B. Kadenbach (✉)
Fachbereich Chemie, Philipps-University Marburg, Marburg 35032, Germany
e-mail: kadenbach@staff.uni-marburg.de

B. Kadenbach (ed.), *Mitochondrial Oxidative Phosphorylation*, Advances in Experimental Medicine and Biology 748, DOI 10.1007/978-1-4614-3573-0_1,

phosphate. For many years the *energy-rich intermediate* of energy transduction between the electron transport in the respiratory chain of mitochondria (and bacteria) and the chemical synthesis of ATP from ADP and phosphate was a matter of controversial discussions until Peter Mitchell's *Chemiosmotic Hypothesis* (Mitchell 1961, 1966) was finally accepted by the leading researchers in the field of oxidative phosphorylation (OxPhos) (Boyer et al. 1977).

Coupled to oxygen consumption in the mitochondria, energy is released in the respiratory chain in three steps and transiently stored as an electrochemical proton gradient across the inner membrane. The gradient is dissipated by the ATP synthase (complex V) to generate ATP (Rees et al. 2009), or converted into heat by passive proton leakage (Jastroch et al. 2010), or through uncoupling proteins (see Chap. 7). The changes of free energy ($\Delta G^{o\prime}$) at the three steps are 52 kJ/mole at complex I (NADH dehydrogenase), 42 kJ/mole at complex III (cytochrome bc_1), and 100 kJ/mole at complex IV (cytochrome oxidase, COX) (Lehninger 1970). In the last step the yield of free energy is twice as high compared to the other two steps, suggesting a regulatory role for COX in the electron transport chain of mitochondria, and thus in OxPhos. In general, enzyme reactions involving large changes of free energy represent regulatory steps of energy metabolism. In fact, COX was found to represent the rate-limiting step of respiration in intact cells (Villani and Attardi 1997, 2001; Piccoli et al. 2006; Dalmonte et al. 2009; Pacelli et al. 2011), but not in isolated mitochondria (Groen et al. 1982). Although many oxidases react with molecular oxygen, only COX converts O_2 into water without forming reactive oxygen species (ROS) like the superoxide radical anion O_2^- or hydrogen peroxide H_2O_2 (Ludwig et al. 2001; Yu et al. 2011) (see Chap. 9). However, significant amounts of ROS can be produced in the mitochondria at complexes I and III (see Chap. 6).

1.2 Mitochondria, a Genetically Independent Cell Organelle

According to the generally accepted endosymbiotic theory (Wallin 1923; Margulis 1975; Zimmer 2009; Richards and Archibald 2011) mitochondria originated from symbiosis of an oxidative α-proteobacterium with an anaerobic pre-eukaryotic (host) cell 1.5–2 billion years ago. The resulting eukaryotic cell generates ATP mainly by OxPhos in the mitochondria, which yields about 15 times more ATP from glucose compared to glycolysis under anaerobic conditions. During evolution most of the genetic material of the ancestral oxidative α-proteobacterium was transferred to the nucleus of the host cell. Furthermore, during evolution eukaryotic cells accumulated additional genes and functions. A genome remained in mitochondria, which is independent of nuclear DNA. Mitochondrial DNA (mtDNA) of mammals codes for 2 ribosomal and 22 transfer RNAs but only for 13 proteins, all representing subunits of the 4 proton pumps of OxPhos (7 of complex I, 1 of complex III, 3 of complex IV, and 2 of complex V). The transcription and translation of mtDNA differs from the nuclear system and requires more than 100 nuclear-encoded proteins which, in addition to about 1,000 other nuclear-encoded proteins, have to be transported into mitochondria by a complex protein transport system (Schmidt et al. 2010) (see Chap. 3). Human mtDNA is inherited only by the mother and has an about ten times higher mutation rate than nuclear DNA, possibly leading to maternally inherited mitochondrial diseases (Shoubridge 2001; Krishnan et al. 2008; DiMauro and Schon 2008) (see Chap. 14). It has been proposed that aging is partly due to somatic mutations of mtDNA, frequently causing defective COX (Müller-Höcker et al. 1992, 1993), concomitant with impaired synthesis of ATP and a decline in cell energetics with increasing age (Wallace 2010).

1.3 The Mitochondrial Proteome

A recent functional classification of the mitochondrial proteome of *Saccharomyces cerevisiae* by Schmidt et al. (2010)—comprising about 850 proteins—have assigned only 15% of the proteome to energy metabolism. Other parts of the proteome involve protein synthesis (13.3%) and genome maintenance and transcription (11.9%). A total of 7.4% are assigned to protein transport and folding, and the knowledge of these transport proteins increased considerably during recent years (Schmidt et al. 2010) (see Chap. 3).

A total of 5.8% of the yeast proteome are proteins involved in the transport of metabolites. Mitochondrial carrier proteins are widely distributed in all eukaryotes and are involved in numerous metabolic pathways. In a recent study, 35 genes for mitochondrial carriers have been identified in yeast and 53 in the human genome, and the number of mitochondrial carrier genes varied in different species between 35 and 125 (Palmieri et al. 2011). Until now, 22 mitochondrial carrier subfamilies have been functionally characterized (Palmieri and Pierri 2010), and several diseases based on mutations of the carrier genes have been described (Palmieri 2008). The first crystal structure of a mitochondrial carrier was presented for the ADP/ATP carrier (Pebay-Peyroula et al. 2003), and it was shown to function as a monomer (Bamber et al. 2007; Kunji and Robinson 2010), in contrast to previous assumptions. The observed dimeric structure of the ADP/ATP carrier in the membrane could have additional regulatory roles (Klingenberg 2008).

Only very few proteins involved in signaling (4.1% of the proteome) have been identified and research in mitochondrial signaling represents a current challenge. Proteins of amino acid (4.2%) and lipid metabolism (3.4%) are also involved in energy metabolism, since they produce the substrates for oxidative metabolism. For 19.3% of the proteome no functions could be assigned yet. Mitochondria also contain proteins involved in other functions such as fission and fusion of the organelle (see Chap. 2), induction of apoptosis (Brenner and Mak 2009; Martinou and Youle 2011; Indran et al. 2011), generation and detoxification of ROS (see Chap. 6), and regulation of cytoplasmic and mitochondrial matrix calcium (Pivovarova and Andrews 2010). In addition, mitochondria are partly involved in the biosynthesis of heme (Schultz et al. 2010) and iron–sulfur proteins (Lill 2009; Sheftel et al. 2010).

1.4 Regulation of Oxidative Phosphorylation

Since the energy requirements of animals vary strongly during their life, a sensitive and rapid regulation of OxPhos is required. This occurs at different levels, as described below, but the details of these regulations are mostly unknown.

1. The number of nuclear-encoded subunits of OxPhos complexes involved in regulation varies in different species (Das et al. 2004). Its number increased during evolution from bacteria to mammalia, for example, in complex I from 14 to 45 (Efremov and Sazanov 2011), and in complex IV from 4 to 13 (Ludwig et al. 2001; Pierron et al., 2012) (see Chap. 8), suggesting additional regulatory properties of OxPhos complexes in mitochondria. In addition, in complex IV (COX) for 5 of the 10 nuclear-encoded subunits different isoforms have been identified (see also Chap. 10) which occur in tissue-specific (Anthony et al. 1990; Hüttemann et al. 2001, 2003a, b), developmental-specific (Van den Bogert et al. 1992; Bonne et al. 1993; Parsons et al. 1996), and species-specific isoforms (Linder et al. 1995). No isoforms have been identified in the other OxPhos complexes.
2. Rapid regulation of OxPhos activity occurs by interaction with allosteric effectors. Noncatalytic (regulatory?) binding sites for ADP and/or ATP have been identified in various enzymes/proteins (Wierenga et al. 1986; Rajagopalan et al. 1999; Robblee et al. 2005; Inoue and Shingyoji 2007) including ATP synthase (Walker et al. 1982). Regulation of enzyme activity by binding of ADP or ATP to the same site, depending on the ATP/ADP ratio, has been demonstrated for COX subunits IV and VIa. Exchange of bound ADP by ATP at the intermembrane domain of subunit IV increases the K_M for cytochrome *c* of the bovine heart enzyme (Napiwotzki and Kadenbach 1998). Exchange of bound ADP by ATP at the matrix domain of COX subunit IV induces an *allosteric ATP-inhibition* (half-maximal at ATP/ADP=28) (Arnold and Kadenbach 1997) (see also Chap. 11), which is abolished by binding of 2,5-diiodothyronine to subunit Va (Arnold et al. 1998). It is also abolished by dephosphorylation of COX (Lee et al. 2001, 2002) or by expression of an isoform of subunit IV (COX subunit IV-2, see Chap. 13). Exchange of bound ADP by ATP at the matrix domain of subunit VIa-H (heart-type isoform) decreases the H^+/e^- stoichiometry of reconstituted bovine heart COX from 1.0 to 0.5 (half-maximal at ATP/ADP=100) (Frank and Kadenbach 1996). The same decrease of H^+/e^- occurs with low concentrations of free palmitate with reconstituted bovine liver COX, but the tissue-specific binding site for palmitate was not determined (Lee and Kadenbach 2001).
3. Regulation of enzyme activity by reversible phosphorylation, first described for the pyruvate dehydrogenase complex (Linn et al. 1969), is expected to occur in many nuclear-encoded mitochondrial enzymes/proteins, although in most cases the specific function of protein phosphorylation remains unknown. All 5 complexes of OxPhos have been described to be phosphorylated (Pagliarini and Dixon 2006; Vogt et al. 2007; Kadenbach et al. 2010). A two- to three-fold stimulation of complex I activity was described by Scacco et al. (2000) in fibroblasts after treatment with dibuturyl-cAMP accompanied by phosphorylation of the

matrix-oriented 18 kDa ESSS subunit (Chen et al. 2004). This phosphorylation is also required for subunit import and contributes to the stability of complex I (De Rasmo et al. 2008). Up to now, 18 phosphorylation sites have been identified in complex IV (Helling et al. 2012). However, the functional significance could only be shown for cAMP-dependent phosphorylation of Tyr-304 of bovine COX subunit I, resulting in inhibition of enzyme activity, which is independent of the ATP/ADP ratio (Lee et al. 2005). The TNF-α (tumor necrosis factor alpha) dependent phosphorylation of this site was suggested to be involved in sepsis, leading in many cases to human death (Samavati et al. 2008) (see Chap. 10).

4. Furthermore, regulation of the activity of OxPhos complexes occurs by reversible formation of *supercomplexes*, also denoted *respirasomes* (Schägger 2002), which could enable enhanced transfer of electrons in the respiratory chain (see Chap. 5). Finally, it has been known for a long time that mitochondria occur in multiple morphological structures and can divide and fuse (see Chap. 2).

The five enzyme complexes of OxPhos in mammalia are composed of multiple subunits, the function of which is mostly unknown: 45 subunits in complex I, 4 subunits in complex II (succinate dehydrogenase), 11 subunits in complex III, 13 subunits in complex IV, and 17 subunits in complex V (ATP synthase) (McKenzie et al. 2009; Carroll et al. 2009). For the assembly of the complexes, which are partly encoded on mtDNA and partly on nuclear DNA (except complex II), multiple assembly factors are required, and their defect can cause various mitochondrial diseases (see Chaps. 4 and 12).

1.5 Crystal Structures of OxPhos Complexes

In order to understand the specific functions of the multiple subunits in the mitochondrial OxPhos complexes, X-ray structures are of great value. From all 5 complexes of OxPhos X-ray structures are available, however, at different resolution. In Fig. 1.1, the X-ray crystal structures of the 5 OxPhos complexes are presented. The X-ray crystallographic analysis of mitochondrial complex I from the aerobic yeast *Yarrowia lipolytica*, composed of 40 subunits, has been determined (Hunte et al. 2010). It shows FMN and 8 iron–sulfur clusters in the peripheral arm which lines up at an angle of 130–140° with respect to the membrane arm. The crystal structure of the hydrophilic domain (peripheral arm) of complex I from *Thermus thermophilus* has been solved at 3.3 Å resolution. It consists of eight subunits and contains all the redox centers of the enzyme, including FMN and nine iron–sulfur clusters (Sazanov and Hinchliffe 2006). The structure of the membrane domain of complex I containing 6 subunits was determined from *Escherichia coli* at 3.0 Å resolution (Efremov and Sazanov 2011). The 14 subunits of bacterial complex I contrast the 45 subunits of mammalian complex I, although both enzymes have the same catalytic activity. The functions of the additional 31 subunits in mammalia remain to be determined.

The crystal structure of mitochondrial complex II has been identified from porcine heart at 2.4 Å resolution (Sun et al. 2005). Complex II is comprised of two

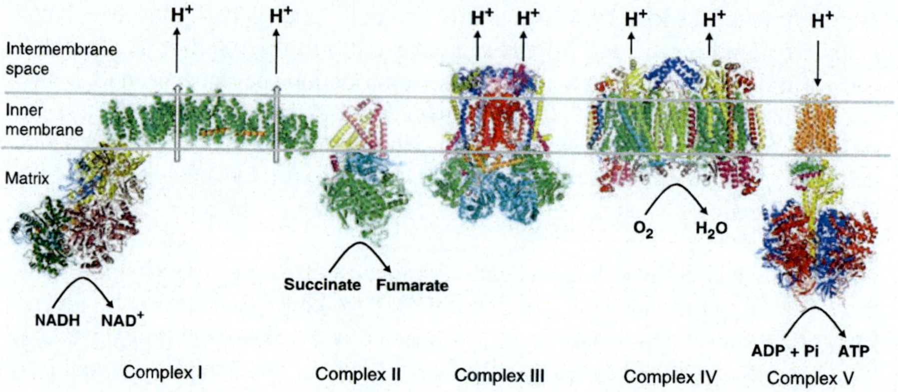

Fig. 1.1 X-ray crystal structures of the 5 complexes of oxidative phosphorylation. The crystal structures were generated using the program PyMOL with the Protein Data Bank (PDB) accession codes 1ZOY (complex II from porcine heart), 3H1H (complex III from chicken), 3ASO (complex IV from bovine heart) and 2XND (complex V (F_1-c-ring subcomplex) from bovine heart). The composite model of complex I combines the peripheral domain from *Thermus thermophilus* (2FUG) and the transmembrane helices of the complex from the yeast *Yarrowia lipolytica* (taken from http://www.bioss.uni-freiburg.de/cms/1015.html). Complexes III and IV were determined as dimers. The electron transfer between complex I and complex II to complex III occurs via ubiquinone (not shown); the electron transfer between complex III and complex IV occurs via cytochrome c (not shown).

hydrophilic proteins, flavoprotein and iron–sulfur protein, and two transmembrane proteins. The X-ray structure of complex III has been determined from bovine (Iwata et al. 1998) and from chicken heart as a dimer (Zhang et al. 1998). The X-ray structure of the dimeric complex III from yeast at 1.9 Å resolution with one molecule of bound cytochrome *c* was published by Nyola and Hunte (2008).

Crystal structures of complex IV, first published in 1995 for COX from bovine heart as a dimer (Tsukihara et al. 1995, 1996) and from *Paracoccus denitrificans* as a monomer (Iwata et al. 1995), have been published repeatedly at increasing resolution for bacterial (Koepke et al. 2009; Buschmann et al. 2010; Liu et al. 2011; Tiefenbrunn et al. 2011) and bovine heart COX (Yoshikawa et al. 1998; Shinzawa-Itoh et al. 2007; Yu et al. 2011). However, up to now, the mechanism of coupling between electron transport and proton pumping remains speculative (Ferguson-Miller et al. 2012; von Ballmoos et al. 2011). An additional proton pathway was identified in COX from bovine heart (H-channel) (Tsukihara et al., 2003; Yoshikawa et al. 2006; Shimokata et al. 2007) (see Chap. 9), which was not found in the bacterial enzyme (Salje et al. 2005).

After the first X-ray structure of bovine F_1-ATPase (Abrahams et al. 1994), higher resolution crystal structures of the stator (Dickson et al. 2006) and the "soluble" F_1-part have been determined at up to 1.9 Å resolution (Bowler et al. 2007). The membrane extrinsic region consists of the spherical catalytic domain made of three α- and three β-subunits and the central stalk (subunits γ, δ, and ε). The X-ray analysis of the whole structure of F_1F_0-ATP synthase was first modeled (Rees et al. 2009) using the membrane domain (F_0) from yeast (Stock et al. 1999). The structure of the F_1-c-ring subcomplex of F_0F_1-ATP synthase from bovine heart was published by Watt et al. (2010), containing in the membrane part a ring of eight c-subunits. The function of most of the nine subunits of the F_0 domain from bovine ATP synthase (subunits a, b, c, d, e, f, g, F6, A6L) (Collinson et al. 1994) remains unknown. Two of them (e and g) are required for ATPase dimer and oligomer formation, which is a key determinant of cristae morphology (Paumard et al. 2002).

Research into the supernumerary (nuclear-encoded) subunits of the OxPhos complexes will help to reveal their specific functions and their possible role in various degenerative human diseases.

Acknowledgement I would really like to thank Rabia Ramzan for preparing Fig. 1.1.

References

Abrahams JP, Leslie AG, Lutter R, Walker JE (1994) Structure at 2.8 Å resolution of F1-ATPase from bovine heart mitochondria. Nature 370(6491):621–628

Anthony G, Stroh A, Lottspeich F, Kadenbach B (1990) Different isozymes of cytochrome c oxidase are expressed in bovine smooth muscle and skeletal or heart muscle. FEBS Lett 277:97–100

Arnold S, Kadenbach B (1997) Cell respiration is controlled by ATP, an allosteric inhibitor of cytochrome c oxidase. Eur J Biochem 249:350–354

Arnold S, Goglia F, Kadenbach B (1998) 3,5-diiodothyronine binds to subunit Va of cytochrome c oxidase and abolishes the allosteric inhibition of respiration by ATP. Eur J Biochem 252:325–330

Bamber L, Harding M, Monne M, Slotboom DJ, Kunji ERS (2007) The yeast mitochondrial ADP/ATP carrier functions as a monomer in mitochondrial membranes. Proc Natl Acad Sci USA 104:10830–10834

Bonne G, Seibel P, Possekel S, Marsac C, Kadenbach B (1993) Expression of humen cytochrome c oxidase subunits during fetal development. Eur J Biochem 217:1099–1107

Bowler MW, Montgomery MG, Leslie AGW, Walker JE (2007) Ground state structure of F1-ATPase from bovine heart mitochondria at 1.9 Å resolution. J Biol Chem 282:14238–14242

Boyer PD, Chance B, Ernster L, Mitchell P, Racker E, Slater EC (1977) Oxidative phosphorylation and photophosphorylation. Annu Rev Biochem 46:955–966

Brenner D, Mak TW (2009) Mitochondrial cell death effectors. Curr Opin Cell Biol 21:871–877

Buschmann S, Warkentin E, Xie H, Langer JD, Ermler U, Michel H (2010) The structure of cbb3 cytochrome oxidase provides insights into proton pumping. Science 329(5989):327–330

Carroll J, Fearnley IM, Wang Q, Walker JE (2009) Measurement of the molecular masses of hydrophilic and hydrophobic subunits of ATP synthase and complex I in a single experiment. Anal Biochem 395:249–255

Chen R, Fearnley IM, Peak-Chew SY, Walker JE (2004) The phosphorylation of subunits of complex I from bovine heart mitochondria. J Biol Chem 279:26036–26045

Collinson IR, Runswick MJ, Buchanan SK, Fearnley IM, Skehel JM, van Raaij MJ, Griffiths DE, Walker JE (1994) Fo membrane domain of ATP synthase from bovine heart mitochondria: purification, subunit composition, and reconstitution with F1-ATPase. Biochemistry 33: 7971–7978

Dalmonte ME, Forte E, Genova ML, Giuffrè A, Sarti P, Lenaz G (2009) Control of respiration by cytochrome c oxidase in intact cells: role of the membrane potential. J Biol Chem 284: 32331–32335

Das J, Miller ST, Stern DL (2004) Comparison of diverse protein sequences of the nuclear-encoded subunits of cytochrome c oxidase suggests conservation of structure underlies evolving functional sites. Mol Biol Evol 21:1572–1582

De Rasmo D, Panelli D, Sardanelli AM, Papa S (2008) cAMP-dependent protein kinase regulates the mitochondrial import of the nuclear encoded NDUFS4 subunit of complex I. Cell Signal 20:989–997

Dickson VK, Silvester JA, Fearnley IM, Leslie AG, Walker JE (2006) On the structure of the stator of the mitochondrial ATP synthase. EMBO J 25:2911–2918

DiMauro S, Schon EA (2008) Mitochondrial disorders in the nervous system. Annu Rev Neurosci 31:91–123

Efremov RG, Sazanov LA (2011) Structure of the membrane domain of respiratory complex I. Nature 476(7361):414–420

Ferguson-Miller S, Hiser C, Liu J (2012) Gating and regulation of the cytochrome c oxidase proton pump. Biochim Biophys Acta 1817:489–494

Frank V, Kadenbach B (1996) Regulation of the H+/e–stoichiometry of cytochrome c oxidase from bovine heart by intraliposomal ATP/ADP ratios. FEBS Lett 382:121–124

Groen AK, Wanders RJA, Westerhoff HV, van der Meer R, Tager JM (1982) Quantification of the contribution of various steps to the control of mitochondrial respiration. J Biol Chem 257:2754–2757

Helling S, Hüttemann H, Ramzan R, Kim SH, Lee I, Müller T, Langenfeld E, Meyer HE, Kadenbach B, Vogt S, Marcus K (2012) Multiple phosphorylations of cytochrome c oxidase and their functions. Proteomics 12:950–959

Hunte C, Zickermann V, Brandt U (2010) Functional modules and structural basis of conformational coupling in mitochondrial complex I. Science 329(5990):448–451

Hüttemann M, Kadenbach B, Grossman LI (2001) Mammalian subunit IV isoforms of cytochrome *c* oxidase. Gene 267:111–123

Hüttemann M, Jaradat S, Grossman LI (2003a) Cytochrome *c* oxidase of mammals contains a testes-specific isoform of subunit VIb – the counterpart to testes-specific cytochrome c? Mol Reprod Dev 66:8–16

Hüttemann M, Schmidt TR, Grossman LI (2003b) A third isoform of cytochrome c oxidase subunit VIII is present in mammals. Gene 312:95–102

Indran IR, Tufo G, Pervaiz S, Brenner C (2011) Recent advances in apoptosis, mitochondria and drug resistance in cancer cells. Biochim Biophys Acta 1807:735–745, Review

Inoue Y, Shingyoji C (2007) The roles of noncatalytic ATP binding and ADP binding in the regulation of dynein motile activity in flagella. Cell Motil Cytoskeleton 64:690–704

Iwata S, Ostermeier C, Ludwig B, Michel H (1995) Structure at 2.8 Å resolution of cytochrome c oxidase from Paracoccus denitrificans. Nature 376:660–669

Iwata S, Lee JW, Okada K, Lee JK, Iwata M, Rasmussen B, Link TA, Ramaswamy S, Jap BK (1998) Complete structure of the 11-subunit bovine mitochondrial cytochrome bc1 complex. Science 281(5373):64–71

Jastroch M, Divakaruni AS, Mookerjee S, Treberg JR, Brand MD (2010) Mitochondrial proton and electron leaks. Essays Biochem 47:53–67

Kadenbach B, Ramzan R, Wen L, Vogt S (2010) New extension of the Mitchell Theory for oxidative phosphorylation in mitochondria of living organisms. Biochim Biophys Acta 1800: 205–212

Klingenberg M (2008) The ADP and ATP transport in mitochondria and its carrier. Biochim Biophys Acta 1778:1978–2021

Koepke J, Olkhova E, Angerer H, Müller H, Peng G, Michel H (2009) High resolution crystal structure of *Paracoccus denitrificans* cytochrome c oxidase: new insights into the active site and the proton transfer pathways. Biochim Biophys Acta 1787:635–645

Krishnan KJ, Reeve AK, Samuels DC, Chinnery PF, Blackwood JK, Taylor RW, Wanrooij S, Spelbrink JN, Lightowlers RN, Turnbull DM (2008) What causes mitochondrial DNA deletions in human cells? Nat Genet 40:275–279

Kunji ER, Robinson AJ (2010) Coupling of proton and substrate translocation in the transport cycle of mitochondrial carriers. Curr Opin Struct Biol 20:440–447

Lee I, Kadenbach B (2001) Palmitate decreases proton pumping of liver-type cytochrome c oxidase. Eur J Biochem 268:6329–6334

Lee I, Bender E, Arnold S, Kadenbach B (2001) New control of mitochondrial membrane potential and ROS-formation. Biol Chem 382:1629–1633

Lee I, Bender E, Kadenbach B (2002) Control of mitochondrial membrane potential and ROS formation by reversible phosphorylation of cytochrome c oxidase. Mol Cell Biochem 234(235):63–70

Lee I, Salomon AR, Ficarro S, Mathes I, Lottspeich F, Grossman LI, Hüttemann M (2005) cAMP-dependent tyrosine phosphorylation of subunit I inhibits cytochrome c oxidase activity. J Biol Chem 280:6094–6100

Lehninger AL (1970) Biochemistry. Worth Publishers, New York

Lill R (2009) Function and biogenesis of iron-sulphur proteins. Nature 460(7257):831–838

Linder D, Freund R, Kadenbach B (1995) Species-specific expression of cytochrome c oxidase isozymes. Comp Biochem Physiol 112B:461–469

Linn TC, Pettit FH, Reed LJ (1969) Alpha-keto acid dehydrogenase complexes. X. Regulation of the activity of the pyruvate dehydrogenase complex from beef kidney mitochondria by phosphorylation and dephosphorylation. Proc Natl Acad Sci USA 62:234–241

Liu J, Qin L, Ferguson-Miller S (2011) Crystallographic and online spectral evidence for role of conformational change and conserved water in cytochrome oxidase proton pump. Proc Natl Acad Sci USA 108(4):1284–1289

Ludwig B, Bender E, Arnold S, Hüttemann M, Lee I, Kadenbach B (2001) Cytochrome c oxidase and the regulation of oxidative phosphorylation. Chembiochem 2:392–403

Margulis L (1975) Symbiotic theory of the origin of eukaryotic organelles; criteria for proof. Symp Soc Exp Biol 29:21–38

Martinou JC, Youle RJ (2011) Mitochondria in apoptosis: Bcl-2 family members and mitochondrial dynamics. Dev Cell 21:92–101

McKenzie M, Lazarou M, Ryan MT (2009) Analysis of respiratory chain complex assembly with radiolabeled nuclear- and mitochondrial-encoded subunits. Methods Enzymol 456:321–339

Mitchell P (1961) Coupling of phosphorylation to electron and hydrogen transfer by a chemiosmotic type of mechanism. Nature 191:144–148

Mitchell P (1966) Chemiosmotic coupling in oxidative and photosynthetic phosphorylation. Biol Rev 41:445–502

Müller-Höcker J, Schneiderbanger K, Stefani FH, Kadenbach B (1992) Progressive loss of cytochrome-c-oxidase in the human extraocular muscles in ageing—a cytochemical-immunohistochemical study. Mutat Res 275:115–124

Müller-Höcker J, Seibel P, Schneiderbanger K, Kadenbach B (1993) Different in situ hybridisation patterns of mitochondrial DNA in cytochrome c oxidase- deficient extraocular muscle fibres in the elderly. Virchows Arch A Pathol Anat Histopathol 422:7–15

Napiwotzki J, Kadenbach B (1998) Extramitochondrial ATP/ADP-ratios regulate cytochrome c oxidase activity via binding to the cytosolic domain of subunit IV. Biol Chem 379:335–339

Nyola A, Hunte C (2008) A structural analysis of the transient interaction between the cytochrome bc1 complex and its substrate cytochrome c. Biochem Soc Trans 36:981–985

Pacelli C, Latorre D, Cocco T, Capuano F, Kukat C, Seibel P, Villani G (2011) Tight control of mitochondrial membrane potential by cytochrome c oxidase. Mitochondrion 11:334–341

Pagliarini DJ, Dixon JE (2006) Mitochondrial modulation: reversible phosphorylation takes center stage? Trends Biochem Sci 31:26–34

Palmieri F (2008) Diseases caused by defects of mitochondrial carriers: a review. Biochim Biophys Acta 1777:564–578

Palmieri F, Pierri CL (2010) Mitochondrial metabolite transport. Essays Biochem 47:37–52

Palmieri F, Pierri CL, De Grassi A, Nunes-Nesi A, Fernie AR (2011) Evolution, structure and function of mitochondrial carriers: a review with new insights. Plant J 66:161–181

Parsons WJ, Williams RS, Shelton JM, Luo Y, Kessler DJ, Richardson JA (1996) Developmental regulation of cytochrome oxidase subunit VIa isoforms in cardiac and skeletal muscle. Am J Physiol 270:H567–H574

Paumard P, Vaillier J, Coulary B, Schaeffer J, Soubannier V, Mueller DM, Brèthes D, di Rago JP, Velours J (2002) The ATP synthase is involved in generating mitochondrial cristae morphology. EMBO J 21:221–230

Pebay-Peyroula E, Dahout-Gonzalez C, Kahn R, Trezeguet V, Lauquin GJ, Brandolin G (2003) Structure of mitochondrial ADP/ATP carrier in complex with carboxyatractyloside. Nature 426:39–44

Piccoli C, Scrima R, Boffoli D, Capitanio N (2006) Control by cytochrome c oxidase of the cellular oxidative phosphorylation system depends on the mitochondrial energy state. Biochem J 396:573–583

Pierron D, Wildman DE, Hüttemann M, Markondapatnaikuni GC, Aras S, Grossman LI (2012) Cytochrome c oxidase: evolution of control via nuclear subunit addition. Biochim Biophys Acta 1817:590–597

Pivovarova NB, Andrews SB (2010) Calcium-dependent mitochondrial function and dysfunction in neurons. FEBS J 277:3622–3636

Rajagopalan K, Watt DS, Haley BE (1999) Orientation of GTP and ADP within their respective binding sites in glutamate dehydrogenase. Eur J Biochem 265:564–571

Rees DM, Leslie AG, Walker JE (2009) The structure of the membrane extrinsic region of bovine ATP synthase. Proc Natl Acad Sci USA 106:21597–21601

Richards TA, Archibald JM (2011) Cell evolution: gene transfer agents and the origin of mitochondria. Curr Biol 21:R112–R114

Robblee JP, Cao W, Henn A, Hannemann DE, De La Cruz EM (2005) Thermodynamics of nucleotide binding to actomyosin V and VI: a positive heat capacity change accompanies strong ADP binding. Biochemistry 44:10238–10249

Salje J, Ludwig B, Richter OM (2005) Is a third proton-conducting pathway operative in bacterial cytochrome c oxidase? Biochem Soc Trans 33:829–831

Samavati L, Lee I, Mathes I, Lottspeich F, Hüttemann M (2008) Tumor necrosis factor alpha inhibits oxidative phosphorylation through tyrosine phosphorylation at subunit I of cytochrome c oxidase. J Biol Chem 283:21134–21144

Sazanov LA, Hinchliffe P (2006) Structure of the hydrophilic domain of respiratory complex I from Thermus thermophilus. Science 311(5766):1430–1433

Scacco S, Vergari R, Scarpulla RC, Technikova-Dobrova Z, Sardanelli A, Lambo R, Lorusso V, Papa S (2000) cAMP-dependent phosphorylation of the nuclear encoded 18-kDa (IP) subunit of respiratory complex I and activation of the complex in serum-starved mouse fibroblast cultures. J Biol Chem 275:17578–17582

Schägger H (2002) Respiratory chain supercomplexes of mitochondria and bacteria. Biochim Biophys Acta 1555:154–159

Schmidt O, Pfanner N, Meisinger C (2010) Mitochondrial protein import: from proteomics to functional mechanisms. Nat Rev Mol Cell Biol 11:655–667

Schultz IJ, Chen C, Paw BH, Hamza I (2010) Iron and porphyrin trafficking in heme biogenesis. J Biol Chem 285:26753–26759

Sheftel A, Stehling O, Lill R (2010) Iron-sulfur proteins in health and disease. Trends Endocrinol Metab 21:302–314

Shimokata K, Katayama Y, Murayama H, Suematsu M, Tsukihara T, Muramoto K, Aoyama H, Yoshikawa S, Shimada H (2007) The proton pumping pathway of bovine heart cytochrome c oxidase. Proc Natl Acad Sci USA 104:4200–4205

Shinzawa-Itoh K, Aoyama H, Muramoto K, Terada H, Kurauchi T, Tadehara Y, Yamasaki A, Sugimura T, Kurono S, Tsujimoto K, Mizushima T, Yamashita E, Tsukihara T, Yoshikawa S (2007) Structures and physiological roles of 13 integral lipids of bovine heart cytochrome c oxidase. EMBO J 26:1713–1725

Shoubridge EA (2001) Cytochrome c oxidase deficiency. Am J Med Genet 106:46–52

Stock D, Leslie AG, Walker JE (1999) Molecular architecture of the rotary motor in ATP synthase. Science 286(5445):1700–1705

Sun F, Huo X, Zhai Y, Wang A, Xu J, Su D, Bartlam M, Rao Z (2005) Crystal structure of mitochondrial respiratory membrane protein complex II. Cell 121:1043–1057

Tiefenbrunn T, Liu W, Chen Y, Katritch V, Stout CD, Fee JA, Cherezov V (2011) High resolution structure of the ba3 cytochrome c oxidase from Thermus thermophilus in a lipidic environment. PLoS One 6(7):e22348

Tsukihara T, Aoyama H, Yamashita E, Tomizaki T, Shinzawa-Itoh H, Shinzawa-Itoh K, Nakashima R, Yaono R, Yoshikawa S (1995) Structures of metal sites of oxidized bovine heart cytochrome coxidase at 2.8 Å. Science 269:1069–1074

Tsukihara T, Aoyama H, Yamashita E, Tomizaki T, Yamaguchi H, Shinzawa-Itoh K, Nakashima R, Yaono R, Yoshikawa S (1996) The whole structure of the 13-subunit oxidized cytochrome c oxidase at 2.8 Å. Science 272:1136–1144

Van den Bogert C, Dekker HL, Cornelissen JC, Van Kuilenburg AB, Bolhuis PA, Muijsers AO (1992) Isoforms of cytochrome c oxidase in tissues and cell lines of the mouse. Biochim Biophys Acta 1099:118–122

Villani G, Attardi G (1997) In vivo control of respiration by cytochrome c oxidase in wild-type and mitochondrial DNA mutation-carrying human cells. Proc Natl Acad Sci USA 94:1166–1171

Villani G, Attardi G (2001) In vivo measurements of respiration control by cytochrome c oxidase and in situ analysis of oxidative phosphorylation. Methods Cell Biol 65:119–131

Vogt S, Rhiel A, Koch V, Kadenbach B (2007) Regulation of oxidative phosphorylation by inhibition of its enzyme complexes via reversible phosphorylation. Curr Enzyme Inhib 3:189–206

von Ballmoos C, Gennis RB, Ädelroth P, Brzezinski P (2011) Kinetic design of the respiratory oxidases. Proc Natl Acad Sci USA 108:11057–11062

Walker JE, Saraste M, Runswick MJ, Gay NJ (1982) Distantly related sequences in the alpha- and beta-subunits of ATP synthase, myosin, kinases and other ATP-requiring enzymes and a common nucleotide binding fold. EMBO J 1:945–951

Wallace DC (2010) Mitochondrial DNA mutations in disease and aging. Environ Mol Mutagen 51:440–450

Wallin IE (1923) The mitochondria problem. Am Nat 57(650):255–261

Watt IN, Montgomery MG, Runswick MJ, Leslie AG, Walker JE (2010) Bioenergetic cost of making an adenosine triphosphate molecule in animal mitochondria. Proc Natl Acad Sci USA 107:16823–16827

Wierenga RK, Terpstra P, Hol WG (1986) Prediction of the occurrence of the ADP-binding beta alpha beta-fold in proteins, using an amino acid sequence fingerprint. J Mol Biol 187: 101–107

Yoshikawa S, Shinzawa-Itoh K, Nakashima R, Yaono R, Yamashita E, Inoue N, Yao M, Fei MJ, Libeu CP, Mizushima T, Yamaguchi H, Tomizaki T, Tsukihara T (1998) Redox-coupled crystal structural changes in bovine heart cytochrome c oxidase. Science 280(5370):1723–1729

Yoshikawa S, Muramoto K, Shinzawa-Itoh K, Aoyama H, Tsukihara T, Shimokata K, Katayama Y, Shimada H (2006) Proton pumping mechanism of bovine heart cytochrome c oxidase. Biochim Biophys Acta 1757:1110–1116

Yu MA, Egawa T, Shinzawa-Itoh K, Yoshikawa S, Yeh SR, Rousseau DL, Gerfen GJ (2011) Radical formation in cytochrome c oxidase. Biochim Biophys Acta 1807:1295–1304

Zhang Z, Huang L, Shulmeister VM, Chi YI, Kim KK, Hung LW, Crofts AR, Berry EA, Kim SH (1998) Electron transfer by domain movement in cytochrome bc1. Nature 392(6677): 677–684

Zimmer C (2009) Origins. On the origin of eukaryotes. Science 325:666–668

Chapter 2
Mitochondrial Dynamics: The Intersection of Form and Function

Andrew Ferree and Orian Shirihai

Abstract Mitochondria within a cell exist as a population in a dynamic morphological continuum. The balance of mitochondrial fusion and fission dictates a spectrum of shapes from interconnected networks to fragmented individual units. This plasticity bestows the adaptive flexibility needed to adjust to changing cellular stresses and metabolic demands. The mechanisms that regulate mitochondrial dynamics, their importance in normal cell biology, and the roles they play in disease conditions are only beginning to be understood. Dysfunction of mitochondrial dynamics has been identified as a possible disease mechanism in Parkinson's disease. This chapter will introduce the budding field of mitochondrial dynamics and explore unique characteristics of affected neurons in Parkinson's disease that increase susceptibility to disruptions in mitochondrial dynamics.

2.1 Introduction

Mitochondrial Dynamics refers to the observation that mitochondria within the individual cell go through fusion and fission events. Visually, this results in a morphological spectrum with contrasting degrees of elongation and fragmentation. Plasticity bestows the adaptive flexibility needed to adjust to changing cellular stresses and metabolic demands. Constant network remodeling also establishes a

A. Ferree
Department of Pharmacology, and Department of Neuroscience,
Boston University School of Medicine, Boston, MA, USA

O. Shirihai (✉)
Department of Medicine, Boston University School of Medicine,
650 Albany Street, Boston, MA 02118, USA
e-mail: Shirihai@bu.edu

B. Kadenbach (ed.), *Mitochondrial Oxidative Phosphorylation*, Advances in Experimental Medicine and Biology 748, DOI 10.1007/978-1-4614-3573-0_2,

mechanism for quality control of the mitochondrial population with important ramifications for long-term function and health.

While our understanding of mitochondrial dynamics is just beginning, descriptions of morphological transitions by mitochondria can be traced to reports dating back nearly a century. In 1914, Lewis and Lewis elegantly describe witnessing fusion and fission events along with an incredible range of structures exhibited by mitochondria in cultured cells (Lewis and Lewis 1914). The body of knowledge surrounding mitochondrial dynamics has expanded greatly from these early studies and this chapter provides a brief introduction into this exciting field.

2.2 Mitochondrial Dynamics Proteins

The state of balance between four dynamin-related proteins essentially controls mitochondrial fusion and fission. From yeast to humans, these highly conserved enzymes share homologous GTPase and transmembrane regions that form complexes and alter the curvature of the mitochondrial membranes. It is the relative activities of oppositional forces that together determine mitochondrial morphology. Complete network fragmentation can result from increased expression or activation of fission proteins. However the mitochondrial network will also fragment if fusion activity is inhibited. Similarly, elongated tubular networks occur with enhanced fusion activity as well as through blockage of fission. These extremes are reminders that antagonism between counteracting enzymatic forces sets the shape of mitochondria and therefore must always be considered simultaneously when deciphering network morphology.

2.3 Fusion

Fusion is categorized into two forms and three mitochondrial localized GTPases control the difference between transient and complete fusion events. Transient fusion involves only outer membranes while complete fusion requires merging of both inner and outer membranes. A complete fusion event occurs with a rapid diffusion of soluble mitochondrial components followed by a more gradual mixing of membrane elements (Twig et al. 2006; Partikian et al. 1998; Karbowski et al. 2004a; Jakobs et al. 2003; Jakobs 2006; Arimura et al. 2004; Busch et al. 2006). This process is believed to bestow complementation between units and increased homogeneity over the network. Complementation is a key mechanism by which mitochondria can rescue a damaged unit within the network. The effect of loss of fusion has been assessed in several model systems. Network fragmentation and susceptibility to apoptosis, decreased mitochondrial membrane potential and oxygen consumption, and increased ROS production are seen with blocking fusion and underscore its importance in maintaining mitochondrial integrity.

Two homologous proteins known as mitofusin 1 and mitofusin 2 (Mfn1 and Mfn2) function together to merge the outer membranes of mitochondria. Both proteins share relevant functional domains and connect adjacent membranes through coiled-coil antiparallel homotypic (Mfn1–Mfn1) and heterotypic (Mfn1–Mfn2) dimers. The GTPase activity of Mfn1 is higher compared to Mfn2, thus the relative proportion of dimer combinations has important functional consequences for fusion rates within the cell (Chen et al. 2003; Koshiba et al. 2004). Turnover occurs in part by polyubiquitination-mediated recruitment of chaperone proteins, such as p97, which mediate retrotranslocation of mitofusins and promote their proteasomal degradation (Tanaka et al. 2010a). Curiously, Mfn2 appears to have other crucial functions in the cell beyond mitochondrial fusion. One such function is tethering mitochondria and endoplasmic reticulum during calcium exchange between the organelles (de Brito and Scorrano 2008). In neurons, Mfn2 has been shown to play a role in motility by connecting mitochondria to the Miro/Milton transport complex (Misko et al. 2010).

Inner mitochondrial membranes are joined via the protein encoded by Optic Atrophy type 1 gene (OPA1) (Song et al. 2009). Expression of OPA1 is highly regulated at the transcriptional level with eight possible isoforms available through alternative splicing (Song et al. 2007). Imported OPA1 protein localizes to the intermembrane space and is further processed by several proteases to produce five additional isoform variations (Ehses et al. 2009). Functional differences between isoforms are not entirely understood but it is known that both long and short forms of OPA1 are needed to maintain fusion capacity (Song et al. 2007; Duvezin-Caubet et al. 2006). In both soluble and membrane associated forms, OPA1 exists in a complex with mitofusins (Cipolat et al. 2004). This interaction is crucial for complete fusion as cleavage of OPA1 disrupts the complex and limits mitochondria to only transient fusion events. In this way, proteolytic removal of long isoforms provides a mechanism for creating network fragmentation in response to stress (Griparic et al. 2007). The specific molecular signals that trigger processing of OPA1 remain largely a mystery but clearly both induction of apoptosis and dissipation of mitochondrial membrane potential induce OPA1 cleavage (Gottlieb 2006; Guillery et al. 2008; Lee et al. 2004; Olichon et al. 2007). This effect may represent a stopgap attempt to limit spread of damaged material within the mitochondrial network by isolating units that pose a risk or have been selected for mitophagy.

2.4 Fission

Fission is crucially involved in numerous important cell pathways including mitochondrial inheritance by daughter cells during cellular division, differentiation of post-mitotic cells such as neurons and cardiomyocytes, mitophagy, and forms of cell death (Lee et al. 2004, 2011a; Yu et al. 2005; Gomes and Scorrano 2008; Mendl et al. 2011; Grohm et al. 2010; Karbowski 2010; Jourdain et al. 2009; Wilkerson and Sankar 2011; Choudhary et al. 2011; Kane and Youle 2010; Shroff et al. 2009; Ishihara et al. 2009; Frank et al. 2001). Loss of fission results in increased mitochondrial connectivity, loss of mtDNA, bioenergetic deficiency, and alterations in apoptosis (Landes and Martinou 2011; Westermann 2010; Sheridan and Martin 2010; Parone et al. 2008). One cytosolic GTPase performs the division of fused mitochondria. Dynamin-related protein 1 (Drp1) translocates to mitochondrial scission sites and polymerizes into structures that surround the perimeter of the organelle (Fukushima et al. 2001). Polymerization activates the GTPase domain of Drp1, which literally causes constriction and pinching of a single unit into two individual daughters (Legesse-Miller et al. 2003). Sub-cellular localization and activity of Drp1 is regulated by several post-translational modifications, such as phosphorylation, ubiquitination, nitrosylation, and sumoylation (Figueroa-Romero et al. 2009; Cho et al. 2009; Wang et al. 2011a; Santel and Frank 2008; Braschi et al. 2009; Taguchi et al. 2007).

Network fragmentation occurs in response to various factors including intracellular calcium levels, mitochondrial membrane potential, and ATP availability (Yoon et al. 2003; Kong et al. 2005). For example, the calcium-sensitive phosphatase calcineurin promotes fission by dephosphorylating cytosolic Drp1, which causes translocation to mitochondria (Scorrano 2005). Two additional proteins that reside on the outer mitochondrial membrane act together as a receptor for organizing Drp1 to sites of fission. Mitochondrial fission 1 protein (hFis1) is a transmembrane protein that marks sites of division (Yu et al. 2005; Koch et al. 2005; Serasinghe and Yoon 2008; Otera et al. 2010; James et al. 2003). Mitochondrial Fission Factor (Mff) interacts with hFis1 and serves as an adaptor that recruits Drp1 to promote polymerization (Otera et al. 2010). Recent studies have demonstrated that Mff is required for fission but hFis1 is dispensible (Otera et al. 2010; Huang et al. 2011a). This surprising finding suggests the existence of other proteins that can supersede hFis1 and act as alternative receptors for Mff and Drp1.

In summary, there are three key steps for mitochondrial fragmentation. First is the localization of fission adaptor proteins, such as hFis1 and Mff, to fission sites. Second Drp1 must be recruited from the cytosol and polymerize at fission sites. Finally, there must be an accompanying inhibition of fusion through cleavage of long Opa1 isoforms within mitochondria. These are the three minimal steps that are required for a continuous network of fused mitochondria to transition towards fragmentation.

2.5 Approaches for Measurement of Mitochondrial Dynamics

While early observational studies describe mitochondrial fusion and fission, direct experimental proof was first obtained using polyethylene glycol (PEG)-mediated cell fusion assays (Legros et al. 2002; Neuspiel et al. 2005). In this method, two separate cultures of cells have their mitochondria labeled with different molecular probes, such as green and red fluorescent proteins (GFP, RFP). Combining the two cell populations in the presence of PEG detergent promotes fusion of plasma membranes and subsequent mixing of mitochondrial populations. Fluorescence heterogeneity is detected in the resultant pool, with cells containing mitochondria purely expressing GFP or RFP while other units display a mixture of both colors. These studies provide direct proof of fusion between individual mitochondria but significant practical limitations of the methodology left ample room for improvement. Recently, more physiologically relevant studies have employed an expanded arsenal of probes in the form of dyes and proteins that allow precise, detailed data collection on a wide range of parameters.

The tracking of dyes that accumulate within mitochondria in a membrane potential dependant manner, such as Tetramethylrhodamine ethyl ester perchlorate (TMRE), was a significant advancement in the study of mitochondrial dynamics. Confocal imaging studies of cells labeled with TMRE reveal stable mitochondrial membrane potentials maintained for a period of 40–80 s followed by a sudden drop of more than 15 mV (Loew et al. 1993). These studies were pioneering in our understanding of mitochondrial biology but limited by an inability to assure that the detected mitochondrion did not fuse and/or divide during the recording time. For example, fission can occur without movement of the two daughter mitochondria or involve only the inner (but not the outer) mitochondrial membrane (Twig et al. 2006; Malka et al. 2005). Therefore fission cannot be reliably identified by observation of separation of a mitochondrion into two segments. Similarly, the repositioning of a mitochondrion to become juxtaposed to another mitochondrion is not an indication that a fusion event occurred (Twig et al. 2006). The use of photoactivatable proteins was a breakthrough because it overcame these technical difficulties of imaging individual organelles that move and change morphology within a complex architecture (Betzig et al. 2006; Patterson and Lippincott-Schwartz 2002).

The creation of tools that allow laser-mediated photoactivation of mitochondrial matrix-targeted GFP (mtPA-GFP) facilitates improved biophysical and morphologic measurements as it is nontoxic for the cells and can therefore be used to make observations over an extended period of time (Karbowski et al. 2004a; Arimura et al. 2004; Busch et al. 2006). Overall fusion rates for a cell can be quantified by activating the mtPA-GFP in a subset of the mitochondrial population and then tracking diffusion of the fluorescence signal over time, as the mtPA-GFP spreads to non-activated fusion partners. While fusion rates vary across cell types and conditions, several studies have shown that when 10–20% of population is activated, the mtPA-GFP equilibrates across the entire network in approximately 45 min (Karbowski et al. 2004a, b; Twig et al. 2008a). This is predicted to result in homogeneity in protein content and function across the mitochondrial population. In addition to analyzing the properties of the entire mitochondrial network within a cell, mtPA-GFP can be used to assess attributes of an individual mitochondrion. These observable characteristics include the size, shape, membrane potential, motility, and temporal properties of fusion. For example, photoactivation of a selected mitochondrion enables real time tracking of that individual. Long-term monitoring of single mitochondrial units with activated mtPA-GFP in INS1 and COS7 cells has allowed for direct quantification of fusion rates. These studies revealed the frequency of fusion to be once every 5–20 min per mitochondrion (Twig et al. 2008a). The duration of fusion events is typically brief, lasting ~100 s and followed by fission (Arimura et al. 2004; Twig et al. 2008a). Thus mitochondria spend most of their time as individual solitary units. These studies provided the groundwork for a concept of the mitochondrial life cycle consisting of two stages, the pre-fusion period (solitary period) and the post-fusion period when mitochondria are connected together (networked period).

The combination of both TMRE and mtPA-GFP has two significant additional benefits in measuring biophysical properties of mitochondria (Twig et al. 2006; Molina and Shirihai 2009). First, it provides means for accurate determination of organelle boundaries that can be easily followed despite movement within a dense mitochondrial network. It is also beneficial because it allows comparison of the fluorescence intensities of the two probes to get a ratiometric value. This offers a tool for quantification of changes in membrane potential that are independent of exact focal plane. By avoiding the need to perform repeated imaging through the entire *z*-axis, monitoring can be done with greatly reduced phototoxicity. The combination approach with TMRE and mtPA-GFP extends the permissible recording periods for tracking mitochondria within a cell from minutes to hours. This advancement helped reveal that mitochondria maintain stable membrane potential during their solitary period for up to 2 h (Twig et al. 2008a; Wikstrom et al. 2007).

One major limitation of direct user-based microscopy studies is they tend to be labor intensive and therefore not amenable to high-throughput screening. Recent description of an innovative cell-free fusion assay addresses this shortcoming with a luciferase-based approach that will allow large-scale screens of modifiers of mitochondrial dynamics (Schauss et al. 2010). Specifically, the assay is based on a bimolecular complementation approach using both mitochondrial targeted yellow fluorescent protein (YFP) and luciferase constructs separated by a leucine zipper. The two split proteins are expressed separately in large cultures of cells from which mitochondrial populations are isolated and purified. During the assay the two populations of mitochondria are mixed and through fusion the split proteins are able to combine to form functional molecules. This new system holds great promise as it provides multiple highly quantifiable readouts. Linking luciferase activity to fusion events provides a much-needed tool for rapid, large-scale screening of conditions that affect mitochondrial dynamics. The methods described here provided important insights into mitochondrial biology yet they likely represent a mere beginning, as creative new approaches expand boundaries in this emerging field.

2.6 Benefits from Mitochondrial Dynamics

Responsive mitochondrial dynamics is an essential part of an array of cellular processes including mitosis, fuel sensing, ATP production, mitophagy, and apoptosis (Arimura et al. 2004; Twig et al. 2008a; Nakada et al. 2001a, b; Skulachev 2001; Liesa et al. 2008; Molina et al. 2009). In some situations, entire network fragmentation is necessary to facilitate autophagic clearance of mitochondria (mitophagy), such as during erythrocyte maturation, sperm mitochondria in oocyte fertilization and apoptosis dependant on PTP opening (Takano-Ohmuro et al. 2000; Shitara et al. 2000; Elmore et al. 2001). On a local level, fusion allows mixing and complementation between two units. In a fused state, exchange of components such as solutes, metabolites, and soluble proteins occurs rapidly (Partikian et al. 1998; Arimura et al. 2004; Chen et al. 2003, 2005; Chen and Chan 2005; Shaw and Nunnari 2002; Griffin et al. 2006) while membrane embedded proteins and mitochondrial DNA spread more slowly (Twig et al. 2006; Legros et al. 2004; Gilkerson et al. 2008; Wikstrom et al. 2009). The ability of mitochondria to fuse together reduces content heterogeneity and thus is a first line of defence against dysfunction (Legros et al. 2002; Chen et al. 2005, 2011; Legros et al. 2004; Hori et al. 2011; Chan 2006; Ono et al. 2001; Mazzoni and Falcone 2011).

Maintaining quality control through mitochondrial dynamics simultaneously optimizes bioenergetic efficiency and reduces risks associated with oxidative phosphorylation by removal of damaged material. For example, inhibition of mitochondrial fission leads to an increase in oxidized proteins along with decreased maximal oxygen consumption rates during uncoupled respiration. These findings suggest the accumulation of oxidized material is due to a loss of clearance rather than increased production of ROS. Failure to properly remove damaged components impairs mitochondrial function and limits reserve capacity. These outcomes are particularly important for long-lived cells with high metabolic demands.

The dependence of quality control on fission may stem from the ability to generate unequal daughter units. Most fission events produce heterogeneous daughters with opposite membrane potential "deflections," usually greater than 5 mV. Oxidized and damaged material is also inequitably distributed and this ability to regularly create uneven fission events suggests a selective mechanism of intra-mitochondrion segregation and separation. The net effect of numerous cycles of asymmetric divisions and selective isolation is the ability to concentrate undesirable material within a minimal number of units. Damage laden mitochondria ultimately get isolated and prevented from fusing with the rest of the network through reduction in fusion proteins (Twig et al. 2008a, b). In various cell types, loss of membrane potential leads to the polyubiquitination and proteasomal degradation of proteins associated with the mitochondrial outer membrane, such as Miro and mitofusins. In addition to degradation of individual proteins, ubiquitination serves to recruit autophagy-related scaffold adaptors, such as p62 and HDAC6 (Huang et al. 2011b; Lee et al. 2011b). These scaffold proteins bind to polyubiquitin chains and serve essentially as receptors for autophagosomes to facilitate lysosomal degradation of the mitochondrial unit through mitophagy. Blocking autophagy is sufficient to cause a buildup of damaged material including mitochondria, particularly in energy intensive tissues such as brain, heart, liver, kidney, and pancreatic beta cells (Twig et al. 2008a; Jung and Lee 2009; Taneike et al. 2010; Kimura et al. 2011). It is worth noting in these cases that dysfunctional mitochondria accumulate without requiring any additional toxins or mitochondrial stressors. Mitochondrial turnover is a major proportion of the basal autophagic processing within cells, especially those with elevated metabolic demands. Long-lived post-mitotic cells with chronic high levels of turnover are inherently vulnerable to disruptions in the quality control pathway (Terman et al. 2010).

2.7 Regulation of Mitochondrial Dynamics

Multiple levels of cell signaling are involved in regulating mitochondrial dynamics. Despite its complexity, the system can be broken down into two simple categories, global and local regulation (Hyde et al. 2010). Figure 2.1 illustrates regulatory elements during the mitochondrial lifecycle and Table 2.1 lists examples of global and local control during fusion, fission, and the solitary period. Control derived from the cellular macroenvironment that affects the entire mitochondrial network is

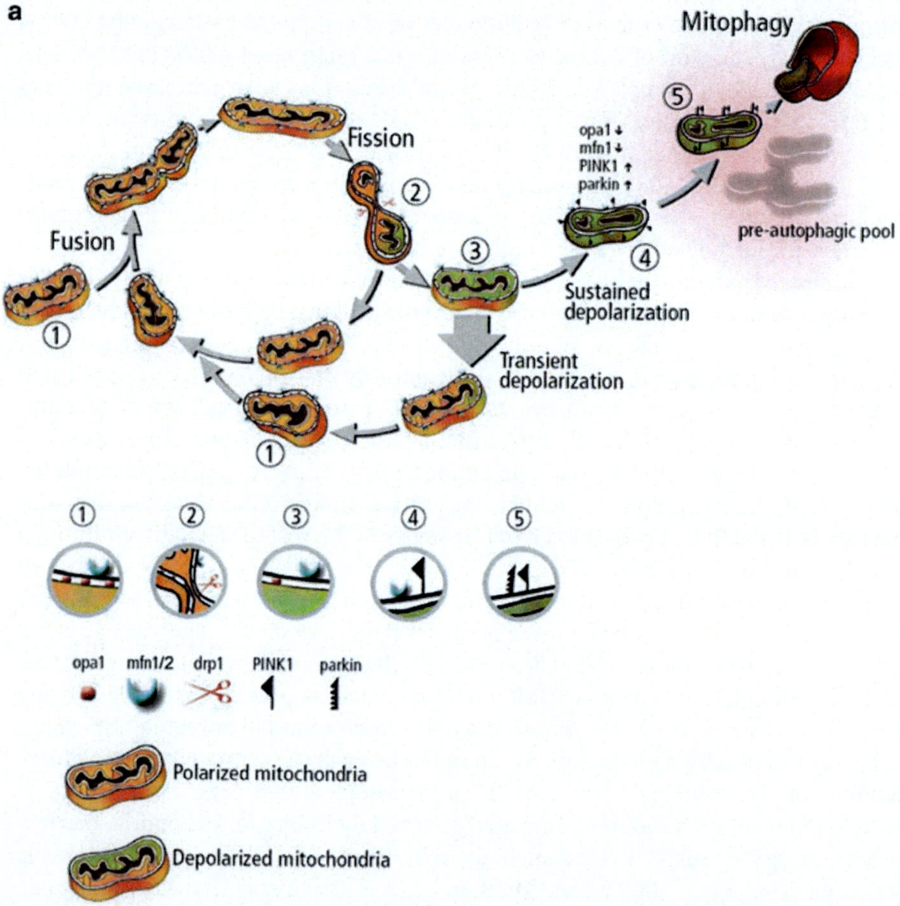

Fig. 2.1 Organellar and cellular controls of the mitochondrial life cycle. The mitochondria life cycle. **(a)** The mitochondria life cycle. Mitochondria go through continuous cycles of fusion and fission. Each cycle last 5–20 min. Fusion is brief (1) and triggers fission events (2). A daughter mitochondrion may maintain intact membrane potential (*orange*) or depolarize (3, *green*). When depolarized a subsequent fusion event is unlikely to occur, unless the mitochondrial re-polarizes. As a result, depolarized daughter mitochondria remain solitary. Depolarized and solitary mitochondria (4) remain for 1–4 h in a pre-autophagic pool before being consumed by the autophagic machinery.

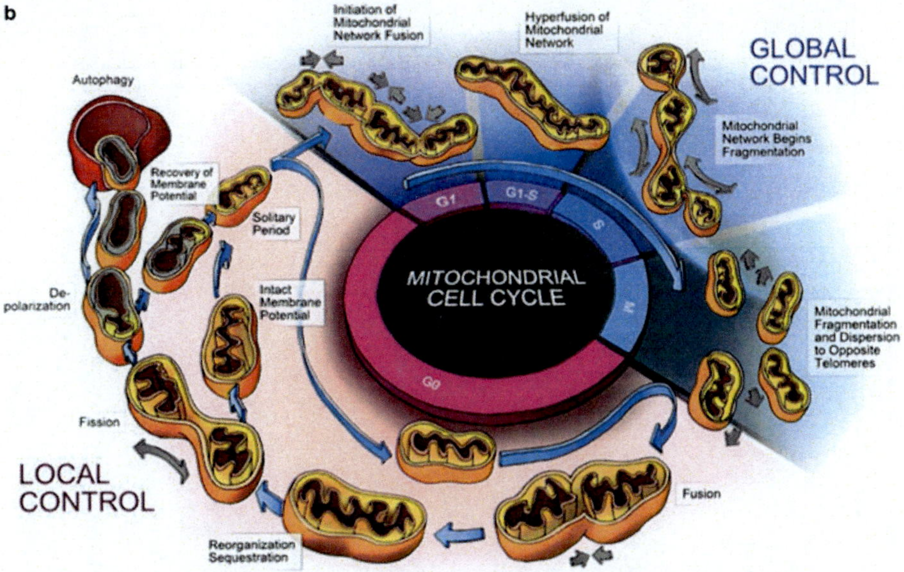

Fig 2.1 (continued) (**b**) The interaction of the mitochondria life cycle with the cell cycle—this diagram depicts the normal life cycle of an individual mitochondrion during the G0 phase of the cell cycle. The mitochondrion undergoes fusion, fission, depolarization, and degradation by autophagy. This process is depicted as one of local control whereby mitochondrial events are largely dictated by the local energetic status and associated local signals. During the cell cycle global signals cause concerted changes in the mitochondrial population, as noted by hyperfusion in the G1-S and fragmentation during the M phase. These global population effects are governed by the cellular demand for energy required by cell division and the need for homogenization and sequestration of cellular components during met-phase. The cell cycle serves as an elegant example of the parities of local and global control

categorized as global regulation of mitochondrial dynamics, such as during the cell cycle (Lee et al. 2004; Taguchi et al. 2007; Scarpulla 2002a; Arakaki et al. 2006). At different stages of mitotic cell division there is transcriptional control of dynamics proteins that lead to opposite extremes in network morphology. A concert of transcription factors mediates increases in mitochondrial mass, respiratory capacity, and energy production that are required during S phase (Scarpulla 2002a, b). Accordingly, there is hyperfusion of the network during G1-S phase while hyperfragmentation occurs in late S and M phases.

Local regulation occurs at the level of the microenvironment of an individual mitochondrion. Fission events are controlled checkpoints for generation of polarized and depolarized mitochondria; therefore changes in membrane potential distribution are a mechanistic example of local regulation (Twig et al. 2008a; Wikstrom et al. 2007). Loss of membrane potential and ATP production causes cleavage and degradation of fusion proteins by mitochondrial proteases and the proteasome (Song et al. 2007; Chan and Chan 2011; Chan et al. 2011). Decreased fusion capacity

Table 2.1 Local (organellar) versus global (cellular) controls of mitochondrial dynamics

Control mechanism	Outcome	References
Fission		
Global cellular control		
Recruitment of Drp1 to mitochondria is calcium dependent and regulated by calcineurin	Elevated cytosolic calcium levels activate calcineurin to dephosphorylate Drp1	Yoon et al. (2003), Kong et al. (2005), Kaddour-Djebbar et al. (2010), Hom et al. (2010), Cribbs and Strack (2007), Cereghetti et al. (2008, 2010), Tan et al. (2011), Wang et al. (2011c)
Prolonged exercise increases transcription and expression of Fis1 while decreasing mitofusins	Acute increases in metabolic demands of skeletal muscle stimulate fission	Ding et al. (2010b)
BH3 only proteins and Bax/Bak induce fission, Bax/Bak in healthy cells control fusion through MFN2	Interact at mitochondrial scission sites to promote fission	Karbowski (2010), Shroff et al. (2009), Karbowski et al. (2002), Wu et al. (2011), Sheridan et al. (2008)
Sumoylation of Drp1 occurs by multiple enzymes and is present at fission sites	Protects Drp1 from degradation and increases fission activity	Figueroa-Romero et al. (2009), Braschi et al. (2009), Harder et al. (2004), Dimmer and Scorrano (2006)
High levels of oxidative stress causes fragmentation of the mitochondrial network	Drp1 phosphorylation and Bid translocation increase fission activity	Grohm et al. (2010), Qi et al. (2011)
Amino acid and other nutrient deprivation causes hyperfusion by downregulation of Drp1	Fused mitochondrial network evades autophagic degradation during starvation	Rambold et al. (2011a, b)
Drp1 expression is transcriptionally activated by p53 protein in response to apoptotic stimuli	The miR-30 family of micro-RNA limit fission by suppressing p53 expression and Drp1 activation	Li et al. (2010)
Inhibition of histone deacetylases induces mitochondrial elongation	Fused networks occur due to decreased Fis1 expression and reduced Drp1 translocation	Lee et al. (2012)
Phosphorylation of Drp1 has opposing effects on fission depending on the kinase	Example: phosphorylation of Drp1 by cAMP kinase increases yet PKC delta decreases fission	Cribbs and Strack (2007), Qi et al. (2011), Kim et al. (2011)
Local organelle control		
Knockdown of Mff promotes elongation of the network and overexpression of Mff promotes fission	Mff recruits Drp-1 to fission sites on the outer membrane independently of hFis1	Otera et al. (2010), Gandre-Babbe and van der Bliek (2008)

Blocking the mitochondrial Na^+/Ca^{2+} exchanger increases interaction between Drp1 and Fis1	Elevated levels of mitochondrial calcium increases fission activity	Kaddour-Djebbar et al. (2010)
Treatment with cysteine-alkylators inhibits fission and fast mitochondrial movement	Loss of movement coincided with microtubule-dependent thin mitochondrial extensions	Bowes and Gupta (2005, 2008)
Conformation specificities and self-interaction dictate the ability of Fis1 to recruit fission machinery	Fis1 activity is regulated by two interaction interfaces and its ability to oligomerize	Serasinghe and Yoon (2008), Zhang and Chan (2007)
Fusion		
Global cellular control		
Activation of PGC1a/PGC-1b/ERRa induces MFN2 mRNA. PGC-1b induces mitochondrial fusion by Mfn2	Increased fusion activity accompanies mitochondrial biogenesis	Liesa et al. (2008), Soriano et al. (2006)
Mitochondrial tubularization and network fusion at G1-S of cell cycle	G1-S stimulates global mitochondrial fusion, mitosis stimulates fission	Lee et al. (2004), Taguchi et al. (2007), Arakaki et al. (2006), Kashatus et al. (2011)
Variation in isoform expression of OPA1 occurs via alternative gene splicing	Expression of the eight splice variants differs across tissues and alter fusion and apoptotic activity	Song et al. (2007), Olichon et al. (2007), Frezza et al. (2006)
The promoter region of MFN2 is a target of the tumor suppressor protein, p53	Mfn2 mRNA and protein levels are up-regulated in a p53-dependant manner	Wang et al. (2010)
Local organelle control		
Functional interaction of OPA1 with MFN1 and physical interaction between mitofusins and OPA1	Protein complex spans the two mitochondrial membranes and permits fusion activity	Cipolat et al. (2004), Guillery et al. (2008)
OPA1 processing by metalloproteases can block fusion activity and alter cristae structure	Provides a mechanism for fusion inhibition at the local level by protease activity	Song et al. (2007, 2009), Ehses et al. (2009), Duvezin-Caubet et al. (2006), Griparic et al. (2007), Guillery et al. (2008), Baricault et al. (2007), Kieper et al. (2010)
Low levels of local GTP induce outer membrane tethering while complete fusion events require high intra-mitochondrial GTP levels	Initial fusion is promoted in energy deficient environments yet complete fusion is regulated by energetic status of the organelles	Meeusen et al. (2006)
G-protein beta2 is enriched in the mitochondrial membrane and interacts with Mfn1 to regulate fusion	Gbeta2 regulates the mobility of Mfn1 within the outer membrane and promotes fusion	Zhang et al. (2010)

(continued)

Table 2.1 (continued)

Control mechanism	Outcome	References
Bcl-x(L) increases rates of fusion and fission with an observed overall network elongation	Bcl-x(L) increases mitochondrial mass concurrent with elevated dynamics cycling	Berman et al. (2009)
Solitary period		
Global cellular control		
Global ADP levels increase mitochondrial movement to synapses	ADP signals mitochondrial motility	Mironov (2009)
G-protein coupled receptor, Ga12, is expressed in mitochondria and regulates motility	GPCRs are sensitive to GDP/GTP levels and can regulate mitochondrial motility	Andreeva et al. (2008)
Bnip3 expression induces Drp1 mediated fission and parkin translocation in adult myocytes	Increased fission activity and parkin translocation enhanced mitophagy	Lee et al. (2011a)
Local organelle control		
Mitochondrial movement along microtubules occurs in an energy-dependent manner	Individual mitochondria move at different rates along microtubules based on ATP levels	Yi et al. (2004), Miller and Sheetz (2004), Guo et al. (2005)
Local redox status of mitochondria impacts membrane potential and velocity of movement	Elevated oxidation leads to depolarization and to increased motility	Gerencser and Nicholls (2008)
PINK1 and Parkin target Miro, mitofusins and other outer membrane proteins for proteasomal degradation and promote mitophagy	Proteasomal degradation of Miro and mitofusins isolate and immobilize mitochondria which increases the pre-autophagic pool	Tanaka et al. (2010a), Wang et al. (2011b), Weihofen et al. (2009), Ziviani et al. (2010), Ziviani and Whitworth (2010), Poole et al. (2008, 2010), Yang et al. (2008), Glauser et al. (2011), Rakovic et al. (2011), Gegg et al. (2010)

results in increased time spent in the solitary phase, and if membrane potential is not recovered, the mitochondrion enters into the pre-autophagic pool (Cipolat et al. 2004, 2006; Baricault et al. 2007; Twig and Shirihai 2011; Rambold et al. 2011a).

Another crucial local regulator of mitochondrial dynamics is the degree of movement of an individual mitochondrion (Twig and Shirihai 2011). Movement greatly increases the chances of fusion perhaps in part because microtubule transport aligns mitochondria (Twig et al. 2010). Alignment facilitates pole interaction between mitochondria and thus increases the likelihood of tethering between mitofusins. Motility depends on the calcium-sensitive mitochondrial Rho GTPase (Miro), which connects mitochondria to the ATP-dependant motor enzymes, dynein and kinesin (Boldogh and Pon 2007; Fehrenbacher et al. 2004; Saotome et al. 2008). Milton is a protein that complexes with Miro and is also required for transport of mitochondria (Glater et al. 2006; Rice and Gelfand 2006). Calcium binding to Miro inhibits motility by causing detachment of the motor protein complex from microtubules (Wang and Schwarz 2009a, b; Wang et al. 2011b). Calcium exchangers, serving a role in calcium buffering, are dependant on ATP, resulting in a dependency of the buffering capacity on mitochondrial ATP synthesis. This is leading to the detachment of mitochondria in calcium rich spots, thus creating a localized mechanism for selective delivery of mitochondria to cellular regions with unmet ATP needs (Yi et al. 2004). Elevated levels of cytosolic calcium inhibit motility by binding to Miro and decreased supply of ATP lowers the activity of ATPase-driven motor proteins. Kinesins themselves are unaffected by calcium levels, so by coupling both ATP availability and local calcium concentrations this local regulation specifically impacts mitochondrial movement and not general microtubule transport.

In neurons, proper distribution of mitochondria is of the upmost importance. The degree of dendritic arborization correlates with mitochondrial content and is dependant on Miro activity (Macaskill et al. 2009; Russo et al. 2009). Mitochondria with high membrane potential and elevated ATP production travel anterogradely to synaptic regions where there is a very high demand for energy (Miller and Sheetz 2004). Appropriately, global elevation of ADP levels in neurons increases delivery of mitochondria to synapses (Mironov 2009). On the contrary, there is fast retrograde transport of depolarized mitochondria with low membrane potential back to the soma to facilitate lysosomal degradation (Boldogh and Pon 2007; Gerencser and Nicholls 2008; Hollenbeck and Saxton 2005). In addition to supplying ATP, mitochondria fulfill a crucial role by buffering cytosolic calcium. The abundant neurotransmitter glutamate activates ionotropic NMDA receptors resulting in local increases in calcium influx. This relationship establishes an important regulatory mechanism for local inhibition of mitochondrial transport by Miro at active synaptic sites (Saotome et al. 2008; Wang and Schwarz 2009b; Macaskill et al. 2009).

The transcriptional and post-translational regulation of OPA1 serves as an excellent final example of both global and local control of mitochondrial dynamics. Through global signaling pathways, alterations in gene transcription can create eight isoforms of OPA1 (Landes et al. 2010). The distinct functions of the different forms of OPA1 are not well understood but clearly they can perform unique activities such as stabilizing cristae and protecting mtDNA (Semenzato et al. 2011;

Merkwirth et al. 2008; Frezza et al. 2006; Elachouri et al. 2011; Yu-Wai-Man et al. 2010). A shift in isoform production affects all mitochondria undergoing protein import and therefore represents a form of global regulation. On the other hand, imported OPA1 is cleaved to produce variants of different lengths by several mitochondrial proteases, including MPP, OMA1, PARL, and Yme1L (Song et al. 2007, 2009; Ehses et al. 2009; Cipolat et al. 2004, 2006; Griparic et al. 2007; Guillery et al. 2008; Ishihara et al. 2004). This proteolytic processing is dependant on membrane potential, metal ion levels, and ATP availability. Both long and short isoforms are required for proper inner membrane fusion and so represents a means of local regulation. For example, stress-induced cleavage of OPA1 long isoforms by OMA1 can disrupt interaction with Mfn1 and thereby block complete fusion (Ehses et al. 2009; Cipolat et al. 2004, 2006; Guillery et al. 2008). In this way, OPA1 is regulated both at the global and local levels to control fusion of mitochondria.

2.8 Mitochondrial Dynamics and Pathology

Alterations to mitochondrial fusion and fission have been demonstrated in several pathological conditions including neurodegeneration, obesity, and type II diabetes. Mutations in genes that encode for fusion proteins provide the clearest connection between mitochondrial dynamics and disease. Charcot-Marie-Tooth (CMT) disease Type 2A is caused by mutations in MFN2 and results in peripheral nervous system dysfunction (Ching et al. 2010; Casasnovas et al. 2010; Ouvrier and Grew 2010; Feely et al. 2011). The most common form of hereditary optic neuropathy is caused by mutations in OPA1 (Ferre et al. 2009; Nochez et al. 2009; Yu-Wai-Man et al. 2011a, b). These diseases confirm the importance of mitochondrial fusion in cell survival and also illustrate the existence of selective susceptibility amongst neuronal subtypes.

Neurons in general are vulnerable to mitochondrial dysfunction due to extreme energy demands coupled with complex, polarized cell structures. Degeneration in Parkinson's disease (PD) occurs selectively in neurons that exemplify these combined susceptibilities and genetic studies strongly implicate defects in mitochondrial dynamics and quality control (Wang et al. 2011b; Braak and Del Tredici 2008; Braak et al. 2004; Narendra and Youle 2011; Dagda and Chu 2009; Whitworth and Pallanck 2009). Specific cellular morphological characteristics create inherent challenges to the networking of mitochondrial populations within PD-sensitive neurons. The A9 dopaminergic (A9-DA) neurons of the substantia nigra elegantly illustrate this principle (Braak and Del Tredici 2008; Braak et al. 2004; Ferrer et al. 2011).

The nigral A9-DA neurons are so polarized and branched that their somas account for less than 1% of total cell volume (Sulzer 2007). Massive neuritic arborization occurs in both axonal and dendritic compartments such that each A9-DA neuron may contain more than 300,000 synapses in its axonal field alone (Arbuthnott and Wickens 2007; Matsuda et al. 2009; Surmeier et al. 2010a, b). Extreme cellular morphology creates a major logistical hurdle and heightens susceptibility to disruptions in mitochondrial transport. Dispersion also limits protective mechanisms of complementation and quality control by decreasing the likelihood of fusion events. Synapses are the most energy-demanding region of the neuron as well as being sites of voltage-gated calcium influx. Proper mitochondrial distribution is therefore critical not only to provide ATP but also to buffer calcium levels (Oliveira 2010; MacAskill et al. 2010).

To reach the pre-synaptic compartment, mitochondria must travel along long, thin, and poorly myelinated axons in A9-DA neurons (Braak et al. 2004). Each of these characteristics increases both metabolic demand and the parallel risk of oxidative stress. Length correlates with surface area and longer axons have increased requirements for ATPase activity by the sodium potassium exchanger (Na^+/K^+ ATPase). Greater distances also increase energy expenditures and travel time for motor proteins bringing cargo back and forth from soma to synapse. Longer retrograde transit times for damaged mitochondria autophagocytosed at synapses likely increases the risk and extent of oxidative damage en route back to lysosomes (Terman et al. 2010; Yue 2007; Yue et al. 2009). Like lanes on a road, width also impacts axonal transport. Thin caliber axons are spatially restrained and this reduces capacity for delivery of both mitochondria and autophagosomes. Thin A9-DA axons have higher surface area to volume ratios and therefore elevated energy demands due to higher basal Na^+/K^+ ATPase activity. Similarly, myelin limits the amount of surface area involved in ion exchange and this insulation dramatically impacts energy demands for maintaining ionic gradients required axonal conductance. Creating specialized sub-domains of the axon permits clustering of mitochondria in energy intensive micro-regions and that may promote mitochondrial fusion (Ohno et al. 2011). Poorly myelinated A9-DA neurons require high Na^+/K^+ ATPase activity over the entire length of the axon are thereby denied the potential benefits of mitochondrial clustering. Collectively, these characteristics of A9-DA neurons likely synergize to heighten sensitivity to disruptions in mitochondrial dynamics, motility, and mitophagy.

Studies of genetic mutations that cause recessive forms of familial PD support these predictions of heightened susceptibility of disruptions in mitochondrial dynamics and quality control in A9-DA neurons. Mutations in DJ-1, PINK1, and Parkin cause parkinsonism and originally these genes were thought to have disparate functions but recently their cellular roles were unified around mitochondrial dynamics and quality control (Narendra and Youle 2011; Dagda and Chu 2009; Whitworth and Pallanck 2009; Chu 2010a; Irrcher et al. 2010; Thomas et al. 2011). Together these genes provide protection against the extremes of mitochondrial membrane potential and ROS production. In this thermostat analogy DJ-1 targets mitochondria that produce excess ROS with normal to high membrane potential. The other extreme is handled by PINK1, which identifies mitochondria with little ROS production due to depolarized membrane potential. Parkin acts as a downstream effecter of both DJ-1 and PINK1 pathways to facilitate selective autophagic clearance of targeted mitochondria.

DJ-1 is a cytosolic chaperone protein that translocates to mitochondria in response to oxidative stress (Canet-Aviles et al. 2004; Moore et al. 2005; Xiong et al. 2009). Loss of DJ-1 function leads to aberrant mitochondrial morphology and function

(Irrcher et al. 2010; Thomas et al. 2011; Goldberg et al. 2005; Krebiehl et al. 2010) as well as increased sensitivity to mitochondrial toxins and ROS (Canet-Aviles et al. 2004; Kim et al. 2004, 2005; Ved et al. 2005; Zhang et al. 2005; Paterna et al. 2007; Taira et al. 2004; Menzies et al. 2005; Meulener et al. 2005). The protective functions of DJ-1 are dependant on a specific cysteine residue located at site 106 and this single amino acid allows DJ-1 to function as a cytoplasmic sensor of mitochondrial oxidative stress (Irrcher et al. 2010; Canet-Aviles et al. 2004; Blackinton et al. 2005, 2009). In support of this role, mitochondria isolated from mice lacking DJ-1 display greater ROS generation and phenotypes associated with loss of DJ-1 are reversed with antioxidants (Irrcher et al. 2010; Thomas et al. 2011). Replacing wild-type protein rescues the phenotypes of DJ-1 knockout cells but not when the replacement protein is mutated at site 106. Exactly how DJ-1 regulates mitochondrial ROS production is not fully clear but two plausible mechanisms include modulation of complex I activity and regulation of uncoupling protein expression (Hayashi et al. 2009; Guzman et al. 2010). Additionally, oxidation of DJ-1 results in binding with the cytosolic E3 ubiquitin ligase Parkin (Moore et al. 2005). DJ-1 functions upstream in this pathway as Parkin overexpression can rescue phenotypes associated with loss of DJ-1 (Irrcher et al. 2010; Thomas and Cookson 2009) but not the reverse (Dodson and Guo 2007).

Maintenance of homeostasis within the mitochondrial population is not complete with only protection from excess ROS by DJ-1. Production of ROS is connected to membrane potential and when a mitochondrion becomes depolarized it decreases ROS output. In this situation the dysfunctional mitochondrion would be undetected by the DJ-1-mediated quality control mechanism. An additional surveillance mechanism is needed to guard against this other potential extreme situation. PTEN-induced kinase 1 (PINK1) is a protein kinase with a mitochondrial targeting signal and a putative transmembrane domain (Chu 2010a, b; Mills et al. 2008). PINK1 is continuously imported into the intermembrane space where it is immediately targeted for degradation by several mitochondrial proteases. This normal turnover is interrupted when mitochondria depolarize and PINK1 accumulates, allowing for kinase signaling to bring about selective mitochondrial removal (Jin et al. 2010; Narendra et al. 2010a). In this way PINK1 provides a mechanism for monitoring mitochondrial function that is not based on ROS production. Loss of PINK1 function leads to a buildup of damaged mitochondrial material along with decreased membrane potential and ATP synthesis (Dagda and Chu 2009; Exner et al. 2007; Liu et al. 2009, 2011; Grunewald et al. 2009; Marongiu et al. 2009; Dagda et al. 2009a, b; Wood-Kaczmar et al. 2008; Gandhi et al. 2009). Mitochondrial dysfunction occurs prior to the onset of any neurodegeneration in mice lacking the PINK1 gene (Gispert et al. 2009; Narendra et al. 2008, 2009, 2010). One downstream event of PINK1 stabilization is the binding and phosphorylation of Miro/Milton mitochondrial transport complexes (Wang et al. 2011b; Weihofen et al. 2009). This PINK1 effect may isolate depolarized mitochondria by limiting their transport and help in the clearance by mitophagy.

Another consequence of depolarization-induced PINK1 accumulation related to mitophagy is the mitochondrial recruitment of Parkin. Cytosolic Parkin is selectively recruited to depolarized mitochondria, which facilitates autophagic elimination of the dysfunctional units (Narendra et al. 2008, 2009, 2010a) and PINK1 is required for this

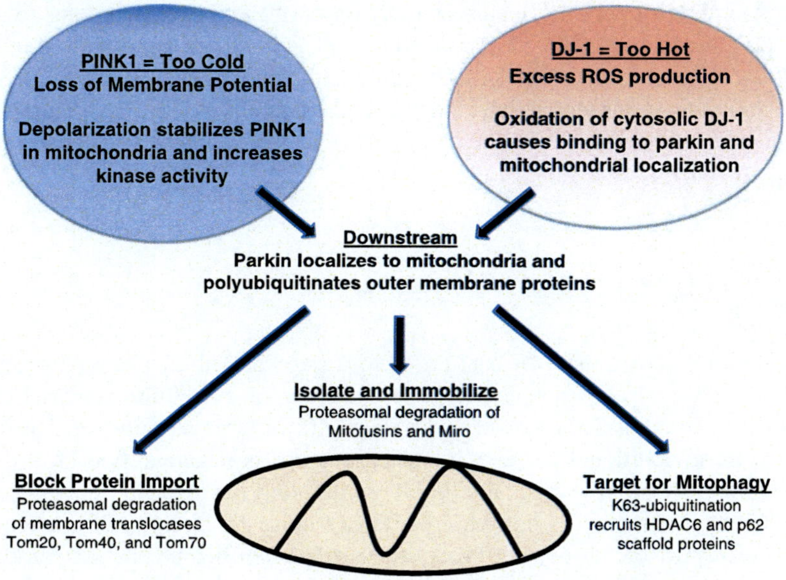

Fig. 2.2 The thermostat model of mitochondrial quality control. Separate pathways maintain mitochondrial homeostasis by safeguarding against functional extremes. In the cold extreme, PINK1 is stabilized within mitochondria upon membrane depolarization leading to increased kinase signaling. Alternatively, oxidative activation of cytosolic DJ-1 occurs in response to the hot extreme of excess ROS production. Downstream of both PINK1 and DJ-1 pathways is the recruitment of the E3 ligase Parkin and attachment of K63 and K48 polyubiquitin chains to mitochondrial outer membrane proteins. Damaged mitochondria are isolated and immobilized by proteasomal degradation of K48-tagged proteins, such as mitofusins and Miro. Proteasomal clearance of mitochondrial translocases prevents repopulation of the outer membrane with newly synthesized replacement proteins. Finally, K63 polyubiquitin chains selectively identifies mitochondria for autophagic clearance by recruitment of scaffold proteins, such as HDAC6 and p62

process (Geisler et al. 2010a, b; Vives-Bauza et al. 2010a, b, c; Vives-Bauza and Przedborski 2010). By attaching lysine 48 (K48) linked polyubiquitin chains, Parkin promotes proteasomal degradation of mitochondrial outer membrane proteins such as Miro, mitofusins, and several transporters (Tanaka et al. 2010a, b; Chan and Chan 2011; Chan et al. 2011; Ziviani et al. 2010; Ziviani and Whitworth 2010; Poole et al. 2008, 2010; Tanaka 2010). In this way K48-mediated proteasomal turnover of outer membrane proteins immobilizes and isolates damaged mitochondria to increase the likelihood of autophagic clearance. In addition, Parkin also creates recruitment signals for mitophagy via lysine 63-linked (K63) ubiquitin chains. Scaffold proteins, such as HDAC6 and p62/SQSTM1, bind to K63-ubiquitin and facilitate localization to the aggresome and clearance by mitophagy (Huang et al. 2011b; Geisler et al. 2010a; Lee et al. 2010; Okatsu et al. 2010; Narendra et al. 2010b; Ding et al. 2010a).

Quality control of mitochondrial performance occurs through two pathways in which either DJ-1 or PINK1 utilize Parkin as a downstream effector, summarized in Fig. 2.2. Age of onset studies from PD patients support this conceptual hierarchy.

Mutations in DJ-1 and PINK1 cause an early onset of Parkinson symptoms relative to the sporadic disease (~30–50 compared to ~60–80 years of age) (Abou-Sleiman et al. 2004; Mizuno et al. 2006). In accordance with being downstream in both pathways, Parkin mutations tend to have very early disease onset with a large number of juvenile cases occurring before the age of 30 (Kitada et al. 1998; Nisipeanu et al. 1999, 2001; Oliveri et al. 2001; Lucking et al. 2000).

2.9 Open Questions and Controversies

1. Studies have shown that Mff is required for fission and hFis1 is not (Otera et al. 2010). This raises the potential for the existence of other mitochondrial outer membrane proteins that can bind and anchor Mff for Drp1 recruitment. The identity of these additional proteins is not known nor is it known how they differ functionally from hFis1. Finally, hFis1 is definitely involved in mitochondrial fission but its function is not required. Could hFis1 be important for a specific form or aspect of fission, such as in the mechanism behind the generation of unequal division?
2. One crucial area of future study is understanding of the regulatory mechanisms for mitochondrial dynamics and how they connect to quality control. These are of particular interest as therapeutic targets since strategies aimed at maximizing upkeep of mitochondrial performance would have broad application across health. Can enhancement of mitochondrial dynamics, quality control, and turnover be a viable therapeutic strategy for treatment of chronic diseases such as diabetes and neurodegeneration?
3. A large portion of publications connecting the DJ-1-PINK1-Parkin pathways were performed with cell lines in nonphysiologic conditions. The degree to which these pathways exist in neurons, surprisingly, remains a matter of debate (Van Laar et al. 2011). It is also worth acknowledging that most patients with Parkin mutations lack Lewy Bodies, the intracellular neuropathological hallmarks of PD (Ahlskog 2009). Assuming the role of Parkin in neurons is to execute mitophagy downstream of DJ-1 and PINK1 signaling, the question of what connects Parkin-mediated mitophagy to Lewy Body formation is another tantalizing question facing the field of PD research.

References

Abou-Sleiman PM, Healy DG, Wood NW (2004) Causes of Parkinson's disease: genetics of DJ-1. Cell Tissue Res 318(1):185–188

Ahlskog JE (2009) Parkin and PINK1 parkinsonism may represent nigral mitochondrial cytopathies distinct from Lewy body Parkinson's disease. Parkinsonism Relat Disord 15(10):721–727

Andreeva AV, Kutuzov MA, Voyno-Yasenetskaya TA (2008) G alpha12 is targeted to the mitochondria and affects mitochondrial morphology and motility. FASEB J 22(8):2821–2831

Arakaki N et al (2006) Dynamics of mitochondria during the cell cycle. Biol Pharm Bull 29(9): 1962–1965

Arbuthnott GW, Wickens J (2007) Space, time and dopamine. Trends Neurosci 30(2):62–69

Arimura S et al (2004) Frequent fusion and fission of plant mitochondria with unequal nucleoid distribution. Proc Natl Acad Sci USA 101(20):7805–7808

Baricault L et al (2007) OPA1 cleavage depends on decreased mitochondrial ATP level and bivalent metals. Exp Cell Res 313(17):3800–3808

Berman SB et al (2009) Bcl-x L increases mitochondrial fission, fusion, and biomass in neurons. J Cell Biol 184(5):707–719

Betzig E et al (2006) Imaging intracellular fluorescent proteins at nanometer resolution. Science 313(5793):1642–1645

Blackinton J et al (2005) Effects of DJ-1 mutations and polymorphisms on protein stability and subcellular localization. Brain Res Mol Brain Res 134(1):76–83

Blackinton J et al (2009) Post-transcriptional regulation of mRNA associated with DJ-1 in sporadic Parkinson disease. Neurosci Lett 452(1):8–11

Boldogh IR, Pon LA (2007) Mitochondria on the move. Trends Cell Biol 17(10):502–510

Bowes TJ, Gupta RS (2005) Induction of mitochondrial fusion by cysteine-alkylators ethacrynic acid and N-ethylmaleimide. J Cell Physiol 202(3):796–804

Bowes T, Gupta RS (2008) Novel mitochondrial extensions provide evidence for a link between microtubule-directed movement and mitochondrial fission. Biochem Biophys Res Commun 376(1):40–45

Braak H, Del Tredici K (2008) Invited Article: Nervous system pathology in sporadic Parkinson disease. Neurology 70(20):1916–1925

Braak H et al (2004) Stages in the development of Parkinson's disease-related pathology. Cell Tissue Res 318(1):121–134

Braschi E, Zunino R, McBride HM (2009) MAPL is a new mitochondrial SUMO E3 ligase that regulates mitochondrial fission. EMBO Rep 10(7):748–754

Busch KB et al (2006) Mitochondrial dynamics generate equal distribution but patchwork localization of respiratory Complex I. Mol Membr Biol 23(6):509–520

Canet-Aviles RM et al (2004) The Parkinson's disease protein DJ-1 is neuroprotective due to cysteine-sulfinic acid-driven mitochondrial localization. Proc Natl Acad Sci USA 101(24):9103–9108

Casasnovas C et al (2010) Phenotypic spectrum of MFN2 mutations in the Spanish population. J Med Genet 47(4):249–256

Cereghetti GM et al (2008) Dephosphorylation by calcineurin regulates translocation of Drp1 to mitochondria. Proc Natl Acad Sci USA 105(41):15803–15808

Cereghetti GM, Costa V, Scorrano L (2010) Inhibition of Drp1-dependent mitochondrial fragmentation and apoptosis by a polypeptide antagonist of calcineurin. Cell Death Differ 17(11):1785–1794

Chan DC (2006) Mitochondria: dynamic organelles in disease, aging, and development. Cell 125(7):1241–1252

Chan NC, Chan DC (2011) Parkin uses the UPS to ship off dysfunctional mitochondria. Autophagy 7(7):771–772

Chan NC et al (2011) Broad activation of the ubiquitin-proteasome system by Parkin is critical for mitophagy. Hum Mol Genet 20(9):1726–1737

Chen H, Chan DC (2005) Emerging functions of mammalian mitochondrial fusion and fission. Hum Mol Genet 14 Spec No. 2:R283–R289

Chen H et al (2003) Mitofusins Mfn1 and Mfn2 coordinately regulate mitochondrial fusion and are essential for embryonic development. J Cell Biol 160(2):189–200

Chen H, Chomyn A, Chan DC (2005) Disruption of fusion results in mitochondrial heterogeneity and dysfunction. J Biol Chem 280(28):26185–26192

Chen Y, Liu Y, Dorn GW 2nd (2011) Mitochondrial fusion is essential for organelle function and cardiac homeostasis. Circ Res 109(12):1327–1331

Ching CK et al (2010) A novel mitofusin 2 gene mutation causing Charcot-Marie-Tooth type 2A disease in a Chinese family. Chin Med J (Engl) 123(11):1466–1469

Cho DH et al (2009) S-nitrosylation of Drp1 mediates beta-amyloid-related mitochondrial fission and neuronal injury. Science 324(5923):102–105
Choudhary V et al (2011) Novel role of androgens in mitochondrial fission and apoptosis. Mol Cancer Res 9(8):1067–1077
Chu CT (2010a) A pivotal role for PINK1 and autophagy in mitochondrial quality control: implications for Parkinson disease. Hum Mol Genet 19(R1):R28–R37
Chu CT (2010b) Tickled PINK1: mitochondrial homeostasis and autophagy in recessive Parkinsonism. Biochim Biophys Acta 1802(1):20–28
Cipolat S et al (2004) OPA1 requires mitofusin 1 to promote mitochondrial fusion. Proc Natl Acad Sci USA 101(45):15927–15932
Cipolat S et al (2006) Mitochondrial rhomboid PARL regulates cytochrome c release during apoptosis via OPA1-dependent cristae remodeling. Cell 126(1):163–175
Cribbs JT, Strack S (2007) Reversible phosphorylation of Drp1 by cyclic AMP-dependent protein kinase and calcineurin regulates mitochondrial fission and cell death. EMBO Rep 8(10): 939–944
Dagda RK, Chu CT (2009) Mitochondrial quality control: insights on how Parkinson's disease related genes PINK1, parkin, and Omi/HtrA2 interact to maintain mitochondrial homeostasis. J Bioenerg Biomembr 41(6):473–479
Dagda RK et al (2009a) Loss of PINK1 function promotes mitophagy through effects on oxidative stress and mitochondrial fission. J Biol Chem 284(20):13843–13855
Dagda RK, Zhu J, Chu CT (2009b) Mitochondrial kinases in Parkinson's disease: converging insights from neurotoxin and genetic models. Mitochondrion 9(5):289–298
de Brito OM, Scorrano L (2008) Mitofusin 2 tethers endoplasmic reticulum to mitochondria. Nature 456(7222):605–610
Dimmer KS, Scorrano L (2006) (De)constructing mitochondria: what for? Physiology (Bethesda) 21:233–241
Ding WX et al (2010a) Nix is critical to two distinct phases of mitophagy, reactive oxygen species-mediated autophagy induction and Parkin-ubiquitin-p62-mediated mitochondrial priming. J Biol Chem 285(36):27879–27890
Ding H et al (2010b) Response of mitochondrial fusion and fission protein gene expression to exercise in rat skeletal muscle. Biochim Biophys Acta 1800(3):250–256
Dodson MW, Guo M (2007) Pink1, Parkin, DJ-1 and mitochondrial dysfunction in Parkinson's disease. Curr Opin Neurobiol 17(3):331–337
Duvezin-Caubet S et al (2006) Proteolytic processing of OPA1 links mitochondrial dysfunction to alterations in mitochondrial morphology. J Biol Chem 281(49):37972–37979
Ehses S et al (2009) Regulation of OPA1 processing and mitochondrial fusion by m-AAA protease isoenzymes and OMA1. J Cell Biol 187(7):1023–1036
Elachouri G et al (2011) OPA1 links human mitochondrial genome maintenance to mtDNA replication and distribution. Genome Res 21(1):12–20
Elmore SP et al (2001) The mitochondrial permeability transition initiates autophagy in rat hepatocytes. FASEB J 15(12):2286–2287
Exner N et al (2007) Loss-of-function of human PINK1 results in mitochondrial pathology and can be rescued by parkin. J Neurosci 27(45):12413–12418
Feely SM et al (2011) MFN2 mutations cause severe phenotypes in most patients with CMT2A. Neurology 76(20):1690–1696
Fehrenbacher KL et al (2004) Live cell imaging of mitochondrial movement along actin cables in budding yeast. Curr Biol 14(22):1996–2004
Ferre M et al (2009) Molecular screening of 980 cases of suspected hereditary optic neuropathy with a report on 77 novel OPA1 mutations. Hum Mutat 30(7):E692–E705
Ferrer I et al (2011) Neuropathology of sporadic Parkinson disease before the appearance of parkinsonism: preclinical Parkinson disease. J Neural Transm 118(5):821–839
Figueroa-Romero C et al (2009) SUMOylation of the mitochondrial fission protein Drp1 occurs at multiple nonconsensus sites within the B domain and is linked to its activity cycle. FASEB J 23(11):3917–3927

Frank S et al (2001) The role of dynamin-related protein 1, a mediator of mitochondrial fission, in apoptosis. Dev Cell 1(4):515–525
Frezza C et al (2006) OPA1 controls apoptotic cristae remodeling independently from mitochondrial fusion. Cell 126(1):177–189
Fukushima NH et al (2001) The GTPase effector domain sequence of the Dnm1p GTPase regulates self-assembly and controls a rate-limiting step in mitochondrial fission. Mol Biol Cell 12(9):2756–2766
Gandhi S et al (2009) PINK1-associated Parkinson's disease is caused by neuronal vulnerability to calcium-induced cell death. Mol Cell 33(5):627–638
Gandre-Babbe S, van der Bliek AM (2008) The novel tail-anchored membrane protein Mff controls mitochondrial and peroxisomal fission in mammalian cells. Mol Biol Cell 19(6):2402–2412
Gegg ME et al (2010) Mitofusin 1 and mitofusin 2 are ubiquitinated in a PINK1/parkin-dependent manner upon induction of mitophagy. Hum Mol Genet 19(24):4861–4870
Geisler S et al (2010a) PINK1/Parkin-mediated mitophagy is dependent on VDAC1 and p62/SQSTM1. Nat Cell Biol 12(2):119–131
Geisler S et al (2010b) The PINK1/Parkin-mediated mitophagy is compromised by PD-associated mutations. Autophagy 6(7):871–878
Gerencser AA, Nicholls DG (2008) Measurement of instantaneous velocity vectors of organelle transport: mitochondrial transport and bioenergetics in hippocampal neurons. Biophys J 95(6):3079–3099
Gilkerson RW et al (2008) Mitochondrial nucleoids maintain genetic autonomy but allow for functional complementation. J Cell Biol 181(7):1117–1128
Gispert S et al (2009) Parkinson phenotype in aged PINK1-deficient mice is accompanied by progressive mitochondrial dysfunction in absence of neurodegeneration. PLoS One 4(6):e5777
Glater EE et al (2006) Axonal transport of mitochondria requires milton to recruit kinesin heavy chain and is light chain independent. J Cell Biol 173(4):545–557
Glauser L et al (2011) Parkin promotes the ubiquitination and degradation of the mitochondrial fusion factor mitofusin 1. J Neurochem 118(4):636–645
Goldberg MS et al (2005) Nigrostriatal dopaminergic deficits and hypokinesia caused by inactivation of the familial Parkinsonism-linked gene DJ-1. Neuron 45(4):489–496
Gomes LC, Scorrano L (2008) High levels of Fis1, a pro-fission mitochondrial protein, trigger autophagy. Biochim Biophys Acta 1777(7–8):860–866
Gottlieb E (2006) OPA1 and PARL keep a lid on apoptosis. Cell 126(1):27–29
Griffin EE, Detmer SA, Chan DC (2006) Molecular mechanism of mitochondrial membrane fusion. Biochim Biophys Acta 1763(5–6):482–489
Griparic L, Kanazawa T, van der Bliek AM (2007) Regulation of the mitochondrial dynamin-like protein Opa1 by proteolytic cleavage. J Cell Biol 178(5):757–764
Grohm J, Plesnila N, Culmsee C (2010) Bid mediates fission, membrane permeabilization and peri-nuclear accumulation of mitochondria as a prerequisite for oxidative neuronal cell death. Brain Behav Immun 24(5):831–838
Grunewald A et al (2009) Differential effects of PINK1 nonsense and missense mutations on mitochondrial function and morphology. Exp Neurol 219(1):266–273
Guillery O et al (2008) Metalloprotease-mediated OPA1 processing is modulated by the mitochondrial membrane potential. Biol Cell 100(5):315–325
Guo X et al (2005) The GTPase dMiro is required for axonal transport of mitochondria to Drosophila synapses. Neuron 47(3):379–393
Guzman JN et al (2010) Oxidant stress evoked by pacemaking in dopaminergic neurons is attenuated by DJ-1. Nature 468(7324):696–700
Harder Z, Zunino R, McBride H (2004) Sumo1 conjugates mitochondrial substrates and participates in mitochondrial fission. Curr Biol 14(4):340–345
Hayashi T et al (2009) DJ-1 binds to mitochondrial complex I and maintains its activity. Biochem Biophys Res Commun 390(3):667–672
Hollenbeck PJ, Saxton WM (2005) The axonal transport of mitochondria. J Cell Sci 118(Pt 23):5411–5419

Hom J et al (2010) Regulation of mitochondrial fission by intracellular Ca^{2+} in rat ventricular myocytes. Biochim Biophys Acta 1797(6–7):913–921
Hori A, Yoshida M, Ling F (2011) Mitochondrial fusion increases the mitochondrial DNA copy number in budding yeast. Genes Cells 16(5):527–544
Huang P, Galloway CA, Yoon Y (2011a) Control of mitochondrial morphology through differential interactions of mitochondrial fusion and fission proteins. PLoS One 6(5):e20655
Huang C et al (2011b) Preconditioning involves selective mitophagy mediated by Parkin and p62/SQSTM1. PLoS One 6(6):e20975
Hyde BB, Twig G, Shirihai OS (2010) Organellar vs cellular control of mitochondrial dynamics. Semin Cell Dev Biol 21(6):575–581
Irrcher I et al (2010) Loss of the Parkinson's disease-linked gene DJ-1 perturbs mitochondrial dynamics. Hum Mol Genet 19(19):3734–3746
Ishihara N, Eura Y, Mihara K (2004) Mitofusin 1 and 2 play distinct roles in mitochondrial fusion reactions via GTPase activity. J Cell Sci 117(Pt 26):6535–6546
Ishihara N et al (2009) Mitochondrial fission factor Drp1 is essential for embryonic development and synapse formation in mice. Nat Cell Biol 11(8):958–966
Jakobs S (2006) High resolution imaging of live mitochondria. Biochim Biophys Acta 1763(5–6):561–575
Jakobs S, Schauss AC, Hell SW (2003) Photoconversion of matrix targeted GFP enables analysis of continuity and intermixing of the mitochondrial lumen. FEBS Lett 554(1–2):194–200
James DI et al (2003) hFis1, a novel component of the mammalian mitochondrial fission machinery. J Biol Chem 278(38):36373–36379
Jin SM et al (2010) Mitochondrial membrane potential regulates PINK1 import and proteolytic destabilization by PARL. J Cell Biol 191(5):933–942
Jourdain I, Gachet Y, Hyams JS (2009) The dynamin related protein Dnm1 fragments mitochondria in a microtubule-dependent manner during the fission yeast cell cycle. Cell Motil Cytoskeleton 66(8):509–523
Jung HS, Lee MS (2009) Macroautophagy in homeostasis of pancreatic beta-cell. Autophagy 5(2):241–243
Kaddour-Djebbar I et al (2010) Specific mitochondrial calcium overload induces mitochondrial fission in prostate cancer cells. Int J Oncol 36(6):1437–1444
Kane LA, Youle RJ (2010) Mitochondrial fission and fusion and their roles in the heart. J Mol Med (Berl) 88(10):971–979
Karbowski M (2010) Mitochondria on guard: role of mitochondrial fusion and fission in the regulation of apoptosis. Adv Exp Med Biol 687:131–142
Karbowski M et al (2002) Spatial and temporal association of Bax with mitochondrial fission sites, Drp1, and Mfn2 during apoptosis. J Cell Biol 159(6):931–938
Karbowski M et al (2004a) Quantitation of mitochondrial dynamics by photolabeling of individual organelles shows that mitochondrial fusion is blocked during the Bax activation phase of apoptosis. J Cell Biol 164(4):493–499
Karbowski M, Jeong SY, Youle RJ (2004b) Endophilin B1 is required for the maintenance of mitochondrial morphology. J Cell Biol 166(7):1027–1039
Kashatus DF et al (2011) RALA and RALBP1 regulate mitochondrial fission at mitosis. Nat Cell Biol 13(9):1108–1115
Kieper N et al (2010) Modulation of mitochondrial function and morphology by interaction of Omi/HtrA2 with the mitochondrial fusion factor OPA1. Exp Cell Res 316(7):1213–1224
Kim AJ, Lee CS, Schlessinger D (2004) Bex3 associates with replicating mitochondria and is involved in possible growth control of F9 teratocarcinoma cells. Gene 343(1):79–89
Kim RH et al (2005) Hypersensitivity of DJ-1-deficient mice to 1-methyl-4-phenyl-1,2,3,6-tetrahydropyrindine (MPTP) and oxidative stress. Proc Natl Acad Sci USA 102(14):5215–5220
Kim H et al (2011) Fine-tuning of Drp1/Fis1 availability by AKAP121/Siah2 regulates mitochondrial adaptation to hypoxia. Mol Cell 44(4):532–544
Kimura T et al (2011) Autophagy protects the proximal tubule from degeneration and acute ischemic injury. J Am Soc Nephrol 22(5):902–913

Kitada T et al (1998) Mutations in the parkin gene cause autosomal recessive juvenile parkinsonism. Nature 392(6676):605–608

Koch A et al (2005) A role for Fis1 in both mitochondrial and peroxisomal fission in mammalian cells. Mol Biol Cell 16(11):5077–5086

Kong D et al (2005) Regulation of Ca^{2+}-induced permeability transition by Bcl-2 is antagonized by Drpl and hFis1. Mol Cell Biochem 272(1–2):187–199

Koshiba T et al (2004) Structural basis of mitochondrial tethering by mitofusin complexes. Science 305(5685):858–862

Krebiehl G et al (2010) Reduced basal autophagy and impaired mitochondrial dynamics due to loss of Parkinson's disease-associated protein DJ-1. PLoS One 5(2):e9367

Landes T, Martinou JC (2011) Mitochondrial outer membrane permeabilization during apoptosis: the role of mitochondrial fission. Biochim Biophys Acta 1813(4):540–545

Landes T et al (2010) OPA1 (dys)functions. Semin Cell Dev Biol 21(6):593–598

Lee YJ et al (2004) Roles of the mammalian mitochondrial fission and fusion mediators Fis1, Drp1, and Opa1 in apoptosis. Mol Biol Cell 15(11):5001–5011

Lee JY et al (2010) Disease-causing mutations in parkin impair mitochondrial ubiquitination, aggregation, and HDAC6-dependent mitophagy. J Cell Biol 189(4):671–679

Lee Y et al (2011a) Mitochondrial autophagy by Bnip3 involves Drp1-mediated mitochondrial fission and recruitment of Parkin in cardiac myocytes. Am J Physiol Heart Circ Physiol 301(5):H1924–H1931

Lee HM et al (2011b) Autophagy negatively regulates keratinocyte inflammatory responses via scaffolding protein p62/SQSTM1. J Immunol 186(2):1248–1258

Lee JS et al (2012) Histone deacetylase inhibitors induce mitochondrial elongation. J Cell Physiol 227(7):2856–2869

Legesse-Miller A, Massol RH, Kirchhausen T (2003) Constriction and Dnm1p recruitment are distinct processes in mitochondrial fission. Mol Biol Cell 14(5):1953–1963

Legros F et al (2002) Mitochondrial fusion in human cells is efficient, requires the inner membrane potential, and is mediated by mitofusins. Mol Biol Cell 13(12):4343–4354

Legros F et al (2004) Organization and dynamics of human mitochondrial DNA. J Cell Sci 117(13):2653–2662

Lewis MR, Lewis WH (1914) Mitochondria in tissue culture. Science 39(1000):330–333

Li J et al (2010) miR-30 regulates mitochondrial fission through targeting p53 and the dynamin-related protein-1 pathway. PLoS Genet 6(1):e1000795

Liesa M et al (2008) Mitochondrial fusion is increased by the nuclear coactivator PGC-1beta. PLoS One 3(10):e3613

Liu W et al (2009) PINK1 defect causes mitochondrial dysfunction, proteasomal deficit and alpha-synuclein aggregation in cell culture models of Parkinson's disease. PLoS One 4(2):e4597

Liu W et al (2011) Pink1 regulates the oxidative phosphorylation machinery via mitochondrial fission. Proc Natl Acad Sci USA 108(31):12920–12924

Loew LM et al (1993) Imaging in five dimensions: time-dependent membrane potentials in individual mitochondria. Biophys J 65(6):2396–2407

Lucking CB et al (2000) Association between early-onset Parkinson's disease and mutations in the parkin gene. N Engl J Med 342(21):1560–1567

Macaskill AF et al (2009) Miro1 is a calcium sensor for glutamate receptor-dependent localization of mitochondria at synapses. Neuron 61(4):541–555

MacAskill AF, Atkin TA, Kittler JT (2010) Mitochondrial trafficking and the provision of energy and calcium buffering at excitatory synapses. Eur J Neurosci 32(2):231–240

Malka F et al (2005) Separate fusion of outer and inner mitochondrial membranes. EMBO Rep 6(9):853–859

Marongiu R et al (2009) Mutant Pink1 induces mitochondrial dysfunction in a neuronal cell model of Parkinson's disease by disturbing calcium flux. J Neurochem 108(6):1561–1574

Matsuda W et al (2009) Single nigrostriatal dopaminergic neurons form widely spread and highly dense axonal arborizations in the neostriatum. J Neurosci 29(2):444–453

Mazzoni C, Falcone C (2011) The importance of mitochondrial fusion in aging. Cell Cycle 10(21):3631
Meeusen S et al (2006) Mitochondrial inner-membrane fusion and crista maintenance requires the dynamin-related GTPase Mgm1. Cell 127(2):383–395
Mendl N et al (2011) Mitophagy in yeast is independent of mitochondrial fission and requires the stress response gene WHI2. J Cell Sci 124(Pt 8):1339–1350
Menzies FM, Yenisetti SC, Min KT (2005) Roles of *Drosophila* DJ-1 in survival of dopaminergic neurons and oxidative stress. Curr Biol 15(17):1578–1582
Merkwirth C et al (2008) Prohibitins control cell proliferation and apoptosis by regulating OPA1-dependent cristae morphogenesis in mitochondria. Genes Dev 22(4):476–488
Meulener M et al (2005) Drosophila DJ-1 mutants are selectively sensitive to environmental toxins associated with Parkinson's disease. Curr Biol 15(17):1572–1577
Miller KE, Sheetz MP (2004) Axonal mitochondrial transport and potential are correlated. J Cell Sci 117(Pt 13):2791–2804
Mills RD et al (2008) Biochemical aspects of the neuroprotective mechanism of PTEN-induced kinase-1 (PINK1). J Neurochem 105(1):18–33
Mironov SL (2009) Complexity of mitochondrial dynamics in neurons and its control by ADP produced during synaptic activity. Int J Biochem Cell Biol 41(10):2005–2014
Misko A et al (2010) Mitofusin 2 is necessary for transport of axonal mitochondria and interacts with the Miro/Milton complex. J Neurosci 30(12):4232–4240
Mizuno Y et al (2006) Progress in familial Parkinson's disease. J Neural Transm Suppl 70: 191–204
Molina AJ, Shirihai OS (2009) Monitoring mitochondrial dynamics with photoactivatable [corrected] green fluorescent protein. Methods Enzymol 457:289–304
Molina AJ et al (2009) Mitochondrial networking protects beta-cells from nutrient-induced apoptosis. Diabetes 58(10):2303–2315
Moore DJ et al (2005) Association of DJ-1 and parkin mediated by pathogenic DJ-1 mutations and oxidative stress. Hum Mol Genet 14(1):71–84
Nakada K, Inoue K, Hayashi J (2001a) Interaction theory of mammalian mitochondria. Biochem Biophys Res Commun 288(4):743–746
Nakada K et al (2001b) Inter-mitochondrial complementation: mitochondria-specific system preventing mice from expression of disease phenotypes by mutant mtDNA. Nat Med 7(8):934–940
Narender T et al (2010) An unprecedented biogenetic-type chemical synthesis of 1(15–>11) abeo-taxanes from normal taxanes. J Nat Prod 73(4):747–750
Narendra DP, Youle RJ (2011) Targeting mitochondrial dysfunction: role for PINK1 and Parkin in mitochondrial quality control. Antioxid Redox Signal 14(10):1929–1938
Narendra D et al (2008) Parkin is recruited selectively to impaired mitochondria and promotes their autophagy. J Cell Biol 183(5):795–803
Narendra D et al (2009) Parkin-induced mitophagy in the pathogenesis of Parkinson disease. Autophagy 5(5):706–708
Narendra DP et al (2010a) PINK1 is selectively stabilized on impaired mitochondria to activate Parkin. PLoS Biol 8(1):e1000298
Narendra D et al (2010b) p62/SQSTM1 is required for Parkin-induced mitochondrial clustering but not mitophagy; VDAC1 is dispensable for both. Autophagy 6(8):1090–1106
Neuspiel M et al (2005) Activated mitofusin 2 signals mitochondrial fusion, interferes with Bax activation, and reduces susceptibility to radical induced depolarization. J Biol Chem 280(26):25060–25070
Nisipeanu P et al (1999) Autosomal-recessive juvenile parkinsonism in a Jewish Yemenite kindred: mutation of Parkin gene. Neurology 53(7):1602–1604
Nisipeanu P et al (2001) Parkin gene causing benign autosomal recessive juvenile parkinsonism. Neurology 56(11):1573–1575
Nochez Y et al (2009) Acute and late-onset optic atrophy due to a novel OPA1 mutation leading to a mitochondrial coupling defect. Mol Vis 15:598–608

Ohno N et al (2011) Myelination and axonal electrical activity modulate the distribution and motility of mitochondria at CNS nodes of Ranvier. J Neurosci 31(20):7249–7258

Okatsu K et al (2010) p62/SQSTM1 cooperates with Parkin for perinuclear clustering of depolarized mitochondria. Genes Cells 15(8):887–900

Olichon A et al (2007) OPA1 alternate splicing uncouples an evolutionary conserved function in mitochondrial fusion from a vertebrate restricted function in apoptosis. Cell Death Differ 14(4):682–692

Oliveira JM (2010) Mitochondrial bioenergetics and dynamics in Huntington's disease: tripartite synapses and selective striatal degeneration. J Bioenerg Biomembr 42(3):227–234

Oliveri RL et al (2001) The parkin gene is not involved in late-onset Parkinson's disease. Neurology 57(2):359–362

Ono T et al (2001) Human cells are protected from mitochondrial dysfunction by complementation of DNA products in fused mitochondria. Nat Genet 28(3):272–275

Otera H et al (2010) Mff is an essential factor for mitochondrial recruitment of Drp1 during mitochondrial fission in mammalian cells. J Cell Biol 191(6):1141–1158

Ouvrier R, Grew S (2010) Mechanisms of disease and clinical features of mutations of the gene for mitofusin 2: an important cause of hereditary peripheral neuropathy with striking clinical variability in children and adults. Dev Med Child Neurol 52(4):328–330

Parone PA et al (2008) Preventing mitochondrial fission impairs mitochondrial function and leads to loss of mitochondrial DNA. PLoS One 3(9):e3257

Partikian A et al (1998) Rapid diffusion of green fluorescent protein in the mitochondrial matrix. J Cell Biol 140(4):821–829

Paterna JC et al (2007) DJ-1 and Parkin modulate dopamine-dependent behavior and inhibit MPTP-induced nigral dopamine neuron loss in mice. Mol Ther 15(4):698–704

Patterson GH, Lippincott-Schwartz J (2002) A photoactivatable GFP for selective photolabeling of proteins and cells. Science 297(5588):1873–1877

Poole AC et al (2008) The PINK1/Parkin pathway regulates mitochondrial morphology. Proc Natl Acad Sci USA 105(5):1638–1643

Poole AC et al (2010) The mitochondrial fusion-promoting factor mitofusin is a substrate of the PINK1/parkin pathway. PLoS One 5(4):e10054

Qi X et al (2011) Aberrant mitochondrial fission in neurons induced by protein kinase C{delta} under oxidative stress conditions in vivo. Mol Biol Cell 22(2):256–265

Rakovic A et al (2011) Mutations in PINK1 and Parkin impair ubiquitination of Mitofusins in human fibroblasts. PLoS One 6(3):e16746

Rambold AS, Kostelecky B, Lippincott-Schwartz J (2011) Together we are stronger: fusion protects mitochondria from autophagosomal degradation. Autophagy 7(12):1568–1569

Rambold AS et al (2011b) Tubular network formation protects mitochondria from autophagosomal degradation during nutrient starvation. Proc Natl Acad Sci USA 108(25):10190–10195

Rice SE, Gelfand VI (2006) Paradigm lost: milton connects kinesin heavy chain to miro on mitochondria. J Cell Biol 173(4):459–461

Russo GJ et al (2009) Drosophila Miro is required for both anterograde and retrograde axonal mitochondrial transport. J Neurosci 29(17):5443–5455

Santel A, Frank S (2008) Shaping mitochondria: the complex posttranslational regulation of the mitochondrial fission protein DRP1. IUBMB Life 60(7):448–455

Saotome M et al (2008) Bidirectional Ca^{2+}-dependent control of mitochondrial dynamics by the Miro GTPase. Proc Natl Acad Sci USA 105(52):20728–20733

Scarpulla RC (2002a) Nuclear activators and coactivators in mammalian mitochondrial biogenesis. Biochim Biophys Acta 1576(1–2):1–14

Scarpulla RC (2002b) Transcriptional activators and coactivators in the nuclear control of mitochondrial function in mammalian cells. Gene 286(1):81–89

Schauss AC et al (2010) A novel cell-free mitochondrial fusion assay amenable for high-throughput screenings of fusion modulators. BMC Biol 8:100

Scorrano L (2005) Proteins that fuse and fragment mitochondria in apoptosis: con-fissing a deadly con-fusion? J Bioenerg Biomembr 37(3):165–170

Semenzato M, Cogliati S, Scorrano L (2011) Prohibitin(g) cancer: aurilide and killing by Opa1-dependent cristae remodeling. Chem Biol 18(1):8–9
Serasinghe MN, Yoon Y (2008) The mitochondrial outer membrane protein hFis1 regulates mitochondrial morphology and fission through self-interaction. Exp Cell Res 314(19):3494–3507
Shaw JM, Nunnari J (2002) Mitochondrial dynamics and division in budding yeast. Trends Cell Biol 12(4):178–184
Sheridan C, Martin SJ (2010) Mitochondrial fission/fusion dynamics and apoptosis. Mitochondrion 10(6):640–648
Sheridan C et al (2008) Bax- or Bak-induced mitochondrial fission can be uncoupled from cytochrome C release. Mol Cell 31(4):570–585
Shitara H et al (2000) Selective and continuous elimination of mitochondria microinjected into mouse eggs from spermatids, but not from liver cells, occurs throughout embryogenesis. Genetics 156(3):1277–1284
Shroff EH et al (2009) BH3 peptides induce mitochondrial fission and cell death independent of BAX/BAK. PLoS One 4(5):e5646
Skulachev VP (2001) Mitochondrial filaments and clusters as intracellular power-transmitting cables. Trends Biochem Sci 26(1):23–29
Song Z et al (2007) OPA1 processing controls mitochondrial fusion and is regulated by mRNA splicing, membrane potential, and Yme1L. J Cell Biol 178(5):749–755
Song Z et al (2009) Mitofusins and OPA1 mediate sequential steps in mitochondrial membrane fusion. Mol Biol Cell 20(15):3525–3532
Soriano FX et al (2006) Evidence for a mitochondrial regulatory pathway defined by peroxisome proliferator-activated receptor-gamma coactivator-1 alpha, estrogen-related receptor-alpha, and mitofusin 2. Diabetes 55(6):1783–1791
Sulzer D (2007) Multiple hit hypotheses for dopamine neuron loss in Parkinson's disease. Trends Neurosci 30(5):244–250
Surmeier DJ, Guzman JN, Sanchez-Padilla J (2010a) Calcium, cellular aging, and selective neuronal vulnerability in Parkinson's disease. Cell Calcium 47(2):175–182
Surmeier DJ et al (2010b) What causes the death of dopaminergic neurons in Parkinson's disease? Prog Brain Res 183:59–77
Taguchi N et al (2007) Mitotic phosphorylation of dynamin-related GTPase Drp1 participates in mitochondrial fission. J Biol Chem 282(15):11521–11529
Taira T et al (2004) DJ-1 has a role in antioxidative stress to prevent cell death. EMBO Rep 5(2):213–218
Takano-Ohmuro H et al (2000) Autophagy in embryonic erythroid cells: its role in maturation. Eur J Cell Biol 79(10):759–764
Tan AR et al (2011) Elevated intracellular calcium causes distinct mitochondrial remodelling and calcineurin-dependent fission in astrocytes. Cell Calcium 49(2):108–114
Tanaka A (2010) Parkin-mediated selective mitochondrial autophagy, mitophagy: Parkin purges damaged organelles from the vital mitochondrial network. FEBS Lett 584(7):1386–1392
Tanaka A et al (2010a) Proteasome and p97 mediate mitophagy and degradation of mitofusins induced by Parkin. J Cell Biol 191(7):1367–1380
Tanaka K, Matsuda N, Okatsu K (2010b) Mechanisms underlying the cause of Parkinson's disease: the functions of Parkin/PINK1. Rinsho Shinkeigaku 50(11):867
Taneike M et al (2010) Inhibition of autophagy in the heart induces age-related cardiomyopathy. Autophagy 6(5):600–606
Terman A et al (2010) Mitochondrial turnover and aging of long-lived postmitotic cells: the mitochondrial-lysosomal axis theory of aging. Antioxid Redox Signal 12(4):503–535
Thomas KJ, Cookson MR (2009) The role of PTEN-induced kinase 1 in mitochondrial dysfunction and dynamics. Int J Biochem Cell Biol 41(10):2025–2035
Thomas KJ et al (2011) DJ-1 acts in parallel to the PINK1/parkin pathway to control mitochondrial function and autophagy. Hum Mol Genet 20(1):40–50
Twig G, Shirihai OS (2011) The interplay between mitochondrial dynamics and mitophagy. Antioxid Redox Signal 14(10):1939–1951

Twig G et al (2006) Tagging and tracking individual networks within a complex mitochondrial web with photoactivatable GFP. Am J Physiol Cell Physiol 291(1):C176–C184

Twig G et al (2008a) Fission and selective fusion govern mitochondrial segregation and elimination by autophagy. EMBO J 27(2):433–446

Twig G, Hyde B, Shirihai OS (2008b) Mitochondrial fusion, fission and autophagy as a quality control axis: the bioenergetic view. Biochim Biophys Acta 1777(9):1092–1097

Twig G et al (2010) Biophysical properties of mitochondrial fusion events in pancreatic beta-cells and cardiac cells unravel potential control mechanisms of its selectivity. Am J Physiol Cell Physiol 299(2):C477–C487

Van Laar VS et al (2011) Bioenergetics of neurons inhibit the translocation response of Parkin following rapid mitochondrial depolarization. Hum Mol Genet 20(5):927–940

Ved R et al (2005) Similar patterns of mitochondrial vulnerability and rescue induced by genetic modification of alpha-synuclein, parkin, and DJ-1 in *Caenorhabditis elegans*. J Biol Chem 280(52):42655–42668

Vives-Bauza C, Przedborski S (2010) PINK1 points Parkin to mitochondria. Autophagy 6(5):674–675

Vives-Bauza C et al (2010a) PINK1/Parkin direct mitochondria to autophagy. Autophagy 6(2):315–316

Vives-Bauza C et al (2010b) Control of mitochondrial integrity in Parkinson's disease. Prog Brain Res 183:99–113

Vives-Bauza C et al (2010c) PINK1-dependent recruitment of Parkin to mitochondria in mitophagy. Proc Natl Acad Sci USA 107(1):378–383

Wang X, Schwarz TL (2009a) Imaging axonal transport of mitochondria. Methods Enzymol 457:319–333

Wang X, Schwarz TL (2009b) The mechanism of Ca^{2+}-dependent regulation of kinesin-mediated mitochondrial motility. Cell 136(1):163–174

Wang W et al (2010) Mitofusin-2 is a novel direct target of p53. Biochem Biophys Res Commun 400(4):587–592

Wang H et al (2011a) Parkin ubiquitinates Drp1 for proteasome-dependent degradation: implication of dysregulated mitochondrial dynamics in Parkinson disease. J Biol Chem 286(13):11649–11658

Wang X et al (2011b) PINK1 and Parkin target Miro for phosphorylation and degradation to arrest mitochondrial motility. Cell 147(4):893–906

Wang JX et al (2011c) miR-499 regulates mitochondrial dynamics by targeting calcineurin and dynamin-related protein-1. Nat Med 17(1):71–78

Weihofen A et al (2009) Pink1 forms a multiprotein complex with Miro and Milton, linking Pink1 function to mitochondrial trafficking. Biochemistry 48(9):2045–2052

Westermann B (2010) Mitochondrial fusion and fission in cell life and death. Nat Rev Mol Cell Biol 11(12):872–884

Whitworth AJ, Pallanck LJ (2009) The PINK1/Parkin pathway: a mitochondrial quality control system? J Bioenerg Biomembr 41(6):499–503

Wikstrom JD et al (2007) beta-Cell mitochondria exhibit membrane potential heterogeneity that can be altered by stimulatory or toxic fuel levels. Diabetes 56(10):2569–2578

Wikstrom JD, Twig G, Shirihai OS (2009) What can mitochondrial heterogeneity tell us about mitochondrial dynamics and autophagy? Int J Biochem Cell Biol 41(10):1914–1927

Wilkerson DC, Sankar U (2011) Mitochondria: a sulfhydryl oxidase and fission GTPase connect mitochondrial dynamics with pluripotency in embryonic stem cells. Int J Biochem Cell Biol 43(9):1252–1256

Wood-Kaczmar A et al (2008) PINK1 is necessary for long term survival and mitochondrial function in human dopaminergic neurons. PLoS One 3(6):e2455

Wu S et al (2011) Bax is essential for Drp1-mediated mitochondrial fission but not for mitochondrial outer membrane permeabilization caused by photodynamic therapy. J Cell Physiol 226(2):530–541

Xiong H et al (2009) Parkin, PINK1, and DJ-1 form a ubiquitin E3 ligase complex promoting unfolded protein degradation. J Clin Invest 119(3):650–660

Yang Y et al (2008) Pink1 regulates mitochondrial dynamics through interaction with the fission/fusion machinery. Proc Natl Acad Sci USA 105(19):7070–7075
Yi M, Weaver D, Hajnoczky G (2004) Control of mitochondrial motility and distribution by the calcium signal: a homeostatic circuit. J Cell Biol 167(4):661–672
Yoon Y et al (2003) The mitochondrial protein hFis1 regulates mitochondrial fission in mammalian cells through an interaction with the dynamin-like protein DLP1. Mol Cell Biol 23(15):5409–5420
Yu T et al (2005) Regulation of mitochondrial fission and apoptosis by the mitochondrial outer membrane protein hFis1. J Cell Sci 118(Pt 18):4141–4151
Yue Z (2007) Regulation of neuronal autophagy in axon: implication of autophagy in axonal function and dysfunction/degeneration. Autophagy 3(2):139–141
Yue Z et al (2009) The cellular pathways of neuronal autophagy and their implication in neurodegenerative diseases. Biochim Biophys Acta 1793(9):1496–1507
Yu-Wai-Man P et al (2010) OPA1 mutations cause cytochrome c oxidase deficiency due to loss of wild-type mtDNA molecules. Hum Mol Genet 19(15):3043–3052
Yu-Wai-Man P, Griffiths PG, Chinnery PF (2011a) Mitochondrial optic neuropathies – disease mechanisms and therapeutic strategies. Prog Retin Eye Res 30(2):81–114
Yu-Wai-Man P et al (2011b) Genetic screening for OPA1 and OPA3 mutations in patients with suspected inherited optic neuropathies. Ophthalmology 118(3):558–563
Zhang Y, Chan DC (2007) Structural basis for recruitment of mitochondrial fission complexes by Fis1. Proc Natl Acad Sci USA 104(47):18526–18530
Zhang L et al (2005) Mitochondrial localization of the Parkinson's disease related protein DJ-1: implications for pathogenesis. Hum Mol Genet 14(14):2063–2073
Zhang J et al (2010) G-protein beta2 subunit interacts with mitofusin 1 to regulate mitochondrial fusion. Nat Commun 1:101
Ziviani E, Whitworth AJ (2010) How could Parkin-mediated ubiquitination of mitofusin promote mitophagy? Autophagy 6(5):660–662
Ziviani E, Tao RN, Whitworth AJ (2010) Drosophila parkin requires PINK1 for mitochondrial translocation and ubiquitinates mitofusin. Proc Natl Acad Sci USA 107(11):5018–5023

Chapter 3
Biogenesis of Mitochondrial Proteins

Johannes M. Herrmann, Sebastian Longen, Daniel Weckbecker, and Matthieu Depuydt

Abstract Depending on the organism, mitochondria consist approximately of 500–1,400 different proteins. By far most of these proteins are encoded by nuclear genes and synthesized on cytosolic ribosomes. Targeting signals direct these proteins into mitochondria and there to their respective subcompartment: the outer membrane, the intermembrane space (IMS), the inner membrane, and the matrix. Membrane-embedded translocation complexes allow the translocation of proteins across and, in the case of membrane proteins, the insertion into mitochondrial membranes. A small number of proteins are encoded by the mitochondrial genome: Most mitochondrial translation products represent hydrophobic proteins of the inner membrane which—together with many nuclear-encoded proteins—form the respiratory chain complexes. This chapter gives an overview on the mitochondrial protein translocases and the mechanisms by which they drive the transport and assembly of mitochondrial proteins.

3.1 Introduction

The mitochondrial proteome shows an extreme complexity. In 2009, the manually validated list of mitochondrial proteins of the MitoP2 database showed 590 entries for yeast, 920 for human and 1,020 for mouse (Elstner et al. 2009). Proteomic studies on the basis of mass spectroscopic analysis of mitochondria isolated from mouse tissues led to the identification of 3,881 different proteins (Pagliarini et al. 2008). Even when it is taken into account that a fraction of these proteins presumably

J.M. Herrmann (✉) • S. Longen • D. Weckbecker • M. Depuydt
Zellbiologie, Universität Kaiserslautern, Erwin-Schrödinger-Strasse 13,
Kaiserslautern 67663, Germany
e-mail: hannes.herrmann@biologie.uni-kl.de

B. Kadenbach (ed.), *Mitochondrial Oxidative Phosphorylation*, Advances in Experimental Medicine and Biology 748, DOI 10.1007/978-1-4614-3573-0_3,

represents contaminations of the samples, this large number is surprising and suggests that mitochondria of eukaryotic cells are more complex than many bacteria. For example, *Rickettsia prowazekii* whose genome was suggested to be closely related to the endosymbiotic ancestor of mitochondria contains only 834 coding sequences (Andersson et al. 1998). Many mitochondrial proteins are phylogenetically related to bacterial proteins nicely showing the prokaryotic origin of mitochondria. However, a large number of proteins were added during evolution of eukaryotic cells. This is obvious from the composition of several mitochondrial protein complexes: e.g., the cytochrome *c* oxidase of bacteria contains four subunits, three of which are catalytically important (Iwata et al. 1995; Stenberg et al. 2007). These three subunits are homologous to the three mitochondrial encoded subunits of mitochondrial cytochrome *c* oxidase. However, the eukaryotic mitochondrial enzyme contains nine additional nuclear-encoded subunits which were acquired during the evolution of the eukaryotic cell (Tsukihara et al. 1996). Similarly, mitochondrial ribosomes share many subunits with bacterial ribosomes but still more than half of all subunits are mitochondrion-specific and were added during the evolution of the eukaryotic cell (Smits et al. 2007). Thus, the mitochondrial proteome represents a mosaic of ancestral components (that are related to bacterial proteins) and components of eukaryotic origin, for which no homologs in prokaryotes are found. Out of the many hundreds of mitochondrial proteins only a very small number (e.g., 8 in yeast, 13 in humans) are synthesized on mitochondrial ribosomes. All other proteins are synthesized in the cytosol from where they are imported into mitochondria.

3.2 Protein Import into Mitochondria

3.2.1 Translocation Across the Outer Membrane

The translocase of the outer membrane (TOM) is the main mitochondrial entrance gate. It plays a central role in mitochondrial preprotein import and makes the initial contact to cytosolic preproteins that are destined to be transported into mitochondria (Fig. 3.1). In fungi and animals, the TOM complex has a size of 490–600 kDa and consists of seven subunits (Tom70, Tom40, Tom22, Tom20, Tom7, Tom6, and Tom5) (Künkele et al. 1998; Model et al. 2008). In general, the subunits can be separated into two distinct groups according to their function (1) Import receptors: Tom20, Tom22 and Tom70 are important for substrate recognition and binding; and (2) the translocation pore: Tom40, Tom5, Tom6 and Tom7 form the membrane-embedded core of the TOM complex.

The import receptors Tom20 and Tom70 are anchored with their N-termini to the outer membrane and expose large, hydrophilic C-terminal receptor domains to the

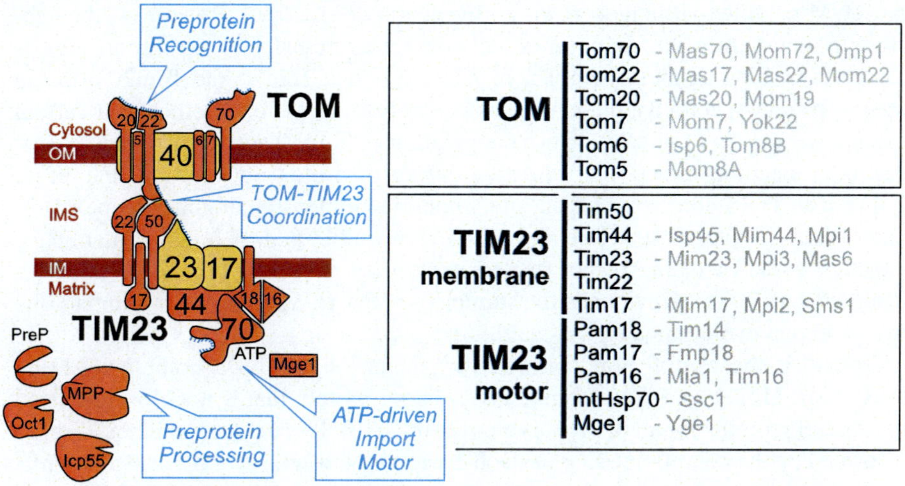

Fig. 3.1 Import of matrix and inner membrane proteins. Proteins with N-terminal presequence are selectively recognized by the Tom20 and Tom22 subunits of the TOM complex and transported through the protein-conducting channel formed by Tom40. Proteins are then passed on to the Tim23 translocase via contacts with the trans site of Tom22 and the IMS domains of Tim50 and Tim23. The membrane-embedded subunits Tim23 and Tim17 form a channel in the inner membrane which allows the transport of preproteins into the matrix. The translocation reaction is driven by the membrane potential across the inner membrane and the hydrolysis of ATP. The matrix chaperone Hsp70 plays an essential role in the vectorial transport of matrix proteins. It interacts with the Tim23 complex in a dynamic fashion; its binding to preprotein is regulated by several subunits of the motor part of the TIM23 complex. Preproteins are proteolytically matured in the matrix by the processing peptidases MPP, Oct1 and Icp55. The presequence peptidase PreP then further degrades the presequences and converts them to amino acids. Channel-forming subunits are shown in *bright orange*. In the *boxes*, the subunits of the translocation complexes are listed. The protein names are shown as they are found in the Saccharomyces Genome Database; alias names are shown in *grey*

cytosol (Kiebler et al. 1993; Schlossmann et al. 1994; Söllner et al. 1989; Waizenegger et al. 2003). Tom22 has an opposite membrane orientation: it consists of an N-terminal cytosolic receptor domain, a central transmembrane span and a C-terminus domain that protrudes into the IMS. The cytosol-exposed receptor domains of Tom20, Tom22 and Tom70 confer substrate recognition and binding (Brix et al. 2000; Hönlinger et al. 1995; Yamano et al. 2008b). Tom20 and Tom22 cooperate and specifically recognize proteins which contain N-terminal matrix-targeting sequences (MTSs) (Shiota et al. 2011). Structural data suggest that a hydrophobic groove in Tom20 is involved in binding of the hydrophobic surface of the amphipatic presequences. In contrast, Tom70 mainly binds to chaperone-bound, hydrophobic precursor proteins that possess internal targeting signals. In the cytosol, preproteins are bound to chaperones of the Hsp70 and Hsp90 family. The cytosolic receptor domain of Tom70 interacts with these chaperones and releases them from preproteins to promote their further translocation into the translocation pore of the TOM complex (Bhangoo et al. 2007; Young et al. 2003). Release of the substrate proteins from the chaperones and subsequent transfer to the import pore is mediated by binding and hydrolysis of ATP. Although Tom70 and Tom20 differ in their substrate specificity both proteins can substitute for each other to a certain extend. Recent data indicate a close cooperation of Tom70 and Tom20 and suggest that both receptors functionally interact during preprotein recognition (Fan et al. 2011). The TOM complex of plant mitochondria employs analogous receptor subunits which in overall structure and function resemble that of fungi or mammals, although some of them are not related in primary sequence (Carrie et al. 2010; Rimmer et al. 2011). Obviously, the similarity of the TOM complexes is in part due to convergent evolution (Perry et al. 2006).

Tom40 is the central pore-forming component of the TOM complex (Ahting et al. 2001; Hill et al. 1998). Tom40 is a β-barrel protein that is not closely related to β-barrel proteins found in prokaryotes (Zeth 2010). It is unclear whether the pore is formed by the central opening of the β-barrel or between several Tom40 subunits in the TOM complex. Tom5, Tom6 and Tom7 are small subunits of the TOM complex that all contain one single transmembrane span. The exact function of these proteins is unclear but they exhibit a stabilizing function on the TOM complex as deletion mutants show defects in the assembly of the TOM complex and/or more labile TOM complexes (Becker et al. 2011; Dembowski et al. 2001; Hönlinger et al. 1996; Model et al. 2001).

Current studies implicate that the activity and the abundance of the TOM complex is regulated by cytosolic kinases. In particular, site-specific phosphorylation of Tom70 by protein kinase A reduces protein import of hydrophobic inner membrane proteins (Rao et al. 2011; Schmidt et al. 2011). Thereby, cells might adapt the levels of protein import to their metabolic needs.

3.2.2 Import of Outer Membrane Proteins

The outer membrane of mitochondria is the only membrane in cells of fungi or animals which contains β-barrel proteins. This reflects the origin of mitochondria from bacterial ancestors. Several β-barrel proteins of mitochondria were identified: VDAC (also referred to as porin), Tom40, Sam50 (also known as Tob55), Mdm10 and Mmm2. These proteins are transported through the protein-conducting channel of the TOM complex into the IMS, from where they are inserted into the outer membrane. This insertion reaction is mediated by a dedicated protein complex: the sorting and assembly machinery (SAM complex), also known as topogenesis of mitochondrial outer membrane β-barrel proteins (TOB) complex. The main component of the SAM complex is Sam50 (Tob55) (Habib et al. 2005; Kozjak et al. 2003; Meisinger et al. 2006; Paschen et al. 2003). Sam50 is a structural (Gentle et al. 2004; Voulhoux et al. 2003) and functional (Muller et al. 2011; Walther et al. 2010) homolog of the bacterial Omp85 protein which mediates the insertion of β-barrel proteins into the outer membrane of gram-negative bacteria. Sam50 consists of two domains: the N-terminal hydrophilic part forms a POTRA (polypeptide translocation associated) domain and faces the IMS. The POTRA domain is important for recognition and further directing of β-barrel precursor proteins to the C-terminal domain. The C-terminal domain of Sam50 is forming a β-barrel and resides in the outer membrane. This domain facilitates the insertion of β-barrel precursor proteins into the outer membrane in a process that is not well understood. In yeast, the SAM complex contains two additional subunits: Sam37/Mas37 and Sam35/Tob38. Individual deletions of Sam50 or Sam35 are lethal due to a blocked protein insertion into the outer membrane. The distinct functions of Sam37 and Sam38 are not known. When the bacterial β-barrel protein PorB is expressed in eukaryotic cells, it was efficiently targeted and inserted into the mitochondrial outer membrane in a Sam50-dependent, but Sam35- and Sam37-independent manner (Jiang et al. 2011). This suggests that the two accessory subunits Sam37 and Sam38 exhibit a mitochondrion-specific function.

Recently, in addition to Sam50, a second paralog of Omp85 was identified in mitochondria of *Trypanosoma brucei* (Pusnik et al. 2011). Interestingly, this protein, called ATOM, serves as import translocase that functions in a TOM-like manner in this parasite. Whether Tom40 is a structurally strongly deviated homolog of ATOM or an unrelated, parallel acquisition of "higher" eukaryotes is not known.

In addition to β-barrel proteins, the outer membrane contains a number of membrane proteins with α-helical transmembrane segments. The mechanisms by which these proteins are integrated into the outer membrane are largely unclear. For some of these proteins a critical role of the receptor domains of the TOM pore was shown while others do not need TOM receptors (Kemper et al. 2008). It remains unclear whether the protein-conducting channel of the TOM complex is used by these proteins since it is difficult to reconcile how the transmembrane domain of a preprotein could be laterally released from a β-barrel protein. Nevertheless, recently published experiments suggest that a model protein consisting of the inner membrane protein Tim23 fused to a cytosol-located folded green fluorescent protein domain can be passed on from the import machinery into the outer membrane; from this observation it was concluded that membrane proteins can be laterally released from the protein-conducting channel of the TOM complex into the outer membrane (Harner et al. 2011).

3.2.3 Import of Matrix Proteins

Nuclear-encoded proteins that are destined for the matrix are translocated in two sequential, though functionally and kinetically coupled reactions across both mitochondrial membranes. Translocation across the outer membrane is mediated by the TOM complex. Subsequently, preproteins are passed on to the TIM23 complex in the inner membrane which threads them into the matrix. Matrix-destined proteins contain N-terminal targeting signals that, in most cases, are removed after translocation into the matrix. These MTSs or presequences consist of 15–60 amino acid residues forming amphiphatic helices with one positively charged and one hydrophobic surface. These signals are both necessary and sufficient to direct proteins into the mitochondrial matrix.

The TIM23 complex consists of two parts: a membrane-embedded part that forms a pore in the inner membrane and a matrix-located motor part that drives the protein translocation process. The membrane-embedded part is made up of three essential subunits (Tim17, Tim23, and Tim50) as well as one non-essential subunit (Tim21). Tim23, potentially together with Tim17, represents the pore-forming subunit of the TIM23 complex (Alder et al. 2008; Truscott et al. 2001). Upon reconstitution in lipid bilayers, Tim23 shows an ion-conducting capacity in electrophysiological measurements (Martinez-Caballero et al. 2007; Meinecke et al. 2006; Truscott et al. 2001) (Fig. 3.1).

The TIM23 complex exposes at least two preprotein binding sites to the IMS. Initially, preproteins are transferred from the *trans* binding site on Tom22 to a receptor domain of Tim50 (Schulz et al. 2011). From there, preproteins are passed on to an IMS domain of Tim23. Preproteins together with the two IMS domains of Tim50 and Tim23 regulate the protein-conducting channel of the TIM23 complex (Donzeau et al. 2000; Geissler et al. 2002; Lohret et al. 1997; Meinecke et al. 2006; Mokranjac et al. 2003a; Yamamoto et al. 2002). Tim17 is structurally and phylogenetically related to Tim23. Its precise function in the import process is still unclear, but it presumably plays a crucial regulatory function in the gating of the TIM23 complex (Martinez-Caballero et al. 2007; Meier et al. 2005a). Tim21 is a non-essential subunit of the TIM23 complex that facilitates the transfer of preproteins from the TOM to the TIM23 complex (Chacinska et al. 2005, 2010; Mokranjac et al. 2005). In addition, it tethers complexes of the respiratory chain to the TIM23 complex presumably to increase the local membrane potential around the translocase of the inner membrane (van der Laan et al. 2006).

The translocation of presequences across the inner membrane depends on the membrane potential. Following translocation into the matrix presequences (and further sequences of the preprotein) are bound by Hsp70 molecules in an ATP-regulated manner (Krayl et al. 2007; Okamoto et al. 2002; Yamano et al. 2008a). The interaction of Hsp70 with the TIM23 complex on the one hand and the presequence on the other hand is an intricate process that is regulated by a number of matrix exposed subunits of the TIM23 complex: Tim44 serves as docking site for Hsp70 (Blom et al. 1993; D'Silva et al. 2004; Schneider et al. 1994), the J-protein Tim14 (Pam18) and the J-like protein Tim16 (Pam16) are co-chaperones that regulate the nucleotide state of Hsp70 (D'Silva et al. 2005; Frazier et al. 2004; Kozany et al. 2004; Mokranjac et al. 2003b, 2006) and Mge1 functions as nucleotide exchange factor (Schneider et al. 1996). Whether the motor subunits are a permanent part of the translocase or only recruited when required is under debate (Chacinska et al. 2005, 2010; Popov-Celeketic et al. 2008, 2011; van der Laan et al. 2007).

In the matrix, most preproteins are proteolytically matured by processing peptidases. The matrix processing peptidase MPP is a dimeric complex that removes the presequences from matrix-destined proteins (Vögtle et al. 2009). A recently published study on the mitochondrial proteome of yeast identified the N-termini of 615 mature proteins indicating that most proteins are processed by MPP. Two additional proteases, Icp55 and Oct1, exhibit additional processing functions and remove one and eight further N-terminal residues from some MPP products, respectively (Naamati et al. 2009; Vögtle et al. 2009).

3.2.4 *Import of Proteins into the IMS*

Despite its small volume, the IMS of mitochondria contains a large variety of proteins. In yeast, more than 50 IMS proteins were identified so far (Herrmann and Riemer 2010) and mammalian mitochondria contain presumably at least twice as many IMS proteins. Proteins of the IMS mediate the transport of metabolites, lipids, metal ions or proteins between both mitochondrial membranes, play roles in the biogenesis of the respiratory chain or regulate apoptosis. The mechanisms by which proteins are transported into the IMS are diverse.

Some IMS proteins contain targeting signals in the form of bipartite presequences. These signals consist of an N-terminal matrix-targeting sequence followed by a hydrophobic sorting domain. The latter serves as stop-transfer signal that arrests the imported protein at the level of the inner membrane. Upon lateral release from the TIM23 complex, these proteins are inserted into the inner membrane. Processing by proteases that cleave in the IMS or the inner membrane releases these proteins into the IMS. The mechanisms by which the TIM23 complex differentiates between proteins that are transferred into the matrix or released into the membrane are only poorly understood. It is likely that a combination of the hydrophobicity of the transmembrane segment, the flanking charges around the hydrophobic region and the content of proline residues in the hydrophobic domain determines the sorting pathway at the level of the TIM23 complex (Bohnert et al. 2010; Botelho et al. 2011; Meier et al. 2005b; Rojo et al. 1998). Proteins that employ bipartite presequences to reach the IMS are for example cytochrome b_2 (Beasley et al. 1993), cytochrome *c* peroxidase (Michaelis et al. 2005; Tatsuta et al. 2007) or Smac/Diablo (Burri et al. 2005).

The yeast protein Mgm1 is present in two isoforms, a soluble form in the IMS and a form tethered to the inner membrane. These two isoforms are generated from an intricated targeting signal (Herlan et al. 2004). Mgm1 contains two consecutive hydrophobic stretches. Insertion of the N-terminal stretch into the inner membrane leads to a longer isoform that is permanently tethered to the inner membrane. Upon translocation of the N-terminal transmembrane span into the matrix, the second transmembrane region is inserted into the inner membrane and cleaved by the membrane-embedded rhomboid protease Pcp1 (Herlan et al. 2003; McQuibban et al. 2003). This shorter isoform is released as soluble protein into the IMS.

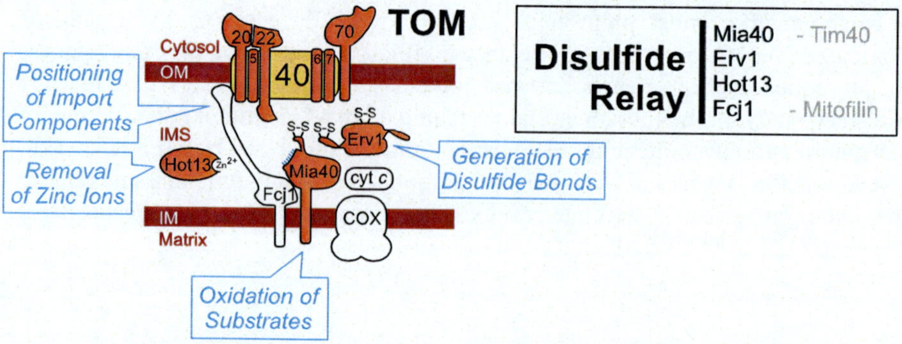

Fig. 3.2 The mitochondrial disulfide relay. The mitochondrial disulfide relay mediates the import of cysteine-containing proteins into the IMS of mitochondria. Substrates of this system reach the IMS through the protein-conducting channel of the TOM complex where they bind to Mia40. Mia40 introduces disulfide bonds into these proteins and drives their folding. Oxidation of Mia40 is mediated by Erv1 which transfers its electrons via cytochrome *c* (cyt *c*) to cytochrome *c* oxidase (COX). Hot13 binds zinc ions to prevent metal binding of Mia40. Fcj1 is a structural inner membrane protein that brings Mia40 into close proximity of the TOM complex to allow the efficient import of substrates of the disulfide relay system

The ATP levels of mitochondria influence the efficiency of the insertion/translocation of the N-terminal transmembrane domain and thereby determine the ratio of both isoforms of Mgm1.

3.2.5 The Mitochondrial Disulfide Relay

A special group of IMS proteins lacks N-terminal presequences. Import of these proteins is facilitated by a dedicated machinery called the mitochondrial disulfide relay (Fig. 3.2). Substrates of this machinery contain critical cysteine residues in their protein sequence. In many cases, these cysteines are arranged in so-called twin Cx_3C or twin Cx_9C motifs, i.e., two pairs of cysteine residues that are separated by three amino acid residues, respectively. Yeast mitochondria contain five twin Cx_3C proteins which are also called small Tim proteins as they play a critical role in the import of carrier proteins (see below in Sect. 3.2.5). The number of proteins with twin Cx_9C motifs is large (14 in yeast and about twice as many in animals) (Cavallaro 2010; Gabriel et al. 2007; Longen et al. 2009). Most twin Cx_9C proteins are required for mitochondrial respiration, but their molecular function is still ill-defined.

The cysteine residues in these proteins are oxidized during the import process. Since only unfolded and linear proteins can traverse the TOM pore, protein oxidation prevents the back-translocation of these proteins into the cytosol (Allen et al. 2003; Lutz et al. 2003; Schwartz and Matouschek 1999). A key component of this import pathway is the oxidoreductase Mia40 (Chacinska et al. 2004; Mesecke et al. 2005; Naoe et al. 2004; Terziyska et al. 2005). In yeast, Mia40 is anchored to the inner membrane by an N-terminal transmembrane segment, whereas in animals and plants Mia40 is a soluble IMS protein. Mia40 contains six cysteine residues, four of which form a structural twin Cx_9C motif. Two redox-active cysteine residues that are arranged in a CPC motif are required for substrate oxidation. Mia40 serves as an intramitochondrial protein receptor that binds substrate proteins via a hydrophobic cleft (Banci et al. 2009, 2010; Kawano et al. 2009; Terziyska et al. 2009). It specifically recognizes hydrophobic motifs in its substrates which are called sequences (Milenkovic et al. 2009; Sideris et al. 2009). Mia40 forms a mixed disulfide with an acceptor cysteine that is in proximity to the MISS signal and finally introduces both disulfide bonds into the substrate protein (Ang and Lu 2009; Bien et al. 2010; Tienson et al. 2009).

Reoxidation of Mia40 is mediated by the sulfhydryl oxidase Erv1 (Allen et al. 2005; Lee et al. 2000; Mesecke et al. 2005; Rissler et al. 2005). Erv1 is an FAD-containing enzyme that transfers electrons from Mia40 to cytochrome *c* of the respiratory chain (Allen et al. 2005; Bihlmaier et al. 2007; Dabir et al. 2007).

The third component of the mitochondrial disulfide relay system is Hot13 which maintains Mia40 in a metal-free conformation that can be efficiently oxidized by Erv1 (Curran et al. 2004; Mesecke et al. 2008; Morgan et al. 2009).

Recently, the inner membrane protein Fcj1was identified as a further component that plays a role in the Mia40-dependent import pathway (von der Malsburg et al. 2011). Fcj1 is a part of a large structural complex in the inner membrane that is important for the formation of cristae and contact sites between the inner and the outer membrane (Alkhaja et al. 2012; Herrmann 2011; Hoppins et al. 2011; Rabl et al. 2009; von der Malsburg et al. 2011). Fcj1 specifically interacts with Mia40 and subunits of the TOM complex thereby improving the interaction of newly imported proteins with Mia40.

3.2.6 The Carrier Import Pathway

The inner membrane of mitochondria contains a large number of transporters to facilitate the exchange of metabolites, nucleotides and other molecules between the cytosol and matrix. Many of these transporters are structurally and phylogenetically related and constitute the family of carrier proteins (Klingenberg 2009; Palmieri et al. 1992). Carrier proteins consist of three structurally similar modules each containing two transmembrane spans (Pebay-Peyroula et al. 2003).

In yeast and animals, carrier proteins lack MTSs. As a consequence, carrier proteins do not embark on the TIM23-pathway. Instead, they employ an alternative inner membrane translocase which is referred to as TIM22 complex (Fig. 3.3). The import route of carrier proteins differs from that of matrix-targeted preproteins already at the level of the outer surface of mitochondria: Whereas matrix-targeting presequences are predominantly recognized by the TOM receptors Tom22 and Tom20 (Abe et al. 2000; Iwata and Nakai 1998; Kiebler et al. 1993; Lithgow et al. 1994a, b; Söllner et al. 1989), carrier proteins bind to the outer membrane protein Tom70 (Hines et al. 1990; Schlossmann et al. 1994; Söllner et al. 1990; Suzuki et al. 2002). In this process, Tom70 recognizes specific regions around the hydrophobic transmembrane

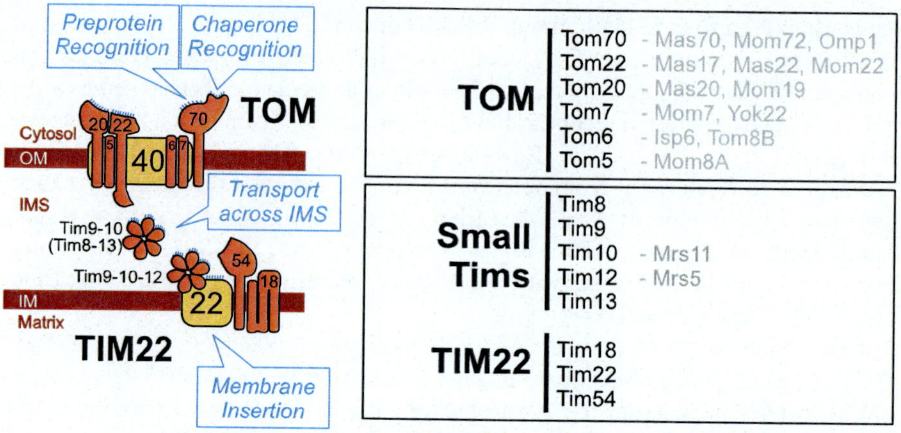

Fig. 3.3 Carrier protein import pathway. In the cytosol, carrier proteins are bound by chaperones which escort them to the Tom70 receptor of the TOM complex. Tom70 promotes the release of chaperones from preproteins so that they can traverse the TOM pore and make contact with the Tim9–Tim10 or Tim8–Tim13 chaperone complexes in the IMS. Carrier proteins are finally passed on to the TIM22 complex in the inner membrane which inserts them into the lipid bilayer in a membrane-dependent manner

domains (Brix et al. 1999, 2000). In the cytosol, carrier precursors are associated with chaperones of the Hsp70 and Hsp90 class. Tom70 triggers the release of the carrier proteins from these chaperones (Komiya et al. 1997; Young et al. 2003) and allows them to enter the protein-conducting channel of the TOM complex.

The carrier proteins traverse the TOM complex as loops. Following translocation into the IMS, carrier proteins bind hexameric complexes of small Tim proteins in the IMS. Small Tim proteins are polypeptides of 8–12 kD which are characterized by a central twin Cx_3C motif. Six of these small proteins form a ring-like core surrounded by the 12 flexible termini of the subunits (Lu et al. 2004; Vergnolle et al. 2005). They resemble jellyfish-like structures with 12 flexible tentacles (Webb et al. 2006). It was suggested that these arms accommodate the hydrophobic regions of carrier proteins in the IMS thereby shielding them from unwanted interactions. Binding of the carrier proteins to the complex of small Tim proteins allows them to fully translocate across the outer membrane and to interact with the TIM22 translocase of the inner membrane. This complex mediates their insertion into the inner membrane in a membrane-potential dependent process.

The TIM22 translocase consists of the three membrane proteins Tim18, Tim22 and Tim54 and of three hydrophilic IMS proteins Tim9, Tim10 and Tim12. Tim22 is the essential core component of the complex which presumably forms the protein-conducting channel (Kerscher et al. 1997; Sirrenberg et al. 1996). Its sequence is homologous to that of Tim17 and Tim23. In the parasite *T. brucei* which has a strongly reduced mitochondrial import machinery only one single Tim17/22/23 protein (*Tb*11.01.4870) is present and presumably facilitates translocation of both matrix and carrier proteins (Schneider et al. 2008). The TIM22 complex forms two connected voltage-activated and signal-gated channels (Kovermann et al. 2002; Peixoto et al. 2007) and therefore is also referred to as twin pore translocase.

The roles of Tim18 and Tim54 are not well understood. Both components contribute to the stability of the TIM22 complex but whether they exhibit a direct role in carrier insertion is not clear (Hwang et al. 2007; Kerscher et al. 1997, 2000; Koehler et al. 2000). Tim9, Tim10 and Tim12 form a hexameric small Tim complex in the IMS that is permanently associated with the TIM22 complex (and distinct from the Tim9–Tim10 hexamer) (Murphy et al. 2001; Sirrenberg et al. 1998). Due to its intrinsic lipid-binding ability Tim12 was suggested to bind directly to the inner membrane (Lionaki et al. 2008). The small Tim proteins Tim8 and Tim13 form a second hexameric chaperone complex in the IMS that presumably partially overlaps in function with the Tim9–Tim10 complex.

3.3 Mitochondrially Encoded Proteins

The number of proteins encoded by the mitochondrial genome differs significantly among eukaryotes: The most complex mitochondrial genomes are found among plants and protists where up to 67 coding sequences are present (Gray et al. 2004; Lang et al. 1997). In contrast, some parasites harbor the most reduced mitochondrial genomes: for example, only three mitochondrial translation products are synthesized in *Plasmodium* (Gray et al. 1999). In most animals and fungi, a characteristic set of about one dozen of proteins is encoded in mitochondria. All these proteins represent membrane-embedded core subunits of complexes I, III, IV and V of the respiratory chain. The human mitochondrial genome contains 13 protein-coding genes (Attardi 1981). That of *Saccharomyces cerevisiae* codes only for eight proteins (Borst and Grivell 1978). Why only hydrophobic proteins are encoded in mitochondria is not clear. It was suggested that these hydrophobic proteins need to be inserted co-translationally in order to prevent their aggregation (hydrophobicity argument). Indeed, some of these proteins cannot be imported into mitochondria as they misfold in the cytosol and jam the import channels (Claros et al. 1995). A second, mutually not exclusive explanation is the regulation of the synthesis of these proteins by mitochondrial signals which require the synthesis and the assembly of the respiratory chain to take place in the same compartment (regulation argument). Also this idea is supported by evidence as regulatory circuits were identified which control the synthesis of translation products in mitochondria and coordinate this directly with their membrane insertion; thereby the accumulation of potentially harmful non-assembled subunits is prevented (Barrientos et al. 2004; Gruschke et al. 2011; Perez-Martinez et al. 2003; Rak and Tzagoloff 2009).

3.3.1 Membrane Insertion of Mitochondrial Translation Products by the Oxa1 Complex

Mitochondrially encoded inner membrane proteins show a characteristic charge distribution: the matrix-exposed segments contain predominantly positively charged residues while domains exposed to the IMS carry a negative net charge (Gavel and von Heijne 1992). This "positive inside rule" was initially described for bacterial inner membrane proteins but also applies—though not as strictly—to eukaryotic membranes (Higy et al. 2004; von Heijne 1989; White and von Heijne 2004). The membrane potential across the inner membrane, i.e., the accumulation of positive charges on the IMS side of the membrane, presumably supports the translocation of negatively charged protein domains from the matrix across the inner membrane. On the other hand, the transport of positive charges seems to be energetically unfavorable so that the distribution of charges that flank transmembrane spans serves as signals that determine protein topology in the inner membrane (Bohnert et al. 2010; Gavel and von Heijne 1992; Herrmann et al. 1995, 1997; Rojo et al. 1999).

The process of membrane integration of mitochondrial translation products is poorly understood. The mitochondrial inner membrane lacks a homolog of the Sec translocase that is found in bacteria and the endoplasmic reticulum (Glick and von Heijne 1996). Protein insertion into the inner membrane is facilitated by the conserved membrane protein Oxa1 (for Oxidase assembly mutant 1). Oxa1 was initially identified in two independent genetic studies as a component involved in the assembly of cytochrome *c* oxidase (Bauer et al. 1994; Bonnefoy et al. 1994). Oxa1-deficient yeast mutants lack cytochrome *c* oxidase activity and show severely reduced levels of other respiratory chain complexes that contain mitochondrial translation products (cytochrome *c* reductase and ATP synthase) (Altamura et al. 1996; Hell et al. 2001). Moreover, Oxa1 plays a role in the biogenesis of complex I of the respiratory chain (which is absent in *S. cerevisiae*) (Nargang et al. 2002; Stiburek et al. 2007). The molecular function of Oxa1 is not clear. Oxa1 is closely related to YidC, a protein that facilitates the insertion and folding of bacterial inner membrane proteins (du Plessis et al. 2006; Houben et al. 2002; Nagamori et al. 2004; Samuelson et al. 2000; Scotti et al. 2000; Wagner et al. 2008). In vitro studies with reconstituted YidC indicated an activity as protein insertase; YidC thereby strongly increased the yield and rate by which membrane proteins integrated into lipid bilayers (Serek et al. 2004; van der Laan et al. 2001, 2004). Proteomic studies suggest that YidC exhibits a rather general role in membrane protein biogenesis in bacteria which is not restricted to the insertion of respiratory chain complexes; rather it appears to play a general role in the insertion and folding of bacterial membrane proteins (Price et al. 2010; Wickström et al. 2011).

3.3.2 Ribosome Binding by the Insertion Machinery

Oxa1 facilitates membrane insertion of mitochondrial translation products. In the absence of Oxa1, newly synthesized membrane proteins accumulate unproductively in the matrix and are rapidly degraded (Fiumera et al. 2007; He and Fox 1997; Hell et al. 1997, 1998; Preuss et al. 2001). However, even in the absence of Oxa1 a basic insertion capacity is still retained, indicating that Oxa1 is not absolutely essential for membrane insertion. Membrane insertion of mitochondrial translation products is kinetically, and presumably also functionally, coupled to their synthesis on mitochondrial ribosomes (van der Klei et al. 1994; Watson 1972). This coupling is achieved by a C-terminal matrix domain of Oxa1 which directly binds to mitochondrial ribosomes (Haque et al. 2010a, b; Jia et al. 2003, 2009; Szyrach et al. 2003). This region contacts the ribosome in proximity to the polypeptide-exit tunnel at which nascent chains emerge (Gruschke et al. 2010; Jia et al. 2003, 2009; Kaur and Stuart 2011; Kohler et al. 2009). It was suggested that an Oxa1 dimer, which contains ten transmembrane spans, forms an insertion pore at the tunnel exit to mediate protein translocation (Kohler et al. 2009). Experimental evidence for a pore-forming capacity of Oxa1 (or YidC) is however lacking.

Oxa1-dependent protein insertion is supported by two additional ribosome-binding inner membrane proteins: Mba1 and Mdm38. The membrane-associated protein Mba1 binds to the large subunit. Mba1-deficient yeast strains still exhibit insertion activity. However, combination mutants lacking both Mba1 and the C-terminal ribosome binding domain of Oxa1 are severely affected (Ott et al. 2006; Preuss et al. 2001). The yeast protein Mdm38 also binds to the large subunit of the mitochondrial ribosome (Frazier et al. 2006; Lupo et al. 2011). It overlaps functionally with Mba1, and *mba1/mdm38* double mutants lack a functional respiratory chain (Bauerschmitt et al. 2010). In addition to its role in mitochondrial protein expression, Mdm38 influences directly or indirectly the ion homeostasis of the matrix (Nowikovsky et al. 2004, 2007). Mdm38 is a conserved protein; mutations in its human homolog LETM1 are associated with the Wolf-Hirschhorn syndrome, a disorder characterized by severe congenital malformations (Bergemann et al. 2005; Tamai et al. 2008).

3.4 Outlook

The import of proteins into mitochondria has been studied for three decades. During this time period the principles of the import reaction were analyzed and many, presumably most, components of the import machinery were identified. Nevertheless, the molecular details by which preproteins are recognized, threaded into the translocation pores and driven across the membranes are unclear. Moreover, it is largely unknown how the import process is regulated to adapt it to the specific (metabolic) needs of a cell. This lack of knowledge might be partially due to the fact that most import studies employed mitochondria isolated from yeast cells as model system. It was recently observed that post-translational modifications regulate cytosol-exposed components of the mitochondrial fission machinery of animal cells. For example, the dynamin-related protein Drp1 is regulated by phosphorylation (Cereghetti et al. 2008; Taguchi et al. 2007), nitrosylation (Nakamura et al. 2010), SUMOylation (Harder et al. 2004) and ubiquitination (Wang et al. 2011). It is tempting to speculate that other mitochondrial outer membrane proteins, like subunits of the TOM complex, are likewise regulated by other modifications. It is high time to move on and investigate import reactions under in vivo conditions in fungi, but even more important, in mammalian cells.

References

Abe Y, Shodai T, Muto T, Mihara K, Torii H, Mishikawa S, Endo T, Kohda D (2000) Structural basis of presequence recognition by the mitochondrial protein import receptor Tom20. Cell 100:551–560

Ahting U, Thieffry M, Engelhardt H, Hegerl R, Neupert W, Nussberger S (2001) Tom40, the pore-forming component of the protein-conducting TOM channel in the outer membrane of mitochondria. J Cell Biol 153:1151–1160

Alder NN, Jensen RE, Johnson AE (2008) Fluorescence mapping of mitochondrial TIM23 complex reveals a water-facing, substrate-interacting helix surface. Cell 134:439–450

Alkhaja AK, Jans DC, Nikolov M, Vukotic M, Lytovchenko O, Ludewig F, Schliebs W, Riedel D, Urlaub H, Jakobs S et al (2012) MINOS1 is a conserved component of mitofilin complexes and required for mitochondrial function and cristae organization. Mol Biol Cell 23(2):247–257

Allen S, Lu H, Thornton D, Tokatlidis K (2003) Juxtaposition of the two distal CX3C motifs via intra-chain disulfide bonding is essential for the folding of Tim10. J Biol Chem 278:38505–38513

Allen S, Balabanidou V, Sideris DP, Lisowsky T, Tokatlidis K (2005) Erv1 mediates the Mia40-dependent protein import pathway and provides a functional link to the respiratory chain by shuttling electrons to cytochrome *c*. J Mol Biol 353:937–944

Altamura N, Capitanio N, Bonnefoy N, Papa S, Dujardin G (1996) The *Saccharomyces cerevisiae OXA1* gene is required for the correct assembly of cytochrome *c* oxidase and oligomycin-sensitive ATP synthase. FEBS Lett 382:111–115

Andersson SG, Zomorodipour A, Andersson JO, Sicheritz-Ponten T, Alsmark UC, Podowski RM, Naslund AK, Eriksson AS, Winkler HH, Kurland CG (1998) The genome sequence of *Rickettsia prowazekii* and the origin of mitochondria. Nature 396:133–140

Ang SK, Lu H (2009) Deciphering structural and functional roles of individual disulfide bonds of the mitochondrial sulfhydryl oxidase Erv1p. J Biol Chem 284:28754–28761

Attardi G (1981) Organisation and expression of the mammalian mitochondrial genome: a lesson in economy. Trends Biochem Sci 6(86–89):100–103

Banci L, Bertini I, Cefaro C, Ciofi-Baffoni S, Gallo A, Martinelli M, Sideris DP, Katrakili N, Tokatlidis K (2009) Mia40 is an oxidoreductase that catalyzes oxidative protein folding in mitochondria. Nat Struct Mol Biol 16:198–206

Banci L, Bertini I, Cefaro C, Cenacchi L, Ciofi-Baffoni S, Felli IC, Gallo A, Gonnelli L, Luchinat E, Sideris D et al (2010) Molecular chaperone function of Mia40 triggers consecutive induced folding steps of the substrate in mitochondrial protein import. Proc Natl Acad Sci USA 107:20190–20195

Barrientos A, Zambrano A, Tzagoloff A (2004) Mss51p and Cox14p jointly regulate mitochondrial Cox1p expression in *Saccharomyces cerevisiae*. EMBO J 23:3472–3482

Bauer M, Behrens M, Esser K, Michaelis G, Pratje E (1994) *PET1402*, a nuclear gene required for proteolytic processing of cytochrome oxidase subunit 2 in yeast. Mol Gen Genet 245:272–278

Bauerschmitt H, Mick DU, Deckers M, Vollmer C, Funes S, Kehrein K, Ott M, Rehling P, Herrmann JM (2010) Ribosome-binding proteins Mdm38 and Mba1 display overlapping functions for regulation of mitochondrial translation. Mol Biol Cell 21:1937–1944

Beasley EM, Muller S, Schatz G (1993) The signal that sorts yeast cytochrome b2 to the mitochondrial intermembrane space contains three distinct functional regions. EMBO J 12:2303–2311

Becker T, Wenz LS, Thornton N, Stroud D, Meisinger C, Wiedemann N, Pfanner N (2011) Biogenesis of mitochondria: dual role of Tom7 in modulating assembly of the preprotein translocase of the outer membrane. J Mol Biol 405:113–124

Bergemann AD, Cole F, Hirschhorn K (2005) The etiology of Wolf-Hirschhorn syndrome. Trends Genet 21:188–195

Bhangoo MK, Tzankov S, Fan AC, Dejgaard K, Thomas DY, Young JC (2007) Multiple 40-kDa heat-shock protein chaperones function in Tom70-dependent mitochondrial import. Mol Biol Cell 18:3414–3428

Bien M, Longen S, Wagener N, Chwalla I, Herrmann JM, Riemer J (2010) Mitochondrial disulfide bond formation is driven by intersubunit electron transfer in Erv1 and proof read by glutathione. Mol Cell 37:516–528

Bihlmaier K, Mesecke N, Terzyiska N, Bien M, Hell K, Herrmann JM (2007) The disulfide relay system of mitochondria is connected to the respiratory chain. J Cell Biol 179:389–395

Blom J, Kübrich M, Rassow J, Voos W, Dekker PJ, Maarse AC, Meijer M, Pfanner N (1993) The essential yeast protein MIM44 (encoded by MPI1) is involved in an early step of preprotein translocation across the mitochondrial inner membrane. Mol Cell Biol 13:7364–7371

Bohnert M, Rehling P, Guiard B, Herrmann JM, Pfanner N, van der Laan M (2010) Cooperation of stop-transfer and conservative sorting mechanisms in mitochondrial protein transport. Curr Biol 20:1227–1232

Bonnefoy N, Chalvet F, Hamel P, Slominski PP, Dujardin G (1994) *OXA1*, a *Saccharomyces cerevisiae* nuclear gene whose sequence is conserved from prokaryotes to eukaryotes controls cytochrome oxidase biogenesis. J Mol Biol 239:201–212

Borst P, Grivell LA (1978) The mitochondrial genome of yeast. Cell 15:705–723

Botelho SC, Osterberg M, Reichert AS, Yamano K, Bjorkholm P, Endo T, von Heijne G, Kim H (2011) TIM23-mediated insertion of transmembrane alpha-helices into the mitochondrial inner membrane. EMBO J 30:1003–1011

Brix J, Rudiger S, Bukau B, Schneider-Mergener J, Pfanner N (1999) Distribution of binding sequences for the mitochondrial import receptors Tom20, Tom22, and Tom70 in a presequence-carrying preprotein and a non-cleavable preprotein. J Biol Chem 274:16522–16530

Brix J, Ziegler GA, Dietmeier K, Schneider-Mergener J, Schulz GE, Pfanner N (2000) The mitochondrial import receptor Tom70: identification of a 25 kDa core domain with a specific binding site for preproteins. J Mol Biol 303:479–488

Burri L, Strahm Y, Hawkins CJ, Gentle IE, Puryer MA, Verhagen A, Callus B, Vaux D, Lithgow T (2005) Mature DIABLO/Smac is produced by the IMP protease complex on the mitochondrial inner membrane. Mol Biol Cell 16:2926–2933

Carrie C, Murcha MW, Whelan J (2010) An in silico analysis of the mitochondrial protein import apparatus of plants. BMC Plant Biol 10:249

Cavallaro G (2010) Genome-wide analysis of eukaryotic twin CX9C proteins. Mol Biosyst 6:2459–2470

Cereghetti GM, Stangherlin A, Martins de Brito O, Chang CR, Blackstone C, Bernardi P, Scorrano L (2008) Dephosphorylation by calcineurin regulates translocation of Drp1 to mitochondria. Proc Natl Acad Sci USA 105:15803–15808

Chacinska A, Pfannschmidt S, Wiedemann N, Kozjak V, Sanjuan Szklarz LK, Schulze-Specking A, Truscott KN, Guiard B, Meisinger C, Pfanner N (2004) Essential role of Mia40 in import and assembly of mitochondrial intermembrane space proteins. EMBO J 23:3735–3746

Chacinska A, Lind M, Frazier AE, Dudek J, Meisinger C, Geissler A, Sickmann A, Meyer HE, Truscott KN, Guiard B et al (2005) Mitochondrial presequence translocase: switching between TOM tethering and motor recruitment involves Tim21 and Tim17. Cell 120:817–829

Chacinska A, van der Laan M, Mehnert CS, Guiard B, Mick DU, Hutu DP, Truscott KN, Wiedemann N, Meisinger C, Pfanner N et al (2010) Distinct forms of mitochondrial TOM-TIM supercomplexes define signal-dependent states of preprotein sorting. Mol Cell Biol 30:307–318

Claros MG, Perea J, Shu YM, Samatey FA, Popot JL, Jacq C (1995) Limitations to in vivo import of hydrophobic proteins into yeast mitochondria – the case of a cytoplasmically synthesized apocytochrome b. Eur J Biochem 228:762–771

Curran SP, Leuenberger D, Leverich EP, Hwang DK, Beverly KN, Koehler CM (2004) The role of Hot13p and redox chemistry in the mitochondrial TIM22 import pathway. J Biol Chem 279:43744–43751

D'Silva P, Liu Q, Walter W, Craig EA (2004) Regulated interactions of mtHsp70 with Tim44 at the translocon in the mitochondrial inner membrane. Nat Struct Mol Biol 11:1084–1091

D'Silva PR, Schilke B, Walter W, Craig EA (2005) Role of Pam16's degenerate J domain in protein import across the mitochondrial inner membrane. Proc Natl Acad Sci USA 102: 12419–12424

Dabir DV, Leverich EP, Kim SK, Tsai FD, Hirasawa M, Knaff DB, Koehler CM (2007) A role for cytochrome c and cytochrome c peroxidase in electron shuttling from Erv1. EMBO J 26:4801–4811
Dembowski M, Kunkele KP, Nargang FE, Neupert W, Rapaport D (2001) Assembly of Tom6 and Tom7 into the TOM core complex of *Neurospora crassa.* J Biol Chem 276:17679–17685
Donzeau M, Kaldi K, Adam A, Paschen S, Wanner G, Guiard B, Bauer MF, Neupert W, Brunner M (2000) Tim23 links the inner and the outer mitochondrial membranes. Cell 101:401–412
du Plessis DJ, Nouwen N, Driessen AJ (2006) Subunit a of cytochrome o oxidase requires both YidC and SecYEG for membrane insertion. J Biol Chem 281:12248–12252
Elstner M, Andreoli C, Klopstock T, Meitinger T, Prokisch H (2009) The mitochondrial proteome database: MitoP2. Methods Enzymol 457:3–20
Fan AC, Kozlov G, Hoegl A, Marcellus RC, Wong MJ, Gehring K, Young JC (2011) Interaction between the human mitochondrial import receptors Tom20 and Tom70 in vitro suggests a chaperone displacement mechanism. J Biol Chem 286:32208–32219
Fiumera HL, Broadley SA, Fox TD (2007) Translocation of mitochondrially synthesized Cox2 domains from the matrix to the intermembrane space. Mol Cell Biol 27:4664–4673
Frazier AE, Dudek J, Guiard B, Voos W, Li Y, Lind M, Meisinger C, Geissler A, Sickmann A, Meyer HE et al (2004) Pam16 has an essential role in the mitochondrial protein import motor. Nat Struct Mol Biol 11:226–233
Frazier AE, Taylor RD, Mick DU, Warscheid B, Stoepel N, Meyer HE, Ryan MT, Guiard B, Rehling P (2006) Mdm38 interacts with ribosomes and is a component of the mitochondrial protein export machinery. J Cell Biol 172:553–564
Gabriel K, Milenkovic D, Chacinska A, Muller J, Guiard B, Pfanner N, Meisinger C (2007) Novel mitochondrial intermembrane space proteins as substrates of the MIA import pathway. J Mol Biol 365:612–620
Gavel Y, von Heijne G (1992) The distribution of charged amino acids in mitochondrial inner-membrane proteins suggests different modes of membrane integration for nuclearly and mitochondrially encoded proteins. Eur J Biochem 205:1207–1215
Geissler A, Chacinska A, Truscott KN, Wiedemann N, Brandner K, Sickmann A, Meyer HE, Meisinger C, Pfanner N, Rehling P (2002) The mitochondrial presequence translocase. An essential role of Tim50 in directing preproteins to the import channel. Cell 111:507–518
Gentle I, Gabriel K, Beech P, Waller R, Lithgow T (2004) The Omp85 family of proteins is essential for outer membrane biogenesis in mitochondria and bacteria. J Cell Biol 164:19–24
Glick BS, von Heijne G (1996) Saccharomyces cerevisiae mitochondria lack a bacterial-type Sec machinery. Protein Sci 5:1–2
Gray MW, Burger G, Lang BF (1999) Mitochondrial evolution. Science 283:1476–1481
Gray MW, Lang BF, Burger G (2004) Mitochondria of protists. Annu Rev Genet 38:477–524
Gruschke S, Grone K, Heublein M, Holz S, Israel L, Imhof A, Herrmann JM, Ott M (2010) Proteins at the polypeptide tunnel exit of the yeast mitochondrial ribosome. J Biol Chem 285:19022–19028
Gruschke S, Kehrein K, Rompler K, Grone K, Israel L, Imhof A, Herrmann JM, Ott M (2011) Cbp3-Cbp6 interacts with the yeast mitochondrial ribosomal tunnel exit and promotes cytochrome b synthesis and assembly. J Cell Biola 193:1101–1114
Habib SJ, Waizenegger T, Lech M, Neupert W, Rapaport D (2005) Assembly of the TOB complex of mitochondria. J Biol Chem 280:6434–6440
Haque ME, Elmore KB, Tripathy A, Koc H, Koc EC, Spremulli LL (2010a) Properties of the C-terminal tail of human mitochondrial inner membrane protein Oxa1L and its interactions with mammalian mitochondrial ribosomes. J Biol Chem 285:28353–28362
Haque ME, Spremulli LL, Fecko CJ (2010b) Identification of protein-protein and protein-ribosome interacting regions of the C-terminal tail of human mitochondrial inner membrane protein Oxa1L. J Biol Chem 285:34991–34998
Harder Z, Zunino R, McBride H (2004) Sumo1 conjugates mitochondrial substrates and participates in mitochondrial fission. Curr Biol 14:340–345

Harner M, Neupert W, Deponte M (2011) Lateral release of proteins from the TOM complex into the outer membrane of mitochondria. EMBO J 30:3232–3241

He S, Fox TD (1997) Membrane translocation of mitochondrially coded Cox2p: distinct requirements for export of N and C termini and dependence on the conserved protein Oxa1p. Mol Biol Cell 8:1449–1460

Hell K, Herrmann J, Pratje E, Neupert W, Stuart RA (1997) Oxa1p mediates the export of the N- and C-termini of pCoxII from the mitochondrial matrix to the intermembrane space. FEBS Lett 418:367–370

Hell K, Herrmann JM, Pratje E, Neupert W, Stuart RA (1998) Oxa1p, an essential component of the N-tail protein export machinery in mitochondria. Proc Natl Acad Sci USA 95:2250–2255

Hell K, Neupert W, Stuart RA (2001) Oxa1p acts as a general membrane insertion machinery for proteins encoded by mitochondrial DNA. EMBO J 20:1281–1288

Herlan M, Vogel F, Bornhovd C, Neupert W, Reichert AS (2003) Processing of Mgm1 by the rhomboid-type protease Pcp1 is required for maintenance of mitochondrial morphology and of mitochondrial DNA. J Biol Chem 278:27781–27788

Herlan M, Bornhovd C, Hell K, Neupert W, Reichert AS (2004) Alternative topogenesis of Mgm1 and mitochondrial morphology depend on ATP and a functional import motor. J Cell Biol 165:167–173

Herrmann JM (2011) MINOS is plus: a Mitofilin complex for mitochondrial membrane contacts. Dev Cell 21:599–600

Herrmann JM, Riemer J (2010) The intermembrane space of mitochondria. Antioxid Redox Signal 13:1341–1358

Herrmann JM, Koll H, Cook RA, Neupert W, Stuart RA (1995) Topogenesis of cytochrome oxidase subunit II – mechanisms of protein export from the mitochondrial matrix. J Biol Chem 270:27079–27086

Herrmann JM, Neupert W, Stuart RA (1997) Insertion into the mitochondrial inner membrane of a polytopic protein, the nuclear encoded Oxa1p. EMBO J 16:2217–2226

Higy M, Junne T, Spiess M (2004) Topogenesis of membrane proteins at the endoplasmic reticulum. Biochemistry 43:12716–12722

Hill K, Model K, Ryan MT, Dietmeier K, Martin F, Wagner R, Pfanner N (1998) Tom40 forms the hydrophilic channel of the mitochondrial import pore for preproteins. Nature 395:516–521

Hines V, Brandt A, Griffiths G, Horstmann H, Brütsch H, Schatz G (1990) Protein import into yeast mitochondria is accelerated by the outer membrane protein MAS70. EMBO J 9:3191–3200

Hönlinger A, Kübrich M, Moczko M, Gärtner F, Mallet L, Bussereau F, Eckerskorn C, Lottspeich F, Dietmeier K, Jacquet M et al (1995) The mitochondrial receptor complex: Mom22 is essential for cell viability and directly interacts with preproteins. Mol Cell Biol 15:3382–3389

Hönlinger A, Bömer U, Alconada A, Eckerskorn C, Lottspeich F, Dietmeier K, Pfanner N (1996) Tom7 modulates the dynamics of the mitochondrial outer membrane translocase and plays a pathway-related role in protein import. EMBO J 15:2125–2137

Hoppins S, Collins SR, Cassidy-Stone A, Hummel E, Devay RM, Lackner LL, Westermann B, Schuldiner M, Weissman JS, Nunnari J (2011) A mitochondrial-focused genetic interaction map reveals a scaffold-like complex required for inner membrane organization in mitochondria. J Cell Biol 195:323–340

Houben EN, Urbanus ML, Van Der Laan M, Ten Hagen-Jongman CM, Driessen AJ, Brunner J, Oudega B, Luirink J (2002) YidC and SecY mediate membrane insertion of a Type I transmembrane domain. J Biol Chem 277:35880–35886

Hwang DK, Claypool SM, Leuenberger D, Tienson HL, Koehler CM (2007) Tim54p connects inner membrane assembly and proteolytic pathways in the mitochondrion. J Cell Biol 178:1161–1175

Iwata K, Nakai M (1998) Interaction between mitochondrial precursor proteins and cytosolic soluble domains of mitochondrial import receptors, Tom20 and Tom70, measured by surface plasmon resonance. Biochem Biophys Res Commun 253:648–652

Iwata S, Ostermeier C, Ludwig B, Michel H (1995) Structure at 2.8 A resolution of cytochrome c oxidase from *Paracoccus denitrificans*. Nature 376:660–669

Jia L, Dienhart M, Schramp M, McCauley M, Hell K, Stuart RA (2003) Yeast Oxa1 interacts with mitochondrial ribosomes: the importance of the C-terminal hydrophilic region of Oxa1. EMBO J 22:6438–6447

Jia L, Kaur J, Stuart RA (2009) Mapping of the *Saccharomyces cerevisiae* Oxa1-mitochondrial ribosome interface and identification of MrpL40, a ribosomal protein in close proximity to Oxa1 and critical for oxidative phosphorylation complex assembly. Eukaryot Cell 8:1792–1802

Jiang JH, Davies JK, Lithgow T, Strugnell RA, Gabriel K (2011) Targeting of Neisserial PorB to the mitochondrial outer membrane: an insight on the evolution of beta-barrel protein assembly machines. Mol Microbiol 82:976–987

Kaur J, Stuart RA (2011) Truncation of the Mrp20 protein reveals new ribosome-assembly subcomplex in mitochondria. EMBO Rep 12:950–955

Kawano S, Yamano K, Naoe M, Momose T, Terao K, Nishikawa S, Watanabe N, Endo T (2009) Structural basis of yeast Tim40/Mia40 as an oxidative translocator in the mitochondrial intermembrane space. Proc Natl Acad Sci USA 106:14403–14407

Kemper C, Habib SJ, Engl G, Heckmeyer P, Dimmer KS, Rapaport D (2008) Integration of tail-anchored proteins into the mitochondrial outer membrane does not require any known import components. J Cell Sci 121:1990–1998

Kerscher O, Holder J, Srinivasan M, Leung RS, Jensen RE (1997) The Tim54p-Tim22p complex mediates insertion of proteins into the mitochondrial inner membrane. J Cell Biol 139: 1663–1675

Kerscher O, Sepuri NB, Jensen RE (2000) Tim18p is a new component of the Tim54p-Tim22p translocon in the mitochondrial inner membrane. Mol Biol Cell 11:103–116

Kiebler M, Keil P, Schneider H, van der Klei IJ, Pfanner N, Neupert W (1993) The mitochondrial receptor complex: a central role of MOM22 in mediating preprotein transfer from receptors to the general insertion pore. Cell 74:483–492

Klingenberg M (2009) Cardiolipin and mitochondrial carriers. Biochim Biophys Acta 1788:2048–2058

Koehler CM, Murphy MP, Bally NA, Leuenberger D, Oppliger W, Dolfini L, Junne T, Schatz G, Or E (2000) Tim18p, a new subunit of the TIM22 complex that mediates insertion of imported proteins into the yeast mitochondrial inner membrane. Mol Cell Biol 20:1187–1193

Kohler R, Boehringer D, Greber B, Bingel-Erlenmeyer R, Collinson I, Schaffitzel C, Ban N (2009) YidC and Oxa1 form dimeric insertion pores on the translating ribosome. Mol Cell 34:344–353

Komiya T, Rospert S, Schatz G, Mihara K (1997) Binding of mitochondrial precursor proteins to the cytoplasmic domains of the import receptors Tom70 and Tom20 is determined by cytoplasmic chaperones. EMBO J 16:4267–4275

Kovermann P, Truscott KN, Guiard B, Rehling P, Sepuri NB, Muller H, Jensen RE, Wagner R, Pfanner N (2002) Tim22, the essential core of the mitochondrial protein insertion complex, forms a voltage-activated and signal-gated channel. Mol Cell 9:363–373

Kozany C, Mokranjac D, Sichting M, Neupert W, Hell K (2004) The J domain-related cochaperone Tim16 is a constituent of the mitochondrial TIM23 preprotein translocase. Nat Struct Mol Biol 11:234–241

Kozjak V, Wiedemann N, Milenkovic D, Lohaus C, Meyer HE, Guiard B, Meisinger C, Pfanner N (2003) An essential role of Sam50 in the protein sorting and assembly machinery of the mitochondrial outer membrane. J Biol Chem 278:48520–48523

Krayl M, Lim JH, Martin F, Guiard B, Voos W (2007) A cooperative action of the ATP-dependent import motor complex and the inner membrane potential drives mitochondrial preprotein import. Mol Cell Biol 27:411–425

Künkele K-P, Heins S, Dembowski M, Nargang FE, Benz R, Thieffry M, Walz J, Lill R, Nussberger S, Neupert W (1998) The preprotein translocation channel of the outer membrane of mitochondria. Cell 93:1009–1019

Lang BF, Burger G, O'Kelly CJ, Cedergren R, Golding GB, Lemieux C, Sankoff D, Turmel M, Gray MW (1997) An ancestral mitochondrial DNA resembling a eubacterial genome in miniature. Nature 387:493–497

Lee J, Hofhaus G, Lisowsky T (2000) Erv1p from *Saccharomyces cerevisiae* is a FAD-linked sulfhydryl oxidase. FEBS Lett 477:62–66

Lionaki E, de Marcos LC, Baud C, Vougioukalaki M, Panayotou G, Tokatlidis K (2008) The essential function of Tim12 in vivo is ensured by the assembly interactions of its C-terminal domain. J Biol Chem 283:15747–15753

Lithgow T, Junne T, Suda K, Gratzer S, Schatz G (1994a) The mitochondrial outer membrane protein Mas22p is essential for protein import and viability of yeast. Proc Natl Acad Sci USA 91:11973–11977

Lithgow T, Junne T, Wachter C, Schatz G (1994b) Yeast mitochondria lacking the two import receptors Mas20p and Mas70p can efficiently and specifically import precursor proteins. J Biol Chem 269:15325–15330

Lohret TA, Jensen RE, Kinnally KW (1997) Tim23, a protein import component of the mitochondrial inner membrane, is required for normal activity of the multiple conductance channel, MCC. J Cell Biol 137:377–386

Longen S, Bien M, Bihlmaier K, Kloeppel C, Kauff F, Hammermeister M, Westermann B, Herrmann JM, Riemer J (2009) Systematic analysis of the twin cx_9c protein family. J Mol Biol 393:356–368

Lu H, Allen S, Wardleworth L, Savory P, Tokatlidis K (2004) Functional TIM10 chaperone assembly is redox-regulated in vivo. J Biol Chem 279:18952–18958

Lupo D, Vollmer C, Deckers M, Mick DU, Tews I, Sinning I, Rehling P (2011) Mdm38 is a 14-3-3-like receptor and associates with the protein synthesis machinery at the inner mitochondrial membrane. Traffic 12:1457–1466

Lutz T, Neupert W, Herrmann JM (2003) Import of small Tim proteins into the mitochondrial intermembrane space. EMBO J 22:4400–4408

Martinez-Caballero S, Grigoriev SM, Herrmann JM, Campo ML, Kinnally KW (2007) Tim17p regulates the twin pore structure and voltage gating of the mitochondrial protein import complex TIM23. J Biol Chem 282:3584–3593

McQuibban GA, Saurya S, Freeman M (2003) Mitochondrial membrane remodelling regulated by a conserved rhomboid protease. Nature 423:537–541

Meier S, Neupert W, Herrmann JM (2005a) Conserved N-terminal negative charges in the Tim17 subunit of the TIM23 translocase play a critical role in the import of preproteins into mitochondria. J Biol Chem 280:7777–7785

Meier S, Neupert W, Herrmann JM (2005b) Proline residues of transmembrane domains determine the sorting of inner membrane proteins in mitochondria. J Cell Biol 170:881–888

Meinecke M, Wagner R, Kovermann P, Guiard B, Mick DU, Hutu DP, Voos W, Truscott KN, Chacinska A, Pfanner N et al (2006) Tim50 maintains the permeability barrier of the mitochondrial inner membrane. Science 312:1523–1526

Meisinger C, Wiedemann N, Rissler M, Strub A, Milenkovic D, Schonfisch B, Muller H, Kozjak V, Pfanner N (2006) Mitochondrial protein sorting: differentiation of beta-barrel assembly by Tom7-mediated segregation of Mdm10. J Biol Chem 281:22819–22826

Mesecke N, Terziyska N, Kozany C, Baumann F, Neupert W, Hell K, Herrmann JM (2005) A disulfide relay system in the intermembrane space of mitochondria that mediates protein import. Cell 121:1059–1069

Mesecke N, Bihlmaier K, Grumbt B, Longen S, Terziyska N, Hell K, Herrmann JM (2008) The zinc-binding protein Hot13 promotes oxidation of the mitochondrial import receptor Mia40. EMBO Rep 9:1107–1113

Michaelis G, Esser K, Tursun B, Stohn JP, Hanson S, Pratje E (2005) Mitochondrial signal peptidases of yeast: the rhomboid peptidase Pcp1 and its substrate cytochrome *c* peroxidase. Gene 354:58–63

Milenkovic D, Ramming T, Muller JM, Wenz LS, Gebert N, Schulze-Specking A, Stojanovski D, Rospert S, Chacinska A (2009) Identification of the signal directing Tim9 and Tim10 into the intermembrane space of mitochondria. Mol Biol Cell 20(10):2530–2539

Model K, Meisinger C, Prinz T, Wiedemann N, Truscott KN, Pfanner N, Ryan MT (2001) Multistep assembly of the protein import channel of the mitochondrial outer membrane. Nat Struct Biol 8:361–370

Model K, Meisinger C, Kuhlbrandt W (2008) Cryo-electron microscopy structure of a yeast mitochondrial preprotein translocase. J Mol Biol 383:1049–1057

Mokranjac D, Paschen SA, Kozany C, Prokisch H, Hoppins SC, Nargang FE, Neupert W, Hell K (2003a) Tim50, a novel component of the TIM23 preprotein translocase of mitochondria. EMBO J 22:816–825

Mokranjac D, Sichting M, Neupert W, Hell K (2003b) Tim14, a novel key component of the import motor of the TIM23 protein translocase of mitochondria. EMBO J 22:4945–4956

Mokranjac D, Popov-Celeketic D, Hell K, Neupert W (2005) Role of Tim21 in mitochondrial translocation contact sites. J Biol Chem 280:23437–23440

Mokranjac D, Bourenkov G, Hell K, Neupert W, Groll M (2006) Structure and function of Tim14 and Tim16, the J and J-like components of the mitochondrial protein import motor. EMBO J 25:4675–4685

Morgan B, Ang SK, Yan G, Lu H (2009) Zinc can play chaperone-like and inhibitor roles during import of mitochondrial small Tim proteins. J Biol Chem 284:6818–6825

Muller JE, Papic D, Ulrich T, Grin I, Schutz M, Oberhettinger P, Tommassen J, Linke D, Dimmer KS, Autenrieth IB et al (2011) Mitochondria can recognize and assemble fragments of a beta-barrel structure. Mol Biol Cell 22:1638–1647

Murphy MP, Leuenberger D, Curran SP, Oppliger W, Koehler CM (2001) The essential function of the small Tim proteins in the TIM22 import pathway does not depend on formation of the soluble 70-kilodalton complex. Mol Cell Biol 21:6132–6138

Naamati A, Regev-Rudzki N, Galperin S, Lill R, Pines O (2009) Dual targeting of Nfs1 and discovery of its novel processing enzyme, Icp55. J Biol Chem 284:30200–30208

Nagamori S, Smirnova IN, Kaback HR (2004) Role of YidC in folding of polytopic membrane proteins. J Cell Biol 165:53–62

Nakamura T, Cieplak P, Cho DH, Godzik A, Lipton SA (2010) S-nitrosylation of Drp1 links excessive mitochondrial fission to neuronal injury in neurodegeneration. Mitochondrion 10:573–578

Naoe M, Ohwa Y, Ishikawa D, Ohshima C, Nishikawa S, Yamamoto H, Endo T (2004) Identification of Tim40 that mediates protein sorting to the mitochondrial intermembrane space. J Biol Chem 279:47815–47821

Nargang FE, Preuss M, Neupert W, Herrmann JM (2002) The Oxa1 protein forms a homooligomeric complex and is an essential part of the mitochondrial export translocase in *Neurospora crassa*. J Biol Chem 277:12846–12853

Nowikovsky K, Froschauer EM, Zsurka G, Samaj J, Reipert S, Kolisek M, Wiesenberger G, Schweyen RJ (2004) The LETM1/YOL027 gene family encodes a factor of the mitochondrial K^+ homeostasis with a potential role in the Wolf-Hirschhorn syndrome. J Biol Chem 279:30307–30315

Nowikovsky K, Reipert S, Devenish RJ, Schweyen RJ (2007) Mdm38 protein depletion causes loss of mitochondrial K^+/H^+ exchange activity, osmotic swelling and mitophagy. Cell Death Differ 14:1647–1656

Okamoto K, Brinker A, Paschen SA, Moarefi I, Hayer-Hartl M, Neupert W, Brunner M (2002) The protein import motor of mitochondria: a targeted molecular ratchet driving unfolding and translocation. EMBO J 21:3659–3671

Ott M, Prestele M, Bauerschmitt H, Funes S, Bonnefoy N, Herrmann JM (2006) Mba1, a membrane-associated ribosome receptor in mitochondria. EMBO J 25:1603–1610

Pagliarini DJ, Calvo SE, Chang B, Sheth SA, Vafai SB, Ong SE, Walford GA, Sugiana C, Boneh A, Chen WK et al (2008) A mitochondrial protein compendium elucidates complex I disease biology. Cell 134:112–123

Palmieri F, Bisaccia F, Iacobazzi V, Indiveri C, Zara V (1992) Mitochondrial substrate carriers. Biochim Biophys Acta 1101:223–227

Paschen SA, Waizenegger T, Stan T, Preuss M, Cyrklaff M, Hell K, Rapaport D, Neupert W (2003) Evolutionary conservation of biogenesis of beta-barrel membrane proteins. Nature 426:862–866

Pebay-Peyroula E, Dahout-Gonzalez C, Kahn R, Trezeguet V, Lauquin GJ, Brandolin G (2003) Structure of mitochondrial ADP/ATP carrier in complex with carboxyatractyloside. Nature 426:39–44

Peixoto PM, Grana F, Roy TJ, Dunn CD, Flores M, Jensen RE, Campo ML (2007) Awaking TIM22, a dynamic ligand-gated channel for protein insertion in the mitochondrial inner membrane. J Biol Chem 282:18694–18701

Perez-Martinez X, Broadley SA, Fox TD (2003) Mss51p promotes mitochondrial Cox1p synthesis and interacts with newly synthesized Cox1p. EMBO J 22:5951–5961

Perry AJ, Hulett JM, Likic VA, Lithgow T, Gooley PR (2006) Convergent evolution of receptors for protein import into mitochondria. Curr Biol 16:221–229

Popov-Celeketic D, Mapa K, Neupert W, Mokranjac D (2008) Active remodelling of the TIM23 complex during translocation of preproteins into mitochondria. EMBO J 27:1469–1480

Popov-Celeketic D, Waegemann K, Mapa K, Neupert W, Mokranjac D (2011) Role of the import motor in insertion of transmembrane segments by the mitochondrial TIM23 complex. EMBO Rep 12:542–548

Preuss M, Leonhard K, Hell K, Stuart RA, Neupert W, Herrmann JM (2001) Mba1, a novel component of the mitochondrial protein export machinery of the yeast Saccharomyces cerevisiae. J Cell Biol 153:1085–1096

Price CE, Otto A, Fusetti F, Becher D, Hecker M, Driessen AJ (2010) Differential effect of YidC depletion on the membrane proteome of *Escherichia coli* under aerobic and anaerobic growth conditions. Proteomics 10:3235–3247

Pusnik M, Schmidt O, Perry AJ, Oeljeklaus S, Niemann M, Warscheid B, Lithgow T, Meisinger C, Schneider A (2011) Mitochondrial preprotein translocase of trypanosomatids has a bacterial origin. Curr Biol 21:1738–1743

Rabl R, Soubannier V, Scholz R, Vogel F, Mendl N, Vasiljev-Neumeyer A, Korner C, Jagasia R, Keil T, Baumeister W et al (2009) Formation of cristae and crista junctions in mitochondria depends on antagonism between Fcj1 and Su e/g. J Cell Biol 185:1047–1063

Rak M, Tzagoloff A (2009) F1-dependent translation of mitochondrially encoded Atp6p and Atp8p subunits of yeast ATP synthase. Proc Natl Acad Sci USA 106:18509–18514

Rao S, Gerbeth C, Harbauer A, Mikropoulou D, Meisinger C, Schmidt O (2011) Signaling at the gate: phosphorylation of the mitochondrial protein import machinery. Cell Cycle 10: 2083–2090

Rimmer KA, Foo JH, Ng A, Petrie EJ, Shilling PJ, Perry AJ, Mertens HD, Lithgow T, Mulhern TD, Gooley PR (2011) Recognition of mitochondrial targeting sequences by the import receptors Tom20 and Tom22. J Mol Biol 405:804–818

Rissler M, Wiedemann N, Pfannschmidt S, Gabriel K, Guiard B, Pfanner N, Chacinska A (2005) The essential mitochondrial protein Erv1 cooperates with Mia40 in biogenesis of intermembrane space proteins. J Mol Biol 353:485–492

Rojo EE, Guiard B, Neupert W, Stuart RA (1998) Sorting of D-lactate dehydrogenase to the inner membrane of mitochondria: analysis of topogenic signal and energetic requirements. J Biol Chem 273:8040–8047

Rojo EE, Guiard B, Neupert W, Stuart RA (1999) N-terminal tail export from the mitochondrial matrix. Adherence to the prokaryotic "positive-inside" rule of membrane protein topology. J Biol Chem 274:19617–19622

Samuelson JC, Chen M, Jiang F, Möller I, Wiedmann M, Kuhn A, Phillips GJ, Dalbey RE (2000) YidC mediates membrane protein insertion in bacteria. Nature 406:637–641

Schlossmann J, Dietmeier K, Pfanner N, Neupert W (1994) Specific recognition of mitochondrial preproteins by the cytosolic domain of the import receptor MOM72. J Biol Chem 269:11893–11901

Schmidt O, Harbauer AB, Rao S, Eyrich B, Zahedi RP, Stojanovski D, Schonfisch B, Guiard B, Sickmann A, Pfanner N et al (2011) Regulation of mitochondrial protein import by cytosolic kinases. Cell 144:227–239

Schneider H-C, Berthold J, Bauer MF, Dietmeier K, Guiard B, Brunner M, Neupert W (1994) Mitochondrial Hsp70/MIM44 complex facilitates protein import. Nature 371:768–774

Schneider H-C, Westermann B, Neupert W, Brunner M (1996) The nucleotide exchange factor MGE exerts a key function in the ATP-dependent cycle of mt-Hsp70-Tim44 interaction driving mitochondrial protein import. EMBO J 15:5796–5803

Schneider A, Bursac D, Lithgow T (2008) The direct route: a simplified pathway for protein import into the mitochondrion of trypanosomes. Trends Cell Biol 18:12–18

Schulz C, Lytovchenko O, Melin J, Chacinska A, Guiard B, Neumann P, Ficner R, Jahn O, Schmidt B, Rehling P (2011) Tim50's presequence receptor domain is essential for signal driven transport across the TIM23 complex. J Cell Biol 195:643–656

Schwartz MP, Matouschek A (1999) The dimensions of the protein import channels in the outer and inner mitochondrial membranes. Proc Natl Acad Sci USA 96:13086–13090

Scotti PA, Urbanus ML, Brunner J, de Gier JW, von Heijne G, van der Does C, Driessen AJ, Oudega B, Luirink J (2000) YidC, the *Escherichia coli* homologue of mitochondrial Oxa1p, is a component of the Sec translocase. EMBO J 19:542–549

Serek J, Bauer-Manz G, Struhalla G, Van Den Berg L, Kiefer D, Dalbey R, Kuhn A (2004) *Escherichia coli* YidC is a membrane insertase for Sec-independent proteins. EMBO J 23:294–301

Shiota T, Mabuchi H, Tanaka-Yamano S, Yamano K, Endo T (2011) In vivo protein-interaction mapping of a mitochondrial translocator protein Tom22 at work. Proc Natl Acad Sci USA 108:15179–15183

Sideris DP, Petrakis N, Katrakili N, Mikropoulou D, Gallo A, Ciofi-Baffoni S, Banci L, Bertini I, Tokatlidis K (2009) A novel intermembrane space-targeting signal docks cysteines onto Mia40 during mitochondrial oxidative folding. J Cell Biol 187:1007–1022

Sirrenberg C, Bauer MF, Guiard B, Neupert W, Brunner M (1996) Import of carrier proteins into the mitochondrial inner membrane mediated by Tim22. Nature 384:582–585

Sirrenberg C, Endres M, Fölsch H, Stuart RA, Neupert W, Brunner M (1998) Carrier protein import into mitochondria mediated by the intermembrane proteins Tim10/Mrs11p and Tim12/Mrs5p. Nature 391:912–915

Smits P, Smeitink JA, van den Heuvel LP, Huynen MA, Ettema TJ (2007) Reconstructing the evolution of the mitochondrial ribosomal proteome. Nucleic Acids Res 35:4686–4703

Söllner T, Griffiths G, Pfaller R, Pfanner N, Neupert W (1989) MOM19, an import receptor for mitochondrial precursor proteins. Cell 59:1061–1070

Söllner T, Pfaller R, Griffiths G, Pfanner N, Neupert W (1990) A mitochondrial import receptor for the ATP/ADP carrier. Cell 62:107–115

Stenberg F, von Heijne G, Daley DO (2007) Assembly of the cytochrome *bo3* complex. J Mol Biol 371:765–773

Stiburek L, Fornuskova D, Wenchich L, Pejznochova M, Hansikova H, Zeman J (2007) Knockdown of human Oxa1l impairs the biogenesis of F(1)F(o)-ATP synthase and NADH:ubiquinone oxidoreductase. J Mol Biol 374:506–516

Suzuki H, Maeda M, Mihara K (2002) Characterization of rat TOM70 as a receptor of the preprotein translocase of the mitochondrial outer membrane. J Cell Sci 115:1895–1905

Szyrach G, Ott M, Bonnefoy N, Neupert W, Herrmann JM (2003) Ribosome binding to the Oxa1 complex facilitates cotranslational protein insertion in mitochondria. EMBO J 22:6448–6457

Taguchi N, Ishihara N, Jofuku A, Oka T, Mihara K (2007) Mitotic phosphorylation of dynamin-related GTPase Drp1 participates in mitochondrial fission. J Biol Chem 282:11521–11529

Tamai S, Iida H, Yokota S, Sayano T, Kiguchiya S, Ishihara N, Hayashi J, Mihara K, Oka T (2008) Characterization of the mitochondrial protein LETM1, which maintains the mitochondrial tubular shapes and interacts with the AAA-ATPase BCS1L. J Cell Sci 121:2588–2600

Tatsuta T, Augustin S, Nolden M, Friedrichs B, Langer T (2007) m-AAA protease-driven membrane dislocation allows intramembrane cleavage by rhomboid in mitochondria. EMBO J 26:325–335

Terziyska N, Lutz T, Kozany C, Mokranjac D, Mesecke N, Neupert W, Herrmann JM, Hell K (2005) Mia40, a novel factor for protein import into the intermembrane space of mitochondria is able to bind metal ions. FEBS Lett 579:179–184

Terziyska N, Grumbt B, Kozany C, Hell K (2009) Structural and functional roles of the conserved cysteine residues of the redox-regulated import receptor Mia40 in the intermembrane space of mitochondria. J Biol Chem 284:1353–1363

Tienson HL, Dabir DV, Neal SE, Loo R, Hasson SA, Boontheung P, Kim SK, Loo JA, Koehler CM (2009) Reconstitution of the mia40-erv1 oxidative folding pathway for the small Tim proteins. Mol Biol Cell 20:3481–3490

Truscott KN, Kovermann P, Geissler A, Merlin A, Meijer M, Driessen AJ, Rassow J, Pfanner N, Wagner R (2001) A presequence- and voltage-sensitive channel of the mitochondrial preprotein translocase formed by Tim23. Nat Struct Biol 8:1074–1082

Tsukihara T, Aoyama H, Yamashita E, Tomizaki T, Yamaguchi H, Shinzawa-Itoh K, Nakashima R, Yaono R, Yoshikawa S (1996) The whole structure of the 13-subunit oxidized cytochrome c oxidase at 2.8 A. Science 272:1136–1144

van der Klei I, Veenhuis M, Neupert W (1994) A morphological view on mitochondrial protein targeting. Microsc Res Tech 27:284–293

van der Laan M, Houben EN, Nouwen N, Luirink J, Driessen AJ (2001) Reconstitution of Sec-dependent membrane protein insertion: nascent FtsQ interacts with YidC in a SecYEG-dependent manner. EMBO Rep 2:519–523

van der Laan M, Bechtluft P, Kol S, Nouwen N, Driessen AJ (2004) F_1F_0 ATP synthase subunit c is a substrate of the novel YidC pathway for membrane protein biogenesis. J Cell Biol 165:213–222

van der Laan M, Wiedemann N, Mick DU, Guiard B, Rehling P, Pfanner N (2006) A role for Tim21 in membrane-potential-dependent preprotein sorting in mitochondria. Curr Biol 16:2271–2276

van der Laan M, Meinecke M, Dudek J, Hutu DP, Lind M, Perschil I, Guiard B, Wagner R, Pfanner N, Rehling P (2007) Motor-free mitochondrial presequence translocase drives membrane integration of preproteins. Nat Cell Biol 9:1152–1159

Vergnolle MA, Baud C, Golovanov AP, Alcock F, Luciano P, Lian LY, Tokatlidis K (2005) Distinct domains of small Tims involved in subunit interaction and substrate recognition. J Mol Biol 351:839–849

Vögtle FN, Wortelkamp S, Zahedi RP, Becker D, Leidhold C, Gevaert K, Kellermann J, Voos W, Sickmann A, Pfanner N et al (2009) Global analysis of the mitochondrial N-proteome identifies a processing peptidase critical for protein stability. Cell 139:428–439

von der Malsburg K, Muller JM, Bohnert M, Oeljeklaus S, Kwiatkowska P, Becker T, Loniewska-Lwowska A, Wiese S, Rao S, Milenkovic D et al (2011) Dual role of mitofilin in mitochondrial membrane organization and protein biogenesis. Dev Cell 21:694–707

von Heijne G (1989) Control of topology and mode of assembly of a polytopic membrane protein by positively charged residues. Nature 341:456–458

Voulhoux R, Bos MP, Geurtsen J, Mols M, Tommassen J (2003) Role of a highly conserved bacterial protein in outer membrane protein assembly. Science 299:262–265

Wagner S, Pop O, Haan GJ, Baars L, Koningstein G, Klepsch MM, Genevaux P, Luirink J, de Gier JW (2008) Biogenesis of MalF and the MalFGK2 maltose transport complex in *Escherichia coli* requires YidC. J Biol Chem 283:17881–17890

Waizenegger T, Stan T, Neupert W, Rapaport D (2003) Signal-anchor domains of proteins of the outer membrane of mitochondria: structural and functional characteristics. J Biol Chem 278:42064–42071

Walther DM, Bos MP, Rapaport D, Tommassen J (2010) The mitochondrial porin, VDAC, has retained the ability to be assembled in the bacterial outer membrane. Mol Biol Evol 27:887–895

Wang H, Song P, Du L, Tian W, Yue W, Liu M, Li D, Wang B, Zhu Y, Cao C et al (2011) Parkin ubiquitinates Drp1 for proteasome-dependent degradation: implication of dysregulated mitochondrial dynamics in Parkinson disease. J Biol Chem 286:11649–11658

Watson K (1972) The organization of ribosomal granules within mitochondrial structures of aerobic and anaerobic cells of *Saccharomyces cerevisae*. J Cell Biol 55:721–726

Webb CT, Gorman MA, Lazarou M, Ryan MT, Gulbis JM (2006) Crystal structure of the mitochondrial chaperone TIM9-10 reveals a six-bladed alpha-propeller. Mol Cell 21:123–133
White SH, von Heijne G (2004) The machinery of membrane protein assembly. Curr Opin Struct Biol 14:397–404
Wickström D, Wagner S, Simonsson P, Pop O, Baars L, Ytterberg AJ, van Wijk KJ, Luirink J, de Gier JW (2011) Characterization of the consequences of YidC depletion on the inner membrane proteome of *E. coli* using 2D Blue Native/SDS-PAGE. J Mol Biol 409:124–135
Yamamoto H, Esaki M, Kanamori T, Tamura Y, Nishikawa S, Endo T (2002) Tim50 Is a subunit of the TIM23 complex that links protein translocation across the outer and inner mitochondrial membranes. Cell 111:519–528
Yamano K, Kuroyanagi-Hasegawa M, Esaki M, Yokota M, Endo T (2008a) Step-size analyses of the mitochondrial Hsp70 import motor reveal the Brownian ratchet in operation. J Biol Chem 283:27325–27332
Yamano K, Yatsukawa Y, Esaki M, Hobbs AE, Jensen RE, Endo T (2008b) Tom20 and Tom22 share the common signal recognition pathway in mitochondrial protein import. J Biol Chem 283:3799–3807
Young JC, Hoogenraad NJ, Hartl FU (2003) Molecular chaperones Hsp90 and Hsp70 deliver preproteins to the mitochondrial import receptor Tom70. Cell 112:41–50
Zeth K (2010) Structure and evolution of mitochondrial outer membrane proteins of beta-barrel topology. Biochim Biophys Acta 1797:1292–1299

Chapter 4
Assembly Factors of Human Mitochondrial Respiratory Chain Complexes: Physiology and Pathophysiology

Daniele Ghezzi and Massimo Zeviani

Abstract Mitochondrial disorders are clinical syndromes associated with abnormalities of the oxidative phosphorylation (OXPHOS) system, the main responsible for the production of energy in the cell. OXPHOS is carried out in the inner mitochondrial membrane by the five enzymatic complexes of the mitochondrial respiratory chain (MRC). The subunits constituting these multimeric complexes have a dual genetic origin, mitochondrial or nuclear. Hence, mitochondrial syndromes can be due to mutations of mitochondrial DNA or to abnormalities in nuclear genes. The biogenesis of the MRC complexes is an intricate and finely tuned process. The recent discovery of several OXPHOS-related human genes, mutated in different clinical syndromes, indicates that the majority of the inherited mitochondrial disorders are due to nuclear genes, and many of them encode proteins necessary for the proper assembly/stability of the MRC complexes. The detailed mechanisms of these processes are not fully understood and the exact function of many such factors remains obscure.

We present an overview on the hypothesized assembly processes of the different MRC complexes, focusing on known assembly factors and their clinical importance.

4.1 Introduction

The mitochondrial respiratory chain (MRC) is composed of five multiheteromeric complexes (complex I, CI; complex II, CII; complex III, CIII; complex IV, CIV or cytochrome *c* oxidase, COX; complex V, CV, or ATP synthase), all embedded in the inner mitochondrial membrane, and two mobile electron shuttles, ubiquinone (Coenzyme Q, CoQ), a lipoidal quinone, and cytochrome *c* (cyt *c*), a heme-containing

D. Ghezzi • M. Zeviani (✉)
Unit of Molecular Neurogenetics, The Foundation "Carlo Besta" Institute of Neurology, Milan, Italy
email: dghezzi@istituto-besta.it; zeviani@istituto-besta.it

B. Kadenbach (ed.), *Mitochondrial Oxidative Phosphorylation*, Advances in Experimental Medicine and Biology 748, DOI 10.1007/978-1-4614-3573-0_4,

small polypeptide. Electron donors including reduced nicotinamide-adenine dinucleotide, $NADH^+$ (via CI), and reduced flavin adenine dinucleotide, $FADH_2$ (via CII and other flavoproteins, including electron transfer flavoprotein ubiquinone reductase or ETF-dehydrogenase (EC 1.5.5.1), the terminal component of fatty acid β-oxidation and ketogenic aminoacid oxidation pathways), supply electrons to CoQ, which donates them to CIII. CIII transfers one electron at a time to cyt *c*, which passes it to COX; COX eventually fixes four electrons to molecular oxygen with the formation of two molecules of water. This process, known as respiration, liberates energy that is partly converted by the proton pumping activity of CI, CIII, and CIV into an electrochemical potential (ΔμH) composed of an electrical gradient (ΔΨ) and a pH gradient, across the inner mitochondrial membrane. ΔμH constitutes the driving proton motive force not only for the phosphorylation of ADP to ATP, operated by CV, but also for a number of other processes, such as heat production, Ca^{++} import inside mitochondria, and protein translocation.

From a genetic standpoint, the MRC is unique, as it is formed through the complementation of two separate genetic systems: the nuclear and the mitochondrial genomes. Four of the five MRC complexes, namely CI, CIII, CIV, and CV, contain subunits encoded by the mitochondrial DNA, mtDNA, and synthesized in situ by the organelle-specific translation machinery. Thus, ad hoc replication, transcription and translation machineries, composed of hundreds of RNA and protein factors, have been maintained through evolution of eukaryotes to carry out the synthesis of a few, but essential, mtDNA-encoded proteins. In humans, seven (ND1, 2, 3, 4, 4L, 5, 6) are components of CI, one (cytochrome *b*) of CIII, three (COI, II, III) of CIV and two (ATPase 6 and 8) of CV.

Specific pathways are required for the assembly of each MRC complex, including the insertion of mtDNA-encoded subunits into the inner membrane of mitochondria, in concert with >80 nuclear DNA encoded subunits; the synthesis and incorporation of several prosthetic groups that form the catalytic redox cores of CI, CII, CIII, and CIV; and the ultimate formation of functionally active holocomplexes. Individual holocomplexes can also organize themselves in respiratory supercomplexes. Additional systems warrant the quality control of protein and non-protein components of the MRC complexes, thus contributing to the maintenance of their structural integrity, functional activity and turnover. Thus, a highly regulated, extremely complex process is at work in mitochondria to control the formation, stability, interactions, function, and plasticity of the MRC. Defects in genes encoding the components of these control and execution systems can compromise the function of the MRC, thus leading to faulty OXPHOS and disease. For instance, many mtDNA mutations affect the mtDNA genes encoding the RNA apparatus for in situ translation (tRNA and rRNA encoding genes) rather than those encoding structural MRC proteins. Likewise, in addition to mutations in nuclear DNA-encoded MRC subunits, a vast number of mitochondrial disorders are due to mutations in "ancillary" gene products controlling their assembly, function and turnover (Ghezzi and Zeviani 2011).

The term "mitochondrial disorders" is referred to (genetic) the defects of OXPHOS. The peculiar genetics of mtDNA, the intergenomic interactions converging on the formation of the MRC, and the structural and functional complexities of the biochemical pathways underpinning OXPHOS, explain the extreme heterogeneity of inherited mitochondrial disorders, which include a vast range of symptoms, severity, age of onset, progression, and outcome (DiMauro and Davidzon 2005). Several modes of transmission have been reported: maternal or Mendelian (autosomal dominant, recessive, X-linked), depending whether the primary genetic defect resides in mtDNA or nuclear DNA. Sporadic conditions, presumably caused by de novo mutations, are also known, as for instance in single mtDNA deletions. Altogether, the prevalence of genetic OXPHOS defects is not less than 1:5,000 (Schaefer et al. 2004; Thorburn et al. 2004; Mancuso et al. 2007), and probably higher (Cree et al. 2009). Clinical manifestations range from lesions in single structures (i.e., the optic nerve in Leber's hereditary optic neuropathy, LHON), tissues or organs (e.g., myopathies, encephalopathies, cardiomyopathies), to multisystem syndromes, with multiple organ involvement, the onset varying from neonatal to adult life. The degree of organ failure is also variable, depending on tissue-specific energy demand, tissue-specific vs. ubiquitous expression of OXPHOS-related genes, mitotic segregation of heteroplasmic mtDNA mutations, the presence of a genetic bottleneck for mtDNA during embryonic development, and other, poorly defined, causes. Because OXPHOS is necessary for energy supply to virtually any cell, any organ can be affected by mitochondrial disease. However, the most common clinical presentations include the involvement of muscle, heart and brain, i.e., postmitotic, specialized tissues, with high metabolic requests (McFarland and Turnbull 2009).

Adhering to the title of our review, the following discussion will focus on factors involved in the assembly of human MRC complexes, mainly those reported to be abnormal in patients, causing faulty OXPHOS and disease (Table 4.1). In broad terms, an "assembly factor" of a given MRC component is any protein (or non-protein) species that plays a role in its formation or stability, but is not stably part of it. Investigation of assembly defects has not only been useful to understand the molecular pathogenesis of several mitochondrial disorders, but has also helped understand the biogenesis of MRC formation in mammals. However, in only a few cases has the detailed mechanism of action been elucidated, so that the definition of "assembly factor" remains largely observational, based on the association between an assembly defect of a given complex with mutations in a particular gene product. For example, BCS1L and SURF1 are important for the assembly of CIII and CIV, but the precise mechanism by which they accomplish this task remains unclear. Moreover, the manifestations of their defects could be due to a multiplicity of functions, some of which unrelated to the MRC. For instance, defects of BCS1L, a CIII assembly factor, are associated with iron overload and multivisceral failure with normal CIII activity in some patients (Visapaa et al. 2002), whereas in others the main abnormality is, as expected, a specific defect of CIII; mutations in SURF1, an assembly factor of CIV, are lethal in humans and flies, but are associated with prolonged longevity and neuroprotection from calcium-induced neuronal excitotoxicity in knockout mice (Dell'Agnello et al. 2007); SCO1 and SCO2, two other CIV assembly factors, are involved in cellular copper homeostasis (Leary et al. 2007), but SCO2 has also been proposed to participate in a p53-dependent checkpoint system that controls the switch from glycolysis to respiration as the main energy

Table 4.1 Assembly factors of the MRC and related mitochondrial disease

Protein	Predicted function/s	Phenotypes
CI assembly factors		
NDUFAF1	CI chaperone interacting transiently with early arm membrane intermediates	Cardiomyoencephalopathy, lactic acidosis
ECSIT	Associated with NDUFAF1	–
NDUFAF2	Stabilizer of late intermediate(s)	Leukoencephalopathy with vanishing white matter, Leigh syndrome
NDUFAF3	Interact with some CI subunits and with NDUFAF4	Variable phenotypes: macrocephaly, severe muscle weakness, myoclonic seizures, brain leukomalacia
NDUFAF4	Interact with some CI subunits and with NDUFAF3	Encephalopathy, antenatal cardiomyopathy
C20ORF7	Probable methyltransferase; early arm membrane assembly	Leigh syndrome, progressive spasticity
C8ORF38	Unclear role but necessary for CI function	Leigh syndrome
NUBPL	Facilitates the effective assembly of Fe–S cofactors and subunits	Leukodystrophy, myopathy, ataxia
ACAD9	Complex I assembler by the interaction with the mitochondrial inner membrane or with NDUFAF1/ECSIT	Cardiomyopathy, encephalopathy, lactic acidosis, exercise intolerance
FOXRED1	Functional role in electrons transfer reactions; aminoacid metabolism	Leigh syndrome
CII assembly factors		
SDHAF1	Fe/S clusters insertion or retention within the CII protein backbone	Encephalomyopathy
SDHAF2	Flavination of SDHA	Hereditary paraganglioma
CIII assembly factors		
BCS1L	Incorporation of RISP	GRACILE syndrome, Bjornstad syndrome, encephalopathy, proximal tubulopathy and liver failure
TTC19	Important in early assembly, interacts with fully assembled CIII	Progressive encephalopathy
HCCS	Synthesis of cyt *c1* and cyt *c*	MIDAS
PTCD2	Maturation and stabilization of cytochrome *b* mRNA	–
CIV assembly factors		
SURF1	Formation of the early subcomplexes of COX	Leigh syndrome
SCO1	Incorporation of copper atoms in the catalytic sites of the nascent complex	Infantile encephalopathy, neonatal hepatopathy, ketoacidotic comas

SCO2	Incorporation of copper atoms in the catalytic sites of the nascent complex	Infantile cardioencephalomyopathy
COX10	Heme A synthesis (conversion of heme *b* to heme *o*)	Leigh syndrome, proximal renal tubulopathy, hypertrophic cardiomyopathy, sensorineural deafness, metabolic acidosis
COX15	Heme A synthesis (conversion of heme *o* to heme *a*)	Infantile cardiomyopathy, Leigh syndrome
COX11	Biosynthesis of heme *a*	–
COX17	Copper recruitment	–
COX19	Copper translocation to mitochondria	–
LPPRC	RNA metabolism, regulation of mtDNA genes	French-Canadian Leigh syndrome
TACO1	COX subunit I translation	Slow progressive Leigh syndrome
C20ORF64	Involved in a very early step of the COX assembly	Fatal neonatal cardiomyopathy
CV assembly factors		
ATPAF1	F1 chaperon; essential for assembly of $\alpha+\beta$ heterooligomer	–
ATPAF2	F1 chaperon; essential for assembly of $\alpha+\beta$ heterooligomer	Degenerative encephalopathy, connatal lactic acidosis, methyl glutaconic aciduria
TMEM70	Assembly of F1; F1 interaction with some of the F0 subunits	Neonatal encephalocardiomyopathy

provider of the cell. Further complexity derives from the recent observation that individual respiratory complexes agglomerate into gigantic supercomplexes.

Studies on the facultative anaerobic yeast *Saccharomyces cerevisiae* have not only led to an understanding of the basic mechanisms underlying the assembly of complexes III, IV, and V, but also to the discovery of many assembly factors, some of which have orthologs in humans.

Because CI is missing in *S. cerevisiae*, other yeasts, such as *Neurospora crassa* and *Yarrowia lipolytica*, or mammals, such as *Bos taurus*, have served as models for the study of CI structure and assembly. Blue native gel electrophoresis (BNGE) is a fundamental technique used to analyze the molecular steps for MRC assembly in different organisms and conditions, including human disease (Schagger and von Jagow 1991; Wittig et al. 2006). For example, the identification of CI subcomplexes in patients has allowed investigators to identify consistent anomalous patterns and relate them to specific genetic defects. Furthermore, the study of subcomplexes in mutant cells, and in normal cells after inhibition of mitochondrial protein synthesis, has led to the delineation of the current models for CI assembly in humans and the refinement of those for CII, CIII, CIV, and CV.

4.2 Complex I

In mammalian mitochondria, CI (NADH ubiquinone oxidoreductase, EC 1.6.5.3) catalyzes the oxidation of reduced NADH by CoQ. CI is the largest complex in the OXPHOS system, consisting of $\approx$45 subunits in mammals, for a total molecular mass of $\approx$1 MDa (Carrol et al. 2006). Seven subunits (ND1–ND6, ND4L) are encoded by the mitochondrial genome, the remaining 38 by nuclear DNA (nDNA) genes. The seven mtDNA encoded subunits and seven nuclear-encoded subunits (NDUFV1, NDUFV2, NDUFS1, NDUFS2, NDUFS3, NDUFS7, NDUFS8) form the catalytic "core" structure of the complex (Koopman et al. 2010), conserved in virtually all organisms that contain a CI, including *Thermus thermophilus*. The remaining nuclear-DNA encoded subunits are possibly involved in CI assembly, stability, and regulation of activity.

Ultrastructural studies of purified CI revealed an "L" shaped object, consisting of two arms, a hydrophilic "peripheral arm" protruding into the matrix, and a hydrophobic "membrane" arm, embedded in the mitochondrial inner membrane, with an angle between the two arms of about 100°. The value of the angle can slightly vary, since the two arms change their reciprocal positions during the catalytic cycle of the complex (Janssen et al. 2006). The two arms harbor three functional modules: (1) the N module, in the peripheral arm, contains the dehydrogenase site, formed by a flavin-mononucleotide (FMN) moiety, responsible for the oxidation of NADH to NAD+; (2) the Q module, in the hinge region between the two arms, contains the CoQ reduction site; (3) the P module, or proton translocation module, constitutes the membrane arm, and includes the seven mtDNA-encoded ND subunits (Vogel et al. 2007). Eight iron–sulfur (Fe–S) clusters are distributed in the N module (n. 5) and the Q module (n. 3); seven clusters form a redox chain, through which electrons flow from FMN to CoQ, whereas the role of the eighth cluster, contained in the N module, is unclear.

The crystal structures of CI from *Thermus thermophilus* (Efremov et al. 2010) and *Yarrowia lipolytica* (Hunte et al. 2010) have provided important information on the mechanisms of electron flow, CoQ reduction, and proton pumping. In particular, the membrane arm, which largely coincides with the P module, includes 63 trans-membrane α-helices distributed among the seven core subunits encoded by mtDNA (Efremov et al. 2010). The P module is composed of two domains of similar size, proximal (P_P) and distal (P_D), connected by a narrow bridge consisting of a long α-helix, which is possibly part of ND5 (Hunte et al. 2010). The driving force for proton translocation is generated by the electrons extracted from NADH, flowing through the seven iron–sulfur cluster centers in the peripheral arm, distributed in the N and Q modules, and eventually converging onto ubiquinone, to form ubiquinol. Energy liberated by the electron flow is transferred from the Q to the P modules where it prompts the P_P–P_D protein bridge to move like a piston through the P_P–P_D domains, thus causing the nearby transmembrane helices to tilt and force proton translocation (Efremov et al. 2010).

4.2.1 *Human Diseases Associated with Complex I Deficiency (MIM 252010)*

Isolated CI deficiency is frequent in mitochondrial disorders (Distelmaier et al. 2009), being responsible for about one-third of all cases (Janssen et al. 2006). The primary genetic defect may be either in mtDNA or in nuclear DNA genes. Given the complexity of this huge enzyme, its dual genetic origin, and the incomplete information about its assembly, turnover, and regulation, it is not surprising that about half of the patients are still genetically undefined (Thorburn et al. 2004), and clinical presentations are so heterogeneous, including, for children, Leigh syndrome (LS), neonatal cardiomyopathy with lactic acidosis, fatal infantile lactic acidosis (FILA), macrocystic leukoencephalopathy, or pure myopathy (Loeffen et al. 2000; Bugiani et al. 2004). CI defects or CI-related mutations, particularly in mtDNA, include Leber's hereditary optic neuropathy (LHON), MELAS (mitochondrial encephalopathy with lactic acidosis and stroke-like episodes), or MELAS/LHON overlap syndromes (Lenaz et al. 2004). Several additional, poorly defined, syndromes include neurological signs, such as ophthalmoparesis, optic atrophy, epilepsy, ataxia and dystonia; gastrointestinal problems, such as pseudo-occlusion and gastroparesis; hepatopathy; cardiomyopathy; and renal dysfunctions, such as tubulopathy or proteinuria (Janssen et al. 2006).

4.2.2 *CI Assembly Model and Assembly Factors*

The complexity of CI assembly reflects its gigantic size, huge number of subunits, double genetics, multiple functions (e.g., redox vs. proton translocation), and numerous prosthetic groups. As mentioned above, there is no CI in the user-friendly yeast *Saccharomyces cerevisiae*. This has been a major hurdle in the definition of the molecular steps of, and players in, CI assembly. Together, these facts explain why CI assembly is still poorly defined; however, observations from CI-defective patients, siRNA experiments in mammalian cells (Remacle et al. 2008; Scheffler et al. 2004; Schneider et al. 2008), and gene targeting in CI-dependent fungi, such as *Neurospora crassa*, have contributed to outline a module-based general model of assembly (Vogel et al. 2007; Lazarou et al. 2009) (Fig. 4.1). Accordingly, the earliest stage consists in the formation of an evolutionarily conserved hydrogenase (Q) module composed of subunits NDUFS2, NDUFS3, NDUFS7, and NDUFS8. After the addition of NDUFA9 (and possibly other subunits), this peripheral-arm intermediate is anchored to the mitochondrial inner membrane, where it is assembled with a membrane arm intermediate containing ND1. The resulting 400 kDa subcomplex is then assembled with a 460 kDa membrane-bound subcomplex, formed by hydrophobic subunits ND2, ND3, ND6, and NDUFB6. Next, subunits ND4 and ND5 are incorporated to form an intermediate, largely consisting of the Q and P modules. Eventually, a pre-formed N module composed of subunits NDUFV1, NDUFV2, NDUFV3, NDUFV6, NDUFS1, NDUFS4, and NDUFS6, join the Q–P intermediate, together with additional peripheral subunits, to form mature CI (McKenzie and Ryan 2010). Accordingly, the composition and size of several CI intermediates, characterized in CI-defective patients, suggest that the peripheral and membrane arms are joined together before the completion of each arm (Lazarou et al. 2007). Many important details are still missing, including, for instance, the role of several single-trans-membrane ancillary subunits, which are supposed to form a scaffold structure "holding" the giant complex during its oscillatory movements; or when and how the several prosthetic groups are incorporated into the complex.

By analogy to complex IV, for which >20 factors cooperate to assemble a 13 subunit complex, at least 10–20 additional CI assembly factors are expected to be found, besides the ones that will be discussed below (Calvo et al. 2010). The list is in fact expanding, thanks to the implementation of a number of strategies, including systematic exome sequencing of CI-defective patients and in silico analysis of interspecific genomic variance. For instance, differential genome subtraction is a powerful approach that exploits what has been considered a drawback, i.e., the absence of CI in *S. cerevisiae,* to select genes that are missing in this organism but present in CI-proficient organisms, including *N. crassa, Yarrowia lipolytica*, and higher eukaryotes, to test the hypothesis that some of them must be functionally related to CI function, including assembly. This strategy has been successfully applied for the identification of a new CI assembly factor, *B17.2* (*NDUFAF2*) (Ogilvie et al. 2005), and has later been implemented to identify 19 proteins that share common ancestry with a large subset of CI proteins. A causative mutation in one of these genes, C8ORF38, has been identified in a family with inherited CI deficiency (Pagliarini et al. 2008).

Some of the known CI assembly factors are physically associated with CI assembly intermediates, e.g., *N. crassa* CIA84 and CIA30 (Kuffner et al. 1998), and their human ortholog NDUFAF1 (Dunning et al. 2007); or ECSIT, a human signaling molecule interacting with NDUFAF1 (Vogel et al. 2007).

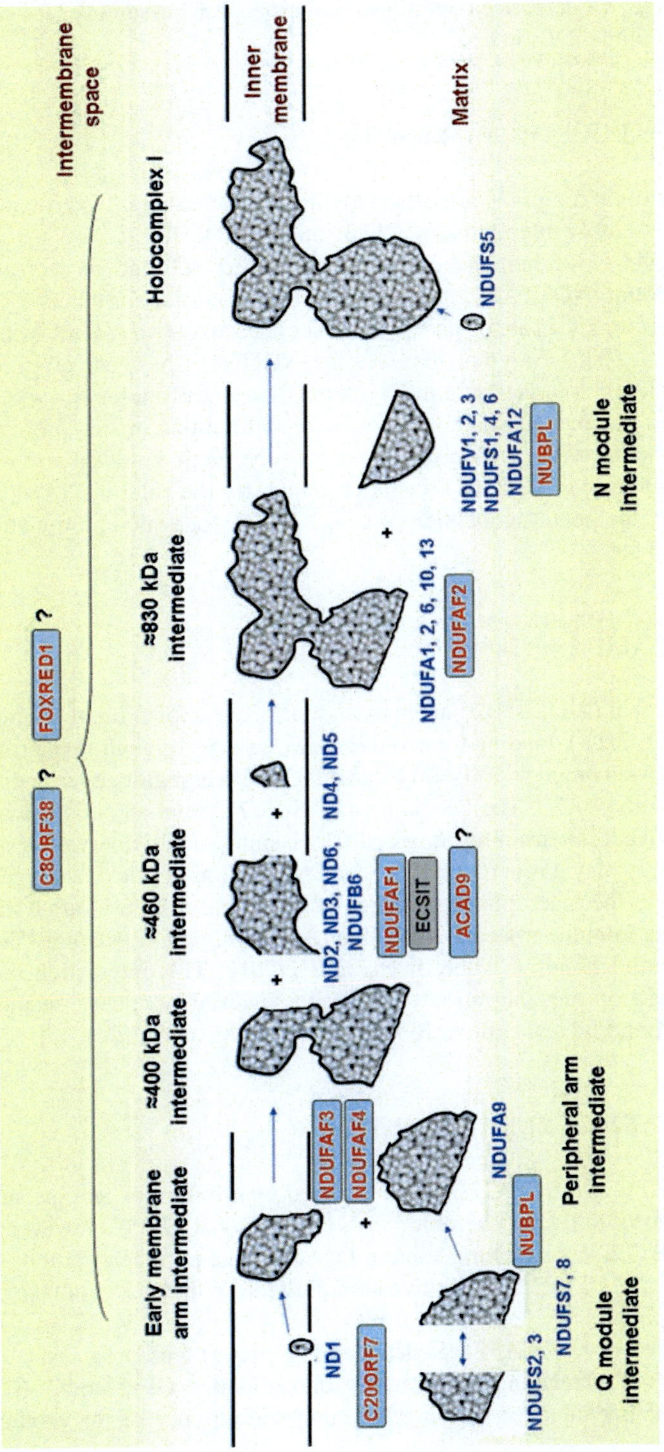

Fig. 4.1 Schematic outline of CI assembly. Assembly factors are *boxed*; in *red* are those associated with human disease, whereas the others are in *grey*. See text for details

Similar to other CI defective conditions, mutations in CI assembly factors cause a wide range of clinical disorders.

4.2.2.1 NDUFAF1/CIA30 (MIM 606934)

Studies in the aerobic fungus *N. crassa* led to the identification of two proteins associated with CI assembly intermediates, CIA30 and CIA84. siRNA-mediated knockdown of *NDUFAF1*, the human homologue of *CIA30*, resulted in decreased CI activity and amount. NDUFAF1 interacts with mitochondrial and nuclear CI subunits (Dunning et al. 2007) and is physically associated with two assembly intermediates (Bych et al. 2008). Based on these results, NDUFAF1 has been proposed as a chaperone transiently interacting with CI intermediates (Kuffner et al. 1998; Vogel et al. 2007), but mechanistic details are missing. Mutations in this protein were reported in a patient with cardiomyoencephalopathy, lactic acidosis and reduced levels of CI (Dunning et al. 2007). Little is known on the role of CIA84 and its human ortholog, the pentatricopeptide protein PTCD1, for which no mutation has been reported so far.

4.2.2.2 ECSIT (MIM 608388)

By two-dimensional (2-D) BNGE analysis and affinity purification, the mitochondrial isoform of ECSIT (standing for evolutionary conserved signaling intermediate in Toll pathway) was found in 500–600 kDa and 830 kDa assembly intermediates of CI, associated with CIA30. The knockdown of *ECSIT* in human cells reduces the levels of NDUFAF1, determining impaired CI assembly and activity (Vogel et al. 2007). ECSIT may play a role in mitochondrial production of reactive oxygen species (ROS) and in the macrophage-mediated antibacterial response: after stimulation of the innate immune system, a Toll-like receptor signaling adaptor (TRAF6) translocates to mitochondria, where it engages ECSIT. This interaction leads to ECSIT ubiquitination and migration to the mitochondrial periphery, resulting in increased mitochondrial and cellular ROS generation (West et al. 2011).

4.2.2.3 NDUFAF2/B17.2L (MIM 609653)

NDUFAF2 (*B17.*2L or *NDUFA12L*) was identified by differential genome subtraction of facultative vs. obligatory aerobic yeasts (Ogilvie et al. 2005). However, while in *Y. lipolytica* B17.2 is a structural subunit located in the peripheral arm of CI, its human ortholog, *NDUFAF2,* is a putative assembly factor that does not take part in the CI holocomplex.

A stop mutation of *NDUFAF2* was detected in a patient with progressive leukoencephalopathy with vanishing white matter, and impaired CI assembly (Ogilvie et al. 2005). Another mutation causing the complete absence of the protein was found in two infants with hypotonia, nystagmus, and ataxia (Barghuti et al. 2008) associated with 30% residual CI activity in muscle mitochondria. Additional homozygous *NDUFAF2* mutations, associated with the complete absence of the protein, were identified in three LS patients (Calvo et al. 2010). In human heart mitochondria, NDUFAF2 co-immunoprecipitates with several CI subunits, and is associated with a late ≈830-kDa CI subassembly intermediate in several CI-deficient patients, but not in control samples (Vogel et al. 2007). Taken together, these studies demonstrate that the complete absence of NDUFAF2 does not prevent the formation of some fully assembled CI holocomplex, indicating that this protein could stabilize late CI intermediate(s) or help late subunits to be incorporated into the nascent complex.

4.2.2.4 NDUFAF3/C3ORF60 (MIM 612911)

Studies on CI-defective patients led to the identification of NDUFAF3 and NDUFAF4 (Saada et al. 2009). Mutations in *NDUFAF3* were found in three families with CI deficiency associated with a spectrum of severe phenotypes: a fulminant syndrome dominated by muscle hypertonia in the first, macrocephaly and severe muscle weakness in the second, and myoclonic epilepsy and leukomalacia in the third. All patients died before 6 months of age. The knockdown of *NDUFAF3* in HeLa cells shows decreased CI amount and activity, and a significant reduction of another assembly factor, NDUFAF4. NDUFAF3 and NDUFAF4 co-migrate in the same CI assembly intermediates and interact with the same set of CI subunits (Saada et al. 2009).

4.2.2.5 NDUFAF4/C6ORF66 (MIM 611776)

A homozygous mutation in *NDUFAF4* was associated with severe CI deficiency in five consanguineous patients presenting with infantile encephalopathy and in one unrelated case of antenatal cardiomyopathy. BNGE analysis of mutant muscle mitochondria revealed reduction of fully assembled CI and accumulation of assembly intermediates (Saada et al. 2008). Similar to NDUFAF3, siRNA knockdown of *NDUFAF4* results in decreased levels of both CI holocomplex and NDUFAF3 (Saada et al. 2009), suggesting physical interaction and functional interdependence.

4.2.2.6 C20ORF7 (MIM 612360)

A homozygous mutation in an anonymous gene, *C20ORF7*, was identified in a lethal neonatal form of CI deficiency by homozygosity mapping followed by candidate gene analysis (Sugiana et al. 2008). Knockdown of *C20ORF7* decreased CI activity in cells and a role for C20ORF7 was demonstrated in the assembly or stability of an early CI assembly intermediate that contains ND1 but not ND2 subunits. A second mutation was later found in affected members from one family, presenting with LS; CI activity was reduced in blood lymphocytes and muscle. However, the genetic etiology was unclear, since a mutation in a second gene, *CRLS1*, encoding cardiolipin synthase 1, was also found in these patients (Gerards et al. 2010). Recently, a *C20ORF7* variant has been reported in children with LS from Ashkenazi Jewish families; in this population the carrier rate is 1:290. Interestingly, these patients show combined deficiency of CI and CIV but expression of the wildtype (wt) gene in mutant cells restored CI activity to normal whereas recovery of CIV was incomplete (Saada et al. 2012). Decreased CIV activity was also observed in human fibroblasts after *C20ORF7* knockdown (Sugiana et al. 2008), suggesting for C20ORF7 an additional role in CIV assembly or in the formation of CI–CIV supercomplexes. For instance, C20ORF7 resembles, and possibly is, a methyltransferase, that could play a role in the posttranscriptional modification of proteins necessary for CI and CIV function and/or expression, or for the stability of the CI–CIV respirasome species (Saada et al. 2012).

4.2.2.7 C8ORF38 (MIM 612392)

Genome subtraction helped identify C8ORF38 as a protein necessary for CI function. A homozygous missense mutation in a conserved residue was associated with LS with isolated CI deficiency. Knockdown of *C8ORF38* resulted in reduced levels and activity of CI, but its exact function remains unclear (Pagliarini et al. 2008).

4.2.2.8 NUBPL/Ind1 (MIM 613621)

Fe–S clusters are present in CI, CII and CIII, and several enzymes and chaperones are required for their biosynthesis (see Sect. 4.6). In addition, Ind1 is a Fe–S cluster binding protein with a specific role in the incorporation of Fe–S centers in CI. Deletion of the *Ind1* gene in *Y. lipolytica* results in a very specific defect of CI, without affecting other mitochondrial Fe–S enzymes (Bych et al. 2008; Sheftel et al. 2009). Knockdown of the human ortholog, *NUBPL*, in cultured cells causes a specific decrease in the amount of CI holocomplex and the formation of an assembly intermediate missing the peripheral arm, which harbors all the Fe–S clusters, but containing the membrane-arm NDUFB6 subunit (Sheftel et al. 2009). These findings suggest that NUBPL/Ind1 specifically facilitates the incorporation of Fe–S clusters into CI peripheral arm.

In a high-throughput screening of CI deficient patients, compound heterozygous mutations in this protein were identified in a single case, presenting with mitochondrial encephalopathy (Calvo et al. 2010).

4.2.2.9 ACAD9 (MIM 611103)

ACAD9 was discovered as an essential CI assembly factor by different approaches: traditional homozygosity mapping and candidate gene screening, exome sequencing, and tandem mass analysis on proteins interacting with known assembly factors. ACAD9 interacts with NDUFAF1 and ECSIT and co-migrates with the same CI assembly intermediates that contain these two factors. Knockdown of ACAD9, NDUFAF1 or ECSIT in cultured cells determines the decrease of all three proteins, and of CI holocomplex as well (Gerards et al. 2011).

Mutations in ACAD9 are associated with infantile hypertrophic cardiomyopathy, encephalopathy and lactic acidosis (Haack et al. 2010; Nouws et al. 2010), with exercise intolerance (Gerards et al. 2011). All patients had a reduction of CI enzymatic activity and assembly. ACAD9 mutant cells and patients respond to riboflavin treatment, with partial correction of CI deficiency and clinical improvement (Haack et al. 2010; Gerards et al. 2011), possibly because ACAD9 is a $FADH_2$ dependent acyl-CoA dehydrogenase. However, although ACAD9 displays a β-oxidative activity in vitro, fatty acid β-oxidation seems normal in ACAD9 mutant patients, the only functional impairment being connected to CI assembly.

4.2.2.10 FOXRED1 (MIM613622)

FOXRED1 was identified by gene screening of CI-defective LS patients (Fassone et al. 2010; Calvo et al. 2010). Knockdown of *FOXRED1* decreases the levels of CI in control fibroblasts. The presence of a FAD-binding domain suggests a role for this protein as a redox enzyme, but its exact function remains elusive (Fassone et al. 2010).

4.3 Complex II

Complex II (succinate dehydrogenase ubiquinone–ubiquinol reductase, E.C. 1.3.5.1) is the only membrane-bound member of the tricarboxylic acid (TCA) cycle, where it functions as a succinate dehydrogenase (SDH), catalyzing the oxidation and dehydration of succinate to fumarate. CII takes part in the MRC by coupling this reaction to the reduction of ubiquinone to ubiquinol, that in turn funnels electrons to CIII (Ernster and Dallner 1995). CII is the smallest MRC complex (123 kDa), consisting of four subunits, encoded by *SDHA*, *SDHB*, *SDHC*, and *SDHD* nuclear genes. SDHA contains a FAD moiety, whereas three Fe–S centers are bound to SDHB. These two hydrophilic subunits are linked to SDHC and SDHD, two small, hydrophobic polypeptides that contain a heme *b* moiety and anchor the complex to the inner mitochondrial membrane (Sun et al. 2005). The crystal structure of porcine heart CII consists of a hydrophilic head protruding into the matrix, a hydrophobic tail embedded in the inner membrane, and a short segment projecting into the intermembrane space (Yankovskaya et al. 2003; Sun et al. 2005). Mitochondrial CII shows close homology with a number of bacterial succinate ubiquinone reductases (SQRs), especially those of α-proteobacteria, from which mitochondria are supposed to derive (Andersson et al. 1998). The aminoacid sequences of the flavin and Fe–S binding domains of CII are highly conserved. The membrane domain is less conserved, although a four-helix bundle motif is ubiquitously present across species.

4.3.1 Human Diseases Associated with CII Deficiency (MIM 252011)

Mitochondrial disease with isolated impairment of CII is rare, encompassing 2–8% of OXPHOS defective cases (Munnich and Rustin 2001; Ghezzi et al. 2009). Two main clinical presentations are known: mitochondrial encephalomyopathy and familial paragangliomas (tumors of chromaffin cells).

In the first group, LS is the most common clinical and neuropathological presentation, but myopathy, encephalopathy, leukodystrophy and isolated cardiomyopathy have also been reported. Mutations in *SDHA*, *SDHD*, and *SDHAF1* genes have been identified in a minority of CII defective patients.

The pathogenesis of CII-associated paragangliomas remains to be explained. The most widely accepted hypothesis is based on induction of the hypoxia program that switches energy metabolism from mitochondrial respiration to glycolysis. This adaptive mechanism would occur through the stabilization of the hypoxia-inducible transcription factor 1 (HIF1) as a result of increased concentration of succinate, the SDH substrate (Bayley and Devilee 2010). The activation of the hypoxic program includes increased cellular uptake of glucose, activation of glycolysis, and promotion of angiogenesis.

The four genetic loci associated with hereditary paragangliomas correspond to *SDHD* (PGL1), *SDHC* (PGL3), *SDHB* (PGL4), and *SDHAF2* (PGL2) genes.

4.3.2 CII Assembly Model and Assembly Factors

The low frequency of human conditions associated with CII deficiency and the absence of a proton pumping activity and mtDNA-encoded subunits may explain the scarcity of information on CII assembly, in spite of the fact that CII is the smallest and simplest MRC complex.

In *E. coli*, an active, soluble SDH is composed of SDHA and SDHB ortholog subunits (SdhA, B), independent of the SDHC and SDHD orthologs (SdhC, D) (Nakamura et al. 1996). However, there are no data in eukaryotes suggesting that the CII hydrophilic module, composed of SDHA and SDHB, builds up independent of the membrane-embedded module, formed by SDHC and SDHD. SDHB seems to play a central role in the stabilization of the human holocomplex, since mutations in each and every SDH subunits cause the loss of SDHB (van Nederveen et al. 2009). CII also contains several prosthetic groups, including one heme *b* moiety, one FAD, and three Fe–S clusters. The function of the single heme *b* moiety is unknown. In *B. subtilis* SQR (Hagerhall 1997; Hederstedt 2002), the absence of heme *b* prevents the hydrophilic subunits to assemble with the membrane-bound subunits. The crystal structure of SQR-related prokaryotic enzymes (e.g., the quinol-fumarate reductase of *W. succinogenes*) shows physical interaction between four of the transmembrane helices and the heme *b* moiety (Lancaster et al. 1999). The histidyl ligands for heme *b* are conserved throughout the species and site-directed mutagenesis of *E. coli* residues involved in heme *b* binding induces the catalytic subunits to dissociate from the membrane-bound subunits. Taken together, these observations support a role for heme *b* in assembly and stability of the complex (Hagerhall and Hederstedt 1996; Nakamura et al. 1996). The current model for CII assembly starts with heme *b* binding to SdhD, followed by linking with SdhC (Lenaz and Genova 2010), to form the membrane-bound module. Heme *b* is also essential to link the hydrophilic module (SdhA + SdhB) to the membrane anchor subunits (SdhC + SdhD). While defects in factors involved in FAD supply (e.g., Flx1) (Tzagoloff et al. 1996) or Fe–S cluster synthesis (e.g., Tcm62) (Dibrov et al. 1998) can impair assembly and activity of CII, together with other Fe–S or FAD-dependent enzymes, only two specific CII assembly factors are presently known, SDHAF1 (Ghezzi et al. 2009) and SDHAF2 (Hao et al. 2009).

4.3.2.1 SDHAF1 (MIM612848)

SDHAF1, standing for SDH Assembly Factor 1, is a small protein containing a tripeptide sequence, LYR, a proposed signature for proteins involved in Fe–S metabolism. Hence, SDHAF1 could play a role in the insertion or retention of the Fe–S clusters within the protein backbone of CII, but this hypothesis needs further experimental evidence.

Although SDHAF1 resides in the mitochondrial matrix, while CII is membrane bound, mutations in this protein are associated with a drastic decrease of CII activity and content in both humans and yeast. A homozygous missense mutation in *SDHAF1* has been identified in related infants affected by leukoencephalopathy with the accumulation of lactate and succinate in the white matter, and severe reduction of CII activity and amount in muscle and fibroblasts (Ghezzi et al. 2009). Complementation assays in both human cells and a yeast model have confirmed the pathogenicity of the mutations. An additional homozygous stop mutation has later been identified in a CII-defective baby girl with leukoencephalopathy, born from consanguineous parents (R. Carrozzo; poster 251 Euromit8 Meeting 2011). To date, no mutation in SDHAF1 has been reported in patients with paraganglioma (Feichtinger et al. 2010).

4.3.2.2 SDHAF2 (MIM 613019)

The function of SDHAF2, standing for SDH assembly factor 2, is likely related to the incorporation of FAD into SDHA (Sdh1 in yeast), since Sdhaf2-less yeast strains, as well as SDHAF2 mutant human fibroblasts, have normal amounts of total Sdh1/SDHA protein but very low levels of the flavinated form (Hao et al. 2009). The binding of FAD to SDHA is probably a self-catalytic process, but requires that the imported SDHA subunit is properly refolded, forming the FAD-binding pouch. Sdhaf2/SDHAF2 could be a chaperone responsible for this step, in cooperation with hsp60 in *S. cerevisiae* (Robinson and Lemire 1996).

A germline missense mutation in SDHAF2, G78R, has been reported in two families with hereditary, multiple head and neck paragangliomas (PGL2). The G78 residue is part of the most conserved region of the protein, and the mutant R78 impairs its interaction with the SDHA subunit (Hao et al. 2009). The same mutation has later been found in a Spanish family, characterized by earlier onset of head and neck PGL. Haplotype analysis indicates that the G78R occurred independently in the two families (Bayley et al. 2010).

4.4 Complex III

CIII (ubiquinol–cytochrome *c* reductase, E.C. 1.10.2.2) catalyzes the electron transfer from reduced CoQH2, (ubiquinol) to cytochrome *c*. CIII is made up of 11 subunits (Iwata et al. 1998), only one (cytochrome *b*, cyt *b*) being encoded by mtDNA. Nuclear genes encode the remaining subunits: apo-cytochrome *c*1, cyt *c*1, the Rieske iron–sulfur protein (RISP or UQCRFS1), two relatively large "core" subunits, Core 1 (UQCRC1) and Core 2 (UQCRC2), and six additional, smaller proteins (UQCR6-11), the functions of which are unknown. In addition to the protein backbone, prosthetic groups of CIII include the two Fe-containing heme moieties of cyt *b* and *c1*, and the Fe–S cluster of RISP; these three subunits form the catalytic redox core of CIII. The CIII monomer is likely a transient form, which quickly converts into a stable, catalytically active homodimer.

4.4.1 Human Diseases Associated with CIII Deficiency (MIM 124000)

CIII defects are rare, compared to those of CI or CIV, the most frequent being caused by mutations in cyt *b*. Most of the cyt *b* mutations are sporadic and cause a mitochondrial myopathy with high plasma levels of creatine kinase and episodic myoglobinuria. CIII deficiency is also found in mutations of two assembly factors: *BCS1L* and *TTC19*.

4.4.2 CIII Assembly Model and Assembly Factors

Studies in *S. cerevisiae* have outlined a module-based assembly model for CIII, which is likely the same in mammals. The assembly starts with the formation of three different modules (1) cyt *b*+Qcr7+Qcr8, (2) cyt *c1*+Qcr6+Qcr9, and (3) Core1+Core2. The three modules assemble together, forming a precomplex to which RISP and Qcr10 are then added, leading to the formation of enzymatically active CIII (Zara et al. 2007). In humans, an additional subunit corresponds to the cleaved presequence of RISP (Brandt et al. 1993), which is inserted in the very last stage of CIII assembly. CIII dimerization occurs in the pre-complex stage, before the incorporation of the last subunits.

More recently, 2D-BNGE analysis of yeast CIII mutant strains has shown the existence of additional *bc1* sub-complexes, suggesting some corrections of the model (Zara et al. 2009) (Fig. 4.2). Accordingly, an initial module, containing cyt *b*+Qcr7+Qcr8, incorporates a second module composed of Core 1+Core 2+cyt *c*1, to form a 500 kDa dimerized sub-complex, to which Qcr6, Qcr9, RISP and Qcr10 are sequentially added.

Only two CIII-specific assembly factors are currently known in humans: BCS1L and TTC19.

4.4.2.1 BCS1L (MIM 603647)

BCS1L is the human ortholog of yeast Bcs1 (standing for b–c synthesis 1). A member of the AAA family, Bcs1 is essential for the incorporation of RISP, and possibly subunit 10, into CIII in the last steps of CIII assembly (Cruciat et al. 1999; Nobrega et al. 1992). Bcs1 seems to interact with the 500 kDa dimeric CIII pre-complex, playing a chaperone-like role to maintain the pre-complex in a conformation suitable for RISP incorporation. A similar function is predicted for BCS1L, since human cells from BCS1L mutant patients show accumulation of the CIII pre-complex, with hardly any trace of either incorporated RISP or fully assembled CIII (Zara et al. 2007). BCS1L is also found in a high molecular weight structure that does not contain any CIII-specific subunit (Fernandez-Vizarra et al. 2007; Cruciat et al. 1999). The latter is probably a membrane integral ring-shaped BCS1L homohexamer, a structure typical of AAA family members.

Several *BCS1L* gene mutations have been reported in CIII deficiency, associated with different clinical presentations: neonatal proximal tubulopathy, hepatopathy and encephalopathy (De Lonlay et al. 2001), and isolated progressive infantile encephalopathy. The acronym GRACILE stands for growth retardation, aminoaciduria, cholestasis, iron overload, lactic acidosis and early death, and designates an infantile condition caused by a specific BCS1L mutation, S78G, which is part of the Finnish disease heritage. A second less-severe disease associated with BCS1L missense mutations is Björnstad syndrome, characterized by neurosensory hearing loss and abnormally curly and brittle hair (*pili torti*). The clinical heterogeneity can

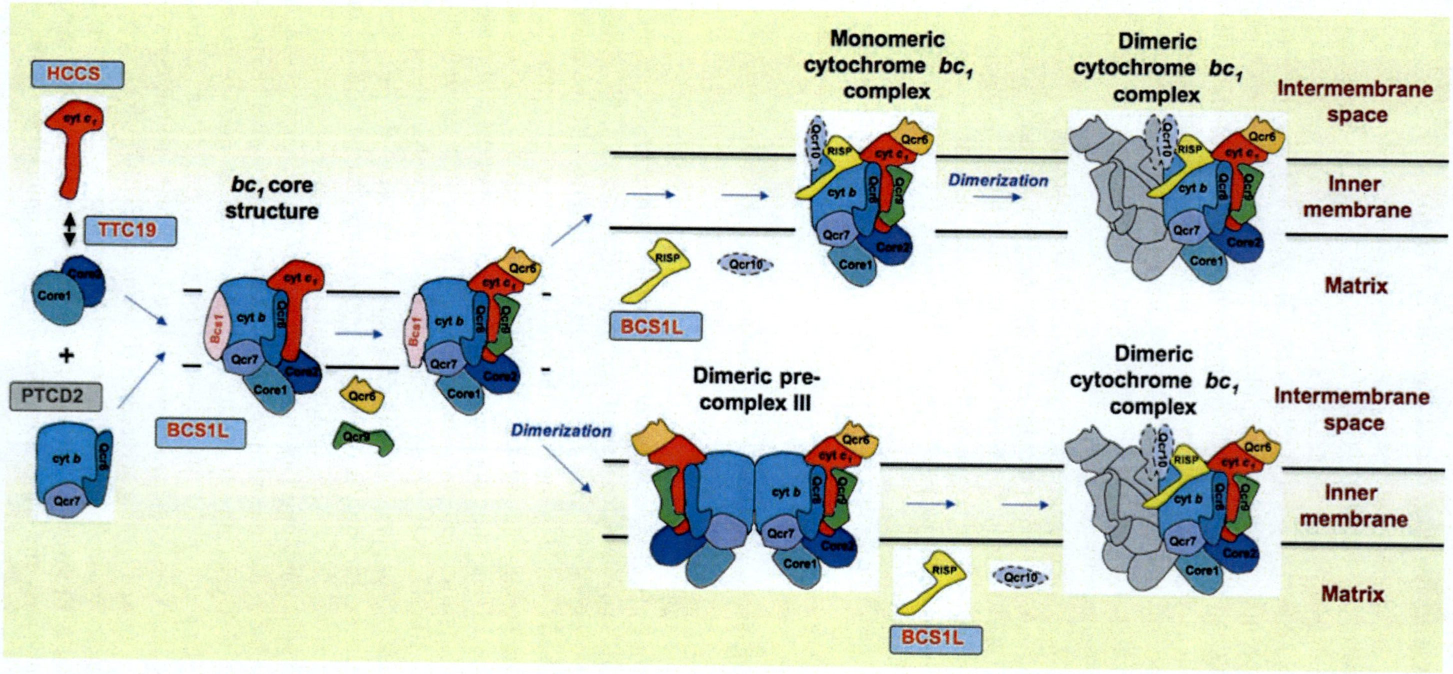

Fig. 4.2 Schematic outline of CIII assembly. Assembly factors are *boxed*; in *red* are those associated with human disease, whereas the others are in *grey*. Two possible pathways are depicted. The top pathway is characterized by monomeric CIII assembly (Zara et al. 2007); the bottom pathway is characterized by early dimerization of CIII pre-complex (Fernandez-Vizarra, 2009). See text for details

be linked to the functional domain affected by the different mutations. In the BCS1L protein there are three different domains: an N-terminal import domain, a BCS1L-specific domain and a C-terminal AAA, ATPase domain. For instance, while mutations have been reported in all domains, those in the ATPase domain are associated with a more severe phenotype (Hinson et al. 2007). A mouse recombinant model carrying the GRACILE mutation reproduces the main clinical features of the human disease, displaying CIII deficiency, with liver and kidney involvement. Interestingly, the clinical manifestations, and the CIII defect, including impaired RISP incorporation, start a few weeks after birth, suggesting the existence of another factor that plays the same role of Bcs1l during embryonic development and neonatal period (Levéen et al. 2011).

4.4.2.2 TTC19 (MIM 613814)

A second, putative CIII assembly factor, TTC19 (tetratricopeptide repeats 19) has recently been discovered, by investigating a CIII-defective, progressive mitochondrial encephalopathy (Ghezzi et al. 2011). All affected individuals carried *TTC19* nonsense mutations, resulting in the absence of the protein. TTC19 protein is localized in the inner mitochondrial membrane, where it takes part in two high molecular weight complexes, 500 kDa and ≥1 MDa in size. While the smaller complex corresponds to $CIII_2$, the larger one is also detected in rho^0 cells (that lack mtDNA and hence CIII) suggesting for TTC19 additional functions, unrelated to the MRC. The TTC19-mutant muscle accumulates CIII assembly intermediates that contain Core 1–2, but not RISP, suggesting that TTC19 is important in early CIII assembly. The tetratricopeptide motif is a protein–protein interaction module found in a number of functionally different proteins that facilitates specific interactions with partner proteins (Blatch and Lassle 1999).

TTC19 has no ortholog in yeast but a *TTC19* knockout model of *D. melanogaster* is characterized by neurological abnormalities associated with CIII deficiency. Similar to what reported in the *Bcs1l* knockin "GRACILE" mouse, normal development and CIII activity of mutant *D. melanogaster* larvae, and the late disease onset of mutant patients, suggest tissue-specific, age-related and developmentally dependent regulation of both TTC19 functions and CIII activity in different cell types and organisms.

4.4.2.3 HCCS (MIM 300056)

In humans, a single heme lyase, holocytochrome *c* synthase (HCCS), catalyzes the covalent binding of heme moieties to both apocytochromes *c* and *c*1 (Bernard et al. 2003). Catalytically active holocytochrome *c*1 is part of CIII, whereas holocytochrome *c* is a mobile cytochrome shuttling electrons from CIII to CIV, and taking part in other important cellular processes, for instance mitochondrial apoptosis (Jiang and Wang 2004).

Mutations in the *HCCS* gene are associated with MIDAS (microphthalmia, dermal aplasia and sclerocornea, MIM 309801), an X-linked dominant trait exclusively found in heterozygous mutant females, while hemizygous mutant males die at the embryonic stage.

4.4.2.4 PTCD2

A factor related to cytochrome *b* was found in a knockout mouse model obtained by random mutagenesis. A member of the pentatricopeptide protein family, PTCD2 (pentatricopeptide repeat domain protein 2) is a mitochondrial RNA-binding protein that participates in the maturation and stabilization of cytochrome *b* mRNA (Xu et al. 2008).

Several steps in human CIII assembly are still missing, including the exact sequence of incorporation for both protein subunits and prosthetic groups, the mechanism of formation of the CIII dimer, the details of BCS1L–RISP interaction and the exact role and interactions of TTC19 with CIII components. Developmental regulation of CIII assembly is suggested by BCS1L and TTC19 animal models. As for other MRC complexes, additional factors are likely involved in the various stages of CIII assembly.

4.5 Complex IV

Complex IV (cytochrome *c* oxidase, COX, E.C. 1.9.3.1), the terminal component of MRC, transfers electrons from reduced cytochrome *c* to molecular oxygen. This reaction is coupled to proton pumping across the inner mitochondrial membrane. Mammalian COX is composed of 13 subunits (Tsukihara et al. 1996; Yoshikawa et al. 1998), the three largest being encoded by mtDNA genes (MTCO1, MTCO2, MTCO3), whereas the remaining subunits are encoded by nuclear genes. The redox catalytic core is composed of subunits MTCO1 and MTCO2, which harbor two iron-containing heme *a* moieties (a and a_3) and two copper centers (CuA and CuB), responsible for the electron transfer. The third mtDNA-encoded subunit, MTCO3, is part of the structural core, being possibly involved in proton pumping. The function of the ten nuclear-encoded subunits (COX4, COX5A, COX5B, COX6A, COX6B, COX6C, COX7A, COX7B, COX7C and COX8) is currently unknown, but they may play regulatory (Arnold and Kadenbach 1997) and stabilization roles. In humans, COX displays multiple tissue-specific isoforms: for instance, COX6A and COX7A have a heart (H) isoform, present in skeletal and cardiac muscle, and a liver (L) isoform, present in extra-muscle tissues (Grossman and Lomax 1997); a testis-specific COX6B2 isoform and a lung-specific COX4 have also been identified. These alternative isoforms confer different kinetic properties to the enzyme, endowing it with remarkable functional plasticity.

4.5.1 Human Diseases Associated with CIV Deficiency (MIM 220110)

In humans, defects of COX activity are nearly as common as CI defects. In infancy, the most frequent manifestation of isolated, profound COX deficiency is LS, but other phenotypes include severe cardiomyopathy and encephalocardiomyopathy. Maternally inherited COX defects are associated with several mutations of mtDNA tRNA genes. Conversely, only a few mutations in the three COX-encoding mtDNA genes and a single mutation in a nuclear-encoded COX subunit, COX6B (Massa et al. 2008), have been reported to date, suggesting that most of the mutations in structural components of CIV are incompatible with extra-uterine life. Accordingly, the most common nuclear-DNA gene defects of COX are due to mutations in the assembly factors.

4.5.2 CIV Assembly Model and Assembly Factors

Many COX assembly factors are conserved between yeast and mammals, although half of them are missing in the latter organisms; in turn, mammalian COX includes subunits that are absent in yeast (i.e., COX7b and COX8). This explains not only the relatively greater wealth of information gained on CIV assembly, compared to other MRC complexes, but also the many details that are still missing.

In summary (Fig. 4.3), COX subunits are sequentially incorporated to form the nascent complex, starting from the insertion of newly synthesized MTCO1 into the inner mitochondrial membrane. This first, crucial step (S1) is followed by the incorporation of subunits COX4 and COX5A, to form a second assembly intermediate, S2. Insertion of heme *a* is likely to occur just after the formation of S1 or during the formation of S2, and proceeds together with the insertion of CuB and heme a_3 into MTCO1. The formation of the MTCO2-associated CuA center is followed by the incorporation of MTCO2 into the S2 intermediate. In yeast, this step requires Cox18 (MIM 610428), a factor also present in humans, catalyzing the insertion of the MTCO2 C-terminal tail into the mitochondrial inner membrane. Next, rapid sequential incorporation of MTCO3 and of smaller nuclear encoded subunits, including COX5b and COX8, to form S3, leads to the formation of a quasi-complete assembly intermediate (also named subcomplex b) (Williams et al. 2004; Stiburek et al. 2005). The addition of a few remaining subunits, including COX6A, COX6B, COX7A and COX7B (Nijtmans et al. 1998; Massa et al. 2008), all converging on the surface of the complex core, results in the formation of a holocomplex monomer (S4). Finally, monomeric COX dimerizes in an active structure that contains the cytochrome *c* binding site (Tsukihara et al. 1996), where two molecules of cytochrome *c* cooperatively bind at the interface between the two COX monomers (Birchmeier et al. 1976; Darley-Usmar et al. 1984) formed by contacts among MTCO1, COX6A, COX6B, and COX5B (Lee et al. 2001).

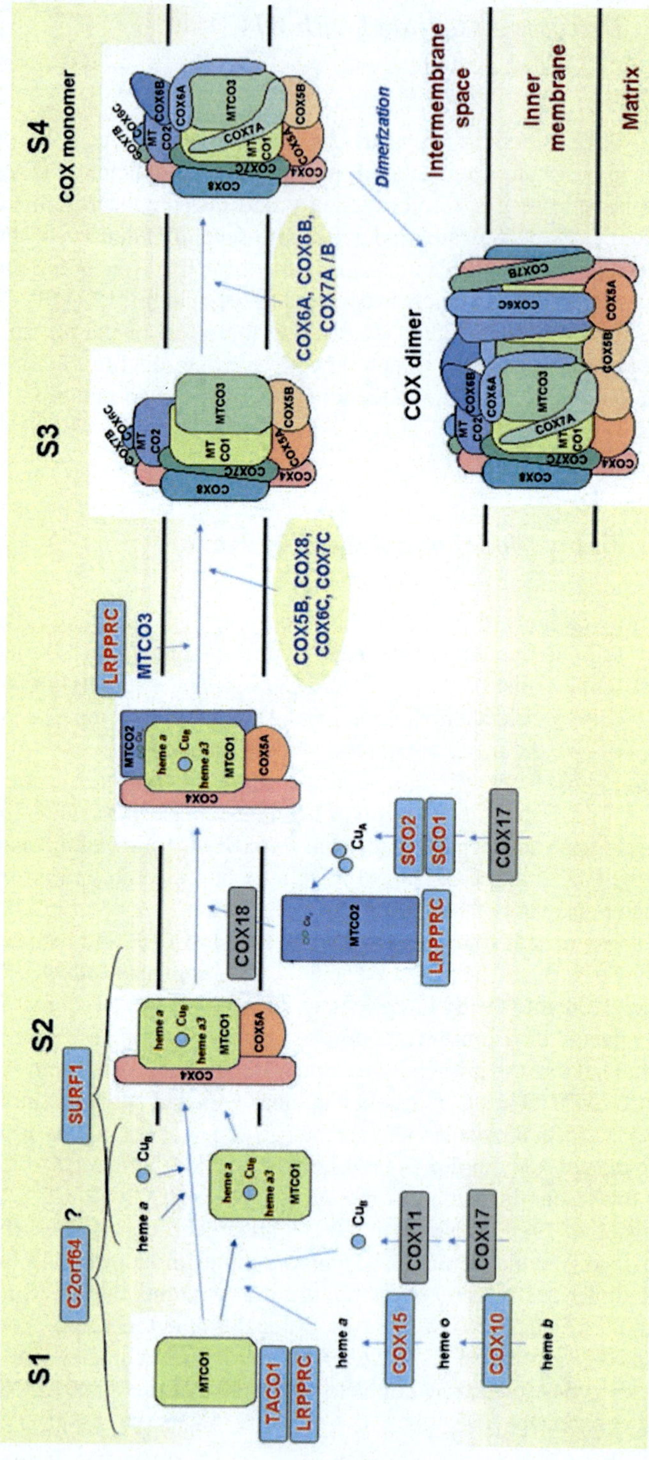

Fig. 4.3 Schematic outline of CIV assembly. Assembly factors are *boxed*; in *red* are those associated with human disease, whereas the others are in *grey*. See text for details

More than 20 nuclear encoded accessory proteins are involved in COX assembly of *S. cerevisiae*, most of which have known human orthologs (Fontanesi et al. 2006). Ancillary gene products are required for incorporation of hemes *a, a3* and copper atoms (CuA, B) into catalytic COX subunits and for maintaining the nascent intermediates in an assembly-competent state (Barrientos et al. 2009; Mick et al. 2007; Stiburek et al. 2006).

Similar to CI, CIII and CV, CIV contains mtDNA encoded subunits, i.e., MTCO1, MTCO2, MTCO3 that are synthesized on mitochondrial ribosomes. In *S. cerevisiae* the insertion into the mitochondrial inner membrane of these in situ translated subunits requires a specific integrase, i.e., OXA1 (standing for oxidase-cytochrome *c* assembly 1), which is located as a homo-oligomeric complex in the inner mitochondrial membrane (Bonnefoy et al. 1994). The function of OXA1 and their OXA1-like orthologs is probably not specific for CIV; for instance, OXA1 depletion in *Neurospora crassa* results in combined reduction of CI and CIV (Nargang et al. 2002). Likewise, knockdown of OXA1L in human cells impairs the biogenesis of CV and CI, whereas CIV seems unaffected (Stiburek et al. 2007). No mutation in human *OXA1L* has been reported in patients with cIV or combined MRC deficiency (Coenen et al. 2005).

We will hereafter focus on those CIV assembly factors that are involved in human disease.

4.5.2.1 SURF1 (MIM 185620)

Although the precise function of SURF1 remains to be elucidated, studies on the yeast ortholog Shy1, and in human mutant cells and tissues, indicate a role for SURF1 in the formation of the early subcomplexes of COX. The mature SURF1 protein is a 30 kDa hydrophobic polypeptide with two transmembrane domains at the N and C termini, which anchor the protein to the mitochondrial inner membrane (Tiranti et al. 1998).

Mutations in *SURF1* are the most common cause of LS associated with COX deficiency. This association is highly specific, and is partly explained by the observation that almost all the SURF1 mutations reported to date cause the complete absence of the protein. Very few missense mutations have been detected (Pecina et al. 2004), sometimes in association with less severe phenotypes (Piekutowska-Abramczuk et al. 2009). *SURF1* mutations determine marked reduction in the amount of fully assembled COX and accumulation of early assembly intermediates S1 and S2 (Tiranti et al. 1999; Stiburek et al. 2005). Although the mechanistic function is still unclear, these results indicate a role for the SURF1 protein in the early stages of COX assembly, most likely before the incorporation of MTCO2 into the S2 assembly intermediate (Nijtmans et al. 1998). In SURF1 null human samples (Tiranti et al. 1998), as well as in yeast strains lacking Shy1 (Mashkevich et al. 1997), fully assembled, functionally active CIV is found in residual amounts, suggesting partial functional redundancy.

Recent studies in bacteria have indicated that two prokaryotic Surf1 putative orthologs (Surf1c and Surf1q) are heme *a* binding proteins, suggesting a role for Surf1 in heme *a* insertion into MTCO1 (Bundschuh et al. 2009).

4.5.2.2 SCO1 (MIM603644) and SCO2 (MIM 604272)

In yeast, two functionally and evolutionarily related COX assembly genes, *SCO1* and *SCO2* (for synthesis of cytochrome oxidase), enable MTCO1 and MTCO2 to be incorporated into nascent CIV by promoting the insertion of Cu^{++} atoms in the catalytic sites CuB and CuA of either subunit. This process requires additional factors, partly conserved in yeast and mammals, including COX17 (MIM 604813) and COX11 (MIM 603648), and is carried out at the very early steps of COX assembly, being necessary to promote the incorporation and to stabilize the assembly of the first protein subunits of CIV, including MTCO1 and MTCO2. In SCO1- and SCO2-deficient cells, S1 and S2 subassemblies accumulate, whereas fully assembled COX is markedly reduced (Williams et al. 2004; Leary et al. 2004; Stiburek et al. 2005); the absence of MTCO2 in the subcomplexes indicates that the formation of the MTCO2-bound CuA center is necessary for the addition of MTCO2 to the S2 intermediate.

Mutations in the human ortholog of *SCO2* were initially found in infants with fatal cardioencephalomyopathy and COX deficiency (Papadopoulou et al. 1999). More than 10 mutations have been reported; however, one particular mutant allele, E140K, is present in almost all affected individuals, either in combination with a second mutant allele, or in homozygosity. The latter genotype is associated with a relatively milder phenotype and delayed disease onset (Joost et al. 2010). Heart hypertrophy in *SCO2*-mutant patients is usually severe, whereas brain involvement may vary, from LS-like to spinal-muscular atrophy-like presentations (Pronicki et al. 2010).

Mutations in *SCO1* have been found in a single large family with multiple cases of neonatal hepatopathy, severe ketoacidosis and isolated, severe COX deficiency in liver and muscle (Valnot et al. 2000). Another *SCO1* case presented with hypertrophic cardiomyopathy, encephalopathy and hepatomegaly with fatal outcome (Stiburek et al. 2009).

4.5.2.3 COX10 (MIM 602125) and COX15 (MIM 603646)

COX10 and COX15 are enzymes involved in the terminal steps of the biosynthesis of hemes *a* and *a3*. *COX10* encodes the heme *a*:farnesyltransferase, which catalyzes the conversion of protoheme (heme *b*) to heme *o*. COX15 converts heme *o* into heme *a* by hydroxylation, with the help of a monooxygenase still uncharacterized in humans. While complete loss of COX10 or COX15 activities is incompatible with extra-uterine life, rare missense mutations have been found in both, resulting in partial COX deficiency. Since heme *a* is incorporated during early COX assembly, mutations in either COX10 or COX15 lead to the arrest and degradation of the complex, with virtually no accumulation of assembly intermediates (Antonicka et al. 2003).

Mutations in *COX10* are associated with a spectrum of conditions including LS; encephalopathy with proximal renal tubulopathy; sensorineural deafness, metabolic acidosis, hypotonia and hypertrophic cardiomyopathy. Likewise, mutations of *COX15* can cause fatal infantile hypertrophic cardiomyopathy, as well as rapidly progressive or protracted LS.

4.5.2.4 COX11 (MIM 602125), COX17 (MIM 603646) and COX19 (MIM 610429)

Studies carried out in yeast have led to the identification of additional genes necessary for the maturation and insertion of the copper prosthetic groups in nascent COX.

The Cox11 protein is a constituent of the inner mitochondrial membrane and, like Cox10 and Cox15, is likely involved in the biosynthesis of heme *a*. Mutations in the *S. cerevisiae* cox11 gene result in a 'petite' phenotype, associated with isolated COX defect.

The Cox17 protein contains a cysteine-rich motif and two additional cysteine residues that are also conserved in copper-binding metallothioneins. Yeast cox17 is a cytosolic protein involved in copper recruitment to mitochondria. Cox17 null mutations cause CIV assembly failure and Cox17 KO mice are embryonic lethal, exhibiting severe COX deficiency. Yeast Cox17 is likely to act downstream from Cox23, a factor that has not been identified in mammals.

Yeast, and human, Cox19 is a cytosolic protein, with a minor fraction being located in the intermembrane space of mitochondria. Its exact function is still unknown, but its structure is similar to that of Cox17, suggesting a role in copper translocation to mitochondria.

Both *COX17* and *COX19* are candidates for CIV defects, but no mutation in either gene has been reported in humans.

4.5.2.5 LRPPRC (MIM 607544) and TACO1 (MIM 612958)

LRPPRC (or LPR130) is a leucine-rich protein of the pentatricopeptide repeat family; its role is still unclear, but it seems to regulate the expression of mtDNA genes (Sasarman et al. 2010), including RNA stability, maturation and processing (Schmitz-Linneweber and Small 2008). While missense LRPPRC mutations are associated with isolated COX deficiency, further decrease of LRPPRC levels determines profound reduction of all mitochondrial transcripts associated with generalized defect of all mtDNA-dependent OXPHOS complexes.

A single LRPPRC mutation, A354V, causes the French-Canadian type of Leigh syndrome (LSFC, MIM 220111), which is restricted to the Charlevoix

Seguenay-Lac Saint-Jean region in Quebec, Canada. The clinical presentation is characterized by mild psychomotor regression and lactic acidosis, with severe COX deficiency in brain and liver. A single patient has been reported to carry an exon deletion, in compound heterozygosity with the A354V (Mootha et al. 2003).

TACO1 (standing for translation activator of CO1; initially named *CCD44*) is a mitochondrial translational activator required for efficient translation of *MTCOI*. A homozygous frameshift mutation in this gene, c.472insC, has been found in patients from a single family, presenting with slowly progressive LS, caused by impaired MTCO1 protein synthesis (Weraarpachai et al. 2009). No other TACO1 mutations have been found to date, suggesting that integrity of this factor is needed for extra-uterine life (Seeger et al. 2010).

4.5.2.6 C2ORF64

PET191, the yeast ortholog of huma C2ORF64, is a COX assembly factor, albeit its mechanistic role is still unknown. A homozygous mutation in *C2ORF64* was recently described in two siblings affected by fatal neonatal cardiomyopathy. 2D-BNGE analysis revealed the accumulation of a small assembly intermediate containing subunit MTCO1 but not MTCO2, COX4, or COX5b, indicating that C2ORF64 is involved in a very early step of COX assembly (Huigsloot et al. 2011).

4.6 Complex V

Complex V (ATP synthase, E.C. 3.6.3.14) dissipates the proton electrochemical gradient generated by the respiratory chain to produce ATP. It comprises an integral membrane cylindrical rotor-like structure, the F0 particle, and a peripheral matrix-facing F1 particle, the catalytic ATP synthase domain (Boyer 1997). F1 has a bulb-like shape, formed by six "cloves". F0 and F1 are physically connected to each other by two additional structures: a centrally located, asymmetrical stalk and an externally tethered stator. F0 is a rotor harboring a proton channel. Following the electrochemical gradient, protons flow through the channel guided by the stator, impressing a rotary motion to F0, which is transmitted to the catalytic head (F1) by the centrally projecting stalk. The sequential tilt of the cloves, impressed by the asymmetrical stalk during each F1 rotation cycle, drives the condensation of ADP and Pi to form three ATP molecules for each cycle (Devenish et al. 2008). All five subunits of F1 (α, β, γ, δ, ε), and most of the F0 subunits (b c, d, e, f, g, OSCP and F6) are nuclear encoded (Collinson et al. 1996). Only two proteins (MTATP6 and 8) are encoded by mtDNA (Boyer 1993). Both MTATP6 and ATP8 are part of the F0, and connect the latter to the stator (Fig. 4.4). Dimeric and higher oligomeric forms of ATP synthase (Arnold et al. 1998; Paumard et al. 2002) seem critical to maintain the shape of mitochondria by promoting the formation of the inner membrane cristae.

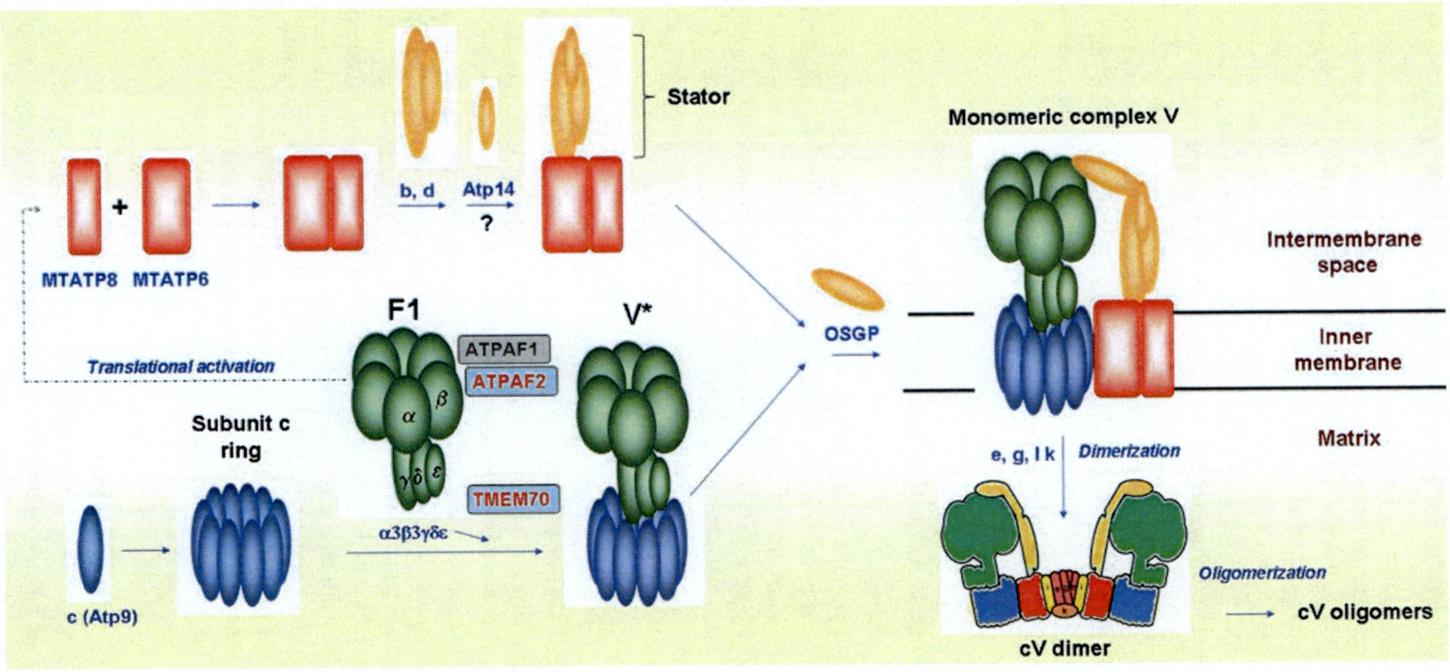

Fig. 4.4 Schematic outline of CV assembly. Assembly factors are *boxed*; in *red* are those associated with human disease, whereas the others are in *grey*. The existence of human orthologs of yeast CV assembly factors follows the model by Rak et al. (2011). See text for details

4.6.1 Human Diseases Associated with CV Deficiency

Maternally transmitted ATP synthase dysfunction can be caused by mutations in *MTATP6* or *MTATP8*, the two mtDNA genes encoding ATP6 and 8 polypeptides. Heteroplasmic missense mutations in MTATP6 (Tatuch and Robinson 1993, Schon et al. 2001) are associated with adult-onset NARP (neuropathy, ataxia and retinitis pigmentosa) or maternally inherited Leigh syndrome (MILS), the clinical severity being proportional to the heteroplasmic mutation load. A single patient with hypertrophic cardiomyopathy carried a nonsense mutation in MTATP8 (Jonckheere et al. 2008). Three disease-causing nuclear genes have been identified so far, two encoding assembly factors (ATPAF2, TMEM70), whereas the third (ATP5E) encodes the epsilon subunit of the F1 domain (Mayr et al. 2010). In other CV deficient cases the genetic cause is still unknown (Sperl et al. 2006).

4.6.2 Complex V Assembly Model and Factors

Experiments based on pulse–chase protein labeling and 2D-BNGE in bacterial, yeast and mammalian cells have delineated the current model for CV assembly. The process starts with the formation of the F1 catalytic core, carried out by specific assembly factors ATPAF1 and ATPAF2 in mammalian mitochondria (corresponding to Atp11 and Atp12 in yeast). Next, the initial F1 intermediate interacts with subunit c (Atp9 in yeast) and the other F0 subunits, forming an assembly intermediate named V*. The two mtDNA-encoded subunits, MTATP6 and MTATP8, are added during the last assembly stage, at least in mammalian cells (Nijtmans et al. 1995), since the V* intermediate and lower order subassemblies build up in MTATP6 or MTATP8 mutant mitochondria (Houstek et al. 2006). In yeast, two distinct assembly intermediates have been characterized, one formed by Atp6, Atp8, at least two stator subunits, and the Atp10 chaperone, the second by the F1 ATPase particle and the Atp9/subunit c ring. This recent result indicates that the assembly process is not a linear addition of single subunits one next to the other, but consists of at least two separate, coordinately regulated pathways eventually converging together at the end stage (Rak et al. 2011). This is in agreement with the notion that the F1 and F0 components seem to derive from functionally unrelated ancestral proteins (Mulkidjanian et al. 2007) that follow independent assembly pathways (Schatz 1968; Tzagoloff 1969). The very final steps in mammalian CV biogenesis include the formation of dimers, coincidental with the addition of subunits e and g (Schagger and Pfeiffer 2000), and the formation of higher order oligomers (V1–V4) (Krause et al. 2005).

4.6.2.1 ATPAF2 (MIM 608918)

Among the several factors known for CV assembly in yeast, only orthologs of ATP11/ATPAF1 and ATP12/ATPAF2 are known in mammals (Wang et al. 2001;

De Meirleir et al. 2004 These proteins are chaperones interacting with subunits β and α of F1, essential for the assembly of the α + β heterooligomer (Ackerman and Tzagoloff 1990, Wang et al. 2000).

To date, only one case of complex V deficiency has been referred to a homozygous missense ATPAF2 mutation associated with degenerative encephalopathy, connatal lactic acidosis and methyl-glutaconic aciduria (De Meirleir et al. 2004). The amount of fully assembled CV was low, but no subassembly intermediates were detected, suggesting that ATPAF2 acts very early during CV assembly (Houstek et al. 1999).

4.6.2.2 TMEM70 (MIM 612418)

Mutations in TMEM70 were found in patients, mostly of Gipsy origin, with neonatal encephalocardiomyopathy and isolated CV deficiency (Cizkova et al. 2008). This is the most frequent cause of ATP synthase deficiency (Honzik et al. 2010; Spiegel et al. 2011). The prevalent homozygous mutation, an A-to-G transition in intron 2 of the *TMEM70* gene, resulting in aberrant splicing and loss of the mRNA transcript, is associated with highly variable clinical severity.

2D-BNGE analysis of samples from TMEM70-mutant patients shows the presence of traces of free F1 ATPase, and small amounts of CV holocomplex, but no F0-F1 subassemblies (Houstek et al. 2009). Following transfection of wild-type *TMEM70* into mutant fibroblasts, CV holoenzyme amount and activity reverse to normal, whereas unassembled F1 disappears, indicating that TMEM70 may be involved in the assembly of F1 itself or in the F1 interaction with some of the F0 subunits.

Ultrastructural studies in TMEM70 mutant mitochondria show loss of invaginations of cristae and formation of concentric membrane rings. These morphological alterations could affect the integrity of mitochondrial nucleoids and hence mtDNA replication and expression (Cameron et al. 2011a), which could explain the variable reduction of other OXPHOS activities in some TMEM70 mutant cells.

Similar to other OXPHOS complexes, and by analogy with yeast, additional assembly factors are likely to concur to the biogenesis of mammalian CV, e.g., FMC1 for F1, and ATP10, ATP23 for F0 assembly. Other factors, such as NCA1-3, NAM1, AEP1-3 ATP22, and ATP25, are involved in mRNA stability, translation, and processing of yeast CV genes (Houstek et al. 2009). Last but not least, assembly of CV dimers and oligomers, which are critical for the formation of tubular cristae (Wagner et al. 2010), is likely to require *ad hoc* chaperones and stabilizing proteins.

4.7 Fe–S Cluster Biosynthesis

Fe–S clusters are essential prosthetic groups for CI, CII and CIII, and for several other mitochondrial and non-mitochondrial enzymes. The biosynthesis of Fe–S clusters and their incorporation into proteins is a complex process that requires numerous factors (Lill and Kispal 2000), including scaffold proteins (i.e., ISCU, mNFU1), as well as cysteine desulfurases (ISCS), iron donors and chaperones (i.e., frataxin) (Li et al. 2009). More than 20 proteins involved in Fe–S cluster biogenesis have been identified in yeast (Lill and Mühlenhoff 2008). In humans, Fe–S clusters are synthesized and assembled into proteins in a series of complex biochemical reactions organized in two parallel pathways (Rouault and Tong 2008). In the first, ISCU is the main scaffold protein, whereas the second has NFU1. The pathways take place in the mitochondrial matrix; a fraction of the Fe–S pool is then exported from mitochondria to supply Fe–S dependent enzymes in the cytosol and nucleus.

Mutations of Fe–S related genes are linked to human disease characterized by multiple MRC deficiency.

4.7.1 ISCU (MIM 611911)

ISCU mutations (MIM 255125) cause hereditary myopathy with lactic acidosis, sometimes complicated by cardiomyopathy or episodes of myoglobinuria. The activities of aconitase and CII, both containing Fe–S clusters, are severely impaired in ISCU mutant patients (Mochel et al. 2008; Kollberg et al. 2009), with less severe reduction of CI and CIII activities as well. Two ISCU isoforms, cytosolic and mitochondrial, are produced by alternative splicing from the same gene transcript (Tong and Rouault 2006; Li et al. 2006), both being active in homeostatic regulation of iron uptake, intracellular iron distribution and mitochondrial iron utilization to form Fe–S clusters.

4.7.2 Frataxin (MIM 606829)

Frataxin interacts with ISCU in the same homeostatic pathway, acting as a mitochondrial iron chaperone that stores and supplies iron in a bioavailable form for the mitochondrial biosynthesis of Fe–S clusters and heme moieties. While the complete absence of frataxin leads to early embryonic lethality, mutations that reduce the amount of frataxin, including the most common one, a $(GAA)_n$ expansion within the first intron of *FXN,* the frataxin-encoding gene, cause Friedreich ataxia (FRDA, MIM 229300), an autosomal recessive, progressive disorder that combines spinocerebellar ataxia and cardiomyopathy (Campuzano et al. 1996). Mutant cells with low frataxin display defective activity of the Fe–S cluster-containing subunits of MRC CI, II and III. A physical interaction of frataxin with SDHA and SDHB subunits of CII has been shown in both human and yeast cell lines (Gonzalez-Cabo et al. 2005).

4.7.3 *NFU1 (MIM 608100)*

A homozygous missense mutation in the scaffold protein NFU1 has been found in affected members from one family, presenting with weakness, lethargy, and severe lactic acidosis. Biochemically, combined deficiency and low amount of the Fe–S MRC complexes were associated with defects of 2-oxoacid dehydrogenases, including pyruvate and 2-oxoglutarate dehydrogenase complexes (PDHc, OGDHc) (Cameron et al. 2011b). Although PDHc and OGDHc do not harbor Fe–S clusters, both covalently bind a lipoate moiety that is the product of a Fe–S cluster enzyme, lipoate synthase. Notably, and different from ISCU patients, the amount and activity of mitochondrial aconitase were normal. The mutation seems to affect the splicing process leading to no immunodetectable mitochondrial NFU1 protein.

4.7.4 *BOLA3 (MIM 613183)*

BolA family members are putative reductases interacting with glutaredoxins (Huynen et al. 2005). The proposed role for BOLA3 is to interact with glutaredoxin 5, which is involved, along with several chaperones, in the insertion of [2Fe–2S] and [4Fe–4S] clusters into apoproteins.

A homozygous single-base insertion in *BOLA3*, predicting the formation of a premature stop codon, was identified in a single subject presenting with cardiomyopathy and epileptic encephalopathy. As for NFU1 mutation, also in *BOLA3* mutant cells the activity and amount of Fe–S containing MRC complexes were reduced, particularly CI (Cameron et al. 2011b).

4.7.5 *GRX5 (MIM 609588)*

Several cochaperones bind to ISCU for the incorporation of [2Fe–2S] clusters into recipient apoproteins, whereas GLRX5 is a specific chaperone for the incorporation of [4Fe–4S] species.

A functional defect in the human glutaredoxin GRX5 was found in a patient with microcytic anemia and iron overload (Camaschella et al. 2007). Mutant cells showed low levels of aconitase, whereas MRC deficiency was not reported. However, the yeast Δgrx5 strain displayed respiratory deficiency, due to the absence of Rip1, the iron–sulfur Rieske protein, that is part of CIII (Bellí et al. 2004) or to the reduction of SDH activity (Kim et al. 2010). The mutant yeast strain also showed a significant decrease in the amount of mtDNA, suggesting a role for grx5 in supporting mitochondrial genome stability (Kim et al. 2010).

4.7.6 ABCB7 (MIM 300135)

The ATP-binding cassette member 7 (ABCB7) is an iron mitochondrial exporter, which controls the supply of Fe–S clusters to cytosolic Fe–S dependent proteins. Mutations in this gene have been found in families with X-linked sideroblastic anemia with ataxia syndrome (XLSA/A; MIM301310). The human ABC7 protein is localized in the inner mitochondrial membrane and belongs to the ATP-binding cassette transporter superfamily. Its yeast ortholog, atm1p, plays a central role in the maturation of cytosolic Fe–S cluster-containing proteins. As expected, deletion of Abcb7 in mouse impairs the activity of cytosolic but not mitochondrial Fe–S enzymes (Pondarré et al. 2006). The presence of cerebellar ataxia in XLSA/A indicates that, as evidenced also in FRDA, spino-cerebellar pathways and cells are exquisitely dependent on Fe–S bioavailability.

4.8 Respiratory Chain Supercomplexes

Integration of the redox reactions occurring in each complex increases the efficiency of the electron flow to molecular oxygen, reducing the electron leak from the MRC. The physical counterpart of this phenomenon is supercomplexes, i.e., supra-molecular structures composed of individual MRC complexes assembled in variable stoichiometric ratios.

The existence of supercomplexes dates back to a "*solid-state*" MRC model consisting of orderly sequences of complex I–complex IV (Chance and Williams 1955), based on the isolation of two or more complexes in consistently defined stoichiometric ratios (Fowler and Richardson 1963; Blair 1967). The "*solid-state*" model was later replaced by a "*fluid*" model, according to which MRC complexes are independently embedded in the lipid bilayer of the inner mitochondrial membrane and electron transfer depends on random collisions between complexes and the mobile carriers, CoQ and cyt *c*. The "*fluid*" model was based on the demonstration that respiratory complexes can be purified individually, retaining their enzymatic activity (Hackenbrock et al. 1986), and by lack of ultrastructural and in vitro evidence of supercomplexes (Capaldi 1982). However, BNGE analysis has later demonstrated the existence of larger sized protein structures (Krause et al. 2004; Schagger 2002; Schagger and Pfeiffer 2000), suggesting that individual MRC complexes exist in dynamic equilibrium with different types of supercomplexes, also referred to as respirasomes (Schagger and Pfeiffer 2000; Acín-Perez et al. 2008; Wittig and Schagger 2009). Kinetic measurements using metabolic flow control analysis have provided further evidence on the existence of functionally relevant, specific supercomplexes (e.g., CI+CIII) (Bianchi et al. 2004). This "*plasticity*" model integrates the two previous ones (Acin-Perez et al. 2008) and includes the following features: (1) supercomplexes are formed by single MRC complexes; (2) are functionally active entities, some of which contain CoQ and cyt *c* ; (3) they can respire, i.e., they can transfer electrons from NADH to O_2; (4) are dynamic structures of variable stoichiometry that (5) optimize electron transfer efficiency to the different energy needs of cells (Wittig et al. 2006; Acín-Pérez et al. 2008; Wittig and Schägger 2009). Finally, a fraction of supercomplexes is associated with CV dimers to bend and fold the inner membrane and form the mitochondrial cristae (Gilkerson et al. 2003; Rabl et al. 2009; Zick et al. 2009).

Although results may vary, depending on concentration and type of detergent, in mitochondria from different species virtually all CI seem to be organized into a supercomplex composed of a CI monomer, a CIII dimer, and up to four COX complexes ($I_1III_2IV_{0-4}$) (Schagger and Pfeiffer 2000). In digitonin-solubilized bovine heart mitochondria, a $CI_1III_2IV_1$ supercomplex has been detected by BNGE, and its three-dimensional structure resolved by electron microscopy (Schäfer et al. 2007). The association into supercomplexes is essential for the stability of CI. In fact, when either complex CIII or CIV fail to get assembled (Acin-Perez et al. 2004; Diaz et al. 2006), leading to impaired formation of the supercomplex, mitochondria display CI deficiency (Lamantea et al. 2002; Saada et al. 2012). Conversely, absence of CI affects the formation of supercomplexes but does not usually impair the activity of the other complexes (Acin-Perez et al. 2008), although rare mutations in CI subunits, for instance NDUFS4 (Ugalde et al. 2004), can be associated with combined CI and CIII deficiency.

The molecular mechanisms and regulation of supercomplex assembly are still unknown. The process is likely to require specific factors, none of which has been identified so far. However, the CI chaperone NDUFAF2 is present in both a 830-kDa CI assembly intermediate and a structure composed of this intermediate bound to a CIII dimer, before completion of CI assembly (Lazarou et al. 2007). This observation supports the hypothesis that supercomplex assembly occurs in conjunction with the formation of individual complexes, in agreement with the experimental evidence showing that critical levels of CIII and CIV are required for supercomplexes to form and to warrant stability and integrity of CI. Together, these results suggest that assembly factors specific to individual complexes may be involved in the formation of supercomplexes. In addition, other components could play a role in supercomplex assembly, for instance cardiolipin. Defects in cardiolipin synthesis and remodeling due to mutations in the X-linked gene tafazzin, encoding an acyltransferase, cause Barth syndrome, a disease characterized by mitochondrial myopathy, cardiomyopathy, growth retardation, and leukopenia (Barth et al. 1999).

Recently, supercomplexes have been proposed as the building blocks for the formation of much larger supramolecular structures, such as respiratory "strings" (Wittig et al. 2006); "core fragments" of respiratory strings have been isolated as respiratory supercomplex multimers, of apparent mass of 35–45 MDa. Likewise, "ATP synthasomes" are supercomplexes formed by adenine nucleotide and inorganic phosphate carriers bound to CV (Ko et al. 2003).

Acknowledgments This work was supported by Fondazione Telethon-Italy (grants GGP11011 and GPP10005 to MZ), Fondazione CARIPLO (grant 2011/0526), the Italian Association of Mitochondrial Disease Patients and Families (Mitocon) and Fondazione Pierfranco e Luisa Mariani (Ricerca 2000 grant to MZ).

References

Acín-Pérez R, Bayona-Bafaluy MP, Fernández-Silva P et al (2004) Respiratory complex III is required to maintain complex I in mammalian mitochondria. Mol Cell 13:805–815

Acín-Pérez R, Fernández-Silva P, Peleato ML, Pérez-Martos A, Enriquez JA (2008) Respiratory active mitochondrial supercomplexes. Mol Cell 32:529–539

Ackerman SH, Tzagoloff A (1990) Identification of two nuclear genes (ATP11, ATP12) required for assembly of the yeast F1-ATPase. Proc Natl Acad Sci USA 87:4986–4990

Andersson SG, Zomorodipour A, Andersson JO et al (1998) The genome sequence of *Rickettsia prowazekii* and the origin of mitochondria. Nature 396:133–140

Antonicka H, Leary SC, Guercin GH et al (2003) Mutations in COX10 result in a defect in mitochondrial heme A biosynthesis and account for multiple, early-onset clinical phenotypes associated with isolated COX deficiency. Hum Mol Genet 12:2693–2702

Arnold S, Kadenbach B (1997) Cell respiration is controlled by ATP, an allosteric inhibitor of cytochrome-c oxidase. Eur J Biochem 249:350–354

Arnold I, Pfeiffer K, Neupert W, Stuart RA, Schägger H (1998) Yeast mitochondrial F1F0-ATP synthase exists as a dimer: identification of three dimer-specific subunits. EMBO J 17:7170–7178

Barghuti F, Elian K, Gomori JM, Shaag A, Edvardson S, Saada A, Elpeleg O (2008) The unique neuroradiology of complex I deficiency due to NDUFA12L defect. Mol Genet Metab 94:78–82

Barrientos A, Gouget K, Horn D, Soto IC, Fontanesi F (2009) Suppression mechanisms of COX assembly defects in yeast and human: insights into the COX assembly process. Biochim Biophys Acta 1793:97–107

Barth PG, Wanders RJ, Vreken P, Janssen EA, Lam J, Baas F (1999) X-linked cardioskeletal myopathy and neutropenia (Barth syndrome) (MIM 302060). J Inherit Metab Dis 22:555–567

Bayley JP, Devilee P (2010) Warburg tumours and the mechanisms of mitochondrial tumour suppressor genes. Barking up the right tree? Curr Opin Genet Dev 20:324–329

Bayley JP, Kunst HP, Cascon A et al (2010) SDHAF2 mutations in familial and sporadic paraganglioma and phaeochromocytoma. Lancet Oncol 11:366–372

Bellí G, Molina MM, García-Martínez J, Pérez-Ortín JE, Herrero E (2004) *Saccharomyces cerevisiae* glutaredoxin 5-deficient cells subjected to continuous oxidizing conditions are affected in the expression of specific sets of genes. J Biol Chem 279:12386–12395

Bernard DG, Gabilly ST, Dujardin G, Merchant S, Hamel PP (2003) Overlapping specificities of the mitochondrial cytochrome c and c1 heme lyases. J Biol Chem 278:49732–49742

Bianchi C, Genova ML, Parenti Castelli G, Lenaz G (2004) The mitochondrial respiratory chain is partially organized in a supercomplex assembly: kinetic evidence using flux control analysis. J Biol Chem 279:36562–36569

Birchmeier W, Kohler CE, Schatz G (1976) Interaction of integral and peripheral membrane proteins: affinity labeling of yeast cytochrome oxidase by modified yeast cytochrome c. Proc Natl Acad Sci USA 73:4334–4338

Blair PV (1967) Preparation and properties of repeating units of electron transfer. Methods Enzymol 10:208–212

Blatch GL, Lassle M (1999) The tetratricopeptide repeat: a structural motif mediating protein-protein interactions. Bioessays 21:932–939

Bonnefoy N, Kermorgant M, Groudinsky O, Minet M, Slonimski PP, Dujardin G (1994) Cloning of a human gene involved in cytochrome oxidase assembly by functional complementation of an oxa1- mutation in *Saccharomyces cerevisiae*. Proc Natl Acad Sci USA 91:11978–11982

Boyer PD (1993) The binding change mechanism for ATP synthase–some probabilities and possibilities. Biochim Biophys Acta 1140:215–250

Boyer PD (1997) The ATP synthase—a splendid molecular machine. Annu Rev Biochem 66:717–749

Brandt U, Yu L, Yu CA, Trumpower BL (1993) The mitochondrial targeting presequence of the Rieske iron–sulfur protein is processed in a single step after insertion into the cytochrome bc1 complex in mammals and retained as a subunit in the complex. J Biol Chem 268:8387–8390

Bugiani M, Invernizzi F, Alberio S et al (2004) Clinical and molecular findings in children with complex I deficiency. Biochim Biophys Acta 1659:136–147

Bundschuh FA, Hannappel A, Anderka O et al (2009) Surf1, associated with Leigh syndrome in humans, is a heme-binding protein in bacterial oxidase biogenesis. J Biol Chem 284:25735–25741

Bych K, Kerscher S, Netz DJA, Pierik AJ, Zwicker K, Huynen MA, Lill R, Brandt U, Balk J (2008) Ind1 is requie for effective complex I assembly. EMBO J 27:1736–1746

Calvo SE, Tucker EJ, Compton AG et al (2010) High-throughput, pooled sequencing identifies mutations in NUBPL and FOXRED1 in human complex I deficiency. Nat Genet 42:851–858

Camaschella C, Campanella A, DeFalco L et al (2007) The human counterpart of zebrafish shiraz shows sideroblastic-like microcytic anemia and iron overload. Blood 110:1353–1358

Cameron JM, Levandovskiy V, Mackay N et al (2011a) Complex V TMEM70 deficiency results in mitochondrial nucleoid disorganization. Mitochondrion 11:191–199

Cameron JM, Janer A, Levandovskiy V et al (2011b) Mutations in iron-sulfur cluster scaffold genes NFU1 and BOLA3 cause a fatal deficiency of multiple respiratory chain and 2-oxoacid dehydrogenase enzymes. Am J Hum Genet 89:486–495

Campuzano V, Montermini L, Moltò MD et al (1996) Friedreich's ataxia: autosomal recessive disease caused by an intronic GAA triplet repeat expansion. Science 271:1423–1427

Capaldi RA (1982) Arrangement of proteins in the mitochondrial inner membrane. Biochim Biophys Acta 694:291–306

Carrol J, Fearnley IM, Skehel JM et al (2006) Bovine complex I is a complex of 45 different subunits. J Biol Chem 281:32724–32727

Chance B, Williams GR (1955) A method for the localization of sites for oxidative phosphorylation. Nature 176:250–254

Cizkova A, Stranecky V, Mayr JA et al (2008) TMEM70 mutations cause isolated ATP synthase deficiency and neonatal mitochondrial encephalocardiomyopathy. Nat Genet 40:1288–1290

Coenen MJ, Smeitink JA, Smeets R, Trijbels FJ, van den Heuvel LP (2005) Mutation detection in four candidate genes (OXA1L, MRS2L, YME1L and MIPEP) for combined deficiencies in the oxidative phosphorylation system. J Inherit Metab Dis 28:1091–1097

Collinson IR, Skehel JM, Fearnley IM, Runswick MJ, Walker JE (1996) The F1F0-ATPase complex from bovine heart mitochondria: the molar ratio of the subunits in the stalk region linking the F1 and F0 domains. Biochemistry 35:12640–12646

Cree LM, Samuels DC, Chinnery PF (2009) The inheritance of pathogenic mitochondrial DNA mutations. Biochim Biophys Acta 1792:1097–1102

Cruciat CM, Hell K, Fölsch H, Neupert W, Stuart RA (1999) Bcs1p, an AAA-family member, is a chaperone for the assembly of the cytochrome bc(1) complex. EMBO J 18:5226–5233

Darley-Usmar VM, Georgevich G, Capaldi RA (1984) Reaction of thionitrobenzoate-modified yeast cytochrome c with monomeric and dimeric forms of beef heart cytochrome c oxidase. FEBS Lett 166:131–135

De Lonlay P, Valnot I, Barrientos A et al (2001) A mutant mitochondrial respiratory chain assembly protein causes complex III deficiency in patients with tubulopathy, encephalopathy and liver failure. Nat Genet 29:57–60

De Meirleir L, Seneca S, Lissens W et al (2004) Respiratory chain complex V deficiency due to a mutation in the assembly gene ATP12. J Med Genet 41:120–124

Dell'agnello C, Leo S, Agostino A, Szabadkai G et al (2007) Increased longevity and refractoriness to Ca(2+)-dependent neurodegeneration in Surf1 knockout mice. Hum Mol Genet 16:431–444

Devenish RJ, Prescott M, Rodgers AJ (2008) The structure and function of mitochondrial F1F0-ATP synthases. Int Rev Cell Mol Biol 267:1–58

Diaz F, Fukui H, Garcia S, Moraes CT (2006) Cytochrome *c* oxidase is required for the assembly/stability of respiratory complex I in mouse fibroblasts. Mol Cell Biol 26:4872–4881

Dibrov E, Fu S, Lemire BD (1998) The *Saccharomyces cerevisiae* TCM62 gene encodes a chaperone necessary for the assembly of the mitochondrial succinate dehydrogenase (complex II). J Biol Chem 273:32042

DiMauro S, Davidzon G (2005) Mitochondrial DNA and disease. Ann Med 37:222–232
Distelmaier F, Koopman WJ, van den Heuvel LP et al (2009) Mitochondrial complex I deficiency: from organelle dysfunction to clinical disease. Brain 132:833–842
Dunning CJ, McKenzie M, Sugiana C et al (2007) Human CIA30 is involved in the early assembly of mitochondrial complex I and mutations in its gene cause disease. EMBO J 1126:3227–3237
Efremov RG, Baradaran R, Sazanov LA (2010) The architecture of respiratory complex I. Nature 465:441–445
Ernster L, Dallner G (1995) Biochemical, physiological and medical aspects of ubiquinone function. Biochim Biophys Acta 1271:195–204
Fassone E, Duncan AJ, Taanman JW et al (2010) FOXRED1, encoding an FAD-dependent oxidoreductase complex-I-specific molecular chaperone, is mutated in infantile onset mitochondrial encephalopathy. Hum Mol Genet 19:4837–4847
Feichtinger RG, Zimmermann F, Mayr JA et al (2010) Low aerobic mitochondrial energy metabolism in poorly- or undifferentiated neuroblastoma. BMC Cancer 10:149
Fernandez-Vizarra E, Bugiani M, Goffrini P et al (2007) Impaired complex III assembly associated with BCS1L gene mutations in isolated mitochondrial encephalopathy. Hum Mol Genet 16:1241–1252
Fernandez-Vizarra E, Tiranti V, Zeviani M (2009) Assembly of the oxidative phosphorylation system in humans: what we have learned by studying its defects. Biochim Biophys Acta 1793:200–211
Fontanesi F, Soto IC, Horn D, Barrientos A (2006) Assembly of mitochondrial cytochrome c-oxidase, a complicated and highly regulated cellular process. Am J Physiol Cell Physiol 291:C1129–C1147
Fowler LR, Richardson SH (1963) Studies on the electron transfer system. On the mechanism of reconstitution of the mitochondrial electron transfer system. J Biol Chem 238:456–463
Gerards M, Sluiter W, van den Bosch BJ et al (2010) Defective complex I assembly due to C20orf7 mutations as a new cause of Leigh syndrome. J Med Genet 47:507–512
Gerards M, van den Bosch BJ, Danhauser K et al (2011) Riboflavin-responsive oxidative phosphorylation complex I deficiency caused by defective ACAD9: new function for an old gene. Brain 134:210–219
Ghezzi D, Zeviani M (2011) Mitochondrial disorders: nuclear gene mutations. In: Encyclopedia of Life Sciences (ELS). John Wiley & Sons, Ltd: Chichester. DOI: 10.1002/9780470015902.
Ghezzi D, Goffrini P, Uziel G et al (2009) SDHAF1, encoding a LYR complex-II specific assembly factor, is mutated in SDH-defective infantile leukoencephalopathy. Nat Genet 41:654–656
Ghezzi D, Arzuffi P, Zordan M et al (2011) Mutations in TTC19 cause mitochondrial complex III deficiency and neurological impairment in humans and flies. Nat Genet 43:259–263
Gilkerson RW, Selker JM, Capaldi RA (2003) The cristal membrane of mitochondria is the principal site of oxidative phosphorylation. FEBS Lett 546:355–358
González-Cabo P, Vázquez-Manrique RP, García-Gimeno MA, Sanz P, Palau F (2005) Frataxin interacts functionally with mitochondrial electron transport chain proteins. Hum Mol Genet 14:2091–2098
Grossman LI, Lomax MI (1997) Nuclear genes for cytochrome c oxidase. Biochim Biophys Acta 1352:174–192
Haack TB, Danhauser K, Haberberger B et al (2010) Exome sequencing identifies ACAD9 mutations as a cause of complex I deficiency. Nat Genet 42:1131–1134
Hackenbrock CR, Chazotte B, Gupte SS (1986) The random collision model and a critical assessment of diffusion and collision in mitochondrial electron transport. J Bioenerg Biomembr 18:331–368
Hägerhäll C (1997) Succinate: quinone oxidoreductases. Variations on a conserved theme. Biochim Biophys Acta 1320:107–141
Hägerhäll C, Hederstedt L (1997) A structural model for the membrane-integral domain of succinate: quinone oxidoreductases. FEBS Lett 389:25

Hao HX, Khalimonchuk O, Schraders M et al (2009) SDH5, a gene required for flavination of succinate dehydrogenase, is mutated in paraganglioma. Science 28:1139–1142

Hederstedt L (2002) Succinate:quinone oxidoreductase in the bacteria *Paracoccus denitrificans* and *Bacillus subtilis*. Biochim Biophys Acta 1553:74–83

Hinson JT, Fantin VR, Schonberger J et al (2007) Missense mutations in the BCS1L gene as a cause of the Bjornstad syndrome. N Engl J Med 356:809–819

Honzík T, Tesarová M, Mayr JA et al (2010) Mitochondrial encephalocardio-myopathy with early neonatal onset due to TMEM70 mutation. J Arch Dis Child 95:296–301

Houstek J, Klement P, Floryk D et al (1999) A novel deficiency of mitochondrial ATPase of nuclear origin. Hum Mol Genet 8:1967–1974

Houstek J, Pickova A, Vojtiskova A, Mracek T, Pecina P, Jesina P (2006) Mitochondrial diseases and genetic defects of ATP synthase. Biochim Biophys Acta 1757:1400–1405

Houstek J, Kmoch S, Zeman J (2009) TMEM70 protein – a novel ancillary factor of mammalian ATP synthase. Biochim Biophys Acta 1787:529–532

Huigsloot M, Nijtmans LG, Szklarczyk R et al (2011) A mutation in C2orf64 causes impaired cytochrome c oxidase assembly and mitochondrial cardiomyopathy. Am J Hum Genet 88:488–493

Hunte C, Zickermann V, Brandt U (2010) Functional modules and structural basis of conformational coupling in mitochondrial complex I. Science 329:448–451

Huynen MA, Spronk CA, Gabaldón T, Snel B (2005) Combining data from genomes, Y2H and 3D structure indicates that BolA is a reductase interacting with a glutaredoxin. FEBS Lett 579:591–596

Iwata S, Lee JW, Okada K et al (1998) Complete structure of the 11-subunit bovine mitochondrial cytochrome bc1 complex. Science 281:64–71

Janssen RJ, Nijtmans LG, van den Heuvel LP, Smeitink JA (2006) Mitochondrial complex I: structure, function and pathology. J Inherit Metab Dis 29:499–515

Jiang X, Wang X (2004) Cytochrome c-mediated apoptosis. Annu Rev Biochem 73:87–106

Jonckheere A, Hogeveen M, Nijtmans L et al (2008) A novel mitochondrial ATP8 (MT-ATP8) gene mutation in a patient with apical hypertrophic cardiomyopathy and neuropathy. J Med Genet 45:129–133

Joost K, Rodenburg R, Piirsoo A et al (2010) A novel mutation in the SCO2 gene in a neonate with early-onset cardioencephalomyopathy. Pediatr Neurol 42:227–230

Kim KD, Chung WH, Kim HJ, Lee KC, Roe JH (2010) Monothiol glutaredoxin Grx5 interacts with Fe-S scaffold proteins Isa1 and Isa2 and supports Fe-S assembly and DNA integrity in mitochondria of fission yeast. Biochem Biophys Res Commun 392:467–472

Ko YH, Delannoy M, Hullihen J, Chiu W, Pedersen PL (2003) Mitochondrial ATP synthasome. Cristae-enriched membranes and a multiwell detergent screening assay yield dispersed single complexes containing the ATP synthase and carriers for Pi and ADP/ATP. J Biol Chem 278:12305–12309

Kollberg G, Tulinius M, Melberg A et al (2009) Clinical manifestation and a new ISCU mutation in iron-sulphur cluster deficiency myopathy. Brain 132:2170–2179

Koopman WJ, Nijtmans LG, Dieteren CE et al (2010) Mammalian mitochondrial complex I: biogenesis, regulation and reactive oxygen species generation. Antioxid Redox Signal 12:1431–1470

Krause F, Scheckhuber CQ, Werner A, Rexroth S, Reifschneider NH, Dencher NA, Osiewacz HD (2004) Supramolecular organization of cytochrome c oxidaseand alternative oxidase-dependent respiratory chains in the filamentous fungus Podospora anserine. J Biol Chem 279:26453–26461

Krause F, Reifschneider NH, Goto S, Dencher NA (2005) Active oligomeric ATP synthases in mammalian mitochondria. Biochem Biophys Res Commun 329:583–590

Kuffner R, Rohr A, Schmiede A, Krull C, Schulte U (1998) Involvement of two novel chaperones in the assembly of mitochondrial NADH: ubiquinone oxidoreductase (complex I). J Mol Biol 283:409–417

Lamantea E, Carrara F, Mariotti C, Morandi L, Tiranti V, Zeviani M (2002) A novel nonsense mutation (Q352X) in the mitochondrial cytochrome b gene associated with a combined deficiency of complexes I and III. Neuromuscul Disord 12:49–52

Lancaster CR, Kroger A, Auer M, Michel H (1999) Structure of fumarate reductase from Wolinella succinogenes at 2.2 A resolution. Nature 402:377–385

Lazarou M, McKenzie M, Ohtake A, Thorburn DR, Ryan MT (2007) Analysis of the assembly profiles for mitochondrial- and nuclear-DNA-encoded subunits into complex I. Mol Cell Biol 27:4228–4237

Lazarou M, Smith SM, Thorburn DR, Ryan MT, McKenzie M (2009) Assembly of nuclear DNA-encoded subunits into mitochondrial complex IV, and their preferential integration into supercomplex forms in patient mitochondria. FEBS J 276:6701–6713

Leary SC, Kaufman BA, Pellecchia G et al (2004) Human SCO1 and SCO2 have independent, cooperative functions in copper delivery to cytochrome c oxidase. Hum Mol Genet 13:1839–1848

Leary SC, Cobine PA, Kaufman BA et al (2007) The human cytochrome c oxidase assembly factors SCO1 and SCO2 have regulatory roles in the maintenance of cellular copper homeostasis. Cell Metab 5:9–20

Lee SJ, Yamashita E, Abe T et al (2001) Intermonomer interactions in dimer of bovine heart cytochrome c oxidase. Acta Crystallogr D: Biol Crystallogr 57:941–947

Lenaz G, Genova ML (2010) Structure and organization of mitochondrial respiratory complexes: a new understanding of an old subject. Antioxid Redox Signal 12:961–1008

Lenaz G, Baracca A, Carelli V, D'Aurelio M, Sgarbi G, Solaini G (2004) Bioenergetics of mitochondrial diseases associated with mtDNA mutations. Biochim Biophys Acta 1658:89–94

Levéen P, Kotarsky H, Mörgelin M, Karikoski R, Elmér E, Fellman V (2011) The GRACILE mutation introduced into Bcs1l causes postnatal complex III deficiency: a viable mouse model for mitochondrial hepatopathy. Hepatology 53:437–447

Li K, Tong WH, Hughes RM, Rouault TA (2006) Roles of the mammalian cytosolic cysteine desulfurase, ISCS, and scaffold protein, ISCU, in iron-sulfur cluster assembly. J Biol Chem 281:12344–12351

Li H, Gakh O, Smith DY, Isaya G (2009) Oligomeric yeast frataxin drives assembly of core machinery for mitochondrial iron-sulfur cluster synthesis. J Biol Chem 284:21971–21980

Lill R, Kispal G (2000) Maturation of cellular Fe-S proteins: an essential function of mitochondria. Trends Biochem Sci 25:352–356

Lill R, Mühlenhoff U (2008) Maturation of iron-sulfur proteins in eukaryotes: mechanisms, connected processes, and diseases. Annu Rev Biochem 77:669–700

Loeffen JL, Smeitink JA, Trijbels JM et al (2000) Isolated complex I deficiency in children: clinical, biochemical and genetic aspects. Hum Mutat 15:123–134

Mancuso C, Scapagini G, Currò D et al (2007) Mitochondrial dysfunction, free radical generation and cellular stress response in neurodegenerative disorders. Front Biosci 12:1107–1123

Mashkevich G, Repetto B, Glerum DM, Jin C, Tzagoloff A (1997) SHY1, the yeast homolog of the mammalian SURF-1 gene, encodes a mitochondrial protein required for respiration. J Biol Chem 272:14356–14364

Massa V, Fernandez-Vizarra E, Alshahwan S et al (2008) Severe infantile encephalomyopathy caused by a mutation in COX6B1, a nucleus-encoded subunit of cytochrome c oxidase. Am J Hum Genet 82:1281–1289

Mayr JA, Havlícková V, Zimmermann F et al (2010) Mitochondrial ATP synthase deficiency due to a mutation in the ATP5E gene for the F1 epsilon subunit. Hum Mol Genet 19:3430–3439

McFarland R, Turnbull DM (2009) Batteries not included: diagnosis and management of mitochondrial disease. J Intern Med 265:210–228

McKenzie M, Ryan MT (2010) Assembly factors of human mitochondrial complex I and their defects in disease. IUBMB Life 62:497–502

Mick DU, Wagner K, van der Laan M et al (2007) Shy1 couples Cox1 translational regulation to cytochrome c oxidase assembly. EMBO J 26:4347–4358

Mochel F, Knight MA, Tong WH et al (2008) Splice mutation in the iron-sulfur cluster scaffold protein ISCU causes myopathy with exercise intolerance. Am J Hum Genet 82:652–660

Mootha VK, Lepage P, Miller K (2003) Identification of a gene causing human cytochrome c oxidase deficiency by integrative genomics. Proc Natl Acad Sci USA 100:605–610

Mulkidjanian AY, Makarova KS, Galperin MY, Koonin EV (2007) Inventing the dynamo machine: the evolution of the F-type and V-type ATPases. Nat Rev Microbiol 5:892–899

Munnich A, Rustin P (2001) Clinical spectrum and diagnosis of mitochondrial disorders. Am J Med Genet 106:4–17

Nakamura K, Yamaki M, Sarada M et al (1996) Two hydrophobic subunits are essential for the heme b ligation and functional assembly of complex II (succinate-ubiquinone oxidoreductase) from *Escherichia coli*. J Biol Chem 271:521–527

Nargang FE, Preuss M, Neupert W, Herrmann JM (2002) The Oxa1 protein forms a homooligomeric complex and is an essential part of the mitochondrial export translocase in *Neurospora crassa*. J Biol Chem 277:12846–12853

Nijtmans LG, Klement P, Houstek J, van den Bogert C (1995) Assembly of mitochondrial ATP synthase in cultured human cells: implications for mitochondrial diseases. Biochim Biophys Acta 1272:190–198

Nijtmans LG, Taanman JW, Muijsers AO, Speijer D, Van den Bogert C (1998) Assembly of cytochrome-c oxidase in cultured human cells. Eur J Biochem 254:389–394

Nobrega FG, Nobrega MP, Tzagoloff A (1992) BCS1, a novel gene required for the expression of functional Rieske iron–sulfur protein in *Saccharomyces cerevisiae*. EMBO J 11:3821–3829

Nouws J, Nijtmans L, Houten SM et al (2010) Acyl-CoA dehydrogenase 9 is required for the biogenesis of oxidative phosphorylation complex I. Cell Metab 12:283–294

Ogilvie I, Kennaway NG, Shoubridge EA (2005) A molecular chaperone for mitochondrial complex I assembly is mutated in a progressive encephalopathy. J Clin Invest 115:2784–2792

Pagliarini DJ, Calvo SE, Chang B et al (2008) A mitochondrial protein compendium elucidates complex I disease biology. Cell 134:112–123

Papadopoulou LC, Sue CM, Davidson MM et al (1999) Fatal infantile cardioencephalomyopathy with COX deficiency and mutations in SCO2, a COX assembly gene. Nat Genet 23:333–337

Paumard P, Vaillier J, Coulary B et al (2002) The ATP synthase is involved in generating mitochondrial cristae morphology. EMBO J 21:221–230

Pecina P, Houstkova H, Hansıkova H, Zeman J, Houstek J (2004) Genetic defects of cytochrome c oxidase assembly. Physiol Res 53:S213–S223

Piekutowska-Abramczuk D, Magner M, Popowska E et al (2009) SURF1 missense mutations promote a mild Leigh phenotype. Clin Genet 76:195–204

Pondarré C, Antiochos BB, Campagna DR et al (2006) The mitochondrial ATP-binding cassette transporter Abcb7 is essential in mice and participates in cytosolic iron-sulfur cluster biogenesis. Hum Mol Genet 15:953–964

Pronicki M, Kowalski P, Piekutowska-Abramczuk D et al (2010) A homozygous mutation in the SCO2 gene causes a spinal muscular atrophy like presentation with stridor and respiratory insufficiency. Eur J Paediatr Neurol 14:253–260

Rabl R, Soubannier V, Scholz R et al (2009) Formation of cristae and crista junctions in mitochondria depends on antagonism between Fcj1 and Su e/g. J Cell Biol 185:1047–1063

Rak M, Gokova S, Tzagoloff A (2011) Modular assembly of yeast mitochondrial ATP synthase. EMBO J 30:920–930

Remacle C, Barbieri MR, Cardol P, Hamel PP (2008) Eukaryotic complex I: functional diversity and experimental systems to unravel the assembly process. Mol Genet Genomics 280: 93–110

Robinson KM, Lemire BD (1996) Covalent attachment of FAD to the yeast succinate dehydrogenase flavoprotein requires import into mitochondria, presequence removal, and folding. J Biol Chem 271:4055–4060

Rouault TA, Tong WH (2008) Iron-sulfur cluster biogenesis and human disease. Trends Genet 24:398–407

Saada A, Edvardson S, Rapoport M et al (2008) C6ORF66 is an assembly factor of mitochondrial complex I. Am J Hum Genet 82:32–38

Saada A, Vogel RO, Hoefs SJ et al (2009) Mutations in NDUFAF3 (C3ORF60), encoding an NDUFAF4 (C6ORF66)-interacting complex I assembly protein, cause fatal neonatal mitochondrial disease. Am J Hum Genet 84:718–727

Saada A, Edvardson S, Shaag A et al (2012) Combined OXPHOS complex I and IV defect, due to mutated complex I assembly factor C20ORF7. J Inherit Metab Dis 35(1):125–131

Sasarman F, Brunel-Guitton C, Antonicka H et al (2010) LRPPRC and SLIRP interact in a ribonucleoprotein complex that regulates posttranscriptional gene expression in mitochondria. Mol Biol Cell 21:1315–1323

Schaefer AM, Taylor RW, Turnbull DM, Chinnery PF (2004) The epidemiology of mitochondrial disorders: past, present and future. Biochim Biophys Acta 1659:115–120

Schäfer E, Dencher NA, Vonck J, Parcej DN (2007) Three-dimensional structure of the respiratory chain supercomplex I1III2IV1 from bovine heart mitochondria. Biochemistry 46:12579–12585

Schägger H (2002) Respiratory chain supercomplexes of mitochondria and bacteria. Biochim Biophys Acta 1555:154–159

Schagger H, Pfeiffer K (2000) Supercomplexes in the respiratory chains of yeast and mammalian mitochondria. EMBO J 19:1777–1783

Schagger H, von Jagow G (1991) Blue native electrophoresis for isolation of membrane protein complexes in enzymatically active form. Anal Biochem 199:223–231

Schatz G (1968) Impaired binding of mitochondrial adenosine triphosphatase in the cytoplasmic "petite" mutant of *Saccharomyces cerevisiae*. J Biol Chem 243:2192–2199

Scheffler IE, Yadava N, Potluri P (2004) Molecular genetics of complex I-deficient Chinese hamster cell lines. Biochim Biophys Acta 1659:160–171

Schmitz-Linneweber C, Small I (2008) Pentatricopeptide repeat proteins: a socket set for organelle gene expression. Trends Plant Sci 13:663–670

Schneider D, Pohl T, Walter J et al (2008) Assembly of the *Escherichia coli* NADH:ubiquinone oxidoreductase (complex I). Biochim Biophys Acta 1777:735–739

Schon EA, Santra S, Pallotti F, Girvin ME (2001) Pathogenesis of primary defects in mitochondrial ATP synthesis. Semin Cell Dev Biol 12:441–448

Seeger J, Schrank B, Pyle A et al (2010) Clinical and neuropathological findings in patients with TACO1 mutations. Neuromuscul Disord 20:720–724

Sheftel AD, Stehling O, Pierik AJ et al (2009) Human ind1, an iron-sulfur cluster assembly factor for respiratory complex I. Mol Cell Biol 29:6059–6073

Sperl W, Jesina P, Zeman J et al (2006) Deficiency of mitochondrial ATP synthase of nuclear genetic origin. Neuromuscul Disord 16:821–829

Spiegel R, Khayat M, Shalev SA et al (2011) TMEM70 mutations are a common cause of nuclear encoded ATP synthase assembly defect: further delineation of a new syndrome. J Med Genet 48:177–182

Stiburek L, Vesela K, Hansikova H et al (2005) Tissue-specific cytochrome c oxidase assembly defects due to mutations in SCO2 and SURF1. Biochem J 392:625–632

Stiburek L, Hansikova H, Tesarova M, Cerna L, Zeman J (2006) Biogenesis of eukaryotic cytochrome c oxidase. Physiol Res 55:S27–S41

Stiburek L, Fornuskova D, Wenchich L, Pejznochova M, Hansikova H, Zeman J (2007) Knockdown of human Oxa1l impairs the biogenesis of F1Fo-ATP synthase and NADH:ubiquinone oxidoreductase. J Mol Biol 374:506–516

Stiburek L, Vesela K, Hansikova H, Hulkova H, Zeman J (2009) Loss of function of Sco1 and its interaction with cytochrome c oxidase. Am J Physiol Cell Physiol 296:C1218–C1226

Sugiana C, Pagliarini DJ, McKenzie M et al (2008) Mutation of C20orf7 disrupts complex I assembly and causes lethal neonatal mitochondrial disease. Am J Hum Genet 83:468–478

Sun F, Huo X, Zhai Y et al (2005) Crystal structure of mitochondrial respiratory membrane protein complex II. Cell 121:1043–1057

Tatuch Y, Robinson BH (1993) The mitochondrial DNA mutation at 8993 associated with NARP slows the rate of ATP synthesis in isolated lymphoblast mitochondria. Biochem Biophys Res Commun 192:124–128

Thorburn DR, Sugiana C, Salemi R et al (2004) Biochemical and molecular diagnosis of mitochondrial respiratory chain disorders. Biochim Biophys Acta 1659:121–128

Tiranti V, Hoertnagel K, Carrozzo R et al (1998) Mutations of SURF-1 in Leigh disease associated with cytochrome c oxidase deficiency. Am J Hum Genet 63:1609–1621

Tiranti V, Galimberti C, Nijtmans L, Bovolenta S, Perini MP, Zeviani M (1999) Characterization of SURF-1 expression and Surf-1p function in normal and disease conditions. Hum Mol Genet 8:2533–2540

Tong WH, Rouault TA (2006) Functions of mitochondrial ISCU and cytosolic ISCU in mammalian iron-sulfur cluster biogenesis and iron homeostasis. Cell Metab 3:199–210

Tsukihara T, Aoyama H, Yamashita E et al (1996) The whole structure of the 13-subunit oxidized cytochrome c oxidase at 2.8 A. Science 272:1136–1144

Tzagoloff A (1969) Assembly of the mitochondrial membrane system. II. Synthesis of the mitochondrial adenosine triphosphatase, F1. J Biol Chem 244:5027–5033

Tzagoloff A, Jang J, Glerum M, Wu M (1996) FLX1 codes for a carrier protein involved in maintaining a proper balance of flavin nucleotides in yeast mitochondria. J Biol Chem 271:7392–7397

Ugalde C, Janssen RJ, van den Heuvel LP, Smeitink JA, Nijtmans LG (2004) Differences in assembly or stability of complex I and other mitochondrial OXPHOS complexes in inherited complex I deficiency. Hum Mol Genet 13:659–667

Valnot I, Osmond S, Gigarel N et al (2000) Mutations of the SCO1 gene in mitochondrial cytochrome c oxidase deficiency with neonatal-onset hepatic failure and encephalopathy. Am J Hum Genet 67:1104–1109

van Nederveen FH, Gaal J, Favier J et al (2009) An immunohistochemical procedure to detect patients with paraganglioma and phaeochromocytoma with germline SDHB, SDHC, or SDHD gene mutations: a retrospective and prospective analysis. Lancet Oncol 10:764–771

Visapää I, Fellman V, Vesa J et al (2002) GRACILE syndrome, a lethal metabolic disorder with iron overload, is caused by a point mutation in BCS1L. Am J Hum Genet 71:863–876

Vogel RO, Smeitink JA, Nijtmans LG (2007) Human mitochondrial complex I assembly: a dynamic and versatile process. Biochim Biophys Acta 1767:1215–1227

Wagner K, Perschil I, Fichter CD, van der Laan M (2010) Stepwise assembly of dimeric F(1)F(o)-ATP synthase in mitochondria involves the small F(o)-subunits k and i. Mol Biol Cell 21:1494–1504

Wang ZG, Sheluho D, Gatti DL, Ackerman SH (2000) The alpha-subunit of the mitochondrial F(1) ATPase interacts directly with the assembly factor Atp12p. EMBO J 19:1486–1493

Wang ZG, White PS, Ackerman SH (2001) Atp11p and Atp12p are assembly factors for the F(1)-ATPase in human mitochondria. J Biol Chem 276:30773–30778

Weraarpachai W, Antonicka H, Sasarman F et al (2009) Mutation in TACO1, encoding a translational activator of COX I, results in cytochrome *c* oxidase deficiency and late-onset Leigh syndrome. Nat Genet 41:833–837

West AP, Brodsky IE, Rahner C et al (2011) TLR signalling augments macrophage bactericidal activity through mitochondrial ROS. Nature 472:476–480

Williams SL, Valnot I, Rustin P, Taanman JW (2004) Cytochrome c oxidase subassemblies in fibroblast cultures from patients carrying mutations in COX10, SCO1, or SURF1. J Biol Chem 279:7462–7469

Wittig I, Schägger H (2009) Supramolecular organization of ATP synthase and respiratory chain in mitochondrial membranes. Biochim Biophys Acta 1787:672–680

Wittig I, Carrozzo R, Santorelli FM, Schägger H (2006) Supercomplexes and subcomplexes of mitochondrial oxidative phosphorylation. Biochim Biophys Acta 1757:1066–1072

Xu F, Ackerley C, Maj MC et al (2008) Disruption of a mitochondrial RNA-binding protein gene results in decreased cytochrome b expression and a marked reduction in ubiquinol-cytochrome *c* reductase activity in mouse heart mitochondria. Biochem J 416:15–26

Yankovskaya V, Horsefield R, Törnroth S et al (2003) Architecture of succinate dehydrogenase and reactive oxygen species generation. Science 299:700–704

Yoshikawa S, Shinzawa-Itoh K, Tsukihara T (1998) Crystal structure of bovine heart cytochrome c oxidase at 2.8 A resolution. J Bioenerg Biomembr 30:7–14

Zara V, Conte L, Trumpower BL (2007) Identification and characterization of cytochrome bc(1) subcomplexes in mitochondria from yeast with single and double deletions of genes encoding cytochrome bc(1) subunits. FEBS J 274:4526–4539

Zara V, Conte L, Trumpower BL (2009) Biogenesis of the yeast cytochrome bc(1) complex. Biochim Biophys Acta 1793:89–96

Zick M, Rabl R, Reichert AS (2009) Cristae formation-linking ultrastructure and function of mitochondria. Biochim Biophys Acta 1793:5–19

Chapter 5
Supramolecular Organisation of the Mitochondrial Respiratory Chain: A New Challenge for the Mechanism and Control of Oxidative Phosphorylation

Giorgio Lenaz and Maria Luisa Genova

Abstract Recent experimental evidence has replaced the random diffusion model of electron transfer with a model of supramolecular organisation based on specific interactions between individual respiratory complexes. These supercomplexes are detected by blue-native electrophoresis and are found to be functionally relevant by flux control analysis; moreover, they have been isolated and characterised by single-particle electron microscopy. The supramolecular association of individual complexes strongly depends on membrane lipid amount and composition and is affected by lipid peroxidation; it also seems to be modulated by membrane potential and protein phosphorylation. Supercomplex association confers several new properties with respect to the non-associated respiratory complexes to the respiratory chain: the most obvious is substrate channelling, specifically addressing Coenzyme Q and cytochrome *c* to interact directly with the partner enzymes without the need of a less efficient random diffusion step; in addition, supramolecular association may provide a further rate advantage by conferring long-range conformational changes to the individual complexes. Additional properties are stabilisation of Complex I, as evidenced by the destabilising effect on Complex I of mutations in either Complex III or Complex IV, and prevention of excessive generation of reactive oxygen species. On the basis of the properties described above, we hypothesise that an oxidative stress acts primarily by disassembling supercomplex associations thereby establishing a vicious circle of oxidative stress and energy failure, ultimately leading to cell damage and disease. We provide evidence that in physiological ageing and in some disease states, characterised by oxidative stress and mitochondrial damage, such as heart failure, neurodegenerative disorders and cancer, a loss of supercomplex association occurs, in line with our working hypothesis.

G. Lenaz • M.L. Genova (✉)
Department of Biochemistry "G. Moruzzi", University of Bologna, Bologna, Italy
e-mail: marialuisa.genova@unibo.it

B. Kadenbach (ed.), *Mitochondrial Oxidative Phosphorylation*, Advances in Experimental Medicine and Biology 748, DOI 10.1007/978-1-4614-3573-0_5,

5.1 Introduction

The accumulation of recent experimental evidence obtained with newly developed techniques has replaced the random diffusion model of electron transfer, based on casual interactions between randomly dispersed components (Hackenbrock et al. 1986), with a model of supramolecular organisation based on specific, though dynamic, interactions between individual components, and leading to the acquisition of new properties (substrate channelling, assembly, morphological organisation) that were unpredictable from the previous reductionist analysis. This change derives from the newly developed holistic approach (Van Regenmortel 2004) that has extended to all areas of molecular and cell biology, where the cell is viewed as an assembly of molecular machines made of proteins interacting with one another to harmonically exert their function in concert (Alberts 1998); within a cell, protein associations are orchestrated and regulated very precisely and the association of more than two binding partners introduces new levels of complexity in their assembly properties (Gavin and Superti-Furga 2003).

A retrospective analysis of the literature reveals that the idea of supramolecular associations between respiratory enzymes was present since the early times: the original view derived from the pioneering studies of Chance and Williams (Chance and Williams 1955) depicted the respiratory chain as a solid-state assembly of flavins and cytochromes in a protein matrix.

Circumstantial evidence against a random distribution of respiratory complexes also came from the early investigations reporting isolation of Complex I–Complex III (Hatefi et al. 1962) and Complex II–Complex III units (Yu et al. 1974), indicating that such units may be preferentially associated in the native membrane. It is not by chance that the investigators that discovered mitochondrial respiratory complexes had in their hands the key for supercomplex associations, but were unaware of it.

Today the evidence of supercomplex association of the respiratory chain is well consolidated (Wittig and Schägger 2009; Lenaz and Genova 2010), although its possible relation with a random distribution of the individual complexes is not completely clarified (Acín-Pérez et al. 2008; Lenaz and Genova 2009). Furthermore, several studies suggest possible pathological implications due to changes in supramolecular structure: this chapter will present a working hypothesis of a vicious circle of oxidative stress and supercomplex disruption at the basis of numerous pathological conditions.

5.2 Evidence for Supramolecular Associations in the Oxidative Phosphorylation System

5.2.1 *Electrophoretic Evidence*

A few reports before the year 2000 on the possible presence of specific associations between respiratory complexes, either fixed (Ozawa et al. 1987) or dynamic (Hochman et al. 1985), were not usually taken with much consideration.

A breakthrough occurred in 2000 when Schägger applied the previously introduced technique of blue-native polyacrylamide gel electrophoresis (BN-PAGE) to digitonin-solubilised yeast and mammalian mitochondria (Schägger and von Jagow 1991; Schägger and Pfeiffer 2000). The newly discovered associations were considered to represent the physiological state of the respiratory complexes and were called *respirasomes*. In the same paper, the authors also described a dimeric state for the ATP-synthase complex.

Since then, BN-PAGE and the closely related milder colourless native PAGE (CN-PAGE) have become a popular experimental strategy for the detection of the respiratory supercomplexes from various biological samples and for the analysis of their protein composition (Krause and Seelert 2008). Respiratory supercomplexes were revealed and characterised in several mammalian tissues, such as rat brain (Dencher et al. 2007; Wernicke et al. 2010; Reifschneider et al. 2006), liver (Dani et al. 2009; Reifschneider et al. 2006), kidney (Reifschneider et al. 2006), skeletal muscle (Lombardi et al. 2009; Reifschneider et al. 2006), heart (Gómez et al. 2009; Reifschneider et al. 2006), in bovine heart (Schägger and Pfeiffer 2001; Schägger 2002; Reifschneider et al. 2006; Krause 2006), in human skin fibroblasts (Colindres et al. 2007) and in many other organisms, e.g. fish (Schäfer et al. 2007a), fungi (Krause et al. 2004a; Marques et al. 2007; Maas et al. 2009; Nübel et al. 2009), *Caenorhabditis elegans* (Brys et al. 2010), Drosophila (Le Pécheur et al. 2009), and also in plants (Eubel et al. 2004; Krause et al. 2004b; Genova et al. 2008) and bacteria (Krause 2006; Schägger 2002; Stroh et al. 2004).

Highly ordered architectures were observed for all the supercomplexes investigated (cf. Sect. 5.3), thus discarding most doubts on artificial protein–protein interactions and supporting the idea that such interactions may also be species- or kingdom-specific (Vonck and Schäfer 2009). The fundamental features of the supramolecular organisation of the standard respiratory complexes I, III and IV as a respirasome are conserved in all higher eukaryotes where such high molecular weight assemblies were demonstrated to exist in multiple forms characterised by different copy numbers of the Complex IV subunit. The supercomplex I–III_2–$IV_{1\text{-}4}$ is one of the most intriguing supercomplexes because it contains all the redox enzymes required for the complete pathway of electron transfer from NADH to molecular oxygen. However, in some cases respirasome-like supercomplexes are detectable but of low abundance compared to the supercomplex I–III_2 where Complex IV is not present. Moreover, respiratory assemblies comprising only Complex III and Complex IV were observed, particularly in some yeast lacking Complex I.

A major I_1–III_2–IV_1 supercomplex was also found in the wild type of the filamentous fungus *P. anserina*, although long-lived mutants deficient in Complex IV and overexpressing the alternative oxidase (AOX) had a different arrangement including a I_2–III_2 supercomplex that revealed to be a major one by a gentle colourless-native PAGE (Krause et al. 2004a, 2006).

Although Complex I is usually a monomer, putative dimeric Complex I has also been described in mitochondria from *Neurospora crassa, Yarrowia lipolitica* and *Solanum tuberosum* where it may be involved in forming string-like and patch-like megacomplexes composed of repeated units of respiratory supercomplexes (Marques et al. 2007; Nübel et al. 2009; Bultema et al. 2009) (cf. Sect. 5.3).

A higher level of complexity in the organisation state of supercomplexes is due to the presence of further interaction partners such as the ADP/ATP carrier, some chaperones for cytochrome *c* oxidase (COX) assembly and the carbonic anhydrase subunits that co-assemble with the OXPHOS supercomplexes (Stuart 2009; Mick et al. 2007; Sunderhaus et al. 2006). Probably several additional proteins are able to interact with cytochrome *c* oxidase, and some examples such as monocarboxylate transporters, mitochondrial nitric oxide synthase and large-conductance Ca^{2+}-activated K^+ channel have already been identified by immunoprecipitation (Hashimoto et al. 2008; Persichini et al. 2005; Ohya et al. 2005). According to Hildebrandt (Hildebrandt 2011) sulphide-quinone oxidoreductase and sulphite oxidase, the two membrane-bound enzymes of the pathway catalysing sulphide oxidation, which transfer electrons into the respiratory chain, might also be associated to a supercomplex in vivo.

Moreover, Wang et al. (2010) provide evidence of a multifunctional fatty acid β-oxidation (FAO) complex within mitochondria that is physically associated with OXPHOS supercomplexes. The paper also shows that metabolic channelling of electrons occurs from FAO directly to the respiratory chain since the reduction of cytochrome *c* is efficiently promoted by the addition of palmitoyl-CoA in a reaction mixture containing excess cytochrome *c*, CoQ and a sucrose gradient fraction of rat liver mitochondrial proteins enriched in respiratory supercomplexes, thus indicating that the necessary FAO components are also present in the fraction to mediate oxidation of acyl-CoAs with no accumulation of pathway intermediates. However, Schönfeld et al. (2010) could not find such an association between Complex I and the electron transferring flavoprotein (ETF) that participates in fatty acid oxidation in rat heart and liver mitochondria.

In few cases (in contrast with most observations), analysis of respiratory chain complexes by BN-PAGE also revealed bands that could point to supercomplex formation between Complex II and other OXPHOS enzymes (Acín-Pérez et al. 2008).

In addition, the F_oF_1 ATP synthase (Complex V) does exist in large proportions as dimers and homo-oligomers in several organisms (Krause et al. 2005; Dudkina et al. 2006; Couoh-Cardel et al. 2010; Wittig and Schägger 2009). It is known that ribbon-like structures composed of ATP synthase dimers can play a major role in shaping the inner mitochondrial membrane because they can exert a bending force on the lipid bilayer and induce the formation of tight bends or ridges often extending for several hundred nanometers in the cristae membrane. Indeed, in all species that were examined, long rows of ATP synthase dimers were found exclusively on the tightly curved edges of lamellar cristae whereas irregularly distributed particles of redox complexes and supercomplexes of the respiratory chain reside predominantly in the flat adjacent membrane regions of the cristae vesicle (Davies et al. 2011). Moreover, it has been observed that the hampering of the ATP synthase dimerisation destabilises the oligomeric architecture, leading concomitantly to anomalous mitochondrial morphologies in the form of onion-like structures (Velours et al. 2009).

5.2.2 *Flux Control Analysis*

The first functional demonstration of the existence of supercomplexes was given by kinetic analysis of the pool function of Coenzyme Q and cytochrome *c* in mitochondria from *Saccharomyces cerevisiae* (Boumans et al. 1998). The finding that these mitochondria did not follow "pool behaviour" unless treated with chaotropic agents was considered a peculiarity of this organism, because pool behaviour had been widely accepted after the pioneering kinetic studies of Kröger and Klingenberg (1973a, b) confirmed by the electron microscopy studies of Hackenbrock et al. (1986) that led to the "random collision model" of electron transfer.

Later on, our studies by flux control analysis confirmed the existence of functional supercomplexes in mammalian and plant mitochondria (Bianchi et al. 2004; Genova et al. 2008). Metabolic flux control analysis allows a quantitative measurement of the control exerted on a composite pathway by its individual enzymatic steps. It is assumed by the theory of this analysis that the individual steps consist of separate enzymes joined by the diffusion of common intermediates. In any such system, the control would be differently exerted by one or more steps in the pathway, but the sum of the control coefficients of all steps would approach 1 and not overcome unity.

The situation would be different, however, in enzymatic supercomplexes where the interactions between active sites are fixed and substrates and intermediates are channelled from one defined site to another one without leaving the protein environment; in such a supercomplex, the metabolic pathway would behave as a single enzyme unit and each component of the catalytic unit would exert the same flux control. In particular, in a system like open non-phosphorylating submitochondrial particles, in which the respiratory chain is totally dissociated from other components of the OXPHOS apparatus (i.e. ATP synthase, membrane potential and carriers), the existence of a supercomplex would elicit a flux control coefficient near unity at any of the respiratory complexes, and the sum of all apparent coefficients would be more than 1 (Kholodenko and Westerhoff 1993). A more detailed explanation and a graphic representation of the two alternatives (random diffusion or channelling) in terms of flux control analysis were reported by Lenaz and Genova (2007).

A strong kinetic evidence of functionally relevant association between Complex I and Complex III in bovine heart submitochondrial particles is constituted by the high rate control of both complexes over NADH oxidation (C_I=1.06, C_{III}=0.99). On the contrary, Complex IV appears to be randomly distributed (C_{IV}=0.26), although it is possible that if any stable interaction with Complex IV exists in mammalian mitochondria, it escaped detection by flux control analysis most likely due to a pronounced abundance of molecules in non-assembled form. Moreover, Complex II is fully rate limiting for succinate oxidation (C_{II}=0.88, C_{III}=0.34, C_{IV}=0.20), clearly indicating the absence of substrate channelling toward Complexes III and IV (Bianchi et al. 2004).

In permeabilised mitochondria from freshly harvested potato tubers, inhibitor titration experiments indicate that Complexes III and IV are involved in the formation of a supercomplex assembly also comprising Complex I (Genova et al. 2008) whereas the alternative dehydrogenases, as well as the molecules of Complex II, act as independent structures within the inner mitochondrial membrane.

5.2.3 *Isolation of Supercomplexes*

Supercomplexes separated by BN-PAGE and related techniques are active for what concerns their component individual complexes, as shown by in-gel catalytic activity assays (Wittig et al. 2007; Schäfer et al. 2006). Since supercomplexes can be isolated after mild solubilisation (Stroh et al. 2004; Braun et al. 2009), they become amenable to investigate their integrated activity. Stroh et al. (2004) first demonstrated that the respirasome isolated from *P. denitrificans* has NADH-cytochrome *c* oxidoreductase and NADH oxidase activities, showing that the supercomplexes are active and presumably exert channelling of the "mobile components" CoQ and cytochrome *c*; indeed CoQ was found enriched tenfold in the supercomplex, whereas the low NADH oxidase activity suggested some loss of cytochrome *c* from the respirasome.

In bovine heart mitochondria solubilised with digitonin, Schäfer et al. (2006) separated the proteins by BN-PAGE and electro-eluted two supercomplexes ($I_1III_2IV_1$ and I_1III_2) whose composition was confirmed by peptide mass fingerprinting. In-gel activity staining showed NADH-dehydrogenase activity of Complex I in both supercomplexes and cytochrome *c* oxidase activity of Complex IV in supercomplex $I_1III_2IV_1$. In addition, the catalytic activity of each enzyme component was tested by spectrophotometric assays using specific inhibitors to confirm the functional integrity of the supercomplexes in the electro-eluted samples. For the first time, the activity of two isolated respiratory supercomplexes was quantitated and a pronounced enzymatic advantage of the $I_1III_2IV_1$ respirasome was demonstrated. Indeed, under the conditions used, Complex I in supercomplex I_1III_2 displayed only about half the activity of that in supercomplex $I_1III_2IV_1$, Complex III was active in supercomplex $I_1III_2IV_1$, but supercomplex I_1III_2 showed only minor cytochrome *c* reductase. In conclusion, both isolated supercomplexes displayed activity, but supercomplex $I_1III_2IV_1$ was significantly more active.

Acin-Perez et al. (2008) confirmed the presence of different forms of supercomplexes after solubilisation of mouse liver mitochondria in different detergents and BN-PAGE; at difference with previous studies, some supercomplexes also contained Complex II and ATP synthase (Complex V). One particular subfraction (band 3) contained all Complexes I, II, III and IV and in addition cytochrome *c* and CoQ_9. The band was excised from the gel and showed full respiratory activity, from either NADH or succinate, that was sensitive to the specific respiratory inhibitors of all involved complexes. Therefore the oxygen consumption shown by the supercomplex bands is not only the consequence of the presence of all the respiratory complexes and "mobile" carriers needed, but also reflects the proper arrangement into a functional structure.

These findings directly confirm that supercomplex organisation is compatible with electron transfer; however, they do not discard the idea that electron transfer is still possible in the membrane in the absence of supercomplex organisation in accordance with the random collision model.

5.3 The Structure of Respiratory Supercomplexes

Partial elucidation of the interaction of the individual respiratory complexes within the supercomplex could be achieved by single-particle electron microscopy.

In 2005, Dudkina and colleagues (2005) presented the 2D electron projection maps of a plant supercomplex consisting of Complex I and dimeric Complex III (III_2). Class averages in single-particle processing showed characteristic F-shaped side views of the particle volume and top views roughly triangular.

Shortly later, Schäfer and colleagues (2007b) provided initial insights into the shape and size of two respiratory chain supercomplexes (I_1III_2 and $I_1III_2IV_1$) isolated from bovine heart mitochondria and they also generated 3D maps by random conical tilt electron microscopy analysis of the negatively stained multicomplex assemblies. The dimensions of the supercomplex in the membrane plane are 28 nm by 24 nm. The spatial organisation and mutual arrangement of all the partner enzymes could be determined quite unambiguously by comparing the tilted view images of the supercomplexes with the known electron microscopic and X-ray structures of the individual components. All the representations show extensive interaction of Complex III with the membrane arm of Complex I while Complex IV, when present, interacts with the distal portion of the same arm. Furthermore, on the basis of the structural information gained by the 3D maps of Schäfer and colleagues (2007b), the ubiquinone and cytochrome *c* binding sites of each complex in the supercomplex $I_1III_2IV_1$ appeared to be in proximity to the binding site of the succeeding complex in the respiratory chain, thus supporting the idea that direct substrate channelling occurs in the supercomplex with short diffusion distances for the mobile electron carriers.

Very recently, new 3D maps at nanoscale resolution allowed interpretation of the architecture of mammalian respirasomes at the level of secondary structure. Interestingly, in the new model by Dudkina and colleagues (2011), the single Complexes I, III_2 and IV appear to be at some distance, suggesting that there is little close contact. In particular, the section through the model on the level of the membrane (Fig. 5.1c) demonstrates gaps between Complex III_2 and Complex IV within the membrane whereas the same two complexes appear to contact each other in their matrix portions close to the membrane (Fig. 5.1b).

The recent work by Althoff and colleagues (2011) also demonstrates that only few points of direct contact are allowed between the three complexes in the mammalian supercomplex $I_1III_2IV_1$ because the average distances exceed 2 nm. Moreover, the same authors indicate that at 19 Å resolution the membrane-embedded part of the supercomplex shows intermediate values of density between that of

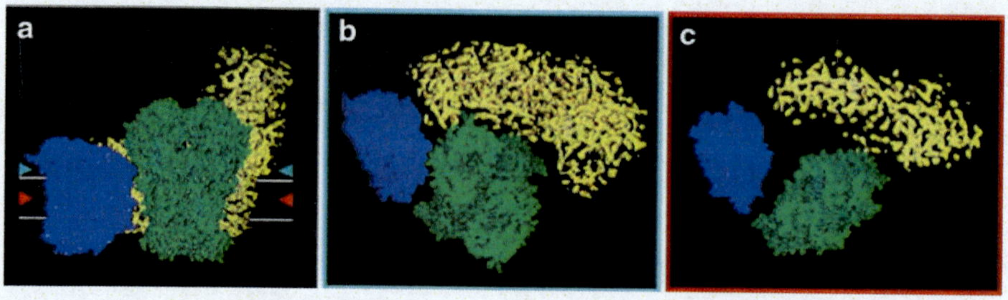

Fig. 5.1 Fitting of the high- and medium-resolution structures of Complexes I, III_2, and IV to the 3D cryo-EM map of $I_1III_2IV_1$ supercomplex: (**a**) space filling model of respirasome, side view from the membrane, *red* and *light-blue arrowheads* show the level of sections in (**b**) and (**c**); (**b**) section through the space-filling model of respirasome on the level of matrix; (**c**) section through the space-filling model of respirasome on the level of membrane, demonstrating gaps between complexes within the supercomplex. In *green*, X-ray structure of the bovine dimeric complex III; in purple, X-ray structure of bovine monomeric complex IV; in *yellow*, the density map of complex I from *Yarrowia lipolytica*. Image taken from (Dudkina et al. 2011), copyright (2011) National Academy of Sciences, USA

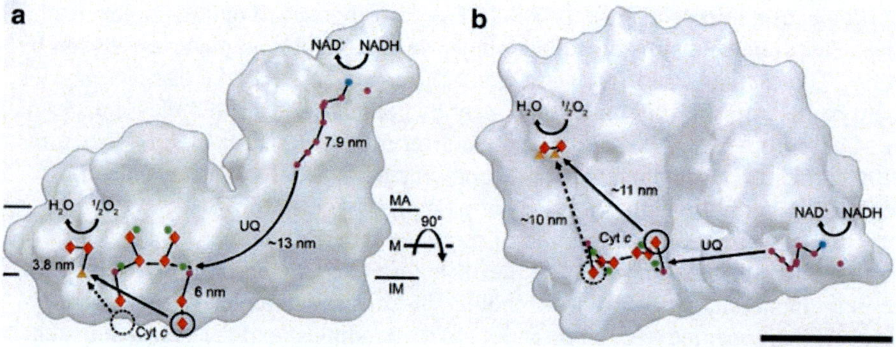

Fig. 5.2 Electron transfer pathways in the supercomplex. Outline of the supercomplex with cofactors active in electron transport marked in *blue* (FMN), *purple* (iron–sulphur-clusters), *green* (quinols/stigmatellins), *red* (hemes) and *orange* (copper atoms), seen from the membrane (**a**) and from the matrix (**b**). Electron trajectories are marked in *black*. The *dashed circle* marks the distal cytochrome *c* binding site, which is unoccupied in the supercomplex. *Straight arrows* in (**a**) indicate the shortest distances from the cytochrome *c* binding sites on complex III to the site of cytochrome *c* oxidation in complex IV. The shorter, proximal branch may be preferred for electron transport. *MA* matrix, *M* membrane, *IM* intermembrane space, *UQ* ubiquinol, *Cyt c* cytochrome *c*. Scale bar, 10 nm. Reprinted by permission from Macmillan Publishers Ltd (Althoff et al. 2011) copyright (2011)

soluble protein and the hydrophobic membrane interior, thus suggesting that the supercomplex is held together at least partly by lipid–protein interactions. Likely, a gap filled with membrane lipid would also facilitate the diffusion of ubiquinol between Complex I and Complex III and it is interesting to note that the CoQ-binding sites in Althoff's model (Fig. 5.2) are conveniently placed for the efficient electron transfer over a short distance of about 13 nm between Complex I and the proximal Complex III monomer. It is proposed that the proximal monomer may be more effective in ubiquinol oxidation while the distal one may be needed to transfer the electron to cytochrome *c* via its flexible Rieske domain, in accordance with the half-of-the-sites reactivity model of Complex III described by the group of Trumpower (Castellani et al. 2010). Furthermore, a trajectory of 10 nm is envisaged in the respirasome along which cytochrome *c* may travel to shuttle electrons towards Complex IV (Fig. 5.2).

Braun and colleagues have also isolated a III_2IV_2 supercomplex from *S. cerevisiae*, a yeast that does not possess a respiratory chain Complex I (Heinemeyer et al. 2007). The paper presents a pseudo-atomic model of the precise interaction of the different components of the supercomplex based on a comparison of the projection maps obtained by electron microscopy with the atomic X-ray structures for Complexes III and IV. Two Complex IV monomers are specifically attached to dimeric Complex III with their convex sides. The opposite sides, which represent the complex IV dimer interface in the X-ray structure, seem to be open for protein–protein interactions. However, no direct evidence was found for the presence of "string"-like structures composed by oligomers of the III_2IV_2 supercomplex.

5.4 Factors that Influence Supramolecular Associations

5.4.1 Lipid Content and Composition

5.4.1.1 General Aspects

These aspects were widely analysed in a previous review (Lenaz and Genova 2007) and will be briefly summarised here.

The fluid mosaic model of membrane structure (Singer and Nicolson 1972) predicted random distribution of integral membrane proteins as a two-dimensional oriented solution in the viscous phospholipid bilayer with high diffusional freedom in the plane of the membrane. Now it is known that this mobility is severely hindered by a great number of restrictions (Jacobson et al. 1995; Vereb et al. 2003), including the existence of hierarchically built supramolecular protein complexes (Damjanovich et al. 1997).

One major determinant of protein organisation in biomembranes is the heterogeneous nature of the lipids in the membrane bilayer (Gil et al. 1998); integral proteins become surrounded by a lipid domain of boundary lipids (Jost et al. 1973) also called lipid annulus (Hesketh et al. 1976), composed of the lipid species providing the least possible mismatch of the protein with the lipids, e.g. with the length of their acyl chains and the lowest distortion of the shape of the membrane.

At low protein concentrations in lipid bilayers, proteins are usually randomly dispersed (Lee 2004), whereas at high concentrations they tend to aggregate. Seen from the protein perspective, protein aggregation would be entropically unfavourable; however, reducing the lipid-exposed protein surface induces a number of bound lipid molecules to return to the more disordered lipid pool, thereby increasing their entropy (Helms 2002).

The possibility that a protein exhibits preference for a given lipid species so that a lipid annulus "wets" the protein surface may be another source of protein aggregation (Lee 2004); the possibility that the wetting layer be shared by two or more proteins has been called capillary condensation and gives rise to aggregation phenomena (Gil et al. 1997) that are at a longer range scale than those involved in direct protein–protein interactions.

5.4.1.2 Lipid–Protein Interactions in Mitochondria

In the mitochondrial inner membrane, integral proteins of the OXPHOS system are densely packed so that the average distance between the complexes may be calculated to be few nanometers (Lenaz 1998); in addition, the presence of different types of lipids would enhance wetting and capillary condensation as well as the possibility of phase separations. Thus, the presence of protein aggregates of the transmembrane respiratory complexes would not be at all unexpected on theoretical grounds. The immobilisation of the proteins would be also favoured by the presence of outer membrane–inner membrane contacts (Brdiczka et al. 2006), by the narrow tubular connections of the cristae (Mannella 2006) and by the high viscosity of the matrix that would exert a slowing effect on the diffusion of proteins spanning the inner membrane, according to the Saffman–Delbruck relation (Saffman and Delbruck 1975).

In addition, direct binding of matrix enzymes with respiratory chain complexes may occur, as proposed for Complex I with several NAD-linked matrix dehydrogenases, including pyruvate dehydrogenase and α-ketoglutarate dehydrogenase complexes, malate dehydrogenase and β-hydroxyacyl-CoA dehydrogenase (Sumegi and Srere 1984; Ovàdi et al. 1994).

Changing the nature of the lipids would dramatically change their matching conditions to the proteins. This could be due to the presence and number of electrostatic charges, different length of the fatty acyl chains, different physical state and different phases existing at the body temperature. Lipid peroxidation removes *cis* double bonds from the fatty acyl chains, inducing profound biophysical changes in the lipid bilayer (Stark 2005), such as increased disorder and possible breakdown of membrane structure and of lipid protein interactions (Megli and Sabatini 2003). Therefore, a change in membrane protein aggregation induced by lipid peroxidation may well be expected on theoretical grounds.

Early experiments reported by Ragan and Heron (1978) provided evidence that purified Complex I and Complex III, when mixed as concentrated solutions in detergent and then co-dialysed, combine reversibly in a 1:1 molar ratio to form a Complex I–Complex III unit (NADH-cytochrome *c* oxidoreductase) that contains equimolar FMN and cytochrome c_1 (clearly at difference with the supercomplexes found by BN-PAGE, where Complex III is present as a dimer, cf. previous sections) and also 2–3 moles of Ubiquinone-10 per mol of protein unit. Activation energy measurements for NADH-cytochrome *c* oxidoreductase activity showed that oxidoreduction of endogenous Ubiquinone-10 proceeds somewhat differently from the oxidation and the reduction of exogenous quinones, supporting the idea that the mobility of Ubiquinone-10 in the Complex I–Complex III unit is highly restricted. However, CoQ-pool behaviour could be restored and Complex I and Complex III could be made to operate independent of each other by raising the concentrations of phospholipid and ubiquinone (approx. a twofold and a sixfold increase, respectively) in the sample (Heron et al. 1978). Inclusion of phospholipid into the reconstituted sample may have a number of effects on the physical state of the system. Heron and co-workers have proposed that, when phospholipid is not in excess of that needed to merely form an annulus, relative mobility of the complexes is lost and Complex I and Complex III are frozen in their I–III super-assembly favouring a stable orientation of the site of reduction of ubiquinone with respect to the site of oxidation.

Heron et al. (1978) also reported that endogenous ubiquinone-10 leaks out of the I–III supercomplex when extra phospholipid is present, causing a decrease in activity that could be alleviated by adding more ubiquinone. It is likely that the function of the large amount of ubiquinone in the natural membrane may be, therefore, to maintain the ubiquinone-10 content in the supercomplex unit when it is formed.

An analogous system, obtained by fusing a crude mitochondrial fraction (R_4B) enriched in Complex I and Complex III (Rieske 1967) with different amounts of phospholipids and CoQ_{10} (Lenaz et al. 1999), was used in our laboratory to discriminate whether the reconstituted protein fraction behaves as individual enzymes (CoQ-pool behaviour) or as assembled supercomplexes depending on the experimental distances between the intramembrane particles. The comparison of the experimentally determined NADH-cytochrome *c* reductase activity with the values expected by theoretical calculation applying the pool equation showed overlapping results at phospholipid dilutions corresponding to distances >50 nm between Complex I and Complex III, whereas at shorter distances that correspond to the mean nearest neighbour distance between respiratory complexes in mitochondria (Vanderkooi 1978; Schwerzmann et al. 1986), pool behaviour was not effective any more (Lenaz et al. 1999; Bianchi et al. 2003). In the two experimental models, kinetic testing according to the Metabolic flux Control Analysis validated the hypothesis of a random organisation and of a functional association between Complex I and Complex III, respectively (Genova et al. 2008).

The formation of the supercomplex I–III is conditioned by the lipid component, but the role played by the lipid environment, in terms of its chemical composition, is not completely known.

All purified preparations of mitochondrial electron transfer complexes are isolated as lipoprotein complexes, the extent of associated lipid depending upon the particular method used for isolation and the phospholipid composition reflecting that for the mitochondrial inner membrane. Predominant phospholipids include cardiolipin, phosphatidylcholine, phosphatidylethanolamine and lesser amounts of neutral lipids and phosphatidylinositol (Fleischer et al. 1962). A dispersive solubilisation effect and a catalytic effect that can be specifically fulfilled only by cardiolipin (Vik and Capaldi 1977; Robinson et al. 1980; Fry and Green 1980, 1981; Lee 2004) are two of the distinguished roles for such membrane lipids. In addition, they can participate in the molecular linkage between the respiratory complexes and, particularly in the case of Complex I and Complex III, provide a sufficiently lipophilic environment for the interaction of the lipophilic electron carrier, ubiquinone.

The phospholipids in closest vicinity to the protein surface, as well as those in the free bilayer, are actually highly mobile and free to exchange, but cardiolipin was indicated as tightly bound being more likely buried within the protein complexes (Kang et al. 1979; Sedlak and Robinson 1999; Lange et al. 2001). The absolute requirement of cardiolipin for the activity of cytochrome oxidase, Complex I and Complex III (Fry and Green 1981), as well as for that of several mitochondrial carriers (Houtkooper and Vaz 2008) suggests that this phospholipid plays a crucial role in the coupled electron transfer process (Fry and Green 1981), but recent results also seem to indicate that cardiolipin stabilises respiratory supercomplexes as well as the individual complexes. The availability of a yeast mutant lacking cardiolipin (Δcrd1 null) provided the opportunity to demonstrate that mitochondrial membranes still contained the III_2–IV_2 supercomplex, but that it was significantly less stable than supercomplexes in the parental strain. Other phospholipids that are increased in the mutant, including phosphatidylethanolamine and phosphatidylglycerol, could not substitute for cardiolipin and could not prevent dissociation of supercomplexes, showing that 90% of the individual homodimers of Complex III and Complex IV are not organised into supercomplex under BN-PAGE conditions (Zhang et al. 2002; Pfeiffer et al. 2003). The putative direct protein–protein interaction of cytochrome oxidase and Complex III in yeast (Pfeiffer et al. 2003) is proposed to also involve one molecule of cardiolipin and one of phosphatidylethanolamine, tightly bound in a cavity of the membrane imbedded domain of Complex III (Lange et al. 2001), suggesting that the two phospholipids can provide a flexible linkage between the interacting subunits of Complex III and Complex IV as well as with the ADP/ATP carrier that is also known to exist in physical association with the III–IV supercomplex (Claypool et al. 2008a; Dienhart and Stuart 2008).

Wenz et al. (2009) studied, by site-directed mutagenesis, the site of specific cardiolipin interaction for supercomplex formation in bc_1 complex and indicate that cardiolipin stabilises supercomplex formation by neutralising the charges of lysine residues in the interaction domain of Complex III with cytochrome oxidase.

In a different study, the stability and assembly of Complex IV was found to be reduced in yeast cells lacking Taz1 (Brandner et al. 2005), the hortologue of human Tafazzin, an acyl transferase involved in the synthesis of mature tetralinoleyl cardiolipin (Neuwald 1997) (see also Sect. 5.6.2.2). It is well documented that exposure of mitochondria to reactive oxygen species (ROS) can affect the respiratory activity via oxidative damage of cardiolipin, which is required for the optimal functioning of the enzyme complexes (Paradies et al. 2000, 2002; Petrosillo et al. 2003). We have demonstrated by flux control analysis that the maintenance of a I–III supercomplex after reconstitution of a protein fraction enriched with Complex I and Complex III (R_4B) into phospholipid vesicles at high protein to lipid ratios (see above) is abolished if lipid peroxidation is induced by 2,2′-azobis-(2-amidinopropane) dihydrochloride (AAPH) before reconstitution (Genova et al. 2008). Evidently, the distortion of the lipid bilayer induced by peroxidation and the alteration of the tightly bound phospholipids determine the dissociation of the supercomplex originally present in the lipid-poor preparation.

5.4.2 *Membrane Potential*

Piccoli et al. (2006) evaluated the impact of the mitochondrial trans-membrane proton electrochemical potential ($\Delta\mu_{H^+}$) on the flux control exerted by cytochrome *c* oxidase on the respiratory activity in intact cells. The results indicate that, under non-phosphorylating conditions mimicking the mitochondrial state 4 respiration, the control strength of the oxidase is low. The authors suggest that such a change in control strength, with respect to state 3 respiration, might be featured in terms of equilibrium between different organisational structures of the enzymatic complexes constituting the mitochondrial OXPHOS (Piccoli et al. 2006). In collaboration with the group of Sarti, we extended the study in actively phosphorylating human hepatoma HepG2 cells under conditions in which the electrical ($\Delta\Psi$) and chemical (ΔpH) components of $\Delta\mu_{H^+}$ were selectively modulated by addition of ionophores. We concluded that $\Delta\Psi$ is predominantly responsible for the tight control exerted by cytochrome *c* oxidase over endogenous respiration whereas ΔpH seems irrelevant in this respect (Dalmonte et al. 2009). Although the driving forces leading to the assembly/disassembly of the supercomplexes have been not clearly defined, it is not inconceivable that, given the membrane-integrated nature of the single complexes, electrostatic/hydrophobic interactions may enter into play in response to $\Delta\Psi$. No direct demonstration of the association state of respiratory complexes as a function of the physiological states of the OXPHOS system is, however, available.

5.4.3 *Possible Role of Post-translational Changes*

Phosphorylation/dephosphorylation events and other post-translational changes of proteins are important factors for protein association/dissociation in signal transduction cascades, cytoskeleton assembly and enzyme activity. For example, glucagon-stimulated phosphorylation of acetyl CoA carboxylase promotes binding of fatty acids to the enzyme and its depolymerisation and inactivation (Clarke and Salati 1985); microtubule-associated protein tau is hyperphosphorylated in Alzheimer's disease inducing its polymerisation to paired helical filaments (Iqbal and Grundke-Iqbal 1996).

It is now established that some of the mitochondrial complexes are subjected to reversible phosphorylation and dephosphorylation. Mitochondria contain protein kinases and phosphatases and both serine/threonine phosphorylation (Thomson 2002; Horbinski and Chu 2005) and tyrosine phosphorylation (Salvi et al. 2005) of mitochondrial proteins are important in the regulation of the activity of these organelles.

Phosphorylation of Complex I has been shown to modify the activity of the enzyme and its ROS generating capacity (Raha et al. 2002; Maj et al. 2004; Scacco et al. 2006; Bellomo et al. 2006; Papa et al. 2008). Cyclic AMP-dependent phosphorylation of the 18 kDa subunit of Complex I, encoded by the nuclear NDUFS4 gene, is required for import of the subunit; modulation of subunit phosphorylation by intramitochondrial protein kinase A and phosphoprotein phosphatase contributes to the stability of Complex I and stimulation of its activity (De Rasmo et al. 2008). It is tempting to speculate that the increase of activity of Complex I and the decrease of ROS generation by phosphorylation may be, in part, the result of enhanced stability of the I–III supercomplex.

Lee et al. (2005) examined the cAMP-dependent phosphorylation of mitochondrial Complex IV isolated from fresh bovine liver and heart in the presence of theophylline, a phosphodiesterase inhibitor that induces high cellular cAMP levels. Under the conditions applied, they were able to demonstrate that phosphorylation of Tyr-304 in subunit I (COX-I) leads to strong decrease of V_{max} in the isolated enzyme while decreasing the K_m for cytochrome *c*, so that the residual oxidase activity is less than 20% at substrate concentrations in the range of the physiological values, compared with saturating substrate concentration. The same effect on cytochrome oxidase kinetics was observed in a cell culture system after treatments that would increase protein phosphorylation (e.g. elevation of cAMP levels by glucagon addition or by forskolin activation of adenylyl cyclase).

Tyrosine 304 is located in the intermembrane space domain of subunit I, on a helix that is in contact with the other catalytic subunit (COX-II), and facing the interface region of the two monomers; therefore a plausible effect of phosphorylation might be the enhancement of the monomer–monomer interaction (Lee et al. 2005). Interestingly, the concave face of Complex IV, which is the dimer interface in the X-ray structure, is also the contact surface of the monomer with its protein partners when it is assembled in the $I_1III_2IV_1$ bovine supercomplex (Schäfer et al. 2007b).

Stable phosphorylation sites in Complex IV are also detectable at tyrosine, serine and threonine in subunit II and III as well as at specific amino acid residues in subunits IV, Vab, VIabc, VIIabc and VIII (Helling et al. 2008) whose function could be to change the binding affinity of Complex IV to specific proteins [i.e. binding of EGFR-pY845 to subunit II, viral protein HBx to subunit III, PKCe to subunit IV, NO synthase to subunit Va, subunit RIa of PKA to subunit Vb and androgen receptor to subunit Vb; for review see (Vogt et al. 2007)].

It is very tempting to speculate that endocrine alterations may affect the assembly state of Complex IV, by hyper- or hypo-phosphorylation of some subunits in the complex. Indeed, cAMP- and PKA-dependent phosphorylation of Complex IV in heart mitochondria (Rosca et al. 2011) was found to be higher in free Complex IV not associated in supercomplex than in the supercomplexed enzyme, suggesting that phosphorylation prevents supercomplex association. Furthermore, changes in the expression of different subunits in Complex IV may affect supercomplex association; it was found that the amount of subunit COX-VIb was increased in heart mitochondrial supercomplexes (which are composed of Complexes I, III and IV) of rats subjected to pharmacological preconditioning before ischemia–reperfusion cycles (Wong et al. 2010).

Kitazoe et al. propose that the increase of threonine and serine composition, found in mitochondrial membrane proteins of terrestrial animals with respect to fishes, enhances protein stability by maximising hydrogen bonds, perhaps involved in the formation of supercomplexes (Kitazoe et al. 2011); since these aminoacids are also largely involved in phosphorylation/dephosphorylation, their increase may contribute to the flexibility of supercomplex organisation.

5.4.4 Is Supercomplex Organisation Fixed or Flexible?

Previous discussions have already considered this aspect. In fact, several data reported in the literature confirm that most or all Complex I are indeed in the form of supercomplex, but most of Complexes II, III and IV appear to be free as isolated complexes. Acin-Perez et al. (2008) propose a *"Plasticity model"* where both types of organisation are possible and functional, depending on the different mitochondrial systems and on the particular physiological states. The plasticity model well suits the information obtained by us by flux control analysis suggesting that electron transfer between Complex I and Complex III is effected *only* by CoQ channelling (at least in beef heart and rat liver mitochondria) whereas that between Complex II and Complex III and between Complex III and Complex IV seems to occur *mostly* by the pools of CoQ and cytochrome *c*, respectively (Bianchi et al. 2004; Genova et al. 2008).

The dynamic character of the supercomplexes is compatible with the factors affecting the association of their protein components and discussed above. Some factors may act in a short or medium time scale, such as the membrane potential (Piccoli et al. 2006; Dalmonte et al. 2009) or protein phosphorylation/dephosphorylation (Rosca et al. 2011), whereas other factors may be operative at longer times, such as the changes in phospholipid composition discussed in Sect. 5.4.1.2.

Nevertheless, a recent study (Muster et al. 2010) suggests that supercomplex dissociation, at least in a random fashion, may not be a fast event. By fusing cells containing mitochondria with respiratory complexes labelled with different fluorescent proteins and resolving their time-dependent re-localisation in living cells, the authors found that a complete reshuffling of respiratory complexes throughout the entire chondriome in single HeLa cells occurs within 2–3 h by organelle fusion and fission. Moreover, polykaryons of fused cells completely re-mixed their complexes in 10–24 h in a progressive manner. Nevertheless, the distribution of respiratory complexes and ATP synthase in fused hybrid mitochondria is not homogeneous but patterned: in co-expressing cells, Complex II is more homogeneously distributed than Complexes I and V, arguing for its higher mobility and less integration in supercomplexes.

5.5 Functional and Structural Consequences for Supramolecular Association

5.5.1 *Kinetic Advantage: Channelling*

The functional consequence of supercomplex assemblies in the respiratory chain is substrate channelling in inter-complex electron transfer. Substrate channelling is the direct transfer of an intermediate between the active sites of two enzymes catalysing consecutive reactions (Ovàdi 1991); in the case of electron transfer, this means direct transfer of electrons between two consecutive enzymes by successive reduction and reoxidation of the intermediate without its diffusion in the bulk medium. In such a case, inter-complex electron transfer becomes indistinguishable from intra-complex electron transfer, so that the so-called mobile intermediates, predicted to exhibit substrate-like behaviour in the classic view of the random collision model (Hackenbrock et al. 1986), would rather be buried in the interface between the two consecutive complexes.

Kinetic analysis allows distinguishing the channelling from the random diffusional encounters; the problem for electron transfer in the respiratory chain was tackled for the first time by the pioneering work of Kröger and Klingenberg (1973a, b). Subsequently flux control analysis was exploited by us with the precise aim of demonstrating channelling (Bianchi et al. 2004).

5.5.1.1 Mechanism of Channelling: Electron Tunnelling or Micro-diffusion?

The fundamental design of electron transfer proteins is two catalytic sites connected by redox chains (Page et al. 2003) that may be entirely within a single protein or belong to different protein subunits. In the respiratory chain of mitochondria, the redox complexes are composed of several subunits containing a trail of cofactors needed to allow sufficiently short distance for electron transfer to occur.

Intra-protein electron transfer is typically limited by tunnelling through the insulating protein medium between the edges of the interacting centres: *electron tunnelling* in protein is reasonably well described by a simple exponential decay with distance, so that the maximal distances allowing physiological electron transfer should not exceed 13–14 Å (Moser et al. 2005).

Inter-protein electron tunnelling obeys to the same exponential rate dependence on distance as intra-protein electron transfer; however, small-scale constrained diffusive motions are sometimes necessary to bring redox centres within the 14 Å tunnelling limit: electron transfer rates reflect diffusional motion of domains of the proteins after a protein–protein complex has been formed (Leys et al. 2003). In the case of supercomplexes formed by apposition of individual complexes connected by potentially mobile cofactors, what is the mechanism of electron transfer? Ideally, we should have a detailed knowledge of the molecular structure of the interacting sites, and this knowledge is still lacking. Obviously, we may have the extremes from close docking of the active sites with real inter-protein tunnelling, up to relatively long distances that may be covered either by important conformation changes or by restricted diffusion (microdiffusion) of the mobile components within the space between the two active sites; all of these alternatives have in common obligate channelling between two fixed sites, so that even the last situation, microdiffusion, would be quite distinguishable from pool behaviour where the interaction of the mobile component may stochastically occur with a great number of possible sites located on several different protein targets which can be reached by random diffusion. In the interaction between Complex I and Complex III within a supercomplex, if the sites are connected by CoQ microdiffusion, it is possible that it takes place within a lipid milieu, although we cannot exclude that the sites are put together by movement of CoQ on the protein or by movement of the protein itself. The previously described kinetic analysis cannot distinguish among different possible mechanisms of channelling. If lipid is involved, then indirect indications may be obtained from studies on the temperature dependence of mitochondrial membrane-bound enzymes (cf. Lenaz and Genova 2010).

In the 3D structure of the mitochondrial supercomplex $I_1III_2IV_1$ which was recently reported by the group of Althoff et al. (2011), a unique arrangement of the three component complexes indicates the pathways along which ubiquinone and cytochrome *c* can travel to shuttle electrons between their respective protein partners. In the above-mentioned model (cf. also Sect. 5.3), one of the Complex III monomers faces the lipid bilayer while the other is surrounded by Complex I. Ubiquinol-binding sites are located between the 49-kDa and the PSST subunits near the first FeS-cluster above the membrane in Complex I and the cytochrome *b* subunit in Complex III. Given that the shortest connection (11.6 nm) between those binding sites would run partly through the aqueous medium, a 13-nm trajectory through a gap within the supercomplex which is filled with membrane lipid is more likely (Fig. 5.2a). Althoff and colleagues also reported the presence of significant amounts of bound phospholipids in the purified supercomplex from mammalian mitochondria and demonstrated that cardiolipin is enriched in the supercomplex compared with bovine heart total lipid. Moreover, HPLC analysis of the lipid extracts indicated that each supercomplex contains at least one molecule of ubiquinol.

Concerning the cytochrome *c* binding sites in the supercomplex, the same authors suggest that a shallow cavity lined by negative charges on the exterior membrane surface near Cu atoms in Complex IV directly faces the similarly shaped binding sites on Complex III, thus making it easy for the small, globular, partly positive cytochrome *c* to pass from one complex to the other like a ball between two cupped hands separated by a trajectory of approximately 11 nm. Residual amounts of cytochrome *c* remaining bound throughout the purification procedure suggest at least partial occupancy of the specific binding sites in the supercomplex (Althoff et al. 2011). This is in accordance with some observations from our laboratory also showing the presence of bound cytochrome *c* by 2D BN/SDS-PAGE analysis and western blot immuno-detection in digitonin-solubilised respirasomes from potato tuber mitochondria (Lenaz et al. 2010).

There are recent indications that supercomplex association may provide further kinetic advantages besides substrate channelling. The study of Schaefer et al. (2006) showed that a supercomplex comprising cytochrome oxidase ($I_1III_2IV_1$) had higher Complex I and Complex III activities than the supercomplex devoid of the terminal oxidase (I_1III_2); evidently, the presence of Complex IV modifies the conformation of the partner complexes in such a way to enhance their catalytic activity. Likewise, Hildebrandt (2011) showed that supercomplex dissociation abolishes the protective effect of dehydroascorbic acid on sulphide toxicity to cytochrome oxidase, suggesting a conformational effect of supramolecular association on the allosteric properties of cytochrome oxidase.

5.5.2 *Stability and Assembly*

The first chromatographic isolation of a complete respirasome ($I_1III_4IV_4$) from digitonin-solubilised membranes of *Paracoccus denitrificans* indicated that Complex I is stabilised by assembly into the NADH oxidase supercomplex since attempts to isolate Complex I from mutant strains lacking Complexes III or IV led to the complete dissociation of Complex I under the conditions of BN-PAGE, which was paralleled by an almost complete loss of NADH-ubiquinone oxidoreductase activity in the mutant strains when the same protocol as for parental strain was applied (Stroh et al. 2004).

Analysis of the state of supercomplexes in human patients with an isolated deficiency of single complexes (Schägger et al. 2004) and in cultured cell models harbouring cytochrome *b* mutations (Acin-Perez et al. 2004; D'Aurelio et al. 2006) also provided evidence that the formation of respirasomes is essential for the assembly/stability of Complex I. Genetic alterations leading to a loss of Complex III prevented respirasome formation and led to secondary loss of Complex I; therefore, primary Complex III assembly deficiencies presented as Complex III/I defects. Conversely, Complex III stability was not influenced by the absence of Complex I.

In particular, D'Aurelio et al. (2006) studied the complementation of mitochondrial DNA (mtDNA) in human cells by fusing two cell lines, one containing a homoplasmic mutation in a subunit of respiratory chain Complex IV, COX-I, and the other with a distinct homoplasmic mutation in a subunit of Complex III, cytochrome *b*. Upon cell fusion, respiration was recovered in hybrids cells, indicating that mitochondria fuse and exchange genetic and protein materials. The recovery of mitochondrial respiration correlated with the presence of supercomplexes containing Complexes I, III and IV; critical amounts of Complexes III or IV are therefore required in order for the supercomplexes to form and provide mitochondrial functional complementation. From these findings, supercomplex assembly emerged as a necessary step for respiration, its defect setting the threshold for respiratory impairment in mtDNA mutant cells.

Several other studies in mutant cells and in human patients having specific defects in a single respiratory complex have pointed out that Complex III and to a lesser extent Complex IV are involved in the assembly and stabilisation of Complex I in mammals (Diaz et al. 2006; McKenzie et al. 2006). This is not the case in fungi, since a *P. anserina* mutant lacking both Complexes III and IV possesses a normal Complex I, presumably as a consequence of special fungal features such as the presence of AOX and a dimerised Complex I (Maas et al. 2009). Conversely, mutations of Complex I had controversial effects, since in some studies they did not affect the amount of other complexes (Schägger et al. 2004; Pineau et al. 2005) while in others they significantly reduced the amounts of Complexes III and IV (Grad and Lemire 2004, 2006; Ugalde et al. 2004). The reason for this discrepancy is not known, but might be related to the specificity of the mutation affecting subunits of Complex I involved in the contacts with the other complexes.

Animal models of the effects of Complex III and Complex IV mutations on Complex I should prove useful for a better understanding of the role of supercomplexes. Two studies in *C. elegans* have been recently published on this aspect.

Suthammarak et al. (2009) knocked down predicted homologues of COX-IV and COX-Va in the nematode *Caenorhabditis elegans* and found that intrinsic Complex I enzymatic activity is dependent on the presence of Complex IV, despite no overall decrease in the amount of complex I. Presumably, the association of Complex I with Complex IV within the supercomplex I–III–IV enhances electron flow through Complex I. In a further study in *C. elegans* the same group (Suthammarak et al. 2010) showed that Complex III defects inhibit Complex I by several different mechanisms involving supercomplex destabilisation. Mutant analysis revealed that Complex III affects supercomplex I–III–IV formation by acting as an assembly/stabilising factor. In addition, a mtDNA mutation affecting Complex III, *ctb-1*, inhibits Complex I function by weakening the interaction of Complex IV in supercomplex I–III–IV. Other Complex III mutations inhibit complex I function either by decreasing the amount of Complex I (*isp-1*), or its assembly as the most active supramolecular form, the I-III-IV supercomplex (*isp-1;ctb-1*). It is suggested that allosteric interactions involve all three complexes within the supercomplex and are necessary for maximal enzymatic activities.

5.5.3 Supercomplexes and ROS Generation

Indirect circumstantial evidence suggests that supercomplex assembly may limit the extent of superoxide generation by the respiratory chain. Panov et al. (2007) reason that the respirasome helps to maintain the redox components of the complexes in the oxidised state through the facilitation of electron flow by channelling, thus limiting ROS formation. Similarly, Seelert et al. (2009) also suggest that facilitation of electron transfer by channelling may limit the detrimental generation of ROS.

Two potential sites for oxygen reduction exist in Complex I, represented by FMN and iron–sulphur cluster N2; controversial results from different laboratories working either on isolated Complex I or on mitochondrial membranes generally indicate that N2 as a source of ROS would be predominant in membrane particles whereas

FMN might become available after Complex I isolation. A reasonable hypothesis is that FMN becomes exposed to oxygen only when Complex I is dissociated from Complex III. Although the molecular structure of the individual complexes does not allow to envisage a close apposition of the matrix arm of Complex I, where FMN is localised, with either Complex III or IV (Dudkina et al. 2005; Schäfer et al. 2007b), the actual shape of the $I_1III_2IV_1$ supercomplex from bovine heart (Schäfer et al. 2007b) suggests a slightly different conformation of Complex I in the supercomplex, showing a smaller angle of the matrix arm with the membrane arm and a higher bending towards the membrane (and presumably Complex III), in line with the notion that Complex I may undergo important conformational changes (Radermacher et al. 2006). Moreover, the observed destabilisation of Complex I in the absence of supercomplex may render the 51 kDa subunit containing the FMN more "loose" allowing it to interact with oxygen. The elevated ROS production observed in *P. anserina* respiring on AOX, where the major form of Complex I is a I_2III_2 supercomplex rather than the usual I_1III_2 supercomplex (Krause et al. 2006), is in line with this reasoning, because it is likely that the Complex I dimer may undergo a less tight interaction than the Complex I monomer with a Complex III dimer.

A direct study conducted in our laboratory on a mitochondrial fraction containing Complexes I and III and reconstituted with different amounts of phospholipids shows that at protein:phospholipids 1:30, when the supercomplex is not formed, the production of superoxide is much higher than at a 1:1 ratio, when most of the protein complexes are assembled as the supercomplex I_1III_2 (E. Maranzana, G. Barbero, G. Lenaz and ML Genova, unpublished).

5.5.4 Supercomplexes and the Mitochondrial Permeability Transition Pore (mtPTP)

The mtPTP is a voltage-dependent non-selective channel in the inner mitochondrial membrane (Bernardi and Forte 2007) that allows solutes of up to 1,500 Da to pass freely across the membrane; the pore opens under conditions of calcium overload and oxidative stress and its state is modulated by several factors; its opening is greatly enhanced by the oxidised state of pyridine nucleotides and of critical dithiols in two different redox sensitive sites, defined P and S, respectively (Fontaine and Bernardi 1999). The molecular structure of the mtPTP is still unknown (Ricchelli et al. 2011).

Belyaeva (2010) reported that mitochondrial inhibitors of Complex I (rotenone) and of Complex III (stigmatellin) exert a protective effect against the mitochondrial swelling evoked by heavy metals such as Ca^{2+}, Hg^{2+}, Zn^{2+} and Cd^{2+}, suggesting an involvement of the respiratory chain in metal-induced permeabilisation. On the basis of differential actions of the metal ions, Belyaeva speculates that the P site is located in Complex I and the S site in Complex III, and suggest that the I–III supercomplex may be a key component of the mtPTP. Although these speculations are highly suggestive, no direct evidence exists yet concerning the participation of the respiratory supercomplex to mtPTP structure.

Interestingly, He and Lemasters (2005) had found that dephosphorylation of the Rieske protein of Complex III by phosphatase treatment induces a shift in its isoelectric point similar to that caused by mtPTP inducers.

5.6 Supercomplexes in Pathology

5.6.1 A Working Hypothesis

An overwhelming body of evidence accumulated in the last decades has demonstrated that mitochondria have a central role in the aetiology and pathogenesis of most major chronic diseases and in ageing itself (cf. Lenaz and Genova 2010; Lenaz 1998; Lenaz et al. 2006; Wallace 2005). The involvement of mitochondria in disease, that has generated the term "Mitochondrial Medicine" (DiMauro et al. 2006), has been largely ascribed to their central role in the production of ROS and to the damaging effect of ROS on these organelles. In particular, damage to mtDNA would induce alterations of the respiratory polypeptides encoded by mtDNA, with consequent decrease of the electron transfer activity of the respiratory chain, leading to further production of ROS, and thus establishing a vicious circle of oxidative stress and energetic decline (Ozawa 1997). This fall of mitochondrial energetic capacity is considered to be the cause of ageing and age-related degenerative diseases (Wallace 2005), although the picture is further complicated (Calabrese et al. 2004) by the complex interplay and cross-talk with nuclear DNA and the rest of the cell (Ryan and Hoogenraad 2007).

Indeed, any lowering of respiratory activity would enhance generation of ROS as in State 4 respiration (controlled state in the absence of ADP) or when Complex I is inhibited, and paradoxically under conditions depending directly or indirectly (HIF1α) upon hypoxia.

ROS may contribute to further mitochondrial dysfunction by several mechanisms: (1) direct alterations of respiratory complexes, e.g. Complex I, particularly on FeS clusters; (2) peroxidation of mitochondrial phospholipids, in particular cardiolipin (Paradies et al. 2000, 2002), that are required for proper assembly and activity of mitochondrial complexes; (3) (further) mtDNA mutations affecting respiratory complexes and (4) stabilisation of HIF1α (Patten et al. 2010).

We now foresee a deep implication of supercomplex organisation as a missing link between oxidative stress and energy failure (Lenaz and Genova 2007). It is tempting to speculate that under conditions of oxidative stress, a dissociation of the Complex I–Complex III assembly occurs, with the loss of facilitated electron channelling and resumption of the less efficient pool behaviour of the free ubiquinone molecules.

As predicted by Lenaz and Genova (2007), dissociation of supercomplexes might have further deleterious consequences, such as disassembly of Complex I and III subunits and loss of intra-complex electron transfer and/or proton translocation; the consequent alteration of electron transfer may elicit further induction of ROS generation.

Following this line of thought, the different susceptibility of different types of cells and tissues to ROS damage may be a consequence of the extent and tightness of supercomplex organisation of their respiratory chains that depend on phospholipid content and composition of their mitochondrial membranes. These changes may have deep metabolic consequences, as depicted in the scheme in Fig. 5.3: an initial enhanced ROS generation due to different possible reasons and originating in different districts of the cell besides mitochondria (Lenaz and Strocchi 2009) would induce supercomplex disorganisation eventually leading to possible decrease of Complex I assembly; both the lack of efficient electron channelling and the loss of Complex I would decrease NAD-linked respiration and ATP synthesis.

5.6.2 *Supercomplex Changes in Pathological Conditions*

5.6.2.1 Ageing

The "Mitochondrial Theory of Ageing" (Linnane et al. 1989) is based on the hypothesis that mitochondrial DNA somatic mutations, caused by accumulation of oxygen radicals damage, induce alterations of the OXPHOS machinery culminating in an energetic failure that is at the basis of cellular senescence. Moreover, a vicious circle (Ozawa 1997) can be established since the accumulated damage to the respiratory chain would enhance ROS generation. Many reports (reviewed in Lenaz et al. 2006) demonstrate that the rate of production of ROS from mitochondria increases with age in mammalian tissues and in fibroblasts during replicative cell senescence, considered to represent a plausible model of in vivo ageing (Hayflick 2003). Trifunovic et al. (2004) showed that expression of a proofreading deficient mtDNA polymerase in a homozygous knock-in mouse strain leads to increased levels of somatic mtDNA mutations causing progressive respiratory chain deficiency; the mice develop symptoms strikingly reminiscent of ageing. Even if these mice do not have an enhanced ROS production (Trifunovic 2006; Trifunovic and Larsson 2008), it is likely that this effect results from the severe extent of mutation in the mutator mice, while the *natural* way to induce mutations is ROS attack. There is overwhelming evidence that the bioenergetic function of mitochondria declines with ageing, especially in post-mitotic tissues (Lenaz et al. 2006). If mtDNA mutations/deletions are at the basis of human ageing, the existence of a biochemical threshold complementing mtDNA mutations is critical, as discussed above. The notion that the respiratory chain is mainly controlled at the level of Complex I suggests that the main alterations due to ageing must be found at the level of this enzyme (Barrientos and Moraes 1999; Lenaz et al. 2000). Flux control analysis in aerobic respiration in coupled liver mitochondria (Ventura et al. 2002) showed that Complex I has little control in young rats but very high control in the old animals, meaning that ageing induces a profound alteration of Complex I that is reflected on the entire OXPHOS.

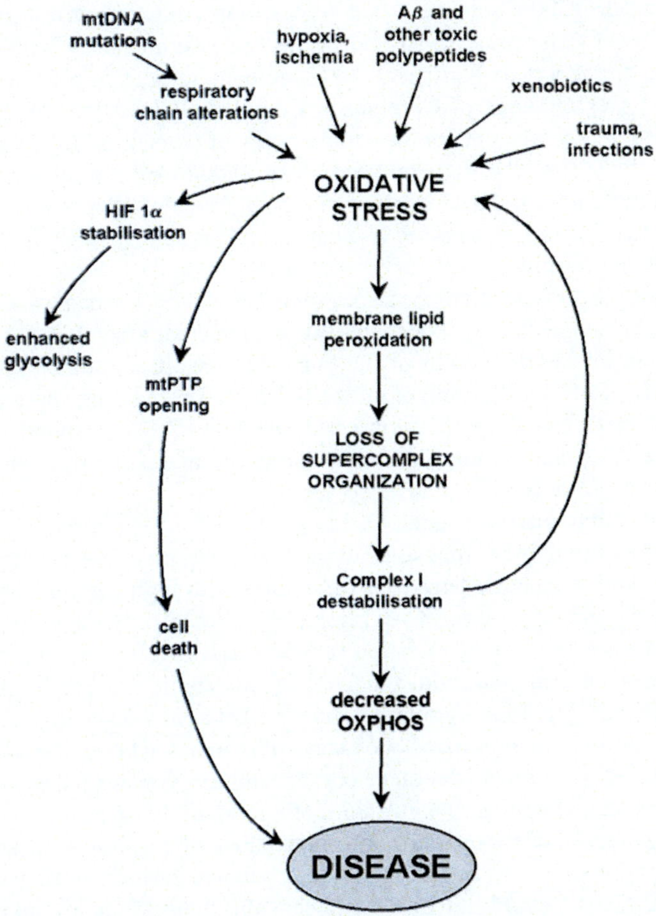

Fig. 5.3 Scheme showing how loss of supercomplex organisation may be involved in a *vicious circle* of oxidative stress and energy failure. Membrane phospholipid peroxidation and consequent loss of supercomplex organisation may occur due to oxidative stress induced by genetic changes or by exogenous factors. The scheme depicts some of the acknowledged causes of mitochondrial oxidative stress (1) mitochondrial DNA mutations occur in mitochondrial cytopathies, in ageing, in neurodegenerative diseases and in cancer and may lead to ROS generation; (2) hypoxia, ischemia (alone or followed by reperfusion) are major causes of oxidative stress; (3) the amyloidogenic peptide Aβ and other proteins increased in neurodegenerative diseases are recognised to enter mitochondria thereby inducing an oxidative stress; (4) xenobiotics may induce redox cycling through Complex I with the production of superoxide (Lenaz and Genova 2010); (5) trauma and infections may act by release of iron or production of cytokines. In all cases, lipid peroxidation destroys supercomplex organisation; the ensuing destabilisation of Complex I results in OXPHOS deficiency and further oxidative stress. A further consequence of oxidative stress may be the mitochondrial permeability transition and cell death. As a consequence of these changes, cells may be forced to rely on glycolysis for energy production, as happens in cancer. See text for explanations

Gomez et al. (2009) showed that supercomplexes composed of various stoichiometries of Complexes I, III and IV declined significantly with age in rat heart. In another study (Lombardi et al. 2009) mitochondria from old rats contained significantly lower amounts of Complex I, Complex III and FoF1-ATP synthase compared with young rats, but contrary to the study of Gomez et al. (2009) the densitometric analysis revealed that supercomplexes displaying the lighter molecular mass were significantly reduced in older mitochondria while the major supercomplex bands were those representing heavier supercomplexes, likely suggesting a compensatory mechanism.

In the mitochondria of rat cortex also, Frenzel et al. (2010) quantified profound age-associated changes in the proportion of supramolecular assemblies of the respiratory chain complexes as well as of the FoF1 ATP synthase; notably, the overall decline with age (40%) in the sum of all the Complex I containing supercomplexes is caused to a large extent by the pronounced decline (58%) of abundance of the supercomplex I_1III_2, supporting the notion and previous results that larger supercomplexes are more stable.

The quoted study by Kitazoe et al. (2011) (cf. Sect. 5.4.3) provides evidence that increase of serine/threonine composition of mitochondrial membrane proteins is crucial for the evolution of large terrestrial vertebrates with high aerobic capacity. An Arrhenius-type equation gave positive correlation between serine/threonine composition and maximum lifespan in terrestrial vertebrates (with a few exceptions relating to the lifestyle of small animals with a high resting metabolic rate), and negative correlation in secondary marine vertebrates such as cetaceans and alligators (which returned from land to water, utilising buoyancy with increased body-size). In particular, marked increases in serine/threonine composition in primates (especially hominoids) were associated with very high values of maximum lifespan.

Although mtDNA deletions and concomitant loss of respiratory activity have been convincingly shown to occur in an age-dependent fashion in the rat kidney, O'Toole et al. (2010) could not find significant differences when comparing the protein abundance of individual respiratory complexes and the supercomplex profiles between young and old mitochondria whereas they suggested that the decrease in endogenous cytochrome *c* may have a contributory role in the age-related decline of OXPHOS capacity.

5.6.2.2 Cardiovascular Disease

The role of loss of supramolecular organisation of the respiratory chain in heart failure has been strongly emphasised (García-Palmer 2008; Rosca and Hoppel 2010).

A study (Rosca et al. 2008) on canine cardiac mitochondria in heart failure induced experimentally by micro-embolisation showed that respiration with NAD-linked substrates, in State 3 or after treatment with uncoupler, was severely affected although the activity of individual Complexes I, III and IV was normal; BN-PAGE showed a severe reduction of supramolecular organisation with particular decrease of the major I-III-IV supercomplex. Clearly the OXPHOS defect was to be ascribed to the supramolecular assembly rather than to the individual components of the respiratory chain; the reason for such diminished assembly was not discussed. Although it was tempting to speculate that enhanced ROS production due to ischemia and reperfusion in the micro-embolised vessels modifies the membrane environment as a consequence of lipid peroxidation, thus disrupting supercomplex assembly, in a subsequent study Rosca et al. (2011) demonstrated that the decrease of supramolecular organisation in heart failure is not due to changes in phospholipid composition since the contents of the main phospholipid species, including cardiolipin, as well as the molecular species of cardiolipin were unchanged. Oxidised cardiolipin molecular species were not observed. On the other hand, in the canine heart mitochondria isolated from heart failure, single enzyme units of Complex IV not incorporated into respirasomes exhibited increased threonine phosphorylation. Since heart failure is associated with increased adrenergic drive to cardiomyocytes, this increased protein phosphorylation might be explained by the involvement of cAMP-activated protein kinase. Indeed, in saponin-permeabilised cardiac fibres, pre-incubation with cAMP decreases oxidative phosphorylation due to a defect localised at Complex IV. The authors propose that phosphorylation of specific Complex IV subunits decreases oxidative phosphorylation either by limiting the incorporation of Complex IV in supercomplexes or by decreasing supercomplex stability (cf. Sect. 5.4.3).

Barth syndrome (BTHS) is a X-linked cardio-skeletal myopathy caused by mutations in the gene Taz1 whose product, tafazzin (Taz1p), is an acyl transferase involved in the synthesis of mature cardiolipin (Neuwald 1997). The mitochondria of BTHS patients exhibit variable respiratory defects (Barth et al. 1996) and abnormal cristae ultrastructure. Interestingly, it was found (Claypool et al. 2008b) that tafazzin physically assembles in several protein complexes of distinct size and composition which include ATP synthase and the adenine nucleotide carrier of the inner mitochondrial membrane; the reduced abundance of such protein clusters, due to instability of mutated tafazzin, induces altered cristae morphology. It was also found (McKenzie et al. 2006) that cardiolipin deficiency in lymphoblasts from patients with Barth syndrome results in the destabilisation of the respiratory supercomplexes by weakening the interactions between Complex I and Complex III.

5.6.2.3 Neurodegeneration

Recent years have seen considerable progress in the allocation of mitochondrial changes in the pathogenesis of Alzheimer disease (AD). There is now little doubt that dementia in AD is associated with the production of the amyloidogenic peptide Aβ. Such effect is likely to be mediated by mitochondrial dysfunction. The toxic form of Aβ has been found not to be the aggregated amyloid plaques but oligomers that are able to enter mitochondria (Reddy et al. 2010; Chen and Yan 2010) where they induce ROS formation and inhibition of oxidative phosphorylation. The interaction of monomeric forms of Aβ with lipid membranes was analysed by neutron diffraction on stacked lipid multilayers (Dante et al. 2002). The use of a selectively deuterated amino acid (i.e. leucine-34, the penultimate amino acid in the C-terminal region) has allowed to unambiguously determine the position of two populations of the peptide, one in the aqueous vicinity of the membrane surface and the second inside the hydrophobic core of the lipid membrane. The deep penetration of the neurotoxic fragment of Aβ(25–35) in the lipid bilayer induces a dramatic perturbation of the membrane structure. Finally, the influence of Aβ(1–42), the most abundant Aβ form in senile plaques, on unilamellar vesicles of phospholipids was investigated by small-angle neutron scattering (Dante et al. 2008). As a result, an increase of the vesicle radii, indicating vesicle fusion was obtained. This effect was particularly enhanced at pH 7.0 and at a high peptide/lipid ratio. At the same time, a thinning of the lipid bilayer occurred. The fusogenic activity of the peptide may have very important consequences and contribute to cytotoxicity by destabilising the cell and mitochondrial membranes (Seelert et al. 2009). The perturbation of the bilayer structure suggests a strong interaction and/or insertion of the longer peptide into the membrane. These results may explain at the molecular level how Aβ may interact with mitochondrial proteins, disrupt the electron transport chain, increase ROS production, cause mitochondrial damage and prevent neurones from functioning normally. It is tempting to suggest that destabilisation of respiratory supercomplexes may be an early step in this course of events.

In a trisomic mouse model (Ts65Dn) of Down syndrome, also characterised by early onset Alzheimer dementia and amyloid deposition (Hattori et al. 2000), we have found by 2D BN/SDS-PAGE disruption of the supramolecular association of Complex III with Complex IV, together with a decrease of total Complex IV (bound plus free), offering a possible explanation of the observed decrease of NADH oxidation accompanied by lower control by Complex I (M. Faccioli, M.L. Genova, G. Lenaz, unpublished observations).

In a mouse model of Cockayne syndrome characterised by neurological and developmental impairment and premature ageing, the mtDNA accumulates oxidative damage. The concomitant bioenergetic alteration parallels a simpler organisation of supercomplexes consisting of Complexes I, III and IV in addition to partially disassembled Complex V in the inner mitochondrial membrane (Osenbroch et al. 2009).

Arthur et al. (2009) investigated mitochondria from post-mortem brains of patients with Parkinson disease and found a generalised decrease of protein levels of all four respiratory complexes and ATP synthase in comparison with aged controls, together with higher levels of 8-oxy-guanosine, but no significant depletion of mtDNA. The results were interpreted in terms of respirasome deficiency, although the decrease of Complex II may not be in line with this interpretation (Arthur et al. 2009).

5.6.2.4 Cancer

In rapidly growing tumours, the mean partial pressure of oxygen declines as distance from the nearest blood vessel increases. Cancer cells exposed to hypoxia are characterised in general by a decrease of mitochondrial respiration and oxidative phosphorylation, together with a strong enhancement of glycolysis (the so-called

Warburg effect, first reported by Otto Warburg (1956) and then confirmed in many reports). Decreased OXPHOS becomes operative during hypoxia because of the stabilisation of HIF1α (Semenza 2003). This factor at normal oxygen tension undergoes hydroxylation of two key proline residues by HIF-prolyl hydroxylases, which allow the recognition of the factor by pVHL, the product of the von Hippel-Lindau gene that addresses the protein to ubiquitin-mediated degradation. During hypoxia, the factor becomes stabilised by inactivation of prolyl hydroxylase, and binds to hypoxia-responsive elements in the DNA, stimulating a large array of genes (Semenza 2007; Kaluz et al. 2008). Besides glycolytic enzymes, HIF1α enhances the expression of PDK (pyruvate dehydrogenase kinase) thus inhibiting pyruvate dehydrogenase (PDH) complex and decreasing the input of reducing equivalents to the respiratory chain (Kim et al. 2006). The decreased input of carbon into the Krebs cycle must necessarily be limited; otherwise the lack of citrate would impair the synthesis of lipids and cholesterol that are required for the formation of new cellular membranes during tumour growth. This is the reason why glutamine becomes a preferred substrate for replenishing the Krebs cycle for both energetic and biosynthetic purposes.

It is of interest that HIF1α is also stabilised by several non-hypoxic stimuli, e.g. by succinate (that inhibits prolyl hydroxylase) (King et al. 2006), but also by pyruvate and oxaloacetate (Lu et al. 2005; McFate et al. 2008) and, interestingly, by ROS (Patten et al. 2010), although the activating effect of ROS on HIF is still a controversial issue (Bell et al. 2008). ROS could act either directly or by stimulating PDK, thus leading to accumulation of pyruvate (Sun et al. 2009; Patten et al. 2010). Thus, derangement of the Krebs cycle may lead to changes in steady-state concentrations of important metabolites that are able to control the activity of HIF1α.

In some cases, the depression of respiratory activity is clearly the consequence of disruptive tumourigenic mtDNA mutations affecting the respiratory chain complexes and enhancing the production of ROS (Brandon et al. 2006); however, the pathological relevance of mtDNA mutations in cancer cells is controversial (Frezza and Gottlieb 2009). Nonetheless, a clear-cut correlation between the occurrence of pathogenic mtDNA mutations and mitochondrial energetic impairment is a well-demonstrated feature of oncocytic tumours (Bonora et al. 2006; Porcelli et al. 2010). Mitochondrial DNA mutations have been consistently found in cancerous cells (Wallace 2005); they have been found to be associated with enhanced ROS production, and ROS act both as mutagens and cellular mitogens (Klaunig and Kamendulis 2004); thus the involvement of mtDNA mutations in cancer may well be of pathogenic importance. It must be underlined that mtDNA mutations leading to increased ROS production are those allowing at least a partial assembly of the respiratory complexes. On the contrary, for instance, oncocytic tumours display homoplasmic mutations disassembling Complex I, which may not be ROS-generating mutations, if one considers that the main ROS production site might be lacking as a whole (Porcelli et al. 2010; Koopman et al. 2007; Moran et al. 2010; Tuppen et al. 2010). The mutant load ought to be considered when analysing functional effects of mtDNA mutations. A cell line carrying a heteroplasmic ND5 mtDNA mutation showed significantly enhanced tumour growth, while cells with homoplasmic form of the same mutation inhibited tumour formation. These results indicate that the mtDNA mutations might play an important role in the early stage of cancer development, possibly through alteration of ROS generation and apoptosis (Park et al. 2009).

Decreased mitochondrial activity is considered to be tumourigenic, mainly due to the enhanced ROS production and also because H_2O_2 exported to the nucleus enhances the transcription of selected genes that favour tumour progression (Wallace 2005). Indeed, ROS act both as mutagens and cellular mitogens (Klaunig and Kamendulis 2004).

We postulate that downregulation of Complex I activity may be a key strategy of tumour cells to depress mitochondrial activity and enhance ROS production. Among the causes of Complex I deactivation, it is tempting to suggest an early involvement of supercomplex disorganisation, with a vicious circle of ROS generation and supercomplex disassembly (Lenaz and Genova 2010) (see also Sect. 5.5.3).

A cell line of a malignant thyroid oncocytoma, characterised by abnormal mitochondrial proliferation, contains a mutation of mtDNA preventing expression of the subunit ND1 of Complex I (Bonora et al. 2006). These cells exhibit a dramatic decline of ATP synthesis supported by NAD-dependent substrates, while in the mitochondria isolated from these cells the Complex I activity is strongly depressed (Bonora et al. 2006). We have also found by BN-PAGE of mitochondrial proteins from the oncocytic cell line, a complete absence of high molecular weight aggregates containing either Complex I or Complex IV, that are instead present in the control cell line (Lenaz, unpublished). In another study (Baracca et al. 2010) we report that down-regulation of the respiratory chain in Ras-transformed fibroblasts is operated through strong decrease of Complex I activity and content, probably due to lack of correct assembly of the subunits as a consequence of altered supercomplex organisation, as demonstrated by the loss of the highest molecular weight $I_1III_2IV_{1\text{-}2}$ supercomplex. Significantly, ROS generation was strongly enhanced in the Ras cells (Lenaz et al. 2010).

5.7 Conclusions

In the recent years, our view of the mitochondrial oxidative phosphorylation system has somewhat changed in a more integrated approach to mitochondrial structure and function within the cell. This new understanding has opened unexpected perspectives and novel research avenues particularly in relation to the established role that mitochondria exert in pathological changes. The field of mitochondrial medicine is now growing exponentially, shedding light on the pathogenesis of diseases in almost every branch of pathology. In this scenario, the supramolecular structure of the respiratory chain and ATP synthase has acquired a strong position, in view of the multiple role that supercomplex association exerts in mitochondria. It is now ascertained that mitochondrial dysfunction and oxidative stress are at the basis of many diseases: in this chapter, we have provided evidence pertaining to the working hypothesis that an oxidative stress acts primarily by disassembling supercomplex associations in the respiratory chain, thereby establishing a vicious circle of oxidative stress and energy failure, ultimately leading to cell damage and disease. The extent and nature of the damage may lead the cell either to death with tissue degeneration or to deregulation of growth and neoplastic transformation. It is certainly true that in the near future, we will experience new exciting developments of mitochondrial medicine brought by our growing understanding of the molecular mechanisms that can affect mitochondrial OXPHOS metabolism and it is easy to predict that this will lead to important therapeutic applications.

References

Acin-Perez R, Bayona-Bafaluy MP, Fernandez-Silva P, Moreno-Loshuertos R, Perez-Martos A, Bruno C, Moraes CT, Enriquez JA (2004) Respiratory complex III is required to maintain complex I in mammalian mitochondria. Mol Cell 13:805–815

Acín-Pérez R, Fernández-Silva P, Peleato ML, Pérez-Martos A, Enriquez JA (2008) Respiratory active mitochondrial supercomplexes. Mol Cell 32(4):529–539

Alberts B (1998) The cell as a collection of protein machines: preparing the next generation of molecular biologists. Cell 92:291–294

Althoff T, Mills DJ, Popot JL, Kühlbrandt W (2011) Arrangement of electron transport chain components in bovine mitochondrial supercomplex I(1)III(2)IV(1). EMBO J 30(22):4652–4664. doi:10.1038/emboj.2011.324

Arthur CR, Morton SL, Dunham LD, Keeney PM, Bennett JP Jr (2009) Parkinson's disease brain mitochondria have impaired respirasome assembly, age-related increases in distribution of oxidative damage to mtDNA and no differences in heteroplasmic mtDNA mutation abundance. Mol Neurodegener 4:37

Baracca A, Chiaradonna F, Sgarbi G, Solaini G, Alberghina L, Lenaz G (2010) Mitochondrial complex I decrease is responsible for bioenergetic dysfunction in K-ras transformed cells. Biochim Biophys Acta 1797:314–323

Barrientos A, Moraes CT (1999) Titrating the effects of mitochondrial complex I impairment in the cell physiology. J Biol Chem 274:16188–16197

Barth PG, Van den Bogert C, Bolhuis PA, Scholte HR, van Gennip AH, Schutgens RB et al (1996) X-linked cardioskeletal myopathy and neutropenia (Barth syndrome): respiratory-chain abnormalities in cultured fibroblasts. J Inherit Metab Dis 19:157–160

Bell EL, Klimova T, Chandel NS (2008) Targeting the mitochondria for cancer therapy: regulation of hypoxia-inducible factor by mitochondria. Antioxid Redox Signal 10(3):635–640

Bellomo F, Piccoli C, Cocco T, Scacco S, Papa F, Gaballo A et al (2006) Regulation by the cAMP cascade of oxygen free radical balance in mammalian cells. Antioxid Redox Signal 8:495–502

Belyaeva EA (2010) Mitochondrial respiratory chain inhibitors modulate the metal-induced inner mitochondrial membrane permeabilization. Acta Biochim Pol 57(4):435–441

Bernardi P, Forte M (2007) The mitochondrial permeability transition pore. Novartis Found Symp 287:157–164, discussion 164–169

Bianchi C, Fato R, Genova ML, Parenti Castelli G, Lenaz G (2003) Structural and functional organization of complex I in the mitochondrial respiratory chain. Biofactors 18:3–9

Bianchi C, Genova ML, Parenti Castelli G, Lenaz G (2004) The mitochondrial respiratory chain is partially organized in a supercomplex assembly: kinetic evidence using flux control analysis. J Biol Chem 279:36562–36569

Bonora E, Porcelli AM, Gasparre G, Biondi A, Ghelli A, Carelli V, Baracca A, Tallini G, Martinuzzi A, Lenaz G, Rugolo M, Romeo G (2006) Defective oxidative phosphorylation in thyroid oncocytic carcinoma is associated with pathogenic mitochondrial DNA mutations affecting complexes I and III. Cancer Res 66:6087–6096

Boumans H, Grivell LA, Berden JA (1998) The respiratory chain in yeast behaves as a single functional unit. J Biol Chem 273:4872–4877
Brandner K, Mick DU, Frazier AE, Taylor RD, Meisinger C, Rehling P (2005) Taz1, an outer mitochondrial membrane protein, affects stability and assembly of inner membrane protein complexes: implications for Barth Syndrome. Mol Biol Cell 16:5202–5214
Brandon M, Baldi P, Wallace DC (2006) Mitochondrial mutations in cancer. Oncogene 25(34):4647–4662
Braun HP, Sunderhaus S, Boekema EJ, Kouril R (2009) Purification of the cytochrome C reductase/cytochrome C oxidase super complex of yeast mitochondria. Methods Enzymol 456:183–190
Brdiczka DG, Zorov DB, Sheu SS (2006) Mitochondrial contact sites: their role in energy metabolism and apoptosis. Biochim Biophys Acta 1762:148–163
Brys K, Castelein N, Matthijssens F, Vanfleteren JR, Braeckman BP (2010) Disruption of insulin signalling preserves bioenergetic competence of mitochondria in ageing *Caenorhabditis elegans*. BMC Biol 8:91
Bultema JB, Braun HP, Boekema EJ, Kouril R (2009) Megacomplex organization of the oxidative phosphorylation system by structural analysis of respiratory supercomplexes from potato. Biochim Biophys Acta 1787(1):60–67
Calabrese V, Scapagnini G, Ravagna A, Colombrita C, Spadaro F, Butterfield DA, Giuffrida Stella AM (2004) Increased expression of heat shock proteins in rat brain during aging: relationship with mitochondrial function and glutathione redox state. Mech Ageing Dev 125:325–335
Castellani M, Covian R, Kleinschroth T, Anderka O, Ludwig B, Trumpower BL (2010) Direct demonstration of half-of-the-sites reactivity in the dimeric cytochrome bc1 complex: enzyme with one inactive monomer is fully active but unable to activate the second ubiquinol oxidation site in response to ligand binding at the ubiquinone reduction site. J Biol Chem 285(1):502–510
Chance B, Williams GR (1955) A method for the localization of sites for oxidative phosphorylation. Nature 176:250–254
Chen JX, Yan SS (2010) Role of mitochondrial amyloid-beta in Alzheimer's disease. J Alzheimers Dis 20(Suppl 2):S569–S578
Clarke SD, Salati IMK (1985) Fatty acid-mediated disaggregatiion of acetyl CoA carboxylase in isolated liver cells. Fed Proc 44:2458–2462
Claypool SM, Oktay Y, Boontheung P, Loo JA, Koehler CM (2008a) Cardiolipin defines the interactome of the major ADP/ATP carrier protein of the mitochondrial inner membrane. J Cell Biol 182(5):937–950
Claypool SM, Boontheung P, McCaffery JM, Loo JA, Koehler CM (2008b) The cardiolipin transacylase, tafazzin, associates with two distinct respiratory components providing insight into Barth syndrome. Mol Biol Cell 19(12):5143–5155
Colindres M, Fournier C, Ritter S, Zahnreich S, Decker H, Dencher N, Frenzel M (2007) Increase of oxidative stress in normal human fibroblasts after irradiation. GSI Sci Rep 356
Couoh-Cardel SJ, Uribe-Carvajal S, Wilkens S, García-Trejo JJ (2010) Structure of dimeric F1F0-ATP synthase. J Biol Chem 285(47):36447–36455
D'Aurelio M, Gajewski CD, Lenaz G, Manfredi G (2006) Respiratory chain supercomplexes set the threshold for respiration defects in human mtDNA mutant cybrids. Hum Mol Genet 15:2157–2169
Dalmonte ME, Forte E, Genova ML, Giuffrè A, Sarti P, Lenaz G (2009) Control of respiration by cytochrome c oxidase in intact cells: role of the membrane potential. J Biol Chem 284(47): 32331–32335
Damjanovich S, Gaspar R Jr, Pieri C (1997) Dynamic receptor superstructures at the plasma membrane. Q Rev Biophys 30:67–106
Dani D, Shimokawa I, Komatsu T, Higami Y, Warnken U, Schokraie E, Schnölzer M, Krause F, Sugawa MD, Dencher NA (2009) Modulation of oxidative phosphorylation machinery signifies a prime mode of anti-ageing mechanism of calorie restriction in male rat liver mitochondria. Biogerontology 11(3):321–334
Dante S, Hauss T, Dencher NA (2002) Beta-amyloid 25 to 35 is intercalated in anionic and zwitterionic lipid membranes to different extents. Biophys J 83(5):2610–2616

Dante S, Hauss T, Brandt A, Dencher NA (2008) Membrane fusogenic activity of the Alzheimer's peptide A beta(1-42) demonstrated by small-angle neutron scattering. J Mol Biol 376(2): 393–404

Davies KM, Strauss M, Daum B, Kief JH, Osiewacz HD, Rycovska A, Zickermann V, Kühlbrandt W (2011) Macromolecular organization of ATP synthase and complex I in whole mitochondria. Proc Natl Acad Sci USA 108(34):14121–14126

De Rasmo D, Panelli D, Sardanelli AM, Papa S (2008) cAMP-dependent protein kinase regulates the mitochondrial import of the nuclear encoded NDUFS4 subunit of complex I. Cell Signal 20:989–997

Dencher NA, Frenzel M, Reifschneider NH, Sugawa M, Krause F (2007) Proteome alterations in rat mitochondria caused by aging. Ann N Y Acad Sci 1100:291–298

Diaz F, Fukui H, Garcia S, Moraes CT (2006) Cytochrome c oxidase is required for the assembly/ stability of respiratory complex I in mouse fibroblasts. Mol Cell Biol 26:4872–4881

Dienhart MK, Stuart RA (2008) The yeast Aac2 protein exists in physical association with the cytochrome bc1-COX supercomplex and the TIM23 machinery. Mol Biol Cell 19(9):3934–3943

DiMauro S, Hirano M, Schon EA (eds) (2006) Mitochondrial medicine. Informa Healthcare, London

Dudkina NV, Eubel H, Keegstra W, Boekema EJ, Braun HP (2005) Structure of a mitochondrial supercomplex formed by respiratory-chain complexes I and III. Proc Natl Acad Sci USA 102(9):3225–3229

Dudkina NV, Sunderhaus S, Braun HP, Boekema EJ (2006) Characterization of dimeric ATP synthase and cristae membrane ultrastructure from Saccharomyces and Polytomella mitochondria. FEBS Lett 580(14):3427–3432

Dudkina NV, Kudryashev M, Stahlberg H, Boekema EJ (2011) Interaction of complexes I, III, and IV within the bovine respirasome by single particle cryoelectron tomography. Proc Natl Acad Sci USA 108(37):15196–15200

Eubel H, Heinemeyer J, Sunderhaus S, Braun HP (2004) Respiratory chain supercomplexes in plant mitochondria. Plant Physiol Biochem 42:937–942

Fleischer S, Brierley G, Klouwen H, Slautterback DB (1962) Studies of the electron transfer system. 47. The role of phospholipids in electron transfer. J Biol Chem 237:3264–3272

Fontaine E, Bernardi P (1999) Progress on the mitochondrial permeability transition pore: regulation by complex I and ubiquinone analogs. J Bioenerg Biomembr 31(4):335–345

Frenzel M, Rommelspacherr H, Sugawa MD, Dencher NA (2010) Ageing alters the supramolecular architecture of OxPhos complexes in rat brain cortex. Exp Gerontol 45:563–572

Frezza C, Gottlieb E (2009) Mitochondria in cancer: not just innocent bystanders. Semin Cancer Biol 19(1):4–11

Fry M, Green DE (1980) Cardiolipin requirement by cytochrome oxidase and the catalytic role of phospholipid. Biochem Biophys Res Commun 93:1238–1246

Fry M, Green DE (1981) Cardiolipin requirement for electron transfer in complex I and III of the mitochondrial respiratory chain. J Biol Chem 256:1874–1880

García-Palmer FJ (2008) Lack of functional assembly in mitochondrial supercomplexes: a new insight into impaired mitochondrial function? Cardiovasc Res 80:3–4

Gavin AC, Superti-Furga G (2003) Protein complexes and proteome organization from yeast to man. Curr Opin Chem Biol 7:21–27

Genova ML, Baracca A, Biondi A, Casalena G, Faccioli M, Falasca AI, Formiggini G, Sgarbi G, Solaini G, Lenaz G (2008) Is supercomplex organization of the respiratory chain required for optimal electron transfer activity? Biochim Biophys Acta 1777:740–746

Gil T, Sabra MC, Ipsen JH, Mouritsen OG (1997) Wetting and capillary condensation as means of protein organization in membranes. Biophys J 73:1728–1741

Gil T, Ipsen JH, Mouritsen OG, Sabra MC, Sperotto MM, Zuckermann MJ (1998) Theoretical analysis of protein organization in lipid membranes. Biochim Biophys Acta 1376:245–266

Gómez LA, Monette JS, Chavez JD, Maier CS, Hagen TM (2009) Supercomplexes of the mitochondrial electron transport chain decline in the aging rat heart. Arch Biochem Biophys 490:30–35

Grad LI, Lamire D (2006) Riboflavin enhances the assembly of mitochondrial cytochrome c oxidase in *C. elegans* NADH ubiquinone reductase mutants. Biochim Biophys Acta 1757:115–122

Grad LI, Lemire D (2004) Mitochondrial complex I mutations in *Caenorhabditis elegans* produce cytochrome c oxidase deficiency, oxidative stress and vitamin-response lactic acidosis. Hum Mol Genet 13:303–314

Hackenbrock CR, Chazotte B, Gupte SS (1986) The random collision model and a critical assessment of diffusion and collision in mitochondrial electron transport. J Bioenerg Biomembr 18:331–368

Hashimoto T, Hussien R, Cho H-S, Kaufer D, Brooks GA (2008) Evidence for the mitochondrial lactate oxidation complex in rat neurons: demonstration of an essential component of brain lactate shuttles. PLoS One 3:e2915

Hatefi Y, Haavik AG, Griffiths DE (1962) Studies on the electron transfer system. XL. Preparation and properties of mitochondrial DPNH-coenzyme Q reductase. J Biol Chem 237:1676–1680

Hattori M, Fujiyama A, Taylor TD et al (2000) Chromosome 21 mapping and sequencing consortium. The DNA sequence of human chromosome 21. Nature 405:311–319

Hayflick L (2003) Living forever and dying in the attempt. Exp Gerontol 38:1231–1241

He L, Lemasters JJ (2005) Dephosphorylation of the Rieske iron-sulfur protein after induction of the mitochondrial permeability transition. Biochem Biophys Res Commun 334(3):829–837

Heinemeyer J, Braun HP, Boekema EJ, Kouril R (2007) A structural model of the cytochrome C reductase/oxidase supercomplex from yeast mitochondria. J Biol Chem 282(16):12240–12248

Helling S, Vogt S, Rhiel A, Ramzan R, Wen L, Marcus K et al (2008) Phosphorylation and kinetics of mammalian cytochrome c oxidase. Mol Cell Proteomics 7:1714–1724

Helms V (2002) Attraction within the membrane. EMBO Rep 3:1133–1138

Heron C, Ragan CI, Trumpower BL (1978) The interaction between mitochondrial NADH-ubiquinone oxidoreductase and ubiquinol-cytochrome c oxidoreductase – restoration of ubiquinone-pool behaviour. Biochem J 174:791–800

Hesketh TR, Smith GA, Houslay MD, McGill KA, Birdsall NJ, Metcalfe JC et al (1976) Annular lipids determine ATPase activity of a calcium transport protein complexes with dipalmitoyllecithin. Biochemistry 15:4145–4151

Hildebrandt TM (2011) Modulation of sulfide oxidation and toxicity in rat mitochondria by dehydroascorbic acid. Biochim Biophys Acta 1807(9):1206–1213

Hochman J, Ferguson-Miller S, Schindler M (1985) Mobility in the mitochondrial electron transport chain. Biochemistry 24:2509–2516

Horbinski C, Chu CT (2005) Kinase signaling cascades in the mitochondrion: a matter of life and death. Free Radic Biol Med 38:2–11

Houtkooper RH, Vaz FM (2008) Cardiolipin, the heart of mitochondrial metabolism. Cell Mol Life Sci 65:2493–2506

Iqbal K, Grundke-Iqbal I (1996) Molecular mechanism of Alzheimer's neurofibrillary degeneration and therapeutic intervention. Ann N Y Acad Sci 777:132–138

Jacobson K, Sheets ED, Simson R (1995) Revisiting the fluid mosaic model of membranes. Science 268:1441–1442

Jost P, Griffith OH, Capaldi RA, Vanderkooi G (1973) Evidence for boundary lipids in membranes. Proc Natl Acad Sci USA 70:480–484

Kaluz S, Kaluzova M, Stanbridge EJ (2008) Rational design of minimal hypoxia-inducible enhancers. Biochem Biophys Res Commun 370(4):613–618

Kang SY, Gutowsky HS, Hsung JC, Jacobs R, King TE, Rice D et al (1979) Nuclear magnetic resonance investigation of the cytochrome oxidase—phospholipid interaction: a new model for boundary lipid. Biochemistry 18:3257–3267

Kholodenko NB, Westerhoff HV (1993) Metabolic channelling and control of the flux. FEBS Lett 320:71–74

Kim JW, Tchernyshyov I, Semenza GL, Dang CV (2006) HIF-1-mediated expression of pyruvate dehydrogenase kinase: a metabolic switch required for cellular adaptation to hypoxia. Cell Metab 3(3):177–185

King A, Selak MA, Gottlieb E (2006) Succinate dehydrogenase and fumarate hydratase: linking mitochondrial dysfunction and cancer. Oncogene 25(34):4675–4682

Kitazoe Y, Kishino H, Hasegawa M, Matsui A, Lane N, Tanaka M (2011) Stability of mitochondrial membrane proteins in terrestrial vertebrates predicts aerobic capacity and longevity. Genome Biol Evol 3:1233–1244

Klaunig JE, Kamendulis LM (2004) The role of oxidative stress in carcinogenesis. Annu Rev Pharmacol Toxicol 44:239–267

Koopman WJ, Verkaart S, Visch HJ, van Emst-de Vries S, Nijtmans LG, Smeitink JA, Willems PH (2007) Human NADH:ubiquinone oxidoreductase deficiency: radical changes in mitochondrial morphology? Am J Physiol Cell Physiol 293(1):C22–C29

Krause F (2006) Detection and analysis of protein-protein interactions in organellar and prokaryotic proteomes by native gel electrophoresis: (membrane) protein complexes and supercomplexes. Electrophoresis 27:2759–2781

Krause F, Seelert H (2008) Detection and analysis of protein-protein interactions of organellar and prokaryotic proteomes by blue native and colorless native gel electrophoresis. Curr Protoc Protein Sci Chapter 14:Unit 14.11

Krause F, Scheckhuber CQ, Werner A, Rexroth S, Reifschneider NH, Dencher NA, Osiewacz HD (2004a) Supramolecular organization of cytochrome c oxidase- and alternative oxidase-dependent respiratory chains in the filamentous fungus *Podospora anserina*. J Biol Chem 279:26453–26461

Krause F, Reifschneider NH, Vocke D, Seelert H, Rexroth S, Dencher NA (2004b) "Respirasome"-like supercomplexes in green leaf mitochondria of spinach. J Biol Chem 279:48369–48375

Krause F, Reifschneider NH, Goto S, Dencher NA (2005) Active oligomeric ATP synthases in mammalian mitochondria. Biochem Biophys Res Commun 329(2):583–590

Krause F, Scheckhuber CQ, Werner A, Rexroth S, Reifschneider NH, Dencher NA, Osiewacz HD (2006) OXPHOS Supercomplexes: respiration and life-span control in the aging model *Podospora anserina*. Ann N Y Acad Sci 1067:106–115

Kröger A, Klingenberg M (1973a) The kinetics of the redox reactions of ubiquinone related to the electron-transport activity in the respiratory chain. Eur J Biochem 34:358–368

Kröger A, Klingenberg M (1973b) Further evidence of the pool function of ubiquinone as derived from the inhibition of the electron transport by antimycin. Eur J Biochem 39:313–323

Lange C, Nett JH, Trumpower BL, Hunte C (2001) Specific roles of protein-phospholipid interactions in the yeast cytochrome bc1 complex structure. EMBO J 20:6591–6600

Le Pécheur M, Morrow G, Kim H.-J, Schäfer E, Dencher N, Tanguay RM (2009) Characterization of OXPHOS complexes in long-lived flies overexpressing Hsp22. Mitochondria in ageing and age-related disease, MiMage final meeting (and LINK-AGE Topic Research) Group Meeting, abstract 16, p 35

Lee AG (2004) How lipids affect the activities of integral membrane proteins. Biochim Biophys Acta 1666:62–87

Lee I, Salomon AR, Ficarro S, Mathes I, Lottspeich F, Grossman LI et al (2005) cAMP-dependent tyrosine phosphorylation of subunit I inhibits cytochrome c oxidase activity. J Biol Chem 280:6094–6100

Lenaz G (1998) Role of mitochondria in oxidative stress and ageing. Biochim Biophys Acta 1366:53–67

Lenaz G, Genova ML (2007) Kinetics of integrated electron transfer in the mitochondrial respiratory chain: random collisions vs. solid state electron channeling. Am J Physiol Cell Physiol 292:C1221–C1239

Lenaz G, Genova ML (2009) Mobility and function of coenzyme Q (ubiquinone) in the mitochondrial respiratory chain. Biochim Biophys Acta 1787(6):563–573

Lenaz G, Genova ML (2010) Structure and organization of mitochondrial respiratory complexes: a new understanding of an old subject. Antioxid Redox Signal 12:961–1008

Lenaz G, Strocchi P (2009) Reactive oxygen species in the induction of toxicity, Chapter 15. In: Ballantyne B, Marrs T, Syversen T (eds) General and applied toxicology. Wiley, Chichester

Lenaz G, Fato R, Di Bernardo S, Jarreta D, Costa A, Genova ML et al (1999) Localization and mobility of coenzyme Q in lipid bilayers and membranes. Biofactors 9:87–93

Lenaz G, D'Aurelio M, Merlo Pich M, Genova ML, Ventura B, Bovina C, Formaggini G, Parenti Castelli G (2000) Mitochondrial bioenergetics in aging. Biochim Biophys Acta 1459:397–404

Lenaz G, Baracca A, Fato R, Genova ML, Solaini G (2006) New insights into structure and function of mitochondria and their role in ageing and disease. Antioxid Redox Signal 8:417–437

Lenaz G, Baracca A, Barbero G, Bergamini C, Dalmonte ME, Del Sole M, Faccioli M, Falasca A, Fato R, Genova ML, Sgarbi G, Solaini G (2010) Mitochondrial respiratory chain super-complex I-III in physiology and pathology. Biochim Biophys Acta 1797(6–7):633–640

Leys D, Basran J, Talfournier F, Sutcliffe MJ, Scrutton NS (2003) Extensive conformational sampling in a ternary electron transfer complex. Nat Struct Biol 10:219–225

Linnane AW, Marzuki S, Ozawa T, Tanaka M (1989) Mitochondrial DNA mutations as an important contributor to ageing and degenerative diseases. Lancet 1:642–645

Lombardi A, Silvestri E, Cioffi F, Senese R, Lanni A, Goglia F, de Lange P, Moreno M (2009) Defining the transcriptomic and proteomic profiles of rat ageing skeletal muscle by the use of a cDNA array, 2D- and Blue native-PAGE approach. J Proteomics 72:708–721

Lu H, Dalgard CL, Mohyeldin A, McFate T, Tait AS, Verma A (2005) Reversible inactivation of HIF-1 prolyl hydroxylases allows cell metabolism to control basal HIF-1. J Biol Chem 280(51):41928–41939

Maas MF, Krause F, Dencher NA, Sainsard-Chanet A (2009) Respiratory complexes III and IV are not essential for the assembly/stability of complex I in fungi. J Mol Biol 387:259–269

Maj MC, Raha S, Myint T, Robinson BH (2004) Regulation of NADH/CoQ oxidoreductase: do phosphorylation events affect activity? Protein J 23:25–32

Mannella CA (2006) The relevance of mitochondrial membrane topology to mitochondrial function. Biochim Biophys Acta 1762:140–147

Marques I, Dencher NA, Videira A, Krause F (2007) Supramolecular organization of the respiratory chain in *Neurospora crassa* mitochondria. Eukaryot Cell 6(12):2391–2405

McFate T, Mohyeldin A, Lu H, Thakar J, Henriques J, Halim ND, Wu H, Schell MJ, Tsang TM, Teahan O, Zhou S, Califano JA, Jeoung NH, Harris RA, Verma A (2008) Pyruvate dehydrogenase complex activity controls metabolic and malignant phenotype in cancer cells. J Biol Chem 283(33):22700–22708

McKenzie M, Lazarou M, Thorburn DR, Ryan MT (2006) Mitochondrial respiratory chain supercomplexes are destabilized in Barth Syndrome patients. J Mol Biol 361(3):462–469

Megli FM, Sabatini K (2003) EPR studies of phospholipid bilayers after lipoperoxidation. 1. Inner molecular order and fluidity gradient. Chem Phys Lipids 125:161–172

Mick DU, Wagner K, van der Laan M, Frazier AE, Perschil I, Pawlas M, Meyer HE, Warscheid B, Rehling P (2007) Shy1 couples Cox1 translational regulation to cytochrome c oxidase assembly. EMBO J 26(20):4347–4358

Moran M, Rivera H, Sanchez-Arago M, Blazquez A, Merinero B, Ugalde C, Arenas J, Cuezva JM, Martin MA (2010) Mitochondrial bioenergetics and dynamics interplay in complex I-deficient fibroblasts. Biochim Biophys Acta 1802(5):443–453

Moser CC, Page CC, Dutton PL (2005) Tunneling in PSII. Photochem Photobiol Sci 4:933–939

Muster B, Kohl W, Wittig I, Strecker V, Joos F, Haase W, Bereiter-Hahn J, Busch K (2010) Respiratory chain complexes in dynamic mitochondria display a patchy distribution in life cells. PLoS One 5(7):e11910

Neuwald AF (1997) Barth syndrome may be due to an acyltransferase deficiency. Curr Biol 7:R465–R466

Nübel E, Wittig I, Kerscher S, Brandt U, Schägger H (2009) Two-dimensional native electrophoretic analysis of respiratory supercomplexes from *Yarrowia lipolytica*. Proteomics 9(9):2408–2418

O'Toole JF, Patel HV, Naples CJ, Fujioka H, Hopple CL (2010) Decreased cytochrome c mediates an age-related decline of oxidative phosphorylation in rat kidney mitochondria. Biochem J 427:105–112

Ohya S, Kuwata Y, Sakamoto K, Muraki K, Imaizumi Y (2005) Cardioprotective effects of estradiol include the activation of large-conductance Ca^{2+}-activated K^+ channels in cardiac mitochondria. Am J Physiol Heart Circ Physiol 289:H1635–H1642

Osenbroch PØ, Auk-Emblem P, Halsne R, Strand J, Forstrøm RJ, van der Pluijm I, Eide L (2009) Accumulation of mitochondrial DNA damage and bioenergetic dysfunction in CSB defective cells. FEBS J 276:2811–2821

Ovàdi J (1991) Physiological significance of metabolic channelling. J Theor Biol 152:135–141

Ovàdi J, Huang Y, Spivey HO (1994) Binding of malate dehydrogenase and NADH channelling to complex I. J Mol Recognit 7:265–272

Ozawa T (1997) Genetic and functional changes in mitochondria associated with aging. Physiol Rev 77:425–464

Ozawa T, Nishikimi M, Suzuki H, Tanaka M, Shimomura Y (1987) Structure and assembly of mitochondrial electron-transfer complexes. In: Ozawa T, Papa S (eds) Bioenergetics: structure and function of energy-transducing systems. Japan Science Society Press, Tokyo, pp 101–119

Page CC, Moser CC, Dutton PL (2003) Mechanism for electron transfer within and between proteins. Curr Opin Chem Biol 7:551–556

Panov A, Dikalov S, Shalbuyeva N, Hemendinger R, Greenamyre JT, Rosenfeld J (2007) Species- and tissue-specific relationships between mitochondrial permeability transition and generation of ROS in brain and liver mitochondria of rats and mice. Am J Physiol Cell Physiol 292:C708–C718

Papa S, De Rasmo D, Scacco S, Signorile A, Technikova-Dobrova Z, Palmisano G et al (2008) Mammalian complex I: a regulable and vulnerable pacemaker in mitochondrial respiratory function. Biochim Biophys Acta 1777:719–728

Paradies G, Petrosillo G, Pistolese M, Ruggiero FM (2000) The effect of reactive oxygen species generated from the mitochondrial electron transport chain on the cytochrome c oxidase activity and on the cardiolipin content in bovine heart submitochondrial particles. FEBS Lett 466:323–326

Paradies G, Petrosillo G, Pistolese M, Ruggiero FM (2002) Reactive oxygen species affect mitochondrial electron transport complex I activity through oxidative cardiolipin damage. Gene 286:135–141

Park JS, Sharma LK, Li H, Xiang R, Holstein D, Wu J, Lechleiter J, Naylor SL, Deng JJ, Lu J, Bai Y (2009) A heteroplasmic, not homoplasmic, mitochondrial DNA mutation promotes tumorigenesis via alteration in reactive oxygen species generation and apoptosis. Hum Mol Genet 18(9):1578–1589

Patten DA, Lafleur VN, Robitaille GA, Chan DA, Giaccia AJ, Richard DE (2010) Hypoxia-inducible factor-1 activation in nonhypoxic conditions: the essential role of mitochondrial-derived reactive oxygen species. Mol Biol Cell 21(18):3247–3257

Persichini T, Mazzone V, Polticelli F, Moreno S, Venturini G, Clementi E, Colasanti M (2005) Mitochondrial type I nitric oxide synthase physically interacts with cytochrome c oxidase. Neurosci Lett 384:254–259

Petrosillo G, Ruggiero FM, Di Venosa N, Paradies G (2003) Decreased complex III activity in mitochondria isolated from rat heart subjected to ischemia and reperfusion: role of reactive oxygen species and cardiolipin. FASEB J 17:714–716

Pfeiffer K, Gohil V, Stuart RA, Hunte C, Brandt U, Greenberg ML et al (2003) Cardiolipin stabilizes respiratory chain supercomplexes. J Biol Chem 278:52873–52880

Piccoli C, Scrima R, Boffoli D, Capitanio N (2006) Control by cytochrome c oxidase of the cellular oxidative phosphorylation system depends on the mitochondrial energy state. Biochem J 396:573–583

Pineau B, Mathieu C, Gérard-Hirne C, De Paepe R, Chétrit P (2005) Targeting the NAD7 subunit in mitochondria restores a functional complex I and a wild type phenotype in the *Nicotiana sylvestris* CMS II mutant lacking nad7. J Biol Chem 280(28):25994–26001

Porcelli AM, Ghelli A, Ceccarelli C, Lang M, Cenacchi G, Capristo M, Pennisi LF, Morra I, Ciccarelli E, Melcarne A, Bartoletti-Stella A, Salfi N, Tallini G, Martinuzzi A, Carelli V, Attimonelli M, Rugolo M, Romeo G, Gasparre G (2010) The genetic and metabolic signature of oncocytic transformation implicates HIF1alpha destabilization. Hum Mol Genet 19(6):1019–1032

Radermacher M, Ruiz T, Clason T, Benjamin S, Brandt U, Zickermann V (2006) The three-dimensional structure of complex I from *Yarrowia lipolytica*: a highly dynamic enzyme. J Struct Biol 154:269–279

Ragan CI, Heron C (1978) The interaction between mitochondrial NADH-ubiquinone oxidoreductase and ubiquinol-cytochrome c oxidoreductase – evidence for stoicheiometric association. Biochem J 174:783–790

Raha S, Myint AT, Johnstone L, Robinson BH (2002) Control of oxygen free radical formation from mitochondrial complex I: roles for protein kinase A and pyruvate dehydrogenase kinase. Free Radic Biol Med 32:421–430

Reddy PH, Manczak M, Mao P, Calkins MJ, Reddy AP, Shirendeb U (2010) Amyloid-beta and mitochondria in aging and Alzheimer's disease: implications for synaptic damage and cognitive decline. J Alzheimers Dis 20(Suppl 2):S499–S512

Reifschneider NH, Goto S, Nakamoto H, Takahashi R, Sugawa M, Dencher NA, Krause F (2006) Defining the mitochondrial proteomes from five rat organs in a physiologically significant context using 2D blue-native/SDS–PAGE. J Proteome Res 5:1117–1132

Ricchelli F, Sileikytė J, Bernardi P (2011) Shedding light on the mitochondrial permeability transition. Biochim Biophys Acta 1807(5):482–490

Rieske JS (1967) Preparation and properties of reduced coenzyme Q-cytochrome c reductase (complex III of the respiratory chain). Methods Enzymol 10:239–245

Robinson NC, Strey F, Talbert L (1980) Investigation of the essential boundary layer phospholipids of cytochrome c oxidase using Triton X-100 delipidation. Biochemistry 19:3656–3661

Rosca MG, Hoppel CL (2010) Mitochondria in heart failure. Cardiovasc Res 88:40–50

Rosca MG, Vazquez EJ, Kerner J, Parland W, Chandler MP, Stanley W, Sabbah HN, Hoppel CL (2008) Cardiac mitochondria in heart failure: decrease in respirasomes and oxidative phosphorylation. Cardiovasc Res 80:30–39

Rosca M, Minkler P, Hoppel CL (2011) Cardiac mitochondria in heart failure: normal cardiolipin profile and increased threonine phosphorylation of complex IV. Biochim Biophys Acta 1807(11):1373–1382

Ryan MT, Hoogenraad NJ (2007) Mitochondrial-nuclear communications. Annu Rev Biochem 76:701–722

Saffman PG, Delbruck M (1975) Brownian motion in biological membranes. Proc Natl Acad Sci USA 72:3111–3113

Salvi M, Brunati AM, Toninello A (2005) Tyrosine phosphorylation in mitochondria: a new frontier in mitochondrial signaling. Free Radic Biol Med 38:1267–1277

Scacco S, Petruzzella V, Bertini E, Luso A, Papa F, Bellomo F et al (2006) Mutations in structural genes of complex I associated with neurological diseases. Ital J Biochem 55:254–262

Schäfer E, Seelert H, Reifschneider NH, Krause F, Dencher NA, Vonck J (2006) Architecture of active mammalian respiratory chain supercomplexes. J Biol Chem 281(22): 15370–15375

Schäfer ER, Cellerino A, Englert C, Frenzel M, Terzibasi E, Dencher NA (2007a) Partial mitochondrial proteome and supramolecular organisation of OXPHOS complexes in the short-lived fish *Nothobranchius furzeri*. Ann Conf German Genetic Soc abstract 43:38

Schäfer E, Dencher NA, Vonck J, Parcej DN (2007b) Three-dimensional structure of the respiratory chain supercomplex I1III2IV1 from bovine heart mitochondria. Biochemistry 46(44):12579–12585

Schägger H (2002) Respiratory chain supercomplexes of mitochondria and bacteria. Biochim Biophys Acta 1555:154–159

Schägger H, Pfeiffer K (2000) Supercomplexes in the respiratory chains of yeast and mammalian mitochondria. EMBO J 19:1777–1783

Schägger H, Pfeiffer K (2001) The ratio of oxidative phosphorylation complexes I-V in bovine heart mitochondria and the composition of respiratory chain supercomplexes. J Biol Chem 276:37861–37867

Schägger H, von Jagow G (1991) Blue native electrophoresis for isolation of membrane protein complexes in enzymatically active form. Anal Biochem 199(2):223–231

Schägger H, de Coo R, Bauer MF, Hofmann S, Godinot C, Brandt U (2004) Significance of respirasomes for the assembly/stability of human respiratory chain complex I. J Biol Chem 279:36349–36353

Schönfeld P, Wieckowski MR, Lebiedzi ska M, Wojtczak L (2010) Mitochondrial fatty acid oxidation and oxidative stress: lack of reverse electron transfer-associated production of reactive oxygen species. Biochim Biophys Acta 1797(6–7):929–938

Schwerzmann K, Cruz-Orive LM, Eggman R, Sänger A, Weibel ER (1986) Molecular architecture of the inner membrane of mitochondria from rat liver: a combined biochemical and stereological study. J Cell Biol 102:97–103

Sedlak E, Robinson NC (1999) Phospholipase A(2) digestion of cardiolipin bound to bovine cytochrome c oxidase alters both activity and quaternary structure. Biochemistry 38:14966–14972

Seelert H, Dani DN, Dante S, Hauss T, Krause F, Schäfer E, Frenzel M, Poetsch A, Rexroth S, Schwassmann HJ, Suhai T, Vonck J, Dencher NA (2009) From protons to OXPHOS supercomplexes and Alzheimer's disease: structure-dynamics-function relationships of energy-transducing membranes. Biochim Biophys Acta 1787:657–671

Semenza GL (2003) Targeting HIF-1 for cancer therapy. Nat Rev Cancer 3(10):721–732

Semenza GL (2007) Oxygen-dependent regulation of mitochondrial respiration by hypoxia-inducible factor 1. Biochem J 405(1):1–9

Singer SJ, Nicolson GL (1972) The fluid mosaic model of the structure of cell membranes. Science 175(23):720–731

Stark G (2005) Functional consequences of oxidative membrane damage. J Membr Biol 205:1–16

Stroh A, Anderka O, Pfeiffer K, Yagi T, Finel M, Ludwig B, Schägger H (2004) Assembly of respiratory complexes I, III, and IV into NADH oxidase supercomplex stabilizes complex I in *Paracoccus denitrificans*. J Biol Chem 279:5000–5007

Stuart RA (2009) Supercomplex organization of the yeast respiratory chain complexes and the ADP/ATP carrier proteins. Methods Enzymol 456:191–208

Sumegi B, Srere PA (1984) Complex I binds several mitochondrial NAD-coupled dehydrogenases. J Biol Chem 259:15040–15045

Sun W, Zhou S, Chang SS, McFate T, Verma A, Califano JA (2009) Mitochondrial mutations contribute to HIF1alpha accumulation via increased reactive oxygen species and up-regulated pyruvate dehydrogenease kinase 2 in head and neck squamous cell carcinoma. Clin Cancer Res 15(2):476–484

Sunderhaus S, Dudkina NV, Jänsch L, Klodmann J, Heinemeyer J, Perales M, Zabaleta E, Boekema EJ, Braun HP (2006) Carbonic anhydrase subunits form a matrix-exposed domain attached to the membrane arm of mitochondrial complex I in plants. J Biol Chem 281(10):6482–6488

Suthammarak W, Yang YY, Morgan PG, Sedensky MM (2009) Complex I function is defective in complex IV-deficient *Caenorhabditis elegans*. J Biol Chem 284(10):6425–6435

Suthammarak W, Morgan PG, Sedensky MM (2010) Mutations in mitochondrial complex III uniquely affect complex I in *Caenorhabditis elegans*. J Biol Chem 285(52):40724–40731

Thomson M (2002) Evidence of undiswcovered cell regulatory mechanisms: phosphoproteins and protein kinases in mitochondria. Cell Mol Life Sci 59:213–219

Trifunovic A (2006) Mitochondrial DNA and ageing. Biochim Biophys Acta 1757:611–617

Trifunovic A, Larsson NG (2008) Mitochondrial dysfunction as a cause of ageing. J Intern Med 263:167–178

Trifunovic A, Wredenberg A, Falkenberg M, Spelbrink JN, Rovio AJ, Bruder CF, Bohlooly YM, Gedlof S, Oldfors A, Wibom R, Jacobs HT, Larsson NG (2004) Premature ageing in mice expressing defective mitochondrial DNA polymerase. Nature 429:357–359

Tuppen HA, Hogan VE, He L, Blakely EL, Worgan L, Al-Dosary M, Saretzki G, Alston CL, Morris AA, Clarke M, Jones S, Devlin AM, Mansour S, Chrzanowska-Lightowlers ZM, Thorburn DR, McFarland R, Taylor RW (2010) The p.M292T NDUFS2 mutation causes complex I-deficient Leigh syndrome in multiple families. Brain 133(10):2952–2963

Ugalde C, Janssen RJ, Van den Heuvel LP, Smeitink JA, Nijtmans LG (2004) Differences in the assembly and stability of complex I and other OXPHOS complexes in inherited complex I deficiency. Hum Mol Genet 13:659–667

Van Regenmortel MHV (2004) Reductionism and complexity in molecular biology. Scientists now have the tools to unravel biological and overcome the limitations of reductionism. EMBO Rep 5:1016–1020

Vanderkooi G (1978) Organization of protein and lipid components in membranes. In: Fleischer S, Hatefi Y, MacLennan D, Tzagoloff A (eds) Molecular biology of membranes. Plenum, New York, pp 25–55

Velours J, Dautant A, Salin B, Sagot I, Brèthes D (2009) Mitochondrial F1F0-ATP synthase and organellar internal architecture. Int J Biochem Cell Biol 41(10):1783–1789

Ventura B, Genova ML, Bovina C, Formiggini G, Lenaz G (2002) Control of oxidative phosphorylation by complex I in rat liver mitochondria: implications for aging. Biochim Biophys Acta 1553:249–260

Vereb G, Szollosi J, Matko J, Nagy P, Farkas T, Vigh L et al (2003) Dynamic, yet structured: the cell membrane three decades after the Singer-Nicolson model. Proc Natl Acad Sci USA 100:8053–8058

Vik SB, Capaldi RA (1977) Lipid requirements for cytochrome c oxidase activity. Biochemistry 16:5755–5759

Vogt S, Rhiel A, Koch V, Kadenbach B (2007) Regulation of oxidative phosphorylation by inhibition of its enzyme complexes via reversible phosphorylation. Curr Enzym Inhib 3:189–206

Vonck J, Schäfer E (2009) Supramolecular organization of protein complexes in the mitochondrial inner membrane. Biochim Biophys Acta 1793(1):117–124

Wallace DC (2005) A mitochondrial paradigm of metabolic and degenerative diseases, aging, and cancer: a dawn for evolutionary medicine. Annu Rev Genet 39:359–407

Wang Y, Mohsen Al-W, Mihalik SJ, Goetzman ES, Vockley J (2010) Evidence for physical association of mitochondrial fatty acid oxidation and oxidative phosphorylation complexes. J Biol Chem 285:29834–29841

Warburg O (1956) On the origin of cancer cells. Science 123(3191):309–314

Wenz T, Hielscher R, Hellwig P, Schägger H, Richers S, Hunte C (2009) Role of phospholipids in respiratory cytochrome bc(1) complex catalysis and supercomplex formation. Biochim Biophys Acta 1787(6):609–616

Wernicke C, Hellmann J, Zi ba B, Kuter K, Ossowska K, Frenzel M, Dencher NA, Rommelspacher H (2010) 9-Methyl-β-carboline has restorative effects in an animal model of Parkinson's disease. Pharmacol Rep 62:1

Wittig I, Schägger H (2009) Supramolecular organization of ATP synthase and respiratory chain in mitochondrial membranes. Biochim Biophys Acta 1787(6):672–680

Wittig I, Karas M, Schägger H (2007) High resolution clear native electrophoresis for in-gel functional assays and fluorescence studies of membrane protein complexes. Mol Cell Proteomics 6:1215–1225

Wong R, Aponte AM, Steenbergen C, Murphy E (2010) Cardioprotection leads to novel changes in the mitochondrial proteome. Am J Physiol Heart Circ Physiol 298(1):H75–H91

Yu A, Yu L, King TE (1974) Soluble cytochrome b-c1 complex and the reconstitution of succinate-cytochrome c reductase. J Biol Chem 249:4905–4910

Zhang M, Mileykovskaya E, Dowhan W (2002) Gluing the respiratory chain together. Cardiolipin is required for supercomplex formation in the inner mitochondrial membrane. J Biol Chem 277:43553–43556

Chapter 6
Molecular Mechanisms of Superoxide Production by the Mitochondrial Respiratory Chain

Stefan Dröse and Ulrich Brandt

Abstract The mitochondrial respiratory chain is a major source of reactive oxygen species (ROS) in eukaryotic cells. Mitochondrial ROS production associated with a dysfunction of respiratory chain complexes has been implicated in a number of degenerative diseases and biological aging. Recent findings suggest that mitochondrial ROS can be integral components of cellular signal transduction as well. Within the respiratory chain, complexes I (NADH:ubiquinone oxidoreductase) and III (ubiquinol:cytochrome *c* oxidoreductase; cytochrome bc_1 complex) are generally considered as the main producers of superoxide anions that are released into the mitochondrial matrix and the intermembrane space, respectively. The primary function of both respiratory chain complexes is to employ energy supplied by redox reactions to drive the vectorial transfer of protons into the mitochondrial intermembrane space. This process involves a set of distinct electron carriers designed to minimize the unwanted leak of electrons from reduced cofactors onto molecular oxygen and hence ROS generation under normal circumstances. Nevertheless, it seems plausible that superoxide is derived from intermediates of the normal catalytic cycles of complexes I and III. Therefore, a detailed understanding of the molecular mechanisms driving these enzymes is required to understand mitochondrial ROS production during oxidative stress and redox signalling. This review summarizes recent findings on the chemistry and control of the reactions within respiratory complexes I and III that result in increased superoxide generation. Regulatory contributions of other components of the respiratory chain, especially complex II (succinate:ubiquinone oxidoreductase) and the redox state of the ubiquinone pool (Q-pool) will be briefly discussed.

S. Dröse (✉) • U. Brandt
Molecular Bioenergetics Group, Medical School, Cluster of Excellence Frankfurt "Macromolecular Complexes", Center for Membrane Proteomics, Johann Wolfgang Goethe-Universität, Theodor-Stern-Kai 7, Frankfurt am Main 60590, Germany
e-mail: droese@zbc.kgu.de; brandt@zbc.kgu.de

B. Kadenbach (ed.), *Mitochondrial Oxidative Phosphorylation*, Advances in Experimental Medicine and Biology 748, DOI 10.1007/978-1-4614-3573-0_6,

6.1 Introduction

Mitochondria are the "power plants" of eukaryotic cells (Scheffler 2008). For this purpose, they oxidize different substrates, mainly pyruvate, the product of glycolysis and fatty acids. During the tricarboxylic acid cycle and the β-oxidation of fatty acids, electrons are transferred from the intermediates being oxidized onto the hydrogen carriers NAD^+ and FAD that feed the reducing equivalents into the respiratory chain of the inner mitochondrial membrane. The respiratory chain complexes I, III and IV guide the electrons through a series of redox reactions utilizing the free energy released to create a proton motive force that drives ATP synthesis by complex V (ATP synthase). Reactive oxygen species (ROS) are produced mainly as superoxide, when electrons leak in unwanted side reactions from prosthetic groups or coenzymes involved in these redox reactions onto molecular oxygen. Mitochondrial ROS have been implicated in a number of neurodegenerative diseases (Lin and Beal 2006) and the aging process (Muller et al. 2007; Navarro and Boveris 2007). Recent data indicate that mitochondrial ROS can also take part in cellular signalling pathways as "second messengers" (Murphy et al. 2011). While the respiratory chain is generally recognized as the main source of ROS (for recent reviews, see Starkov 2008; Kowaltowski et al. 2009; Brand 2010), other mitochondrial enzymes including matrix NADH dehydrogenases have been reported to produce ROS in vitro. These enzymes are not within the scope of this review. On the other hand, mitochondria comprise a very effective antioxidative defence and it has been shown that isolated mitochondria are rather a sink than a net producer of ROS (Zoccarato et al. 2004; Drechsel and Patel 2010). Also the defence system will not be broadly covered in this book chapter. For an excellent review putting emphasis on this topic, see Starkov (2008).

Within the respiratory chain, complexes I and III have been identified as major ROS generators (Kowaltowski et al. 2009; Brand 2010). It can be assumed that the production rate of deleterious ROS was kept low by evolution and it can be calculated that under normal conditions less than 0.1% of the electrons passing through the respiratory chain leak onto O_2 to form superoxide (Brand 2010; Tahara et al. 2009). Therefore, understanding the molecular mechanism, how these enzymes work, is prerequisite to understand how superoxide is generated. We will highlight recent progress in the understanding of the molecular mechanisms of the complexes involved in ROS generation, before current models for ROS production by these respiratory chain complexes are briefly presented and discussed. Subsequently, the modulating effects of other respiratory chain components will be addressed.

6.2 Some General Considerations on the Investigation of ROS Production by Respiratory Chain Complexes

Much of the vast amount of published data on ROS production by the respiratory chain is controversial. Specifically, there is a long-standing and ongoing discussion, whether ROS from complex I or complex III are more pivotal (Chen et al. 2003).

It has been concluded from thermodynamic considerations that only the FMN/$FMNH_2$ couple of complex I with its rather negative midpoint potential E_{m7} of about −360 mV is capable to reduce molecular oxygen to superoxide (E_{m7} (O_2/$O_2^{\bullet-}$)=−160 mV) at significant rates, while the respective reduced redox-active groups of complex III (heme b_L; $UQH^{\bullet-}$/UQ) are much less potent electron donors (E_{m7}=−180 to +30 mV). However, considering the actual concentrations of the redox intermediates, it can be calculated that also the Q_o site of complex III is capable of effective superoxide production (Murphy 2009). This is supported by ample experimental evidence.

Mitochondrial ROS production has been studied with isolated respiratory chain complexes, submitochondrial particles and intact mitochondria from different species and tissues. Substrate- and tissue-specificity of ROS production has been demonstrated (Tahara et al. 2009) and seems plausible considering different expression of metabolic pathways resulting in differences with respect to the utilization of substrates. Preparations of mitochondria from different organs and tissues vary in their contamination with "microsomal membranes" which include fragments of peroxisomes with high amounts of peroxidases and catalases. These enzymes produce large and variable background rates with most dyes and assays available for ROS detection. This is especially the case for mitochondria isolated from liver and yeast cells. Thus, the purity of the sample, the selection of the dyes and the particular assay protocol for ROS detection have huge impact on the reliability of the experimental data. Discussing this matter is beyond the scope of this review and we therefore refer to the excellent review by Gomes et al. (2005).

Another general problem linked in particular to experiments with intact mitochondria has to be considered: none of the available assays detects ROS release to the intermembrane space and to the matrix with the same efficiency. Superoxide released to the matrix or the intermembrane space is readily dismutated into H_2O_2 and O_2 by the activities of the Mn-SOD (SOD2) and Cu/Zn-SOD (SOD1), respectively. However, the widely used Amplex Red/HRP assay (Zhou et al. 1997), in general a robust and reliable test for the detection of H_2O_2 production, detects only H_2O_2 that reaches the mitochondrial intermembrane space ("outside"). Thus, with intact mitochondria a significant fraction of the H_2O_2 released into the matrix is not monitored by this assay, because it is scavenged by the highly effective antioxidative defence system including thioredoxin-reductase/thioredoxin/peroxiredoxin-3 and -5 system, glutathione peroxidase (GPx) and GSH. In contrast, H_2O_2 released directly into the intermembrane space is not effectively scavenged in isolated mitochondria, since regeneration of the GSH-pool in the intermembrane space depends on the activity of cytoplasmic proteins (Herrmann and Riemer 2010). On the other hand, it has to be considered that oxidized cytochrome *c* still present in the intermembrane space can act as a powerful scavenger of superoxide and competes with SOD1 (Pasdois et al. 2011). The direct reaction of GSH with H_2O_2 is slow and ineffective under physiological conditions, but this reaction is greatly accelerated by the activity of mitochondrial GPx1 and GPx4 (for review, see Winterbourn and Hampton 2008). Brand and coworkers (Treberg et al. 2010) have recently reported that the antioxidative defence can be attenuated by depleting the GSH-pool of mitochondria with 1-chloro-2,4-dinitrobenzene (CDNB). There are contradictory data and interpretations in the literature, whether CDNB mainly attenuates the GSH-pool (Zoccarato et al. 2004; Treberg et al. 2010; Stanley et al. 2011) or whether it primarily affects the peroxiredoxin/thioredoxin system (Drechsel and Patel 2010), which might be kinetically more relevant for H_2O_2 removal (Winterbourn and Hampton 2008; Cox et al. 2010). Yet all studies showed that CDNB largely abolishes the H_2O_2 scavenging capacity of respiring mitochondria. However, CDNB pretreatment has to be done with great care, since it has been shown to potentially increase ROS production by complex I directly (Treberg et al. 2010). It was also shown that the efficiency of the antioxidative defence is reduced with succinate as compared to NADH-regenerating substrates like malate/glutamate (Zoccarato et al. 2004; Drechsel and Patel 2010). Moreover, uncoupling reduces the scavenging capacity by impeding the activity of the ΔpH-dependent transhydrogenase (Rydström 2006) that is responsible for keeping the $NADP^+$-pool in the mitochondrial matrix reduced. Indeed dissipation of the proton motive force by uncoupling in the presence of respiratory chain inhibitors generally increases the observed rates of ROS generation (Cadenas and Boveris 1980; Votyakova and Reynolds 2001; Dröse et al. 2009a, 2011a).

Also in investigations with submitochondrial particles (SMP) or isolated respiratory chain complexes one has to carefully consider side and background reactions that interfere with the particular assay system employed. For example, high background rates have been reported with substrates like NADH or decylubiquinol (DBH) for the Amplex Red/HRP assay (Chen and Schopfer 1999; Votyakova and Reynolds 2004; Dröse and Brandt 2008). Furthermore, especially artificial short chain ubiquinone analogues like DBH can produce superoxide by autooxidation of the reduced or semiquinone form directly (James et al. 2005).

6.3 Complex I

6.3.1 New Findings on the Structure and Molecular Mechanism of Complex I

With more than 40 subunits (45 in bovine heart (Carroll et al. 2006); 42 in the obligate aerobic yeast *Yarrowia lipolytica* (Morgner et al. 2008; Dröse et al. 2011b)), NADH:ubiquinone oxidoreductase (complex I) is by far the largest and most complicated complex of the mitochondrial respiratory chain (overview in Brandt 2006). Bacteria contain a "minimal form" of complex I with the 14 essential or "central" subunits. Electron microscopy revealed that bacterial and mitochondrial complexes share a common L-shaped structure and can be divided into a hydrophilic peripheral arm and a hydrophobic membrane arm of almost equal size (Clason et al. 2010). Functionally and structurally, complex I can be further subdivided into four modules (Brandt 2006; Hunte et al. 2010). The peripheral arm comprises the N- and

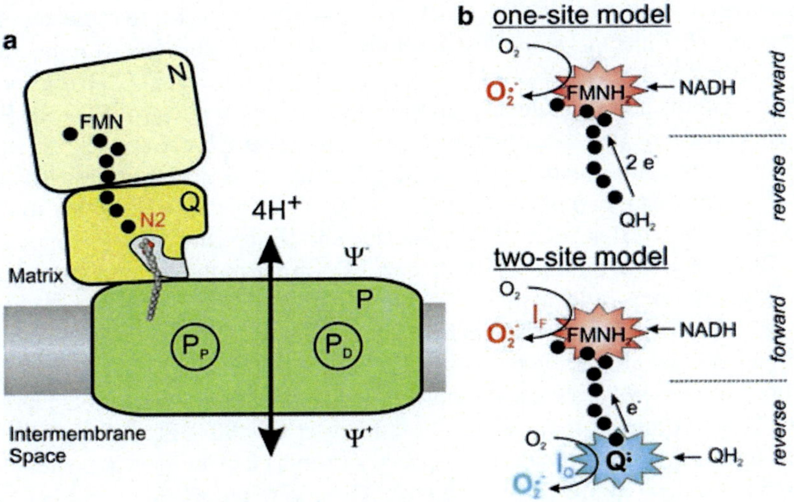

Fig. 6.1 Modular arrangement and superoxide production by complex I (NADH: ubiquinone oxidoreductase). (**a**) Schematic model of complex I illustrating the dissection into functional modules according to Brandt (2006) and the location of functional sites involved in superoxide production. N, N-module comprising the NADH binding site and the tightly bound FMN; Q, Q-module harbouring the ubiquinone binding pocket (*grey*); P, P-module carrying proton-pumping sites in its proximal (P_P) and its distal (P_D) parts; N2, iron–sulphur cluster N2. *Black spheres* indicate the chain of eight iron–sulphur clusters. (**b**) Models for the superoxide production of complex I. In the one-site model (Galkin and Brandt 2005; Kussmaul and Hirst 2006), superoxide is exclusively generated from the tightly bound $FMNH_2$ during both forward (electrons supplied by NADH) and reverse (electrons supplied by QH_2) operation. In the two-site model (Treberg et al. 2011), superoxide is generated from $FMNH_2$ (I_F site) only during forward electron transfer, while during reverse electron transfer a semiquinone in the Q-binding pocket (I_Q site) is the source of superoxide. In any case, superoxide is produced in the peripheral arm of complex I which protrudes into the matrix. See text for further details

Q-modules (for NADH- and ubiquinone (Q) binding, respectively) while the membrane arm comprises the proximal (P_P) and distal (P_D) parts of the P-module that contain the proton-pumping sites (Fig. 6.1a). The mitochondrial complex has acquired a substantial number of "accessory subunits" of largely unknown function (Brandt 2006). Recently, it was proposed that the membrane-associated accessory subunits are important for the stability of the membrane arm and form a scaffold around the central subunits of the proton-pumping modules (Angerer et al. 2011). So far, it has been shown only for mitochondrial complex I that it pumps protons with a stoichiometry of four protons per two electrons transferred from NADH via FMN and seven FeS clusters onto ubiquinone (Wikström 1984; Galkin et al. 1999; Galkin et al. 2006). For a long time, structure and mechanism of complex I had to be considered the "black box" of the respiratory chain (Matsuno-Yagi and Yagi 2001). However, recent progress in X-ray structure determination revealed some of the secrets of complex I and has greatly stimulated the discussion about its molecular mechanism (Brandt 2011; Mourier and Larsson 2011).

X-ray crystallographic analysis of the complex I from the thermophilic eubacterium *Thermus thermophilus* (Sazanov and Hinchliffe 2006; Efremov et al. 2010; Efremov and Sazanov 2011) and *Y. lipolytica* (Hunte et al. 2010) provided important insights into its molecular architecture. All prosthetic groups—one FMN and eight conserved FeS—clusters are located in the peripheral arm. Seven of the eight FeS clusters are organized in a chain allowing direct electron transfer from the FMN in the distal end of the peripheral arm over a distance of ~90 Å towards the membrane arm (Hunte et al. 2010; Sazanov and Hinchliffe 2006; Efremov et al. 2010). Strikingly, the terminal FeS cluster N2 which is the immediate electron donor to ubiquinone is located ~30 Å above the membrane surface (Fig. 6.1a; Hunte et al. 2010). A funnel-like cavity leading from the N-terminal β-sheet of the 49-kDa subunit toward a fully conserved tyrosine residue (Tyr-144 in *Y. lipolytica*, Tyr-87 in *T. thermophilus*) located in the immediate vicinity of cluster N2 has been mapped by site-directed mutagenesis (Tocilescu et al. 2007; Fendel et al. 2008). With its headgroup ubiquinone binds directly to this tyrosine that is essential for enzymatic activity (Tocilescu et al. 2010). The P-module of the membrane arm contains three large hydrophobic subunits exhibiting homology to bacterial Mrp-type Na^+/H^+-antiporters that have been therefore discussed as candidates for harbouring the proton pumps (Mathiesen and Hägerhäll 2002). Mutations of conserved residues in these subunits confirm their essential function for proton translocation (Torres-Bacete et al. 2007; Euro et al. 2008; Nakamaru-Ogiso et al. 2010; Michel et al. 2011). The recently published high-resolution structure of the membrane arm of complex I from *Escherichia coli* indeed revealed putative channels in the three antiporter-like subunits L, M and N (ND2, ND4 and ND5) that could be part of the proton translocation machinery. The three subunits contain 14 structurally conserved helices that share a unique fold with two inverted structural repeats of five transmembrane helices each unusually arranged in a face-to-back manner (Efremov and Sazanov 2011). There are indications for a fourth putative proton-translocation channel at the interface of subunits N, K, J and A (ND2, ND3, ND4L and ND6). However, the functionality of the proposed proton translocation pathways will have to be established by experimental evidence from functional studies. ND4 and ND5 are located in the P_D-module while ND2 and the other central ND subunits are located in the P_P-module (Dröse et al. 2011b; Angerer et al. 2011). The two P-modules are connected by a long helical element (also called helix H_L) oriented in parallel to the long axis of the membrane arm at the matrix facing surface (Hunte et al. 2010; Efremov et al. 2010; Efremov and Sazanov 2011). This lateral helix represents the major part of a C-terminal extension of subunit ND5. It has been proposed that this "rod" may be involved in conformational coupling of the two modules. Mutated complex I from *E. coli* that contained a truncated version of this α-helix pumps protons with reduced stoichiometry supporting its putative role in energy transmission (Steimle et al. 2011). However, this was challenged by a more extensive mutagenesis study in *E. coli* suggesting that H_L may not be critically involved in proton pumping (Belevich et al. 2011). Our detailed functional analysis of a catalytically active subcomplex from *Y. lipolytica* that lacks the entire distal arm including ND4 and ND5 demonstrated proton pumping by both, the P_P and P_D

module since removal of the latter still allowed proton pumping with a stoichiometry of $2H^{+}/2e^{-}$ (Dröse et al. 2011b). This clearly showed that half of the proton-pumping sites are located in either module, but awaiting more structural and functional evidence their actual number is still open (Brandt 2011; Efremov and Sazanov 2011). On the other hand, it seems clear from the recent structural and functional investigations that the proton translocation is driven via long range conformational changes that are generated by the redox reactions in the peripheral arm and transmitted to the membrane arm. Any additional contribution by direct coupling, as again proposed recently (Ohnishi et al. 2010; Treberg and Brand 2011), seems highly unlikely.

6.3.2 Superoxide Production by Complex I: A One- or a Two-Site Issue?

In intact mitochondria, complex I generates superoxide in two different situations: (1) Under conditions where electrons back up in the chain of FeS clusters; this occurs when NADH is present and the downstream respiratory chain is blocked, e.g. in the presence of inhibitors like rotenone or DQA (2-*n*-decyl-quinazolin-4-yl-amine, SAN 549) that bind to the Q-binding site or in presence of inhibitors of complex III like stigmatellin or antimycin A or of complex IV like KCN (Chen et al. 2003; Dröse et al. 2009a; Galkin and Brandt 2005). (2) Under conditions of so-called reverse electron transfer (RET), when electrons flow back from complex II via ubiquinone to complex I, which requires a high membrane potential as the driving force (Votyakova and Reynolds 2001; Grivennikova and Vinogradov 2006).

There is still an ongoing controversy whether superoxide is produced from one site, i.e. the flavin site, under both conditions (Galkin and Brandt 2005; Pryde and Hirst 2011), or whether this site produces ROS only in the "forward mode" when electrons are delivered from NADH, while in the "reverse mode" superoxide is instead generated from a semiquinone in the Q-binding site (Fig. 6.1b; Brand 2010; Treberg et al. 2011). Superoxide production by complex I has been investigated with different preparations: purified detergent-solubilized or reconstituted enzymes (Galkin and Brandt 2005; Kussmaul and Hirst 2006; Esterhazy et al. 2008; Dröse et al. 2009b), submitochondrial particles (Grivennikova and Vinogradov 2006; Pryde and Hirst 2011; Vinogradov and Grivennikova 2005) or intact mitochondria (Votyakova and Reynolds 2001; Kushnareva et al. 2002; St Pierre et al. 2002; Liu et al. 2002; Lambert and Brand 2004a) from different sources (mammalian mitochondria from heart, skeletal muscle or brain; *Y. lipolytica*; *E. coli*). All preparations have their advantages and disadvantages. With purified enzymes and SMP, electrons can be directly supplied by the addition of NADH. In case of SMP from yeast and fungi (Galkin and Brandt 2005; Dröse et al. 2009b) it may be required to use deamino-NADH that is not turned over by alternative or type-2 NADH dehydrogenases (Kerscher et al. 2008). However, if H_2O_2 production is measured by the commonly used Amplex Red/HRP assay (Zhou et al. 1997), a substantial background reaction with pyridine nucleotides has to be considered (Chen and Schopfer 1999; Votyakova and Reynolds 2004) that can be reduced by the addition of superoxide dismutase and by keeping the NADH concentration as low as possible (Kussmaul and Hirst 2006). On the other hand, one has to assure that—when comparing uninhibited and inhibited rates—the catalytic activity is constant over the entire experimental period in all samples. Furthermore, we and others observed that some short-chain ubiquinone analogues that have to be added in investigations with purified complex I can act as redox mediators between molecular oxygen and the enzyme, mostly at a non-physiological site in the hydrophilic domain of complex I (Galkin and Brandt 2005; Cadenas et al. 1977), presumably the flavin (King et al. 2009). Since this reaction decreases with increasing hydrophobicity of the analogue, decylubiquinone is recommended over other short-chain ubiquinones like Q_1 (Dröse et al. 2009b). On the other hand, purified complex I preparations and coupled SMPs have the advantage that ROS production and detection are not affected by the antioxidative defence and that a contribution of matrix NADH dehydrogenases, e.g. α-ketoglutarate dehydrogenase (Starkov et al. 2004) can be excluded. The antioxidative defence in the mitochondrial matrix, especially the mitochondrial peroxiredoxins Prx3 and Prx5 and the combined activity of glutathione (GSH) and glutathion-peroxidase (Winterbourn and Hampton 2008), can scavenge a large portion of ROS produced in or entering into the mitochondrial matrix (Treberg et al. 2010; see above). This has to be considered especially for complex I which releases superoxide exclusively into the matrix (St Pierre et al. 2002; Muller et al. 2004) or when the Amplex Red/HRP assay is used that detects H_2O_2 only outside the mitochondria.

Investigations with intact mitochondria (Treberg et al. 2011), submitochondrial particles (Grivennikova and Vinogradov 2006; Pryde and Hirst 2011) and purified reconstituted or lipid-activated complex I (Galkin and Brandt 2005; Kussmaul and Hirst 2006) indicate that ROS in the forward mode, i.e. when electrons are delivered by NADH, are generated at the isoalloxazine moiety of the FMN cofactor. By directly comparing the superoxide production rate detected with acetylated cytochrome *c* and the H_2O_2 production rate detected with the Amplex Red/HRP assay, Hirst et al. deduced that bovine complex I generates at least 95% superoxide (Kussmaul and Hirst 2006). In contrast, the purified *E. coli* enzyme produces 80% H_2O_2 and 20% superoxide (Esterhazy et al. 2008). Redox titrations and electron paramagnetic resonance spectroscopic studies with purified bovine complex I (Kussmaul and Hirst 2006) and coupled SMP (Pryde and Hirst 2011) suggest that the fully reduced flavin is the source of superoxide. These measurements also excluded the flavin radical and the iron–sulphur clusters including N2 that have also been proposed as direct electron donors for oxygen (Genova et al. 2001; Fato et al. 2009). A similar conclusion was reached by analysing the superoxide production of pure complex I from *Y. lipolytica* (Galkin and Brandt 2005). With purified enzymes from bovine heart and *Y. lipolytica*, neither the turnover nor the inhibition by Q site inhibitors increased the rate of superoxide production (Galkin and Brandt 2005; Kussmaul and Hirst 2006; Dröse et al. 2009b). This is in stark contrast to several reports showing that complex I inhibitors stimulate ROS generation in mitochondrial membranes and intact mitochondria (Chen et al. 2003; Votyakova and Reynolds

2001; Lambert and Brand 2004b; Ohnishi et al. 2005). There are several possible explanations for this discrepancy: (1) in intact mitochondria, ROS production may be partly due to activity of NAD(P)$^+$-dependent matrix enzymes that are stimulated by increased availability of reducing equivalents; (2) the application of hydrophilic analogues of the substrate ubiquinone facilitates nonphysiologic side reactions that do not occur with the endogenous, much more hydrophobic ubiquinones Q_9 and Q_{10}; (3) fast consumption of NADH in submitochondrial particles or mitochondrial membranes is prevented by complex I inhibition, thereby affecting ROS production by complex I itself that is controlled by the NADH/NAD$^+$ ratio (Kussmaul and Hirst 2006). The latter is confirmed by investigations with SMP (Grivennikova and Vinogradov 2006; Pryde and Hirst 2011; Vinogradov and Grivennikova 2005) and intact mitochondria (Kushnareva et al. 2002; Starkov and Fiskum 2003). Using the reconstituted enzyme from *Y. lipolytica* (Dröse et al. 2005), we could show that the membrane potential has only a small effect on the rate of ROS generation in the forward mode (Dröse et al. 2009b). This finding is in agreement with investigations using coupled SMP from bovine heart mitochondria (Pryde and Hirst 2011) and intact mitochondria from guinea-pig brain (Tretter and Adam-Vizi 2007) and rat heart (Dröse et al. 2009a). On the other hand, an increase of NADH-dependent ROS production has been observed in rat brain mitochondria at very high $\Delta\Psi$ values (Starkov and Fiskum 2003). But this could not be unambiguously linked to complex I and may as well have originated from matrix NADH dehydrogenases (Starkov and Fiskum 2003) or complex III (see Sect. 6.3.2).

In contrast, a high membrane potential or proton-motive force (Δp) is absolutely required for superoxide production during RET, as several investigations with coupled SMPs (Grivennikova and Vinogradov 2006; Pryde and Hirst 2011; Vinogradov and Grivennikova 2005) and intact mitochondria (Votyakova and Reynolds 2001; Dröse et al. 2009a; Lambert and Brand 2004b) have shown. It has been reported that especially the ΔpH portion of Δp is responsible for the increased superoxide production during RET (Brand 2010; Lambert and Brand 2004a, b; Zoccarato et al. 2007). However, the observed stimulating effect may be due to a concomitant increase in matrix pH (Selivanov et al. 2008). Furthermore, Pryde and Hirst (2011) found no effect of ΔpH on superoxide production during RET in coupled SMP from bovine heart. As mentioned earlier, there is an ongoing debate whether superoxide is produced during RET from a second site, namely a semiquinone radical in the Q-binding site of complex I (Fig. 6.1b). Indeed, a Δp-sensitive semiquinone radical was detected by EPR measurements in tightly coupled SMP from bovine heart mitochondria (Magnitsky et al. 2002). Further arguments in favour of a two-site model were recently summarised by Brand et al. (Brand 2010; Treberg et al. 2011) (1) superoxide production in intact mitochondria is considerably higher in RET than in forward mode, even in the presence of Q-site inhibitors; (2) this higher rate is not influenced by a greater NADH/NAD$^+$ ratio that was shown to determine the superoxide production from the flavin site (see above) and (3) the higher rate is highly sensitive to Δp. However, Pryde and Hirst (2011) showed in their study with coupled bovine heart SMP that, when set by the NADH/NAD$^+$ redox couple, the potential dependence of NADH-induced superoxide production during the forward mode matches that of RET-induced superoxide production set by the succinate/fumarate redox couple and Δp and demonstrated that both match the potential dependence of the flavin. Importantly, superoxide production during RET was abolished not only by the Q-site inhibitor rotenone, but also by the flavin-site inhibitors NADH-OH, ADP-ribose and diphenyleneiodonium (DPI). The authors concluded that in forward mode, as well as during RET, ROS are produced at the flavin site by the same molecular mechanism. On the other hand, Brand et al. (Treberg et al. 2011) could show in investigations with coupled mitochondria from rat skeletal muscle that at constant membrane potential superoxide generation during RET was determined by the Q-pool redox state and did not correspond to the $NADH/NAD^+$ ratio, while the latter determined superoxide production during forward electron transfer in the presence of the Q-site inhibitor rotenone. These authors proposed a two-site model of complex I superoxide production (Fig. 6.1b, lower panel): one site ($FMNH_2$; site I_F) is in equilibrium with the NAD-pool while the other (semiquinone in the Q-binding site; site I_Q) is not only dependent on the NAD redox state, but also on Δp and the redox state of the Q-pool. However, this interpretation is based on the assumption that also during RET the $NADH/NAD^+$ ratio is always in equilibrium with $FMNH_2$/FMN and vice versa which has not been demonstrated experimentally. It seems more plausible that during RET also the redox state of FMN directly correlates with the redox state of the Q-pool and the applied ΔΨ. Therefore, this study does not disprove the data of Pryde and Hirst (2011). A final decision, whether complex I contains one or two sites of superoxide production, has to await further experimental evidence.

6.4 Complex III

6.4.1 The Protonmotive Q-Cycle

The molecular mechanism that drives H^+ pumping of the cytochrome bc_1 complex (complex III) has been first proposed by Peter Mitchell as the proton-motive Q-cycle (Mitchell 1975). The general ideas of this model have been approved by a substantial body of experimental evidence, although some modifications and refinements had to be introduced (Brandt and Trumpower 1994; Brandt 1996; Osyczka et al. 2005). Crystal structures of mitochondrial cytochrome bc_1 complexes in various conformations and from different organisms clearly define the positions of cofactors involved in electron transfer and allow insight into possible proton conduction pathways and ubiquinone binding sites. Still, the exact position of the substrate in the ubiquinol oxidation site has not been resolved yet (overviews in Hunte et al. 2003, 2008). Three transmembrane subunits of complex III contain redox prosthetic groups, the diheme cytochrome *b*, cytochrome c_1 and the so-called Rieske iron–sulphur protein. Mitochondrial cytochrome bc_1 complex can contain up to eight additional accessory subunits that are not essential for catalysis. Two ubiquinone reaction centres on opposite sides of the membranes are mainly formed by

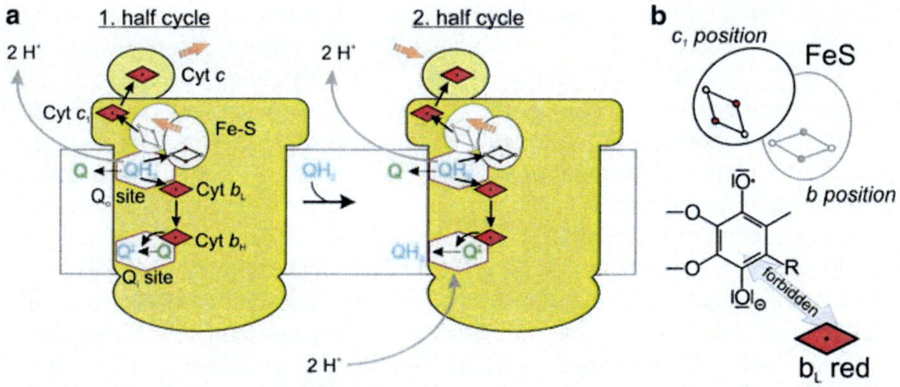

Fig. 6.2 The proton-motive Q-cycle of the cytochrome bc_1-complex. (**a**) Schematic model highlighting only the prosthetic groups/coenzymes of one monomer of the functional dimer. The hemes of the cytochromes (cyt) are indicated by a *red* rhombus, ubiquinone binding sites (Q_o site, ubiquinol oxidation site; Q_i site, ubiquinone reduction site) are shown as white hexagons, the positions of the flexible head of the Rieske FeS-protein with its [2Fe-2S] cluster as *white ovals*. Electron transfer is indicated by *solid black arrows*, protonation/deprotonation by *grey solid arrows*, movement of protein or domains by *orange arrows*. Bifurcated ubiquinol oxidation at the Q_o site directs one electron to the high-potential chain (FeS, Cyt c_1, Cyt *c*) and the other to the low-potential chain (Cyt b_L, Cyt b_H, Q). In the first half cycle, one cytochrome *c* is reduced, in addition to the reduction of one ubiquinone to ubisemiquinone in the Q_i site. During the second half cycle, a second cytochrome *c* is reduced in addition to the reduction of ubisemiquinone in the Q_i site to ubiquinol. The complete cycle translocates two charges across the inner membrane. (**b**) Destabilization due to excess of negative charge prevents that a negatively charged semiquinone and reduced heme b_L are present at the same time in the Q_o site. See text for further details

cytochrome *b*: the ubiquinol-oxidation centre (called centre P or Q_o site) on the positive side of the membrane and the ubiquinone-reduction centre (centre N or Q_i site) on the negative side of the membrane. The crystal structures revealed further that the cytochrome bc_1 complex is an obligate homodimer (Hunte et al. 2003). Functional implications of this dimeric organization were supported by extensive kinetic (reviewed in Covian and Trumpower 2008) and mutational studies with yeast and bacterial complex III indicating that electrons rapidly equilibrate between the cytochrome *b* subunits (Gong et al. 2005; Covian and Trumpower 2005; Castellani et al. 2010; Swierczek et al. 2010; Lanciano et al. 2011) and that there is conformational communication between centre P and centre N (Brandt et al. 1991; Covian and Trumpower 2006).

The proton-motive Q-cycle (Fig. 6.2a) starts with the oxidation of ubiquinol to ubiquinone at the Q_o site, where the two electrons are fed into a high- and low-potential chain in a bifurcated pathway (Brandt and Trumpower 1994; Osyczka et al. 2005). One electron is transferred onto the iron–sulphur cluster of the Rieske protein followed by a movement of its flexible protein domain from the "b-position" (close to cytochrome *b*) to the "c_1-position" (close to cytochrome c_1) that allows further electron transfer to the heme of cytochrome c_1 and finally to soluble cytochrome *c*. The second electron enters the low-potential chain at heme b_L (low potential heme *b*) and is transferred via heme b_H (high potential heme *b*) onto ubiquinone bound to the Q_i site which is reduced to a stabilized semiquinone species that is detectable by EPR spectroscopy (Ohnishi and Trumpower 1980). During oxidation of ubiquinol, two protons are released on the positive side of the membrane (intermembrane space in the case of mitochondria). In the second round or half cycle, two more protons are released upon ubiquinol oxidation at the Q_o site to the positive side of the membrane. The electron entering the high-potential chain reduces a second cytochrome *c* and the electron entering the low-potential chain eventually reduces the semiquinone waiting in the Q_i site to ubiquinol. This is accompanied by the uptake of two protons from the negative side of the membrane (the matrix in the case of mitochondria). Hence, sided uptake and release of protons due to topologically segregated oxidation of ubiquinol and reduction of quinone at opposite sides of the membrane together with the vectorial transport of electrons across the membrane through cytochrome *b* overall result in the net translocation of $2H^+/2e^-$ (Fig. 6.2a).

The bifurcated oxidation of ubiquinol at the Q_o site is the most critical step in the proton-motive Q-cycle since it constitutes the reaction, where the actual chemistry driving the vectorial proton translocation takes place. It is also the reaction which is most controversially discussed and several models for its detailed redox chemistry have been proposed (Brandt and Trumpower 1994; Brandt 1996, 1998; Osyczka et al. 2005; Trumpower 2002; Crofts 2004). Since each partial reaction of the proton-motive Q cycle is completely reversible (Osyczka et al. 2004), a mechanistic constraint or gating must exist that suppresses a number of potential short circuits, which would result in unproductive or even deleterious aberrant electron transfer (for a review, see Osyczka et al. 2005). Notably, it seems that the production of superoxide is the result of one of these short circuits (see below). Obligate bifurcation of electron flow at the Q_o site (centre P) can be explained by a combination of a chemical control of ubiquinol oxidation and a "catalytic switch" of the Rieske iron–sulphur protein between two positions (Brandt 1998). In this model, obligate bifurcation is enforced by the fact that a (negatively charged) semiquinone species and reduced heme b_L cannot be present in the Q_o site at the same time (Fig. 6.2b). Similarly, Osyczka et al. (2004, 2005) suggested a "double-gating" mechanism which incorporates a semiquinone as intermediate that allows ubiquinol oxidation when the Rieske FeS cluster and heme b_L are both oxidized and that permits ubiquinone reduction in the reverse mode, when FeS cluster and heme b_L are both reduced, but forbids quinone electron transfer when FeS is oxidized and heme b_L is reduced. Also other models include a semiquinone as a true intermediate in a sequential mechanism of ubiquinol oxidation (Crofts 2004). Even such models imply that ubisemiquinone is formed only transiently at the ubihydroquinone oxidation site, and never accumulates to significant amounts in the functional enzyme. It follows that the occupancy for this redox intermediate is expected to be extremely low and thus formation of a semiquinone associated with the Q_o site has been proven very difficult to show experimentally (de Vries et al. 1981; Zhang et al. 2007; Cape et al. 2007). In models proposing a concerted electron transfer from ubiquinol onto both redox centres, the participation of a semiquinone intermediate is completely excluded (Osyczka et al. 2004, 2005; Trumpower 2002;

Zhu et al. 2007). The fundamental question, whether a semiquinone is a true intermediate during ubiquinol oxidation, has immediate implications for the mechanism of superoxide formation at the Q_o site.

6.4.2 *Mechanism of Superoxide Production at the Q_o Site*

It has been shown a long time ago (Cadenas et al. 1977; Boveris et al. 1976) that superoxide is formed at the ubiquinol oxidation centre of the cytochrome bc_1 complex. It was shown that complex III releases ROS to both sides of the membrane (St Pierre et al. 2002; Muller et al. 2004). The native enzyme does not produce measurable amounts of superoxide, but the rate is strongly increased under conditions of so-called oxidant-induced reduction, i.e. in the presence of the specific Q_i site inhibitor antimycin A, sufficient amounts of reducing equivalents and an oxidized downstream respiratory chain (Dröse and Brandt 2008; Cape et al. 2007; Muller et al. 2002, 2003; Forquer et al. 2006). Also a high membrane potential can greatly enhance superoxide production from the Q_o site (Rottenberg et al. 2009; Liu 2010). The latter can be explained by the slowdown of electron transfer from heme b_L to heme b_H when a positive outside membrane potential is applied. Increased superoxide production also has been observed in a cytochrome bc_1 complex from *Rhodobacter capsulatus* carrying a point mutation in the cytochrome *b* gene (Lee et al. 2011). Mutation of a conserved Tyrosine residue (Tyr302 in *R. capsulatus*) located near the Q_o site decreased catalytic activity and largely increased the superoxide production to rates that were comparable to the antimycin A-induced rates of the wild-type enzyme. The superoxide production of the mutant enzymes did not increase further upon addition of the specific Q_i site inhibitor. The authors concluded that the loss of Tyr302 led to electron leakage from the Q_o site to O_2 in a way independent of the antimycin A effect (Lee et al. 2011). This may indicate that this conserved tyrosine has a protective role in suppressing superoxide production at the Q_o site by a yet unknown molecular mechanism.

It is generally assumed in the ROS field that inhibiting reduction of ubiquinone at the Q_i site by antimycin A or slowing it down by a high membrane potential results in a backup of electrons in cytochrome *b* that leads to an accumulation of a semiquinone radical at the Q_o site, which can transfer its electron to oxygen forming superoxide (Muller et al. 2002, 2003; Muller 2000; Cape et al. 2009). However, as discussed earlier such a semiquinone radical at the Q_o site occurs, if at all, at very low occupancy even in the presence of antimycin A (see Sect. 6.3.1). When we analysed the ROS production of succinate-fuelled bovine heart SMP, we observed that the rate of antimycin A-induced superoxide production increases when the complex II activity is inhibited by the competitive inhibitors malonate or oxaloacetate (Dröse and Brandt 2008). A more than threefold increase is observed when approximately 75% of the succinate oxidase activity is inhibited. ROS generation is completely abolished in the presence of stigmatellin indicating that under these conditions superoxide is produced at the Q_o site. Such a stimulating effect of complex II inhibitors on the antimycin A-induced superoxide generation had been observed before (Trumpower and Simmons 1979; Ksenzenko et al. 1983, 1984) and an effect of the Q-pool redox state was proposed. Similarly, titrating the Q-pool redox state by the succinate:fumarate ratios resulted in a bell-shaped curve for the rate of antimycin A-induced ROS generation reaching a maximum at a ratio of 1:10 (Starkov and Fiskum 2001). Also measurements with intact mitochondria point towards a major impact of the redox state of the Q-pool on the rate of antimycin A-induced superoxide production (Dröse et al. 2009a, 2011a). To test whether indeed partial oxidation of the ubiquinone pool stimulated ROS production by antimycin-inhibited cytochrome bc_1 complex, we directly titrated the ubiquinone/ubiquinol ratio by mixing *n*-decylubihydroquinol with its oxidized counterpart *n*-decylubihydroquinone (Dröse and Brandt 2008). We found in experiments with SMP and with the purified lipid-activated cytochrome bc_1 complex from bovine heart that the rate of ROS production increases proportionally until a maximal value is reached at about 25–30% ubiquinone. We concluded that superoxide is generated at the Q_o site by reverse electron transfer from reduced heme b_L onto molecular oxygen and that oxidized ubiquinone serves as a redox mediator. While our model also includes a semiquinone as electron donor for superoxide production, this semiquinone is not the intermediate formed during normal turnover at the Q_o site of the cytochrome bc_1 complex. Rather ubiquinone serves as a redox mediator catalysing the transfer of an electron from heme b_L to oxygen. Our model was independently confirmed by Osyczka et al., who have carefully analysed the superoxide production of different cytochrome bc_1 mutants of *R. capsulatus* (Borek et al. 2008; Sarewicz et al. 2010). They concluded from their experimental analysis and kinetic models that the steady-state level of superoxide-generating semiquinone is predominantly controlled by reverse electron transfer from heme b_L to quinone. However, this interpretation has been recently challenged by Brand and coworkers (Quinlan et al. 2011). They confirmed in experiments with intact mitochondria from rat skeletal muscle mitochondria that the antimycin A-induced ROS generation was maximal at an intermediate reduction state of the Q-pool and could further show that it corresponded with the redox state of heme b_L, in that maximal production occurred at 70–80% reduction of heme b_L. Furthermore, when they applied a membrane potential, they found that ROS production also correlated with the redox state of heme b_H. In a kinetic model they predicted a semiquinone intermediate that is maximally stabilized by the fully reduced low-potential chain. However, this model does not take into account some important mechanistic restrictions, e.g. that a reduced heme b_L and a semiquinone anion cannot coexist in the Q_o site (Fig. 6.2b). Hence, superoxide production by reverse electron transfer from reduced heme b_L via oxidized ubiquinone as proposed by us and Osyczka still appears more likely.

6.4.3 Are Complex III ROS Involved in Redox Signalling?

There is increasing evidence that mitochondria-derived ROS are not always deleterious, but can also be an integral part of the cellular signalling machinery (for reviews, see Murphy et al. 2011; Hamanaka and Chandel 2010). Especially superoxide generated at the Q_o site of the cytochrome bc_1 complex has been implicated in redox signalling in physiological processes like hypoxic adaptation and—related to this—cardioprotection by ischemic preconditioning (for reviews, see Hamanaka and Chandel 2010; Guzy and Schumacker 2006). Initially, Schumacker et al. observed increased ROS production during simulated ischemic preconditioning in a cardiomyocyte cell culture model that was suppressed by the Q_o site inhibitor myxothiazol (Vanden Hoek et al. 1998). Subsequently, they showed that ROS generated at complex III are both required and sufficient to initiate HIF-1α stabilization during hypoxia (Chandel et al. 2000; Guzy et al. 2005). These findings are still controversially discussed, since a decrease and not an increase of mitochondrial ROS production was observed in isolated mitochondria at reduced pO_2 (Hoffman et al. 2007; Hoffman and Brookes 2009). The role of mitochondrial ROS in stabilization of HIF-1α has also been questioned, whereas mitochondria might affect HIF-1α stability by altering the cellular oxygen availability (Chua et al. 2010).

So far a general problem in pinpointing the actual generators of ROS under certain conditions in cellular or animal models has been that specific inhibitors or a knockdown of essential complex subunits have to be used to see an effect (Guzy et al. 2005; Bell et al. 2007). In case of complex III this is especially problematic, since there is no bypass for electrons in the respiratory chain and inhibition of complex III—either by knockdown or inhibitors—will always have effects other than ROS generation, like a reduced mitochondrial membrane potential, a reduced cellular energy charge or a metabolic shift to glycolysis. Besides, interpretations of some knockdown studies are hardly firm from a bioenergetic and structural point of view, since it was proposed that cells deficient in cytochrome *b* like the cybrid cell line used by Rana et al. (2000) were able to generate ROS at the Q_o site (Bell et al. 2007). However, these ROS cannot originate from the cytochrome bc_1 complex, since the X-ray structures clearly show that cytochrome *b* harbours the Q_o site (Hunte et al. 2003, 2008). Furthermore, it is quite often not considered that addition of antimycin A to a complex cellular system, where substrate availability cannot be controlled, will not only increase superoxide production at the Q_o site, but also at complex I.

On the other hand, a recent study revealed that terpestacin, a small microbial molecule that binds to the 13.4 kDa subunit (UQCRB in human; Subunit 6 in bovine), attenuated hypoxia-induced ROS production (Jung et al. 2010) without inhibiting mitochondrial respiration. In vivo, terpestacin blocked HIF-1α activation and tumour angiogenesis. This study may not only have important implications for the selective suppression of tumour progression, it also shows from a mechanistic point of view that a modulation of superoxide production at the Q_o site may be controlled by the accessory subunits of complex III. To fully understand the role of complex III-derived ROS in hypoxic signalling and in other cellular processes, it would be desirable to have reliable biomarkers that allowed discriminating between different generator sites of mitochondrial superoxide production. For this purpose, we have recently started to use redox-DIGE (difference gel electrophoresis) to monitor changes in the redox status of protein thiols under different ROS producing regimes. These investigations with isolated mitochondria did not only show that ROS generated at complex I and complex III target different proteins, it also revealed potential candidates for such biomarkers (unpublished results).

6.5 Other Factors Modulating ROS Production by the Respiratory Chain

It is reasonable to assume that also other components of the respiratory chain and the strict control of coupling in oxidative phosphorylation would have an impact on the mitochondrial ROS production, even when not directly involved in producing ROS. The important functional modulation of cytochrome *c* oxidase (complex IV) for the regulation of respiratory activity has been reviewed extensively elsewhere (Kadenbach et al. 2009; Arnold 2012) and is also covered in this book by Chaps. 10, 11, and 13. Cytochrome *c* has also to be considered, since it is a powerful superoxide scavenger and its loss, e.g. during ischaemia, is a major determinant of ROS production by mitochondria under pathophysiological conditions (Pasdois et al. 2011). ATP synthase (complex V) affects ROS production by the respiratory chain complexes via its role in controlling the membrane potential. While elevated levels of $\Delta\Psi$ stimulate superoxide production at both complex I and III during normal "downhill" or "forward" electron transfer, the highest values of ROS generation are observed especially under conditions of RET (see above). Thus, the absence of adenosine nucleotides or the presence of oligomycin promotes ROS production of succinate-fuelled mitochondria (Dröse et al. 2009a).

Recently, the impact of complex II (succinate:ubiquinone oxidoreductase) on mitochondrial ROS production has received more attention (Ralph et al. 2011). Complex II is unique among the respiratory chain complexes since it does not pump protons. It is the only membrane-bound component of the TCA cycle, thereby forming a direct interface between respiratory chain and acetyl-CoA oxidation (Cecchini 2003). It is well known that inhibition of complex II in intact mitochondria attenuates superoxide generation during RET (Dröse et al. 2009a, 2011a; Treberg et al. 2011). A fundamental question arising from these observations is whether the high succinate concentrations needed for RET indeed occur under physiological conditions. The group of Starkov has shown that succinate accumulates in mitochondria under hypoxic conditions (Starkov 2008). Furthermore, investigations with isolated mitochondria revealed that succinate-dependent superoxide generation by RET occurred also in the concurrent presence of NAD-linked substrates like malate/glutamate (Zoccarato et al. 2007; Muller et al. 2008). Therefore, it seems plausible that the oxidative damage occurring upon reperfusion after hypoxic periods is due to RET and that the cardioprotective effect of complex II inhibitors can be explained by an attenuation of ROS production (Dröse et al. 2009a, 2011a). It has been shown that diazoxide, a known K_{ATP} channel opener that also inhibits complex II (Schäfer et al. 1969; Hanley et al. 2002; Dröse et al. 2006), can attenuate the ROS generation that occurs after and during an ischemic period and upon reperfusion in isolated rat heart mitochondria (Ozcan et al. 2002) and isolated Langendorff perfused rat hearts (Pasdois et al. 2008). However, we found with succinate fuelled isolated rat heart mitochondria that complex II inhibition can also stimulate superoxide production by the Q_o site of complex III (Dröse et al. 2009a, 2011a). This can be explained by the resulting increase in oxidation of the Q-pool, which favours superoxide production under conditions of oxidant-induced reduction (Dröse and Brandt 2008; see Sect. 6.3.2). This could explain mechanistically how "signalling ROS" are generated at complex III that have been implicated in ischemic and pharmacological preconditioning (Vanden Hoek et al. 1998; Forbes et al. 2001). Our finding that complex II inhibitors (e.g. diazoxide, oxaloacetate, atpenin A5) have ambivalent effects on mitochondrial ROS generation under specific conditions have been confirmed by independent investigations (Liu et al. 2010; Hirata et al. 2011). Furthermore, Dos Santos et al. (Pasdois et al. 2008) observed a transient increase in ROS generation in perfused hearts during preconditioning with diazoxide, although it also decreased ROS produced during reperfusion. In addition to this direct modulation of superoxide generation by complexes I and III, complex II may also affect ROS generation at other mitochondrial sites via its unique connection to the TCA cycle (for a recent comprehensive review, see Ralph et al. 2011).

6.6 An Integrated Model of ROS Generation by the Respiratory Chain

We propose an integrated model of ROS generation by the respiratory chain that assigns distinct roles to complexes I–III (Dröse et al. 2009a; Fig. 6.3). Since complex I releases superoxide completely into the mitochondrial matrix (St Pierre et al. 2002; Muller et al. 2004), the risk that it is deleterious by inflicting damage

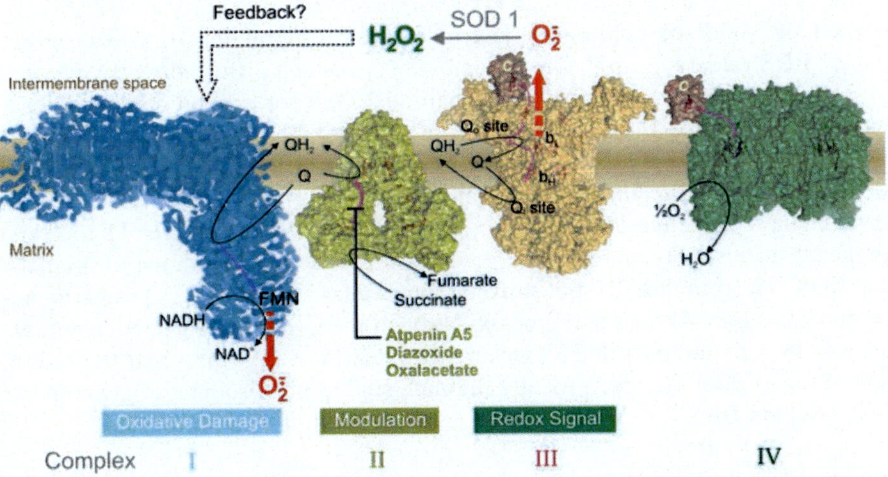

Fig. 6.3 A comprehensive model of ROS generation by the respiratory chain. This overview summarizes the distinct roles of respiratory chain complex I–complex III in mitochondrial ROS production as detailed in the text. Modulation of ROS production at complex I and III by inhibitors of complex II occurs only under specific conditions, i.e. when succinate contributes substantially to the electron supply of the respiratory chain

on mitochondrial DNA (Costa et al. 2011) or by inducing opening of the mitochondrial permeability pore (Halestrap and Pasdois 2009) is high. Superoxide production by complex I can occur in the forward or during reverse mode. In the forward mode, superoxide generation is mainly regulated by the NADH/NAD^+ ratio (Kussmaul and Hirst 2006) and to a lesser extent by the membrane potential, while in the reverse mode the membrane potential (Votyakova and Reynolds 2001) and the redox state of the Q-pool (Treberg et al. 2011) seem to be the main determinants. A further regulation by the active/deactive transition of complex I (reviewed in Vinogradov 1998) may have to be taken into account under physiological and pathophysiological conditions, e.g. hypoxia (Maklashina et al. 2002; Galkin et al. 2009).

Since the recent X-ray data on mitochondrial complex I have shown that the ubiquinone binding site is located in the peripheral arm at a distance of about 30 Å above the membrane surface (Hunte et al. 2010), superoxide formed by complex I will be released always into the matrix, even if this would occur not only at the FMN site (Pryde and Hirst 2011) but during RET also at the Q-binding site (Treberg et al. 2011). In contrast, superoxide generated at the Q_o site of complex III can be expected to be released into the intermembrane space (St Pierre et al. 2002; Muller et al. 2004). This superoxide production is promoted by an elevated membrane potential (Rottenberg et al. 2009) and controlled by the redox states of the Q-pool and of heme b_L (Dröse and Brandt 2008; Quinlan et al. 2011). We suggest that reverse electron transfer from reduced heme b_L onto oxidized ubiquinone is the molecular mechanism, which transiently generates a semiquinone radical as direct source for superoxide (Dröse and Brandt 2008; Borek et al. 2008; Sarewicz et al. 2010). Hence, ROS production at the Q_o site of complex III is controlled by a delicate balance between membrane potential, availability of reduced and oxidized ubiquinone and redox state of complex III (Dröse et al. 2009a). ROS released into the intermembrane space can diffuse directly into the cytosol, where they may serve as "signalling ROS" for several stress-induced signalling pathways. Finally, inhibition of complex II can modulate the ROS generation at complexes I and III under specific conditions (Dröse et al. 2009a, 2011a), e.g. when succinate is accumulated in the mitochondrial matrix. This may attenuate the deleterious ROS production due to RET at complex I by lowering the electron supply and $\Delta\Psi$, while it may stimulate the generation of "signalling ROS" at complex III by shifting the redox state of the Q-pool to an intermediate state. At the same time oxidation of the Q-pool is expected to reduce superoxide generation by RET (Treberg et al. 2011) suggesting that the redox state of the Q-pool may reciprocally regulate superoxide production at complex I and complex III.

Acknowledgements We thank Lea Bleier, Volker Zickermann and Klaus Zwicker for careful and critical reading of our manuscript and for helpful discussion. Funding by the Deutsche Forschungsgemeinschaft, SFB815 Project A2, and the BMBF project 0315584A: GerontoMitoSys, is gratefully acknowledged. This study was also supported by the Excellence Initiative of the German Federal and State Governments (EXC 115).

References

Angerer H, Zwicker K, Wumaier Z, Sokolova L, Heide H, Steger M, Kaiser S, Nübel E, Brutschy B, Radermacher M, Brandt U, Zickermann V (2011) A scaffold of accessory subunits links the peripheral arm and the distal proton pumping module of mitochondrial complex I. Biochem J 437:279–288

Arnold S (2012) The power of life-cytochrome *c* oxidas takes center stage in metabolic control, cell signalling and survival. Mitochondrion 12:46–56

Belevich G, Knuuti J, Verkhovsky MI, Wikström M, Verkhovskaya M (2011) Probing the mechanistic role of the long alpha-helix in subunit L of respiratory complex I from *Escherichia coli* by site-directed mutagenesis. Mol Microbiol 82:1086–1095

Bell EL, Klimova TA, Eisenbart J, Moraes CT, Murphy MP, Budinger GRS, Chandel NS (2007) The Q_o site of the mitochondrial complex III is required for the transduction of hypoxic signaling via reactive oxygen species production. J Cell Biol 177:1029–1036

Borek A, Sarewicz M, Osyczka A (2008) Movement of the iron-sulfur head domain of cytochrome bc_1 transiently opens the catalytic Q_o site for reaction with oxygen. Biochemistry 47:12365–12370

Boveris A, Cadenas E, Stoppani AO (1976) Role of ubiquinone in the mitochondrial generation of hydrogen peroxide. Biochem J 156:435–444

Brand MD (2010) The sites and topology of mitochondrial superoxide production. Exp Gerontol 45:466–472

Brandt U (1996) Bifurcated ubihydroquinone-oxidation in the cytochrome bc_1 complex by proton-gated charge-transfer. FEBS Lett 387:1–6

Brandt U (1998) The chemistry and mechanics of ubihydroquinone oxidation at center P (Q_o) of the cytochrome *bc1* complex. Biochim Biophys Acta 1365:261–268

Brandt U (2006) Energy converting NADH:quinone oxidoreductase (complex I). Annu Rev Biochem 75:69–92

Brandt U (2011) A two-state stabilization-change mechanism for proton-pumping complex I. Biochim Biophys Acta 1807:1364–1369

Brandt U, Trumpower BL (1994) The protonmotive Q cycle in mitochondria and bacteria. CRC Crit Rev Biochem 29:165–197

Brandt U, Haase U, Schägger H, von Jagow G (1991) Significance of the "Rieske" iron-sulfur protein for formation and function of the ubiquinol oxidation pocket of mitochondrial cytochrome *c* reductase (bc_1 complex). J Biol Chem 266:19958–19964

Cadenas E, Boveris A (1980) Enhancement of hydrogen-peroxide formation by protophores and ionophores in antimycin-supplemented mitochondria. Biochem J 188:31–37

Cadenas E, Boveris A, Ragan CI, Stoppani AO (1977) Production of superoxide radicals and hydrogen peroxide by NADH-ubiquinone reductase and ubiquinol-cytochrome *c* reductase from beef-heart mitochondria. Arch Biochem Biophys 180:248–257

Cape JL, Bowman MK, Kramer DM (2007) A semiquinone intermediate generated at the Q_o site of the cytochrome *bc1* complex: importance for the Q-cycle and superoxide production. Proc Natl Acad Sci USA 104:7887–7892

Cape JL, Aidasani D, Kramer DM, Bowman MK (2009) Substrate redox potential controls superoxide production kinetics in the cytochrome bc_1 complex. Biochemistry 48:10716–10723

Carroll J, Fearnley IM, Skehel JM, Shannon RJ, Hirst J, Walker JE (2006) Bovine complex I is a complex of 45 different subunits. J Biol Chem 281:32724–32727

Castellani M, Covian R, Kleinschroth T, Anderka O, Ludwig B, Trumpower BL (2010) Direct demonstration of half-of-the-sites reactivity in the dimeric cytochrome bc_1 complex. J Biol Chem 285:502–510

Cecchini G (2003) Function and structure of complex II of the respiratory chain. Annu Rev Biochem 72:77–109

Chandel NS, McClintock DS, Feliciano CE, Wood TM, Melendez JA, Rodriguez AM, Schumacker PT (2000) Reactive oxygen species generated at mitochondrial complex III stabilize hypoxia-inducible factor-1α during hypoxia. A mechanism of O_2 sensing. J Biol Chem 275:25130–25138

Chen SX, Schopfer P (1999) Hydroxyl-radical production in physiological reactions - A novel function of peroxidase. Eur J Biochem 260:726–735

Chen Q, Vazquez EJ, Moghaddas S, Hoppel CL, Lesnefsky EJ (2003) Production of reactive oxygen species by mitochondria: central role of complex III. J Biol Chem 278:36027–36031

Chua YL, Dufour E, Dassa EP, Rustin P, Jacobs HT, Taylor CT, Hagen T (2010) Stabilization of hypoxia-inducible factor-1 alpha protein in hypoxia occurs independently of mitochondrial reactive oxygen species production. J Biol Chem 285:31277–31284

Clason T, Ruiz T, Schägger H, Peng G, Zickermann V, Brandt U, Michel H, Radermacher M (2010) The structure of eukaryotic and prokaryotic complex I. J Struct Biol 169:81–88

Costa RAP, Romagna CD, Pereira JL, Souza-Pinto NC (2011) The role of mitochondrial DNA damage in the cytotoxicity of reactive oxygen species. J Bioenerg Biomembr 43:25–29

Covian R, Trumpower BL (2005) Rapid electron transfer between monomers when the cytochrome bc_1 complex dimer is reduced through center N. J Biol Chem 280:22732–22740

Covian R, Trumpower BL (2006) Regulatory interactions between ubiquinol oxidation and ubiquinone reduction sites in the dimeric cytochrome bc_1 complex. J Biol Chem 281:30925–30932

Covian R, Trumpower BL (2008) Regulatory interactions in the dimeric cytochrome bc_1 complex: the advantages of being a twin. Biochim Biophys Acta 1777:1079–1091

Cox AG, Winterbourn CC, Hampton MB (2010) Mitochondrial peroxiredoxin involvement in antioxidant defence and redox signalling. Biochem J 425:313–325

Crofts AR (2004) Proton-coupled electron transfer at the Q_o-site of the bc_1 complex controls the rate of ubihydroquinone oxidation. Biochim Biophys Acta 1655:77–92

de Vries S, Albracht SPJ, Berden JA, Slater EC (1981) A new species of bound ubisemiquinone anion in QH_2: cytochrome *c* oxidoreductase. J Biol Chem 256:11996–11998

Drechsel DA, Patel M (2010) Respiration-dependent H_2O_2 removal in brain mitochondria via the thioredoxin/peroxiredoxin system. J Biol Chem 285:27850–27858

Dröse S, Brandt U (2008) The mechanism of mitochondrial superoxide production by the cytochrome bc_1 complex. J Biol Chem 283:21649–21654

Dröse S, Galkin A, Brandt U (2005) Proton pumping by complex I (NADH:ubiquinone oxidoreductase) from *Yarrowia lipolytica* reconstituted into proteoliposomes. Biochim Biophys Acta 1710:87–95

Dröse S, Brandt U, Hanley PJ (2006) K^+-independent actions of diazoxide question the role of inner membrane K_{ATP} channels in mitochondrial cytoprotective signaling. J Biol Chem 281:23733–23739

Dröse S, Hanley PJ, Brandt U (2009a) Ambivalent effects of diazoxide on mitochondrial ROS production at respiratory chain complexes I and III. Biochim Biophys Acta 1790:558–565

Dröse S, Galkin A, Brandt U (2009b) Measurement of superoxide formation by mitochondrial complex I of *Yarrowia lipolytica*. Methods Enzymol 456:475–490

Dröse S, Bleier L, Brandt U (2011a) A common mechanism links differently acting complex II inhibitors to cardioprotection: modulation of mitochondrial reactive oxygen species production. Mol Pharmacol 79:814–822

Dröse S, Krack S, Sokolova L, Zwicker K, Barth HD, Morgner N, Heide H, Steger M, Nübel E, Zickermann V, Kerscher S, Brutschy B, Radermacher M, Brandt U (2011b) Functional dissection of the proton pumping modules of mitochondrial complex I. PLoS Biol 9:e1001128

Efremov RG, Sazanov LA (2011) Structure of the membrane domain of respiratory complex I. Nature 476:414–420

Efremov RG, Baradaran R, Sazanov LA (2010) The architecture of respiratory complex I. Nature 465:441–445

Esterhazy D, King MS, Yakovlev G, Hirst J (2008) Production of reactive oxygen species by complex I (NADH:ubiquinone oxidoreductase) from *Escherichia coli* and comparison to the enzyme from mitochondria. Biochemistry 47:3964–3971

Euro L, Belevich G, Verkhovsky MI, Wikström M, Verkhovskaya M (2008) Conserved lysine residues of the membrane subunit NuoM are involved in energy conversion by the proton-pumping NADH:ubiquinone oxidoreductase (Complex I). Biochim Biophys Acta 1777:1166–1172

Fato R, Bergamini C, Bortolus M, Maniero AL, Leoni S, Ohnishi T, Lenaz G (2009) Differential effects of mitochondrial complex I inhibitors on production of reactive oxygen species. Biochim Biophys Acta 1787:384–392
Fendel U, Tocilescu MA, Kerscher S, Brandt U (2008) Exploring the inhibitor binding pocket of respiratory complex I. Biochim Biophys Acta 1777:660–665
Forbes RA, Steenbergen C, Murphy E (2001) Diazoxide-induced cardioprotection requires signaling through a redox-sensitive mechanism. Circ Res 88:802–809
Forquer I, Covian R, Bowman MK, Trumpower BL, Kramer DM (2006) Similar transition states mediate the Q-cycle and superoxide production by the cytochrome bc_1 complex. J Biol Chem 281:38459–38465
Galkin A, Brandt U (2005) Superoxide radical formation by pure complex I (NADH:ubiquinone oxidoreductase) from *Yarrowia lipolytica*. J Biol Chem 280:30129–30135
Galkin AS, Grivennikova VG, Vinogradov AD (1999) $\rightarrow H^+/2\ e^-$ stoichiometry in NADH-quinone reductase reactions catalyzed by bovine heart submitochondrial particles. FEBS Lett 451:157–161
Galkin A, Dröse S, Brandt U (2006) The proton pumping stoichiometry of purified mitochondrial complex I reconstituted into proteoliposomes. Biochim Biophys Acta 1757:1575–1581
Galkin A, Abramov AY, Frakich N, Duchen MR, Moncada S (2009) Lack of oxygen deactivates mitochondrial complex I. J Biol Chem 284:36055–36061
Genova ML, Ventura B, Giuliano G, Bovina C, Formiggini G, Parenti Castelli G, Lenaz G (2001) The site of production of superoxide radical in mitochondrial complex I is not a bound ubisemiquinone but presumably iron-sulfur cluster N2. FEBS Lett 505:364–368
Gomes A, Fernandes E, Lima JLFC (2005) Fluorescence probes used for detection of reactive oxygen species. J Biochem Biophys Methods 65:45–80
Gong X, Yu L, Xia D, Yu CA (2005) Evidence for electron equilibrium between the two hemes *b*(L) in the dimeric cytochrome bc_1 complex. J Biol Chem 280:9251–9257
Grivennikova VG, Vinogradov AD (2006) Generation of superoxide by the mitochondrial complex I. Biochim Biophys Acta 1757:553–561
Guzy RD, Schumacker PT (2006) Oxygen sensing by mitochondria at complex III: the paradox of increased reactive oxygen species during hypoxia. Exp Physiol 91:807–819
Guzy RD, Hoyos B, Robin E, Chen H, Liu LP, Mansfield KD, Simon MC, Hammerling U, Schumacker PT (2005) Mitochondrial complex III is required for hypoxia-induced ROS production and cellular oxygen sensing. Cell Metab 1:401–408
Halestrap AP, Pasdois P (2009) The role of the mitochondrial permeability transition pore in heart disease. Biochim Biophys Acta 1787:1402–1415
Hamanaka RB, Chandel NS (2010) Mitochondrial reactive oxygen species regulate cellular signaling and dictate biological outcomes. Trends Biochem Sci 35:505–513
Hanley PJ, Mickel M, Löffler M, Brandt U, Daut J (2002) K_{ATP} channel-independent targets of diazoxide and 5-hydroxydecanoate in the heart. J Physiol 542:735–741
Herrmann JM, Riemer J (2010) The intermembrane space of mitochondria. Antioxid Redox Signal 13:1341–1358
Hirata N, Shim YH, Pravdic D, Lohr NL, Pratt PF, Weihrauch D, Kersten JR, Warltier DC, Bosnjak ZJ, Bienengraeber M (2011) Isoflurane differentially modulates mitochondrial reactive oxygen species production via forward versus reverse electron transport flow. Anesthesiology 115:531–540
Hoffman DL, Brookes PS (2009) Oxygen sensitivity of mitochondrial reactive oxygen species generation depends on metabolic conditions. J Biol Chem 284:16236–16245
Hoffman DL, Salter JD, Brookes PS (2007) Response of mitochondrial reactive oxygen species generation to steady-state oxygen tension: implications for hypoxic cell signaling. Am J Physiol Heart Circul Physiol 292:H101–H108
Hunte C, Palsdottir H, Trumpower BL (2003) Protonmotive pathways and mechanisms in the cytochrome bc_1 complex. FEBS Lett 545:39–46
Hunte C, Solmaz S, Palsdóttir H, Wenz T (2008) A structural perspective on mechanism and function of the cytochrome bc_1 complex. Results Probl Cell Differ 45:253–278

Hunte C, Zickermann V, Brandt U (2010) Functional modules and structural basis of conformational coupling in mitochondrial complex I. Science 329:448–451

James AM, Cocheme HM, Smith RAJ, Murphy MP (2005) Interactions of mitochondria-targeted and untargeted ubiquinones with the mitochondrial respiratory chain and reactive oxygen species - Implications for the use of exogenous ubiquinones as therapies and experimental tools. J Biol Chem 280:21295–21312

Jung HJ, Shim JS, Lee J, Song YM, Park KC, Choi SH, Kim ND, Yoon JH, Mungai PT, Schumacker PT, Kwon HJ (2010) Terpestacin inhibits tumor angiogenesis by targeting UQCRB of mitochondrial complex III and suppressing hypoxia-induced reactive oxygen species production and cellular oxygen sensing. J Biol Chem 285:11584–11595

Kadenbach B, Ramzan R, Vogt S (2009) Degenerative diseases, oxidative stress and cytochrome *c* oxidase function. Trends Mol Med 15:139–147

Kerscher S, Dröse S, Zickermann V, Brandt U (2008) The three families of respiratory NADH dehydrogenases. Results Problems Cell Different 45:185–222

King MS, Sharpley MS, Hirst J (2009) Reduction of hydrophilic ubiquinones by the flavin in mitochondrial NADH:ubiquinone oxidoreductase (complex I) and production of reactive oxygen species. Biochemistry 48:2053–2062

Kowaltowski AJ, Souza-Pinto NC, Castilho RF, Vercesi AE (2009) Mitochondria and reactive oxygen species. Free Radic Biol Med 47:333–343

Ksenzenko M, Konstantinov AA, Khomutov GB, Tikhonov AN, Ruuge EK (1983) Effect of electron transfer inhibitors on superoxide generation in the cytochrome bc_1 site of the mitochondrial respiratory chain. FEBS Lett 155:19–24

Ksenzenko M, Konstantinov AA, Khomutov GB, Tikhonov AN, Ruuge EK (1984) Relationships between the effects of redox potential, alpha-thenoyltrifluoroacetone and malonate on O_2^- and H_2O_2 generation by submitochondrial particles in the presence of succinate and antimycin. FEBS Lett 175:105–108

Kushnareva Y, Murphy AN, Andreyev A (2002) Complex I-mediated reactive oxygen species generation: modulation by cytochrome *c* and $NAD(P)^+$ oxidation-reduction state. Biochem J 368:545–553

Kussmaul L, Hirst J (2006) The mechanism of superoxide production by NADH: ubiquinone oxidoreductase (complex I) from bovine heart mitochondria. Proc Natl Acad Sci USA 103:7607–7612

Lambert AJ, Brand MD (2004a) Superoxide production by NADH:ubiquinone oxidoreductase (complex I) depends on the pH gradient across the mitochondrial inner membrane. Biochem J 382:511–517

Lambert AJ, Brand MD (2004b) Inhibitors of the quinone-binding site allow rapid superoxide production from mitochondrial NADH:ubiquinone oxidoreductase (complex I). J Biol Chem 279:39414–39420

Lanciano P, Lee DW, Yang HH, Darrouzet E, Daldal F (2011) Intermonomer electron transfer between the low-potential *b* hemes of cytochrome bc_1. Biochemistry 50:1651–1663

Lee DW, Selamoglu N, Lanciano P, Cooley JW, Forquer I, Kramer DM, Daldal F (2011) Loss of a conserved tyrosine residue of cytochrome *b* induces reactive oxygen species production by cytochrome bc_1. J Biol Chem 286:18139–18148

Lin MT, Beal MF (2006) Mitochondrial dysfunction and oxidative stress in neurodegenerative diseases. Nature 443:787–795

Liu SS (2010) Mitochondrial Q cycle-derived superoxide and chemiosmotic bioenergetics. Ann N Y Acad Sci 1201:84–95

Liu Y, Fiskum G, Schubert D (2002) Generation of reactive oxygen species by the mitochondrial electron transport chain. J Neurochem 80:780–787

Liu B, Zhu XH, Chen CL, Hu KL, Swartz HM, Chen YR, He GL (2010) Opening of the mitoK(ATP) channel and decoupling of mitochondrial complex II and III contribute to the suppression of myocardial reperfusion hyperoxygenation. Mol Cell Biochem 337:25–38

Magnitsky S, Toulokhonova L, Yano T, Sled VD, Hägerhall C, Grivennikova VG, Burbaev DS, Vinogradov AD, Ohnishi T (2002) EPR characterization of ubisemiquinones and iron-sulfur

cluster N2, central components of the energy coupling in the NADH- ubiquinone oxidoreductase (complex I) in situ. J Bioenerg Biomembr 34:193–208
Maklashina E, Sher Y, Zhou HZ, Gray MO, Karliner JS, Cecchini G (2002) Effect of anoxia/reperfusion on the reversible active/de-active transition of NADH-ubiquinone oxidoreductase (complex I) in rat heart. Biochim Biophys Acta 1556:6–12
Mathiesen C, Hägerhäll C (2002) Transmembrane topology of the NuoL, M and N subunits of NADH:quinone oxidoreductase and their homologues among membrane-bound hydrogenases and bona fide antiporters. Biochim Biophys Acta 1556:121–132
Matsuno-Yagi A, Yagi T (2001) Introduction: complex I - an L-shaped black box. J Bioenerg Biomembr 33:155–157
Michel J, DeLeon-Rangel J, Zhu ST, Van Ree K, Vik SB (2011) Mutagenesis of the L, M, and N subunits of complex I from *Escherichia coli* indicates a common role in function. PLoS One 6:e17420
Mitchell P (1975) The protonmotive Q cycle: a general formulation. FEBS Lett 59:137–139
Morgner N, Zickermann V, Kerscher S, Wittig I, Abdrakhmanova A, Barth HD, Brutschy B, Brandt U (2008) Subunit mass fingerprinting of mitochondrial complex I. Biochim Biophys Acta 1777:1384–1391
Mourier A, Larsson NG (2011) Tracing the trail of protons through complex I of the mitochondrial respiratory chain. PLoS Biol 9:e1001129
Muller F (2000) The nature and mechanism of superoxide production by the electron transport chain: its relevance to aging. J Am Aging Assoc 23:227–253
Muller F, Crofts AR, Kramer DM (2002) Multiple Q-cycle bypass reactions at the Q_o site of the cytochrome bc_1 complex. Biochemistry 41:7866–7874
Muller FL, Roberts AG, Bowman MK, Kramer DM (2003) Architecture of the Q_o site of the cytochrome bc_1 complex probed by superoxide production. Biochemistry 42:6493–6499
Muller FL, Liu YH, Van Remmen H (2004) Complex III releases superoxide to both sides of the inner mitochondrial membrane. J Biol Chem 279:49064–49073
Muller FL, Lustgarten MS, Jang Y, Richardson A, Van Remmen H (2007) Trends in oxidative aging theories. Free Radic Biol Med 43:477–503
Muller FL, Liu YH, Abdul-Ghani MA, Lustgarten MS, Bhattacharya A, Jang YC, Van Remmen H (2008) High rates of superoxide production in skeletal-muscle mitochondria respiring on both complex I- and complex II-linked substrates. Biochem J 409:491–499
Murphy MP (2009) How mitochondria produce reactive oxygen species. Biochem J 417:1–13
Murphy MP, Holmgren A, Larsson NG, Halliwell B, Chang CJ, Kalyanaraman B, Rhee SG, Thornalley PJ, Partridge L, Gems D, Nystrom T, Belousov V, Schumacker PT, Winterbourn CC (2011) Unraveling the biological roles of reactive oxygen species. Cell Metab 13:361–366
Nakamaru-Ogiso E, Kao MC, Chen H, Sinha SC, Yagi T, Ohnishi T (2010) The membrane subunit NuoL(ND5) is involved in the indirect proton pumping mechanism of *Escherichia coli* complex I. J Biol Chem 285:39070–39078
Navarro A, Boveris A (2007) The mitochondrial energy transduction system and the aging process. Am J Physiol Cell Physiol 292:C670–C686
Ohnishi T, Trumpower BL (1980) Differential effects of antimycin on ubisemiquinone bound in different environments in isolated succinate:cytochrome *c* reductase complex. J Biol Chem 255:3278–3284
Ohnishi ST, Ohnishi T, Muranaka S, Fujita H, Kimura H, Uemura K, Yoshida K, Utsumi K (2005) A possible site of superoxide generation in the complex I segment of rat heart mitochondria. J Bioenerg Biomembr 37:1–15
Ohnishi T, Nakamaru-Ogiso E, Ohnishi ST (2010) A new hypothesis on the simultaneous direct and indirect proton pump mechanisms in NADH-quinone oxidoreductase (complex I). FEBS Lett 584:4131–4137
Osyczka A, Moser CC, Daldal F, Dutton PL (2004) Reversible redox energy coupling in electron transfer chains. Nature 427:607–612
Osyczka A, Moser CC, Dutton PL (2005) Fixing the Q cycle. Trends Biochem Sci 30:176–182

Ozcan C, Bienengraeber M, Dzeja PP, Terzic A (2002) Potassium channel openers protect cardiac mitochondria by attenuating oxidant stress at reoxygenation. Am J Physiol Heart Circ Physiol 282:H531–H539

Pasdois P, Beauvoit B, Tariosse L, Vinassa B, Bonoron-Adele S, Dos Santos P (2008) Effect of diazoxide on flavoprotein oxidation and reactive oxygen species generation during ischemia-reperfusion: a study on Langendorff-perfused rat hearts using optic fibers. Am J Physiol Heart Circ Physiol 294:H2088–H2097

Pasdois P, Parker JE, Griffiths EJ, Halestrap AP (2011) The role of oxidized cytochrome *c* in regulating mitochondrial reactive oxygen species production and its perturbation in ischaemia. Biochem J 436:493–505

Pryde KR, Hirst J (2011) Superoxide is produced by the reduced flavin in mitochondrial complex I. J Biol Chem 286:18056–18065

Quinlan CL, Gerencser AA, Treberg JR, Brand MD (2011) The mechanism of superoxide production by the antimycin-inhibited mitochondrial Q-cycle. J Biol Chem 286:31361–31372

Ralph SJ, Moreno-Sanchez R, Neuzil J, Rodriguez-Enriquez S (2011) Inhibitors of succinate:quinone reductase/complex II regulate production of mitochondrial reactive oxygen species and protect normal cells from ischemic damage but induce specific cancer cell death. Pharm Res 28:2695–2730

Rana M, de Coo I, Diaz F, Smeets H, Moraes CT (2000) An out-of-frame cytochrome *b* gene deletion from a patient with parkinsonism is associated with impaired complex III assembly and an increase in free radical production. Ann Neurol 48(5):774–781

Rottenberg H, Covian R, Trumpower BL (2009) Membrane potential greatly enhances superoxide generation by the cytochrome bc_1 complex reconstituted into phospholipid vesicles. J Biol Chem 284:19203–19210

Rydström J (2006) Mitochondrial NADPH, transhydrogenase and disease. Biochim Biophys Acta 1757:721–726

Sarewicz M, Borek A, Cieluch E, Swierczek M, Osyczka A (2010) Discrimination between two possible reaction sequences that create potential risk of generation of deleterious radicals by cytochrome bc1. Implications for the mechanism of superoxide production. Biochim Biophys Acta 1797:1820–1827

Sazanov LA, Hinchliffe P (2006) Structure of the hydrophilic domain of respiratory complex I from *Thermus thermophilus*. Science 311:1430–1436

Schäfer G, Wegener C, Portenhauser R, Bojanovski D (1969) Diazoxide, an inhibitor of succinate oxidation. Biochem Pharmacol 18:2678–2681

Scheffler IE (2008) Mitochondria, 2nd edn. Wiley, Hoboken, NJ, pp 1–484

Selivanov VA, Zeak JA, Roca J, Cascante M, Trucco M, Votyakova TV (2008) The role of external and matrix pH in mitochondrial reactive oxygen species generation. J Biol Chem 283:29292–29300

St Pierre J, Buckingham JA, Roebuck SJ, Brand MD (2002) Topology of superoxide production from different sites in the mitochondrial electron transport chain. J Biol Chem 277:44784–44790

Stanley BA, Sivakumaran V, Shi S, McDonald I, Lloyd D, Watson WH, Aon MA, Paolocci N (2011) Thioredoxin reductase-2 is essential for keeping low levels of H_2O_2 emission from isolated heart mitochondria. J Biol Chem 286:33669–33677

Starkov AA (2008) The role of mitochondria in reactive oxygen species metabolism and signaling. Ann N Y Acad Sci 1147:37–52

Starkov AA, Fiskum G (2001) Myxothiazol induces H_2O_2 production from mitochondrial respiratory chain. Biochem Biophys Res Commun 281:645–650

Starkov AA, Fiskum G (2003) Regulation of brain mitochondrial H_2O_2 production by membrane potential and NAD(P)H redox state. J Neurochem 86:1101–1107

Starkov AA, Fiskum G, Chinopoulos C, Lorenzo BJ, Browne SE, Patel MS, Beal MF (2004) Mitochondrial alpha-ketoglutarate dehydrogenase complex generates reactive oxygen species. J Neurosci 24:7779–7788

Steimle S, Bajzath C, Dorner K, Schulte M, Bothe V, Friedrich T (2011) Role of subunit NuoL for proton translocation by respiratory complex I. Biochemistry 50:3386–3393

Swierczek M, Cieluch E, Sarewicz M, Borek A, Moser CC, Dutton PL, Osyczka A (2010) An electronic bus bar lies in the core of cytochrome bc_1. Science 329:451–454

Tahara EB, Navarete FDT, Kowaltowski AJ (2009) Tissue-, substrate-, and site-specific characteristics of mitochondrial reactive oxygen species generation. Free Radic Biol Med 46:1283–1297

Tocilescu MA, Fendel U, Zwicker K, Kerscher S, Brandt U (2007) Exploring the ubiquinone binding cavity of respiratory complex I. J Biol Chem 282:29514–29520

Tocilescu MA, Fendel U, Zwicker K, Dröse S, Kerscher S, Brandt U (2010) The role of a conserved tyrosine in the 49-kDa subunit of complex I for ubiquinone binding and reduction. Biochim Biophys Acta 1797:625–632

Torres-Bacete J, Nakamaru-Ogiso E, Matsuno-Yagi A, Yagi T (2007) Characterization of the NuoM (ND4) subunit in *Escherichia coli* NDH-1 - Conserved charged residues essential for energy-coupled activities. J Biol Chem 282:36914–36922

Treberg JR, Brand MD (2011) A model of the proton translocation mechanism of complex I. J Biol Chem 286:17579–17584

Treberg JR, Quinlan CL, Brand MD (2010) Hydrogen peroxide efflux from muscle mitochondria underestimates matrix superoxide production - a correction using glutathione depletion. FEBS J 277:2766–2778

Treberg JR, Quinlan CL, Brand MD (2011) Evidence for two sites of superoxide production by mitochondrial NADH-ubiquinone oxidoreductase (complex I). J Biol Chem 286:27103–27110

Tretter L, Adam-Vizi V (2007) Moderate dependence of ROS formation on $\Delta\Psi m$ in isolated brain mitochondria supported by NADH-linked substrates. Neurochem Res 32:569–575

Trumpower BL (2002) A concerted, alternating sites mechanism of ubiquinol oxidation by the dimeric cytochrome bc_1 complex. Biochim Biophys Acta 1555:166–173

Trumpower BL, Simmons Z (1979) Diminished inhibition of mitochondrial electron transfer from succinate to cytochrome *c* by thenoyltrifluoroacetone induced by antimycin. J Biol Chem 254:4608–4616

Vanden Hoek TL, Becker LB, Shao Z, Li C, Schumacker PT (1998) Reactive oxygen species released from mitochondria during brief hypoxia induce preconditioning in cardiomyocytes. J Biol Chem 273:18092–18098

Vinogradov AD (1998) Catalytic properties of the mitochondrial NADH-ubiquinone oxidoreductase (complex I) and the pseudo-reversible active/inactive enzyme transition. Biochim Biophys Acta 1364:169–185

Vinogradov AD, Grivennikova VG (2005) Generation of superoxide-radical by the NADH:ubiquinone oxidoreductase of heart mitochondria. Biochemistry (Mosc) 70:120–127

Votyakova TV, Reynolds IJ (2001) $\Delta\psi_m$-Dependent and -independent production of reactive oxygen species by rat brain mitochondria. J Neurochem 79:266–277

Votyakova TV, Reynolds IJ (2004) Detection of hydrogen peroxide with Amplex Red: interference by NADH and reduced glutathione auto-oxidation. Arch Biochem Biophys 431:138–144

Wikström MKF (1984) Two protons are pumped from the mitochondrial matrix per electron transferred between NADH and ubiquinone. FEBS Lett 169:300–304

Winterbourn CC, Hampton MB (2008) Thiol chemistry and specificity in redox signaling. Free Radic Biol Med 45:549–561

Zhang H, Osyczka A, Dutton PL, Moser CC (2007) Exposing the Complex III Q_o semiquinone radical. Biochim Biophys Acta 1767:883–887

Zhou MJ, Diwu ZJ, Panchuk Voloshina N, Haugland RP (1997) A stable nonfluorescent derivative of resorufin for the fluorometric determination of trace hydrogen peroxide: applications in detecting the activity of phagocyte NADPH oxidase and other oxidases. Anal Biochem 253:162–168

Zhu J, Egawa T, Yeh SR, Yu LD, Yu CA (2007) Simultaneous reduction of iron-sulfur protein and cytochrome b_L during ubiquinol oxidation in cytochrome bc_1 complex. Proc Natl Acad Sci USA 104:4864–4869

Zoccarato F, Cavallini L, Alexandre A (2004) Respiration-dependent removal of exogenous H_2O_2 in brain mitochondria - Inhibition by Ca^{2+}. J Biol Chem 279:4166–4174

Zoccarato F, Cavallini L, Bortolami S, Alexandre A (2007) Succinate modulation of H_2O_2 release at NADH:ubiquinone oxidoreductase (complex I) in brain mitochondria. Biochem J 406:125–129

Chapter 7
Studies on the Function and Regulation of Mitochondrial Uncoupling Proteins

Richard K. Porter

Abstract Mitochondrial uncoupling proteins are members of the SLC25 family of solute carriers. Models of mitochondrial transporter function predict that uncoupling proteins are solute carriers. Evidence in the literature suggests that uncoupling proteins can transport protons, fatty acid anions, chloride anions, and recently the dicarboxylate succinate. Studies have also demonstrated that UCPs can be covalently modified and in some instances this covalent modification is needed to affect uncoupling function. The current evidence from functional analyses of mammalian uncoupling proteins is summarized in this chapter.

7.1 Introduction

Great advances have been made in characterizing the function of solute carriers across the mitochondrial inner membrane (Palmieri 2004, 2008). The mitochondrial solute carrier family are small proteins with molecular masses that range from 30 to 34 kDa and are defined by the SLC25 genes in humans. Twenty-two members of the mitochondrial solute carrier family have been characterized so far and they transport dicarboxylate, tricarboxylate, keto acids, amino acids, nucleotides, and coenzymes/cofactor. A key feature of the structure of this mitochondrial solute carrier family is their tripartite structure consisting of three internally repeated sequences. The uncoupling proteins represent a sub-family of the mitochondrial solute carrier family and include *UCP1* (SLC25A7), *UCP2* (SLC25A8), *UCP3* (SLC25A9), *UCP4* (SLC25A27), and *UCP5* (SLC25A14) (Ricquier and Bouillaud 2000; Bouillaud et al. 2001; Krauss et al. 2005). Modelling of mitochondrial solute carriers, partially based on the crystal structure of the adenine nucleotide translocator (Pebay-Peyroula

R.K. Porter (✉)
School of Biochemistry and Immunology, Trinity College Dublin, Dublin 2, Ireland
e-mail: rkporter@tcd.ie

B. Kadenbach (ed.), *Mitochondrial Oxidative Phosphorylation*, Advances in Experimental Medicine and Biology 748, DOI 10.1007/978-1-4614-3573-0_7,

et al. 2003), predicts that theses transporters share the same structural fold, consisting of six transmembrane α-helices and three matrix helices, arranged with threefold pseudo-symmetry (Kunji and Robinson 2006; Robinson et al. 2008; Kunji and Robinson 2010). Based on analysis of the structures proposed by Kunji, all uncoupling protein would be predicted to be solute, possibly ketoacid, transporters. The chapter summarizes the results of functional studies on mitochondrial UCPs, reported in the literature, with a particular emphasis on mammalian forms of the protein.

7.2 Location of Mitochondrial Uncoupling Proteins

7.2.1 Location of UCP1 in Mammals

Uncoupling protein 1 (UCP1), originally termed thermogenin then UCP, is the carrier protein that defines uncoupling protein transporter sub-family. UCP1 is a 33 kDa, 307 amino acid, mitochondrial inner membrane protein and is classically associated with brown adipose tissue (BAT) of mammals including human infants (for reviews, see Nicholls and Locke 1984; Nicholls 2001, 2006; Cannon and Nedergaard 2004). BAT has also been confirmed to be present in adult humans as a result of positron emission tomography scan analysis and subsequent analysis of tissue for UCP1 and other indices of BAT function (Virtanen et al. 2009; Cypess et al. 2009; van Marken Lichtenbelt et al. 2009; Zingaretti et al. 2009). Sympathetic innervation of BAT, usually as a result of cold adaptation/acclimation, results in activation of BAT. The net result of noradrenalin on brown adipocytes is to mobilize intracellular fatty acids which overcome purine nucleotide inhibition of UCP1 dissipating the proton electrochemical gradient (Δp) across the mitochondrial inner membrane by UCP1. The resulting cycle, of proton pumping and leaking, increases electron transport chain activity. The heat generated from the consequent increase in metabolic flux defines the molecular basis of non-shivering thermogenesis in mammals. Extensive blood supply to the brown adipose tissue ensures that the mammal is kept warm under cold conditions (Nicholls and Locke 1984; Nicholls 2001, 2006; Cannon and Nedergaard 2004; Souza et al. 2007). Interestingly, UCP1 has been demonstrated to be present in mitochondria of rat and mouse thymus/thymocytes (Carroll et al. 2004, 2005; Porter 2006). The thymus is the site of T-helper and cytotoxic T-cell maturation and selection (Ritter and Crispe 1992). Early, evidence for the existence of UCP1 in thymus includes detection of RNA transcripts for UCP1 in whole thymus and in isolated thymocytes of rats and mice and UCP1 specific immunoblot data for mitochondria isolated from whole thymus and thymocytes of rats and mice, but not in thymus mitochondria from UCP1 knock-out mice. This early data was challenged by Frontini et al. (2007) who attributed UCP1 detection with BAT in the vicinity of thymus tissue. However, subsequent confocal microscopy demonstrating UCP1 protein associated with mitochondria in situ in thymocytes from wild type but not UCP1 knock-out mice confirms its presence in thymus (Adams et al. 2008a, b). The role for UCP1 in thymus would not appear to be for thermogenesis (Brennan et al. 2006) but for the maturation and fate of developing T-cells (Adams et al. 2010). In addition, there have been two reports of constitutive UCP1 protein expression in tissues other than BAT. Reports of UCP1 in uterine longitudinal smooth muscle cells by Nibbelink et al. (2001) were challenged by Rousset et al. (2003) based on the non-specificity of antibodies to detect UCP1 and uncontested reports exist for the expression of UCP1 in human islet cells (Sale et al. 2007)

7.2.2 *Location of UCP2 Protein in Mammals*

The UCP2 protein shares 59% amino acid sequence identity with UCP1 (Fleury et al. 1997). UCP2 was originally thought to be a regulator of basal metabolism (Fleury et al. 1997), due to the fact that basal proton leak in mitochondria is a significant contributor to basal metabolism (Brand et al. 1994; Porter 2001) and due to the ubiquitous tissue distribution of UCP2 transcripts. Investigating UCP2 function is only meaningful in the context of UCP2 protein expression, the possession of an antibody that was sensitive and selective to UCP2 was a priority. Pecqueur et al. (2001) made great progress in this regard by selecting out polyclonal antibodies that can discriminate UCP2 from other mitochondrial proteins, including other UCPs and transporters. It soon became apparent that UCP2 protein was not present in the wide variety of tissues where UCP2 transcripts were found. A further complication, accounting for the discrepancies in the level of UCP2 mRNA in tissues/cells and the UCP2 protein level, is due to the open-reading-frame controlling translation of UCP2. Thus, UCP2 protein has only been definitively found in a more limited range of tissues or more specifically mitochondrial fractions from those tissues: brain (subcortical brain regions such as the hypothalamus), spleen, lung, stomach, white adipose tissue, pancreatic β-cells, BAT cells, kidney, WAT cells, circulating macrophages, erythroid cells, thymocytes and Kupffer cells, dendritic cells, mastocytes and neutrophils, B-lymphocytes and T-lymphocytes (Richard et al. 1998; Arsenijevic et al. 2000; Joseph et al. 2002; Krauss et al. 2002; Mattiasson and Sullivan 2006; Flachs et al. 2007; Affourtit and Brand 2008; Azzu et al. 2008; Friederich et al. 2008). Although UCP2 has a high concentration in spleen mitochondria it has been estimated that UCP2 is ~16-fold less abundant in spleen mitochondria than UCP1 is in BAT mitochondria (Pecqueur et al. 2001). UCP2 protein also has a short half-life estimated to be 30 min, compared to 30 h for UCP1 (Rousset et al. 2007), and UCP2 has been demonstrated to be degraded via the cytosolic-ubiquitin proteasome system (Azzu and Brand 2010).

7.2.3 *Location of UCP3 in Mammals*

Uncoupling protein 3 (UCP3) was first discovered by Boss et al. (1997) and was predicted to be involved in heat production in muscle (Samec et al. 1998) due to its homology (57%) with the well-characterized uncoupling protein, UCP1. UCP3 is predominantly found associated with skeletal muscle mitochondria (Boss et al. 1997; Cunningham et al. 2003; Carroll and Porter 2004) and has a short half-life (hours) associated with the cytosolic-ubiquitin proteasome system (Azzu and Brand 2010). A consistent observation shows that UCP3 levels in skeletal muscle mitochondria are increased (2–3-fold) on starvation or following treatment with pharmacological doses of triiodothyronine (Cadenas et al. 1999; Cunningham et al. 2003). A further, interesting observation demonstrates that UCP3 is essential for the 3,4-methylenedioxymethamphetamine (MDMA aka Ecstasy) induced hyperthermia in skeletal muscle of mice (Mills et al. 2003). UCP3 protein has also consistently been detected in thymus and spleen where it appears to have a role in determining thymocyte development into mature T-cells (Kelly and Porter 2011). UCP3 protein has also been detected in brown adipose tissue mitochondria (Cunningham et al. 2003) and skin cells (Mori et al. 2008). UCP3 protein also appears to be inducible ectopically in liver following treatment with the pan-peroxisome proliferator-activated receptor (PPAR) ligand and fatty acid analogue tetradecylthioacetic acid (Wensaas et al. 2009).

7.2.4 Location of Other UCPs in Mammals

UCP4 and UCP5 (aka brain-derived mitochondrial carrier 1, BMCP1) are primarily expressed in the central nervous system (Sanchis et al. 1998; Mao et al. 1999; Bouillaud et al. 2001). Mammalian UCP4 and UCP5 have 30% homology to UCP1. UCP4 lacks key residues present reported to be important for proton translocation and has been postulated to be the ancestral uncoupling protein (Hanák and Ježek 2001) although that conclusion has been challenged (Sokolova and Sokolov 2005). Interestingly, nematodes express a UCP ortholog, ceUCP4, which has 46% homology to human UCP4 (Pfeiffer et al. 2011).

7.3 Mechanism of Action of UCPs

7.3.1 Mechanism of Action of UCP1

Understanding how UCP1 functions is crucial to elucidating the mechanism of non-shivering thermogenesis and probably the mechanism of action of other UCPs. UCP1 function has been investigated at the cellular and sub-cellular levels, as well as at the level of UCP1 native and expressed protein reconstituted into liposomes membranes. Mitochondria isolated under standard conditions from active BAT are essentially uncoupled, but on addition of purine nucleotides, oxygen consumption due to the UCP1 catalysed proton leak is inhibited (reviewed in Nicholls and Locke 1984; Cannon and Nedergaard 2004). Addition of nanomolar concentrations of long chain free fatty acids can alleviate the purine nucleotide inhibition of isolated BAT mitochondria (Nicholls and Rial 1999). The nature of this fatty acid dependency and purine nucleotide inhibition of proton leak through UCP1 is still a matter on investigation and not without controversy. The evidence for fatty acid-dependent alleviation of purine nucleotide inhibition of UCP1 activity, at the mitochondrial level, is consistent with observations at the cellular level for UCP1 activation. The most convincing studies demonstrating fatty acid-dependent UCP1 activity in brown adipocytes is summarized in Cannon and Nedergaard (2004). An increase in oxygen consumption rate can be demonstrated in brown adipocytes from UCP1 wild-type mice on direct addition of noradrenaline to those cells, whereas there is no increase in oxygen consumption rates on addition of noradrenaline to brown adipocytes from UCP1 knock-out mice. Furthermore, the fatty acid dependency of UCP1 is highlighted by the observation that brown adipocytes from UCP1 wild-type mice clearly display an increase in oxygen consumption on addition of the long chain fatty acid, oleate, whereas brown adipocytes from UCP1 knock-out mice do not respond to oleate to the same extent.

The nature by which free fatty acids are involved in the UCP uncoupling process is one source of controversy. Two models for the role of fatty acids in the mechanism of action of UCP1 have been proposed: The "fatty acid protonophore model" and the "buffering model". The "fatty acid protonophore model" proposes that protonated fatty acids non-enzymatically flip across the mitochondrial inner membrane, and deprotonate thus uncoupling the mitochondria. Thus, UCP1 functions as a "flippases" translocating the resulting anionic fatty acids back across the bilayer leaflets of the mitochondrial inner membrane and completing the uncoupling process (reviewed Garlid et al. 2000, 2001). Thus, in the "fatty acid protonophore model" UCP1 does not transport protons but facilitates a cycle of uncoupling by free fatty acids. The key evidence for the "fatty acid protonophore model" comes from the many observations that native UCP1 reconstituted into liposome membranes can transfer charge across the liposome membrane using fatty acids. However, the key observations are the fact that native UCP1 reconstituted into liposome membranes can transfer charge in the presence of undecanesulphonate ($C_{11}SO_3^-$) a non-protonatable (pK ~2) fatty acid analogue (reviewed in Garlid et al. 2000). A requirement for the "buffering model" is a protonatable and deprotonatable carboxyl group provided by the α-carboxyl of long-chain fatty acids. The only conclusion for the observed (GDP-sensitive) charge transfer in the presence of undecanesulphonate is that UCP1 flipped this anionic fatty acid analogue across the lipid bilayer. In our laboratory we were able to confirm that undecanesulphonate could catalyse charge transfer across liposome membranes, in accordance with the key evidence for the "fatty acid protonophore model" (Breen et al. 2006).

The "buffering model", sometimes called the "cofactor-activation model", proposes that UCP1 acts as a proton conduit across the mitochondrial inner membrane and importantly that fatty acids act as cofactors/activators providing an addition carboxyl group at a key intra-membrane site facilitating proton movement. Two reviews by Klingenberg summarize the data for the "buffering model" (Klingenberg and Huang 1999; Klingenberg et al. 1999; Nicholls and Rial 1999). It is clear that long-chain fatty acids, as opposed to short-chain fatty acids or very long-chain fatty acids, are optimal for activity of native UCP1 reconstituted into liposome membranes. The most convincing data for the "buffering model" though comes from the observations made using a fatty acid chemically modified at the omega carbon but which is unmodified at the α-carboxyl. The covalent coupling of the hydrophobic glucose molecule to the omega end of palmitate makes for an unflippable fatty acid that potentially can still "activate" UCP1. However, if UCP1 is a flippase, then glucose pyranoside-O-ω-palmitate should not facilitate UCP1 activity, as the hydrophilic glucose on the ω-end would impair flipping of the fatty acid. Thus when the experiment was performed, it was shown that the rate of proton translocation through UCP1 in the presence of glucose pyranoside-O-ω-palmitate was twice that recorded for palmitate alone, thus providing convincing evidence for the "buffering model". We re-investigated this key evidence for the "buffering model" and synthesized glucose pyranoside-O-ω-palmitate (aka glucose-O-ω-palmitate) de novo (Gouin et al. 2005). In contrast to the data reported by Klingenberg in his reviews, we showed that glucose-O-ω-palmitate could *not* "activate" proton translocation in liposomes containing native reconstituted UCP1 nor could it "activate" oxygen consumption by mitochondria from brown adipose tissue in the presence of purine nucleotides (Breen et al. 2006). We would argue that our observation with glucose-O-ω-palmitate undermines the key evidence for the "buffering model" model for UCP1.

Controversy also surrounds the nature of the interaction between the fatty acids and the purine nucleotides on UCP1. As mentioned earlier, like all mitochondrial inner membrane transporters so far characterized, UCP1 is predicted to have a tripartite structure (Klingenberg et al. 1999). Photoaffinity studies suggest that residues in the third domain and the C-terminus are required for purine nucleotide binding (Klingenberg et al. 1999) and chimeric studies have predicted that the central domain is required for fatty acid binding (Jimenez-Jimenez et al. 2006). Thus, data for isolated BAT mitochondria which demonstrate that nanomolar concentrations of fatty acids can alleviate purine nucleotide inhibition of UCP1, intuitively suggests an allosteric interaction between two separate binding sites, one for purine nucleotides and one for fatty acids. On the other hand, Shabalina et al. (2004) demonstrated that oleate competes with GDP for the purine nucleotide binding site in isolated BAT mitochondria, which is at odds with the aforementioned observations.

Another area of controversy, which relates to the mechanism of UCP1 was a report that ubiquinone was required for UCP1 activity. As already mentioned, native UCP1 from rat or hamster BAT can be reconstituted into liposomes (Klingenberg and Huang 1999; Nicholls and Rial 1999; Garlid et al. 2001) and using a fluorimeter and pH-sensitive fluorescent dyes, GDP sensitive, fatty acid-dependent proton, and indeed chloride, transport through UCP1 can be demonstrated. In addition to natively purified UCP1, *E. coli* expressed hamster, rat, and human UCP1 (Breen et al. 2006) have been reconstituted in liposomes where proton and chloride transport studies were undertaken. Data using the UCP1 reconstitution and assay systems by Klingenberg laboratory show that proton flux is fatty acid (micromolar) dependent, ubiquinone dependent, and inhibited by nanomolar purine nucleotide concentrations (Echtay et al. 2000). However, data from the Garlid laboratory using a different reconstitution and assay systems gave different results. In this instance, native UCP1 or expressed UCP1 (from *E. coli* or yeast) were reconstituted into liposomes and measurements of proton flux and charge movement, to assay flippase activity/uncoupling activity, were undertaken using fluorometric methods. Proton flux was shown to be dependent on micromolar fatty acids and to be nucleotide sensitive (micromolar) (Jabůrek et al. 1999; Jabůrek and Garlid 2003). Chloride flux was not measured. Essentially Garlid demonstrated that UCP1 flippase activity was not ubiquinone dependent. The discrepancy over ubiquinone dependency may reflect methodological difference in UCP1 reconstitution and assay systems between the two laboratories.

Interestingly, patch-clamp studies on reconstituted UCP1 also indicate sensitivity to anion channel inhibitors, endorsing the observation in liposomes that UCP1 can transport anions (Huang and Klingenberg 1996) and chloride transport by UCP1 has also been observed in mitochondria (Ježek and Garlid 1990; Nicholls 2006).

More recently, covalent modification of UCP1 has come from our laboratory (Carroll et al. 2008). Mass spectrometry has identified phosphorylation on serine 51 in UCP1 purified from cold-acclimated rats. Furthermore, immunoblot analysis of UCP1 purified from BAT mitochondria demonstrates that there is a greater proportion of serine phosphorylation associated with UCP1 purified from cold-acclimated rats when compared to rats kept at room temperature. Serine 51 is in the first domain, so the role of this domain or the significance associated with its phosphorylation has yet to be determined.

7.3.2 *Mechanism of Action of UCP2 and UCP3*

Evidence of a role for UCP2 and UCP3 in proton leak is ambiguous but has been demonstrated in various studies. In light of the fact that mitochondrial proton leak in a significant contributor to basal metabolism (Brand et al. 1994). One obvious consequence of UCP2 being a catalyst for basal mitochondrial proton leak might be that knocking-out UCP2 in whole animals would result in a fatter phenotype, but in fact there was no reported difference in adiposity in UCP2 knock-out mice compared to wild-type controls (Arsenijevic et al. 2000) or in UCP3 knock-out mice compared to wild types (Vidal-Puig et al. 2000). Work with transgenic mice overexpressing human UCP3 in mouse skeletal muscle resulted in increased proton leak with marked physiological consequences of weight loss and hyperphagia (Clapham et al. 2000). These observations may be due to a functioning UCP3 or may be due to overabundance of protein in the inner membrane. Similarly overexpression of the combination of UCP2 and UCP3 in mice reduces body fat (Horvath et al. 2003). By contrast, neither UCP2 nor UCP3 can compensate for the lack of thermogenic capacity in UCP1 knock-out mice (Enerbäck et al. 1997). However, Krauss et al. (2002) have demonstrated that in situ proton leak is reduced in thymocytes isolated from UCP2−/− mice when compared to those from wild-type mice and Azzu et al. (2008) have demonstrated that UCP2 protein levels correlate with proton leak in pancreatic beta cells.

The picture with respect to reconstituted UCP2 and UCP3 and their ability to facilitate proton leak is a bit clearer. Reconstituted UCP2 can transport protons in planar lipid bilayers (Beck et al. 2007) and UCP2 and UCP3 can facilitate proton transport in liposomes in a fatty acid dependent, GDP inhibitable fashion (Echtay et al. 2000; Jabůrek and Garlid 2003). Further evidence to support a direct role for UCP2 in proton leak comes from the observations with isolated mitochondria. There is an inducible (GDP-sensitive) UCP2-dependent uncoupling of kidney, spleen, and pancreatic β-cell mitochondria under conditions of continuous intra-mitochondrial and extra-mitochondrial superoxide production (Echtay et al. 2002a, b, 2003; Murphy et al. 2003). A UCP2- and UCP3-dependent catalysed proton leak has also been shown to be activated in isolated mitochondria by direct addition of phospholipid fatty acid breakdown products (e.g. 4-hydroxynonenals). Such oxidation products usually occur as a result of superoxide damage to (membrane) polyunsaturated fatty acids (Jabůrek and Garlid 2003; Esteves and Brand 2005). Consequentially, a model for UCPs in regulating free radical production by mitochondria, as a result of a negative feedback loop, has been proposed (Esteves and Brand 2005). A role for UCP2 in regulating ROS production in the thymus was first demonstrated by Nègre-Salvayre et al. (1997), as a result of observing GDP-sensitive ROS production by thymus mitochondria. However, it may well be that some of the GDP-sensitive ROS production observed by Nègre-Salvayre et al. (1997) for thymus mitochondria could be due to the presence of UCP1 in thymus (Carroll et al. 2005). Certainly, the absence of UCP2 in spleen mitochondria results in increased ROS production when compared to mitochondria isolated from spleens of wild-type mice (Bai et al. 2005), although the involvement of superoxide in UCP2 activation in spleen and lung mitochondria could not be demonstrated by Couplan et al. (2002). The idea of a physiological mechanism to alleviate ROS production by mitochondria was proposed by Skulachev (1996) and is termed mild uncoupling. Mild uncoupling is based on the observations that uncoupling lowers Δp, decreases the degree of reduction in the electron transport chain, and thus will reduce ROS production from the electron transport chain. Hence it has been proposed that UCPs act as sensor of ROS production by mitochondria, with the result that their activation relieves ROS production by mitochondria (Esteves and Brand 2005). Direct evidence to support this model, came from the recent observations that UCP2 and UCP3 are glutathionylated, with specific sites being identified for UCP3 (Cys25 and Cys259), and data was presented to suggest that ROS can overcome this covalent modification to activate UCP2 and UCP3 (Mailloux et al. 2011; Mailloux and Harper 2011). Other evidence for in vivo activation of UCP3 was borne out of the observation of Mills et al. (2003) who demonstrated that UCP3 was essential for 3,4-methylenedioxymethamphetamine (MDMA aka Ecstasy) induced hyperthermia in skeletal muscle of mice. It was subsequently demonstrated in our laboratory that preservation of in vivo phosphorylation of UCP3 on serine and tyrosine sites, from mice/rats treated with MDMA, results in increased proton leak through UCP3 (Kelly et al. 2012). We concluded that the MDMA-induced hyperthermia in skeletal muscle is due to increased proton leak in vivo as a result of activation of UCP3 through phosphorylation.

Other stresses such as starvation have also resulted in changes in UCP abundance. Most noteworthy is the consistent observation of increased UCP3 (2–3-fold) abundance in muscle mitochondria of starved animals (Cadenas et al. 1999; Cunningham et al. 2003). However, the increase in UCP3 protein abundance does not result in an increase in proton leak (Cadenas et al. 1999) although it does result in increased fatty acid transport and oxidation rate in skeletal muscle (Bezaire et al. 2005). Consistent with a role in fatty acid metabolism is the data from Jabůrek et al. (2004) and Lombardi et al. (2010) who suggest that the physiological role for UCP2 and UCP3 may be to transport oxidized polyunsaturated fatty acid anions (hydroperoxy fatty acid anions).

On a very different tack, other researchers have provided circumstantial evidence using UCP overexpression and knock-down experiments in single endothelial, HeLa cells (human cervical epithelium), HEK293 cells (human embryonic kidney cells), and AtT20 cells (mouse anterior pituitary-derived cell line) that UCP2 and UCP3 are involved in mitochondrial calcium transport and speculate that UCP2 and UCP3 may be conductive subunits of the Ca^{2+} selective ion channel (Trenker et al. 2007; Graier et al. 2007). However, there is no doubt that the interpretation of the data is contentious (Brookes et al. 2008; Trenker et al. 2008).

7.3.3 Mechanism of Action of UCP4 and UCP5

As already mentioned the mitochondrial solute carrier family are involved in transporting dicarboxylate, tricarboxylate, keto acids, amino acids, nucleotides, and coenzymes/cofactors across the mitochondrial inner membrane and from distillation of the data reviewed here so far, UCPs would appear capable of transporting protons, fatty acid anions, and inorganic anions. On first investigation UCP4 also appeared to be an uncoupler, as ectopic expression of UCP4 decreased mitochondrial membrane in HEK293T cells (Mao et al. 1999). Similarly, overexpression of UCP5 in a neuroblastoma cell line localized to mitochondria decreased mitochondrial membrane potential, reduced ATP production, and increased overall oxygen consumption indicating uncoupling activity (Kwok et al. 2010). Interestingly, recent evidence on the function of UCP4 comes from the nematode *Caenorhabditis elegans* which expresses a single UCP ortholog ceUCP4. Pfeiffer et al. (2011) have demonstrated that ceUCP4 knock-outs have decreased mitochondrial succinate-driven (complex II) respiration and that ceUCP4 inhibition blocked succinate respiration and import in mitochondria. The data suggest that ceUCP4 is in fact a succinate transporter, and thus in this instance the data would be consistent with the models of UCP function proposed by Kunji and Robinson (2006), Robinson et al. (2008), and Kunji and Robinson (2010).

Acknowledgement R.K. Porter would like to thank Bernhard Kadenbach for the invitation to write this chapter.

References

Adams AE, Hanrahan O, Nolan DN, Voorheis HP, Fallon PG, Porter RK (2008a) Images of mitochondrial UCP 1 in mouse thymocytes using confocal microscopy. Biochim Biophys Acta 1777:115–117

Adams AE, Carroll AM, Fallon PG, Porter RK (2008b) Mitochondrial uncoupling protein 1 expression in thymocytes. Biochim Biophys Acta 1777:772–776

Adams AE, Kelly OM, Porter RK (2010) Absence of mitochondrial uncoupling protein 1 affects apoptosis in thymocytes, thymocyte/T-cell profile and peripheral T-cell number. Biochim Biophys Acta 1797(6–7):807–816

Affourtit C, Brand MD (2008) On the role of uncoupling protein-2 in pancreatic beta cells. Biochim Biophys Acta 1777:973–979

Arsenijevic D, Onuma H, Pecqueur C, Raimbault S, Manning B, Miroux B, Couplan E, Alves-Guerra M, Goubern M, Surwit R, Bouillaud F, Richard D, Collins S, Ricquier D (2000) Distribution of the uncoupling protein-2 gene in mice reveals a role in immunity and reactive oxygen species production. Nat Genet 26:435–439

Azzu V, Brand MD (2010) Degradation of an intra-mitochondrial protein by the cytosolic proteasome. J Cell Sci 123:578–585

Azzu V, Affourtit C, Breen EP, Parker N, Brand MD (2008) Dynamic regulation of uncoupling protein 2 content in INS-1E insulinoma cells. Biochim Biophys Acta 1777:1378–1383

Bai Y, Onuma H, Bai X, Medvedev AV, Misukonis M, Weinberg JB, Cao W, Robidoux J, Floering LM, Daniel KF, Collins S (2005) Persistent NF-κB activation in UCP2–/– mice leads to enhanced nitric oxide and inflammatory cytokine production. J Biol Chem 280:19062–19069

Beck V, Jabůrek M, Demina T, Rupprecht A, Porter RK, Ježek P, Pohl EE (2007) High efficiency of polyunsaturated fatty acids in the activation of human uncoupling protein 1 and 2 reconstituted in planar lipid bilayers. FASEB J 4:1137–1144

Bezaire V, Spriet LL, Campbell S, Sabet N, Gerrits M, Bonen A, Harper ME (2005) Constitutive UCP3 overexpression at physiological levels increases mouse skeletal muscle capacity for fatty acid transport and oxidation. FASEB J 19:977–979

Boss O, Samec S, Dulloo AG, Seydoux J, Muzzin P, Giacobino JP (1997) Uncoupling protein 3: a new member of the mitochondrial carrier family with tissue specific expression. FEBS Lett 408:39–42

Bouillaud F, Couplan E, Pecqueur C, Ricquier D (2001) Homologues of the uncoupling protein from brown adipose tissue (UCP1): UCP2, UCP3, BMCP1 and UCP4. Biochim Biophys Acta 1504:107–119

Brand MD, Chien LF, Ainscow EK, Rolfe DF, Porter RK (1994) The causes and functions of mitochondrial proton leak. Biochim Biophys Acta 1187:132–139

Breen EP, Gouin S, Brennan CM, Murphy AF, Haines LR, Pearson AM, Jackson TW, Murphy PV, Porter RK (2006) On the mechanism of mitochondrial uncoupling protein 1 function. J Biol Chem 281:2114–2119

Brennan CM, Breen EP, Porter RK (2006) Cold acclimation and oxygen consumption in the thymus. Biochim Biophys Acta 1757:1463–1468

Brookes PS, Parker N, Buckingham JA, Vidal-Puig A, Halestrap AP, Gunter TE, Nicholls DG, Bernardi P, Lemasters JL, Brand MD (2008) UCPs-unlikely calcium porters. Nat Cell Biol 10:1235–1237

Cadenas S, Buckingham JA, Samec S, Seydoux J, Din N, Dulloo AG, Brand MD (1999) UCP2 and UCP3 rise in starved rat skeletal muscle but mitochondrial proton conductance is unchanged. FEBS Lett 462:257–260

Cannon B, Nedergaard J (2004) Brown adipose tissue: function and physiological significance. Physiol Rev 84:277–359

Carroll AM, Porter RK (2004) Starvation sensitive UCP 3 expression in rat thymus and spleen. Biochim Biophys Acta 1700:145–150

Carroll AM, Haines LR, Pearson TW, Brennan C, Breen EP, Porter RK (2004) Immunodetection of UCP1 in rat thymocytes. Biochem Soc Trans 32:1066–1067

Carroll AM, Haines LR, Pearson TW, Fallon PG, Walsh C, Brennan C, Breen EP, Porter RK (2005) Identification of a functioning mitochondrial uncoupling protein 1 in thymus. J Biol Chem 280:15534–15543

Carroll AM, Porter RK, Morrice NA (2008) Identification of serine phosphorylation in mitochondrial uncoupling protein 1. Biochim Biophys Acta 1777:1060–1065

Clapham JC, Arch JR, Chapman H, Haynes A, Lister C, Moore GB, Piercy V, Carter SA, Lehner I, Smith SA, Beeley LJ, Godden RJ, Herrity N, Skehel M, Changani KK, Hockings PD, Reid DG, Squires SM, Hatcher J, Trail B, Latcham J, Rastan S, Harper AJ, Cadenas S, Buckingham JA, Brand MD, Abuin A (2000) Mice overexpressing human uncoupling protein-3 in skeletal muscle are hyperphagic and lean. Nature 406:415–418

Couplan E, del Mar Gonzalez-Barroso M, Alves-Guerra MC, Ricquier D, Goubern M, Bouillaud F (2002) No evidence for a basal, retinoic, or superoxide-induced uncoupling activity of the uncoupling protein 2 present in spleen or lung mitochondria. J Biol Chem 277:26268–26275

Cunningham O, McElligott AM, Carroll AM, Breen E, Reguenga C, Oliveira MEM, Azevedo JE, Porter RK (2003) Selective detection of UCP 3 expression in skeletal muscle: effect of thyroid status and temperature acclimation. Biochim Biophys Acta 1604:170–179

Cypess AM, Lehman S, Williams G, Tal I, Rodman D, Goldfine AB, Kuo FC, Palmer EL, Tseng YH, Doria A, Kolodny GM, Kahn CR (2009) Identification and importance of brown adipose tissue in adult humans. N Engl J Med 360:1509–1517

Echtay KS, Winkler E, Klingenberg M (2000) Coenzyme Q is an obligatory cofactor for uncoupling protein function. Nature 408:609–613

Echtay KS, Roussel D, St-Pierre J, Jekabsons MB, Cadenas S, Stuart JA, Harper JA, Roebuck SJ, Morrison A, Pickering S, Clapham JC, Brand MD (2002a) Superoxide activates mitochondrial uncoupling proteins. Nature 415:96–99

Echtay KS, Murphy MP, Smith RA, Talbot DA, Brand MD (2002b) Superoxide activates mitochondrial uncoupling protein 2 from the matrix side. Studies using targeted antioxidants. J Biol Chem 277:47129–47135

Echtay KS, Esteves TC, Pakay JL, Jekabsons MB, Lambert AJ, Portero-Otín M, Pamplona R, Vidal-Puig AJ, Wang S, Roebuck SJ, Brand MD (2003) A signalling role for 4-hydroxy-2-nonenal in regulation of mitochondrial uncoupling. EMBO J 22:4103–4110

Enerbäck S, Jacobsson A, Simpson EM, Guerra C, Yamashita H, Harper ME, Kozak LP (1997) Mice lacking mitochondrial uncoupling protein are cold-sensitive but not obese. Nature 387:90–94

Esteves TC, Brand MD (2005) The reactions catalysed by the mitochondrial uncoupling proteins UCP2 and UCP3. Biochim Biophys Acta 1709:35–44

Flachs P, Sponarova J, Kopecky P, Horvath O, Sediva A, Nibbelink M, Casteilla L, Medrikova D, Neckar J, Kolar F, Kopecky J (2007) Mitochondrial uncoupling protein 2 gene transcript levels are elevated in maturating erythroid cells. FEBS Lett 581:1093–1097

Fleury C, Neverova M, Collins S, Raimbault S, Champigny O, Levi-Meyrueis C, Bouillaud F, Seldin MF, Surwit RS, Ricquier D, Warden CH (1997) Uncoupling protein-2: a novel gene linked to obesity and hyperinsulinemia. Nat Genet 15:269–272

Friederich M, Fasching A, Hansell P, Nordquist L, Palm F (2008) Diabetes-induced up-regulation of uncoupling protein-2 results in increased mitochondrial uncoupling in kidney proximal tubular cells. Biochim Biophys Acta 1777:935–940

Frontini A, Rousset S, Cassard-Doulcier AM, Zingaretti C, Ricquier D, Cinti S (2007) Thymus uncoupling protein 1 is exclusive to typical brown adipocytes and is not found in thymocytes. J Histochem Cytochem 55:183–189

Garlid KD, Jabůrek M, Ježek P, Vařecha M (2000) How do uncoupling proteins uncouple? Biochim Biophys Acta 1459:383–389

Garlid KD, Jabůrek M, Ježek P (2001) Mechanism of uncoupling protein action. Biochem Soc Trans 29:803–806

Gouin S, Pilgrim W, Porter RK, Murphy PV (2005) Synthesis of a glycolipid for studying mechanisms of mitochondrial uncoupling proteins. Carb Res 340:1547–1552

Graier WF, Frieden M, Malli R (2007) Mitochondria and Ca^{2+} signaling: old guests, new functions. Pflugers Arch 455:375–396

Hanák P, Ježek P (2001) Mitochondrial uncoupling proteins and phylogenesis–UCP4 as the ancestral uncoupling protein. FEBS Lett 495:137–141

Horvath TL, Diano S, Miyamoto S, Barry S, Gatti S, DAlberati D, Flivak F, Lombardi A, Moreno M, Goglia F, Mor G, Hamilton J, Kachinskas D, Warden CH Horwitz B (2003) Uncoupling proteins-2 and 3 influence obesity and inflammation in transgenic mice. Int J Obes Relat Metab Disord 27:433–442

Huang SG, Klingenberg M (1996) Chloride channel properties of the uncoupling protein from brown adipose tissue mitochondria: a patch clamp study. Biochemistry 35:16806–16814

Jabůrek M, Garlid KD (2003) Reconstitution of recombinant uncoupling proteins: UCP1, -2, and -3 have similar affinities for ATP and are unaffected by coenzyme Q10. J Biol Chem 278:25825–25831

Jabůrek M, Vařecha M, Gimeno RE, Dembski M, Ježek P, Zhang M, Burn P, Tartaglia LA, Garlid KD (1999) Transport function and regulation of mitochondrial uncoupling proteins 2 and 3. J Biol Chem 274:26003–26007

Jabůrek M, Miyamoto S, Di Mascio P, Garlid KD, Ježek P (2004) Hydroperoxy fatty acid cycling mediated by mitochondrial uncoupling protein UCP2. J Biol Chem 279:53097–530102

Ježek P, Garlid KD (1990) New substrates and competitive inhibitors of the Cl^- translocating pathway of the uncoupling protein of brown adipose tissue mitochondria. J Biol Chem 265:19303–19311

Jimenez-Jimenez J, Ledesma A, Zaragoza P, Gonzalez-Barroso MM, Rial E (2006) Fatty acid activation of the uncoupling proteins requires the presence of the central matrix loop from UCP1. Biochim Biophys Acta 1757:1292–1296

Joseph JW, Koshkin V, Zhang CY, Wang J, Lowell BB, Chan CB, Wheeler MB (2002) Uncoupling protein 2 knockout mice have enhanced insulin secretory capacity after a high-fat diet. Diabetes 51:3211–3219

Kelly OM, Porter RK (2011) Absence of mitochondrial uncoupling protein 3: effect on thymus and spleen in the fed and fasted mice. Biochim Biophys Acta 1807:1064–1074

Kelly OM, McNamara YM, Manzke LH, Meegan MJ, Porter RK (2012) The preservation of in vivo phosphorylated and activated uncoupling protein 3 (UCP3) in isolated skeletal muscle mitochondria following administration of 3,4-methylenedioxymethamphetamine (MDMA aka ecstasy) to rats/mice. Mitochondrion 12(1):110–119

Klingenberg M, Huang S-G (1999) Structure and function of the uncoupling protein from brown adipose tissue. Biochim Biophys Acta 1415:271–296

Klingenberg M, Echtay KS, Bienengraeber M, Winkler E, Huang SG (1999) Structure-function relationship in UCP1. Int J Obes Relat Metab Disord 23:S24–S29

Krauss S, Zhang CY, Lowell BB (2002) A significant portion of mitochondrial proton leak in intact thymocytes depends on expression of UCP2. Proc Natl Acad Sci USA 99:118–122

Krauss S, Zhang CY, Lowell BB (2005) The mitochondrial uncoupling-protein homologues. Nat Rev Mol Cell Biol 6:248–261

Kunji ERS, Robinson AJ (2006) The conserved substrate binding site of mitochondrial carriers. Biochim Biophys Acta 175:1237–1248

Kunji ERS, Robinson AJ (2010) Coupling of proton and substrate translocation in the transport cycle of mitochondrial carriers. Curr Opin Struct Biol 20:440–447

Kwok KH, Ho PW, Chu AC, Ho JW, Liu HF, Yiu DC, Chan KH, Kung MH, Ramsden DB, Ho SL (2010) Mitochondrial UCP5 is neuroprotective by preserving mitochondrial membrane potential, ATP levels, and reducing oxidative stress in MPP+ and dopamine toxicity. Free Radic Biol Med 49:1023–1035

Lombardi A, Busiello RA, Napolitano L, Cioffi F, Moreno M, de Lange P, Silvestri E, Lanni A, Goglia F (2010) UCP3 translocates lipid hydroperoxide and mediates lipid hydroperoxide-dependent mitochondrial uncoupling. J Biol Chem 285:16599–16605

Mailloux RJ, Harper ME (2011) Uncoupling proteins and the control of mitochondrial reactive oxygen species production. Free Radic Biol Med 51:1106–1115
Mailloux RJ, Seifert EL, Bouillaud F, Aguer C, Collins S, Harper ME (2011) Glutathionylation acts as a control switch for uncoupling proteins UCP2 and UCP3. J Biol Chem 286:21865–21875
Mao W, Yu XX, Zhong A, Li W, Brush J, Sherwood SW, Adams SH, Pan G (1999) UCP4, a novel brain-specific mitochondrial protein that reduces membrane potential in mammalian cells. FEBS Lett 443:326–330
Mattiasson G, Sullivan PG (2006) The emerging functions of UCP2 in health, disease, and therapeutics. Antioxid Redox Signal 8:1–38
Mills EM, Banks ML, Sprague JE, Finkel T (2003) Pharmacology: uncoupling the agony from ecstasy. Nature 426:403–404
Mori S, Yoshizuka N, Takizawa M, Takema Y, Murase T, Tokimitsu I, Saito M (2008) Expression of uncoupling proteins in human skin and skin-derived cells. J Invest Dermatol 128:1894–1900
Murphy MP, Echtay KS, Blaikie FH, Asin-Cayuela J, Cocheme HM, Green K, Buckingham J, Taylor ER, Hurrell F, Hughes G, Miwa S, Cooper CE, Svistunenko DA, Smith RA, Brand MD (2003) Superoxide activates uncoupling proteins by generating carbon-centered radicals and initiating lipid peroxidation: studies using a mitochondria-targeted spin trap derived from alpha-phenyl-N-tert-butylnitrone. J Biol Chem 278:48534–48545
Nègre-Salvayre A, Hirtz C, Carrera G, Cazenave R, Troly M, Salvayre R, Penicaud L, Casteilla L (1997) A role for uncoupling protein-2 as a regulator of mitochondrial hydrogen peroxide generation. FASEB J 11:809–815
Nibbelink M, Moulin K, Arnaud E, Duval C, Penicaud L, Casteilla L (2001) Brown fat UCP1 is specifically expressed in uterine longitudinal smooth muscle cells. J Biol Chem 276:47291–47295
Nicholls DG (2001) A history of UCP1. Biochem Soc Trans 29:751–755
Nicholls DG (2006) The physiological regulation of uncoupling proteins. Biochim Biophys Acta 1757:459–466
Nicholls DG, Locke RM (1984) Thermogenic mechanisms in brown fat. Physiol Rev 64:1–64
Nicholls DG, Rial E (1999) A history of the first uncoupling protein, UCP1. J Bioenerg Biomembr 5:399–406
Palmieri F (2004) The mitochondrial transporter family (SLC25): physiological and pathological implications. Pflugers Arch 447:689–709
Palmieri F (2008) Diseases caused by defects of mitochondrial carriers: a review. Biochim Biophys Acta 1777:564–578
Pebay-Peyroula E, Dahout-Gonzalez C, Kahn R, Trézéguet V, Lauquin GJ, Brandolin G (2003) Structure of mitochondrial ADP/ATP carrier in complex with carboxyatractyloside. Nature 426:39–44
Pecqueur C, Alves-Guerra MC, Gelly C, Levi-Meyrueis C, Couplan E, Collins S, Ricquier D, Bouillaud F, Miroux B (2001) Uncoupling protein 2, in vivo distribution, induction upon oxidative stress, and evidence for translational regulation. J Biol Chem 276:8705–8712
Pfeiffer M, Kayzer EB, Yang X, Abramson E, Kenaston MA, Lago CU, Lo HH, Sedensky MM, Lunceford A, Clarke CF, Wu SJ, McLeod C, Finkel T, Morgan PG, Mills EM (2011) Caenorhabditis elegans UCP4 protein controls complex II-mediated oxidative phosphorylation through succinate transport. J Biol Chem 286:37712–37720
Porter RK (2001) Allometry of mammalian cellular oxygen consumption. Cell Mol Life Sci 58:815–822
Porter RK (2006) A new look at UCP 1. Biochim Biophys Acta 1757:446–448
Richard D, Rivest R, Huang Q, Bouillaud F, Sanchis D, Champign O, Ricquier D (1998) Distribution of the uncoupling protein 2 mRNA in the mouse brain. J Comp Neurol 397:549–560
Ricquier D, Bouillaud F (2000) The uncoupling protein homologues: UCP1, UCP2, UCP3, StUCP and AtUCP. Biochem J 345:161–179
Ritter MA, Crispe IN (1992) In: Rickwood D, Male D (eds) The thymus. IRL Press, Oxford University Press, Oxford
Robinson AJ, Overy C, Kunji ERS (2008) The mechanism of transport by mitochondrial carriers based on analysis of symmetry. Proc Natl Acad Sci USA 105:17766–17771

Rousset S, Alves-Guerra MC, Ouadghiri-Bencherif S, Kozak LP, Miroux B, Richard D, Bouillaud F, Ricquier D, Cassard-Doulcier AM (2003) Uncoupling protein 2, but not uncoupling protein 1, is expressed in the female mouse reproductive tract. J Biol Chem 278:45843–45847

Rousset S, Mozo J, Dujardin G, Emre Y, Masscheleyn S, Ricquier D, Cassard-Doulcier AM (2007) UCP2 is a mitochondrial transporter with an unusual very short half-life. FEBS Lett 581:479–482

Sale MS, Hsu FC, Palmer ND, Gordon CJ, Keene KL, Borgerink HM, Sharma AJ, Bergman RN, Taylor KD, Saad MF, Norris JM (2007) The uncoupling protein 1 gene, UCP1, is expressed in mammalian islet cells and associated with acute insulin response to glucose in African American families from the IRAS Family Study. BMC Endocr Disord 7:1. doi:10.1186/1472-6823-7-1

Samec S, Seydoux J, Dulloo AG (1998) Role of UCP homologues in skeletal muscles and brown adipose tissue: mediators of thermogenesis or regulators of lipids as fuel substrate? FASEB J 12:715–724

Sanchis D, Fleury C, Chomiki N, Goubern M, Huang Q, Neverova M, Grégoire F, Easlick J, Raimbault S, Lévi-Meyrueis C, Miroux B, Collins S, Seldin M, Richard D, Warden C, Bouillaud F, Ricquier D (1998) BMCP1, a novel mitochondrial carrier with high expression in the central nervous system of humans and rodents, and respiration uncoupling activity in recombinant yeast. J Biol Chem 273:34611–34615

Shabalina IG, Jacobsson A, Cannon B, Nedergaard J (2004) Native UCP1 displays simple competitive kinetics between the regulators purine nucleotides and fatty acids. J Biol Chem 279:38236–38248

Skulachev VP (1996) Role of uncoupled and non-coupled oxidations in maintenance of safely low levels of oxygen and its one-electron reductants. Q Rev Biophys 29:169–202

Sokolova IM, Sokolov EP (2005) Evolution of mitochondrial uncoupling proteins: novel invertebrate UCP homologues suggest early evolutionary divergence of the UCP family. FEBS Lett 579:313–317

Souza SC, Christoffolete MA, Ribeiro MO, Miyoshi H, Strissel KJ, Stancheva ZS, Rogers NH, D'Eon TM, Perfield JW, Imachi H, Obin MS, Bianco AC, Greenberg AS (2007) Perilipin regulates the thermogenic actions of norepinephrine in brown adipose tissue. J Lipid Res 48:1273–1279

Trenker M, Malli R, Fertschai I, Levak-Frank S, Graier WF (2007) Uncoupling proteins 2 and 3 are fundamental for mitochondrial Ca^{2+} uniport. Nat Cell Biol 9:445–452

Trenker M, Fertschai I, Malli R, Graier WF (2008) UCP2/3-likely to be fundamental for mitochondrial Ca^{2+} uniport. Nat Cell Biol 10:1237–1240

van Marken Lichtenbelt WD, Vanhommerig JW, Smulders NM, Drossaerts JM, Kemerink GJ, Bouvy ND, Schrauwen P, Teule GJ (2009) Cold-activated brown adipose tissue in healthy men. N Engl J Med 360:1500–1508

Vidal-Puig AJ, Grujic D, Zhang CY, Hagen T, Boss O, Ido Y, Szczepanik A, Wade J, Mootha V, Cortright R, Muoio DM, Lowell BB (2000) Energy metabolism in uncoupling protein 3 gene knockout mice. J Biol Chem 275:16258–16266

Virtanen KA, Lidell ME, Orava J, Heglind M, Westergren R, Niemi T, Taittonen M, Laine J, Savisto NJ, Enerbäck S, Nuutila P (2009) Functional brown adipose tissue in healthy adults. N Engl J Med 360:1518–1525

Wensaas AJ, Rustan AC, Rokling-Andersen MH, Caesar R, Jensen J, Kaalhus O, Graff BA, Gudbrandsen OA, Berge RK, Drevon CA (2009) Dietary supplementation of tetradecylthioacetic acid increases feed intake but reduces body weight gain and adipose depot sizes in rats fed on high-fat diets. Diabetes Obes Metab 11:1034–1049

Zingaretti MC, Crosta F, Vitali A, Guerrieri M, Frontini A, Cannon B, Nedergaard J, Cinti S (2009) The presence of UCP1 demonstrates that metabolically active adipose tissue in the neck of adult humans truly represents brown adipose tissue. FASEB J 23:3113–3120

Chapter 8
Evolution of the Couple Cytochrome *c* and Cytochrome *c* Oxidase in Primates

Denis Pierron, Derek E. Wildman, Maik Hüttemann, Thierry Letellier, and Lawrence I. Grossman

Abstract Mitochondrial energy metabolism has been affected by a broad set of ancient and recent evolutionary events. The oldest example is the endosymbiosis theory that led to mitochondria and a recently proposed example is adaptation to cold climate by anatomically modern human lineages. Mitochondrial energy metabolism has also been associated with an important area in anthropology and evolutionary biology, brain enlargement in human evolution. Indeed, several studies have pointed to the need for a major metabolic rearrangement to supply a sufficient amount of energy for brain development in primates.

D. Pierron
Center for Molecular Medicine and Genetics, Wayne State University School of Medicine, Detroit, MI 48201, USA

Laboratoire de Physiopathologie Mitochondriale, INSERM, Université Victor Segalen Bordeaux 2, 146, rue Léo Saignat, Bordeaux 33076, France

D.E. Wildman
Center for Molecular Medicine and Genetics, Wayne State University School of Medicine, Detroit, MI 48201, USA

M. Hüttemann • L.I. Grossman (✉)
Center for Molecular Medicine and Genetics, and Cardiovascular Research Institute, Wayne State University School of Medicine, Detroit, MI, USA

Karmanos Cancer Institute, Detroit, MI, USA
e-mail: mhuttema@wayne.edu; l.grossman@wayne.edu

T. Letellier
Laboratoire de Physiopathologie Mitochondriale, INSERM, Université Victor Segalen Bordeaux 2, 146, rue Léo Saignat, Bordeaux 33076, France
e-mail: tletel@u-bordeaux2.fr

B. Kadenbach (ed.), *Mitochondrial Oxidative Phosphorylation*, Advances in Experimental Medicine and Biology 748, DOI 10.1007/978-1-4614-3573-0_8,

The genes encoding for the coupled cytochrome *c* (Cyt *c*) and cytochrome *c* oxidase (COX, complex IV, EC 1.9.3.1) seem to have an exceptional pattern of evolution in the anthropoid lineage. It has been proposed that this evolution was linked to the rearrangement of energy metabolism needed for brain enlargement. This hypothesis is reinforced by the fact that the COX enzyme was proposed to have a large role in control of the respiratory chain and thereby global energy production.

After summarizing major events that occurred during the evolution of COX and cytochrome *c* on the primate lineage, we review the different evolutionary forces that could have influenced primate COX evolution and discuss the probable causes and consequences of this evolution. Finally, we discuss and review the co-occurring primate phenotypic evolution.

8.1 Introduction

Mitochondrial energy metabolism has been associated with a broad set of both ancient and recent evolutionary events. The oldest example would be the endosymbiosis event in which an ancient bacterium was incorporated into another cell (Mereschkowski 1905; Sagan 1967); indeed, the increase by more than 15× of available ATP from a glucose molecule due to mitochondrial oxidative phosphorylation (OxPhos) is proposed as a necessary step for the emergence of pluricellular organisms (Lane and Martin 2010). A more recent example would be the adaptation to cold climate by human populations. In this latter proposal, several OxPhos polymorphisms are suspected to increase heat production by decoupling oxygen consumption from energy production and would have been positively selected in the coldest parts of the world (Mishmar et al. 2003; Ruiz-Pesini et al. 2004; Pierron et al. 2008).

Mitochondrial energy metabolism has also been associated with one of the main topics of anthropology and evolutionary biology, brain enlargement during primate evolution (Grossman et al. 2001, 2004; Uddin et al. 2008; Williams et al. 2010). Indeed, several authors have pointed out the need for a major rearrangement of metabolism to supply a sufficient amount of energy necessary for brain development in primates (Leonard et al. 2007; Isler and Van Schaik 2009). In parallel, numerous studies have shown that the genes encoding the constitutive proteins of the OxPhos complexes have been through an adaptive evolutionary process during primate evolution, and specifically on the anthropoid lineages (Old World monkey, New World monkey, ape, and human) (Adkins and Honeycutt 1994; Adkins et al. 1996; Schmidt et al. 1997, 1999, 2002, 2005; Wu et al. 1997, 2000; Andrews and Easteal 2000; Grossman et al. 2001, 2004; Goldberg et al. 2003; Doan et al. 2004; Uddin et al. 2008).

The fourth OxPhos complex (cytochrome *c* oxidase, COX, EC 1.9.3.1) and cytochrome *c* (Cyt *c*) seem to have evolved exceptionally during the anthropoid descent (Doan et al. 2004; Schmidt et al. 2005). Furthermore, this evolution has been proposed to be linked to the energy metabolism rearrangement needed for brain enlargement

(Grossman et al. 2004; Schmidt et al. 2005; Goodman et al. 2009). This hypothesis is reinforced by the fact that the COX enzyme was proposed to have a large role in the control of the respiratory chain and so of global energy production (Pacelli et al. 2011). Indeed, it has been shown that in vivo COX has a high control coefficient on OxPhos activity and its activity is tightly regulated by several cellular mechanisms (Follmann et al. 1998; Fontanesi et al. 2006). Furthermore, COX and Cyt *c* are the only components of primate OxPhos with known tissue-specific isoforms (reviewed in Hüttemann et al. 2008). Arguably, the control function of the COX/Cyt *c* couple could have been a specific evolution target affecting global primate metabolism.

In this chapter, we will trace the evolution of Cyt *c* and COX genes among the primate lineages and discuss the probable causes and consequences of this evolution. First we will present a general view of the biochemical Cyt *c*/COX mechanisms within a macroevolutionary perspective. Then we will review the COX/Cyt *c* differences both across primates and compared to nonprimate mammals, as well as the timeframe of the appearance of these differences. We will also review the different evolutionary forces that have been suggested to influence primate COX gene evolution. Finally, we will discuss the possible causes and consequences of this evolution and attempt to integrate COX/Cyt *c* evolution into a broader view of primate evolution.

8.2 COX Function and Macroevolution in OxPhos

8.2.1 OxPhos System

OxPhos consists of the electron transport chain (ETC) and ATP synthase. The ETC transfers electrons derived from the breakdown of food that are captured in reduced molecules such as NADH. The electrons of the latter enter the ETC via NADH dehydrogenase (complex I). The Krebs (i.e., Citric Acid) cycle feeds additional electrons into the ETC through succinate dehydrogenase (complex II). Both complex I- and II-derived electrons are then transferred to ubiquinone (coenzyme Q or CoQ), which is thereby reduced to ubiquinol. The bc_1-complex (complex III) is an electron relay, converting packages of two electrons delivered from ubiquinol into single electron packages taken up by Cyt *c*. Finally, Cyt *c* binds to COX (complex IV), which transfers electrons to molecular oxygen.

Electron transfer reactions catalyzed by the ETC are coupled to the pumping of protons across the inner mitochondrial membrane by complexes I, III, and IV, but not complex II. The pumped protons generate the mitochondrial proton motive force (Δp_m), which consists of a chemical component, the *p*H difference across the inner membrane, and an electrical component, the mitochondrial membrane potential ($\Delta\Psi_m$), which constitutes the major part of Δp_m in mitochondria. In the last step of OxPhos, Δp_m is utilized by ATP synthase (complex V) to generate ATP from ADP and phosphate, which is coupled to the backflow of protons from the mitochondrial intermembrane space (IMS) to the matrix.

8.2.2 *Electron Transfer from Cytochrome c via Cytochrome c Oxidase to Oxygen*

The precise mechanism of how electron transfer is coupled to proton pumping is still a matter of controversy. Several models exist, some of which involve the heme propionate side chains as the sites where protons formally leave the matrix space and enter the intermembrane space within COX. The electron pathway is well established: after Cyt *c* binds to COX subunit II, a single electron leaves the Cyt *c* heme group and enters COX subunit II via the binuclear Cu_A site. From there the electron enters subunit I and reaches the heme *a* group located near the middle of the membrane and horizontally transfers to the heme a_3-Cu_B binuclear site where oxygen binds. The heme groups are perpendicular to the membrane at an angle of 108° to each other, and their closest edge-to-edge distance is only 5 Å (Ludwig et al. 2001). Four subsequent electron transfer reactions in addition to four protons, which are taken up from the matrix space, are required to reduce dioxygen to water:

$$4\text{Cyt}c^{2+} + \text{O}_2 + (n_{\text{pumped}} + 4_{\text{chemical}})\text{H}^+_{\text{matrix}} \rightarrow 4\text{Cyt}c^{3+} + 2\text{H}_2\text{O} + n\text{H}^+_{\text{cytosol}}.$$

8.2.3 *Regulatory Mechanisms of Cytochrome c Oxidase*

The last step of the ETC, the reaction catalyzed by COX, is the proposed rate-limiting step of the ETC in intact mammalian cells (Villani and Attardi 1997; Villani et al. 1998) but not in isolated mitochondria, likely due to loss of regulatory properties of the OxPhos complexes during mitochondrial isolation (Hüttemann et al. 2011). The essentially irreversible terminal reaction catalyzed by COX has to be well regulated and adapted to temporal and tissue-specific energy requirements. This adaptation is mediated by the presence of tissue-specific and developmentally regulated isoforms only found in COX and Cyt *c* among OxPhos complexes, further supporting the important role of COX in the regulation of overall ETC flux. To date six isoforms have been reported for the nuclear encoded subunits of COX, which are encoded by separate genes (Hüttemann et al. 2011). Three are liver- and heart-type isoform pairs of subunits VIa, VIIa, and VIII. Liver-type COX is found in organs such as liver, brain, and kidney. In contrast, the heart-type isozyme is expressed in heart and skeletal muscle. In addition, there is a lung-specific isoform of COX subunit IV (*COX4i2*), a testis-specific isoform of subunit VIb (*COX6B2*), and a third isoform of subunit VIII (*COX8C*) (Hüttemann et al. 2003b). Of those enzymes studied, their basal activities follow the sequence heart type < liver type < lung type, which inversely correlates with the mitochondrial capacity of the corresponding tissues.

Several additional regulatory mechanisms act on COX including allosteric regulation by a built-in energy sensor, an ATP/ADP binding pocket located on subunit IV that senses the energetic state of the cell and adjusts energy production to demand (see Chap. 11). Other modulators that affect COX activity include nitric oxide (NO), which competes with oxygen for binding at the binuclear heme *a*-Cu_B center and inhibits COX, and thyroid hormone 3,5-diiodothyronine, which binds to subunit Va and releases the allosteric ATP inhibition, thus allowing higher ETC flux even in the presence of high ATP/ADP ratios. This finding was proposed to explain the short-term activating action of thyroid hormones on metabolism and it may also be involved in nonshivering thermogenesis (Ludwig et al. 2001).

COX is also targeted by cell signaling pathways through phosphorylation of specific serine, threonine, and tyrosine residues, which in some instances decisively regulate enzyme activity (see Chap. 10). Therefore, the multitude of regulatory properties found in COX underlines its unique position as the oxygen burning engine of the ETC. Analyzing COX evolution can thus provide insight into important mechanistic and regulatory aspects of aerobic energy metabolism, and it may also help to answer unsolved basic questions, such as the identification of proton exit pathways and oxygen channels within the enzyme that are presumably evolutionarily conserved.

8.2.4 COX Macroevolution

It is now broadly accepted that mitochondria and OxPhos complexes are the relic of an endosymbiosis between a pre-eukaryotic cell and an α-proteobacterium (Sagan 1967; Zimmer 2009; Richards and Archibald 2011). Interestingly, mammalian COX activity is mainly regulated by the nuclear subunits, which are absent in bacterial COX. Genomic data have shown that nuclear subunits accumulated on the various eukaryotic lineages after the endosymbiosis (for review see Pierron et al. 2012). Because the catalytic activity of human COX is not higher than the proteobacterial one, the macroevolutionary process leading to the doubling of COX's size seems not caused by the selection of a more active enzyme. Instead, the appearance of nuclear subunits could have responded to the need for tight regulation of the COX enzyme by the nuclear genome of eukaryotic pluricellular organisms. We speculate that because endosymbiosis has increased energy production by about 15-fold, selective pressures acting on the new eukaryotic cells were not acting to produce more energy but to produce it when and where necessary, and otherwise to reduce the production of toxic products and by-products of accompanying free radicals.

Recent studies have highlighted that the mammalian respiratory chain has one component less than the eukaryotic ancestral respiratory chain: an alternative oxidase (Mcdonald et al. 2009). This enzyme was a substitute for complexes III and IV (COX), performing the oxidation of CoQ and the reduction of oxygen, but without any proton translocation, and was probably lost at the vertebrate stem (Mcdonald et al. 2009). The function of this enzyme is not clear but it has been shown that in plants this enzyme responded to oxygen concentration stress (Szal et al. 2003). The loss of this COX substitute has probably modified the evolutionary constraint acting on COX. Interestingly, whole genome duplication was another concomitant key event modifying the early vertebrate respiratory chain (Wotton and Shimeld 2006). Indeed, it has been shown that several COX subunits duplicated at this point have been retained in the subsequent vertebrate genome, allowing the specialization of different isoforms for the same subunit (i.e., the *COX4i1-COX4i2* pair, which are regulated based on the oxygen concentration (Hüttemann et al. 2007). Such gene duplication has been a continuous process throughout mammalian genome evolution, promoting organ-specific isoforms for the same subunits (i.e., liver or heart isoforms) to allow tight regulation of COX activity based on organ needs (Pierron et al. 2012).

8.3 Evolutionary Events of Primate COX and Cytochrome *c*

Despite their relative evolutionary stability, both Cyt *c* and the majority of subunits of COX have accumulated amino acid replacements in the primate lineage at a faster rate than would be predicted from either their rate during previous mammalian descent or by the slowdown in the mutation rate in primates (Wu and Li 1985; Bailey et al. 1991; Li et al. 1996). COX appears to have developed the role of its subunits further than the other complexes, adapting them both during development and differentiation. We review here their roles in these processes and their evolution.

8.3.1 Expression of Subunits

The overall timing of subunit expression reflects the metabolic program of the embryo. In mouse, a Cyt *c* null mutant is able to develop until mid-gestation, signaling the time of demand for oxidative metabolism (Li et al. 2000). After the start of oxidative metabolism, the earliest expression of COX subunits in tissues that are destined after differentiation to produce tissue-specific forms is of the ubiquitously expressed isoforms. In an early study, Kadenbach et al. compared human fetal tissues from liver, skeletal muscle, heart, and intestine with those of the corresponding adult tissues (Bonne et al. 1993). They found for the contractile tissues heart and skeletal muscle that both COX6A and COX7A transcription between fetal weeks 20–28 were predominantly of the L isoform (*COX6A1* and *COX7A2*), turning on the H isoform (*COX6A2* and *COX7A1*) after birth. Similar timing was seen for subunit 8 (switching from *COX8A* to *COX8B*) (Bonne et al. 1993) and subunit 4 (*COX4I1* to *COX4I2*) (Hüttemann et al. 2001). Depending on the organism and tissue, the developmental isoform is transcribed in addition to, rather than in place of, the initially expressed isoform. For example, there was a complete switch of isoform expression for subunit 6A and a partial one for 7A in human heart and skeletal muscle (Bonne et al. 1993) as well as a partial one for subunit 4 in rodent lung (Hüttemann et al. 2001).

Determining the functional difference between tissue-specific isoforms of the same subunit has proven challenging. Little is known about the functions of COX7A and COX8. In the case of subunit 4, the holoenzyme containing COX4I2 was shown to be at least twice as active as that containing COX4I1 (Hüttemann et al. 2007). The transcription of *COX4I2* has been shown to respond to oxygen concentration (Horvat et al. 2006; Fukuda et al. 2007; Hüttemann et al. 2007), supporting the original suggestion that a pair of cysteines in COX4I2, which are absent from COX4I1, could serve as a hypoxia sensor (Hüttemann et al. 2001). The COX6A isoforms have been associated with energy transduction efficiency. The contractile muscle specific subunit COX6A2 contains an adenine nucleotide binding site whose occupancy can result in a change of H^+/e^- efficiency. At low ATP/ADP the H^+/e^- ratio is close to 1, contributing to maximum efficiency of energy transduction, whereas at a high ATP/ADP ratios it is 0.5, which may contribute to thermogenesis (Frank and Kadenbach 1996; Kadenbach et al. 1998). For COX6A1, binding of palmitate can produce a similar reduction of the H^+/e^- stoichiometry to 0.5, an effect not produced by other fatty acids tested (Lee and Kadenbach 2001). Thus, in this way, depending on the tissue COX can respond to a readout of either energy supply or metabolic cues and appropriately parse potential energy into heat for nonshivering thermogenesis and chemical energy as ATP. Although these isoforms are also present in invertebrates (with less confidence for COX8) (Little et al. 2010), no information is available about whether comparable isoform-specific functions are present.

8.3.2 Subunit Duplication

COX is the only one of the electron transport complexes to contain subunit isoforms,[1] presumably reflecting its central role in regulation. Each of these appears to have arisen by gene duplication, although in almost all cases the duplicates reside on different chromosomes. The subunits currently known to have more than one form are 4, 6A, 6B, 7A, and 8 (Table 8.1).

Subunit 4 was shown to have a second form in mammals with the discovery of a lung-specific isoform (Hüttemann et al. 2001). As had been previously shown for the muscle-specific isoforms, *COX4I2* is also developmentally induced. Its likely origin during a whole genome duplication (Wotton and Shimeld 2006) was previously discussed. Its current function may have been molded in part by subsequent increases in the oxygen content of the earth's atmosphere to a level about 50% higher than the present, a level highest in at least the last billion years. The molding of this isoform during a maximum in atmospheric oxygen concentration suggests a role in dealing with high oxygen concentration, a suggestion reinforced by its presence in the lung (and, as subsequently found, in the placenta, the embryonic gas exchange tissue). Its precise function, however, is not yet clear: although functioning in the body's highest oxygen concentration, its transcription is stimulated at 4% oxygen (Hüttemann et al. 2007) and it has been said to adjust respiration in

[1] We note that possible paralogs in complex I have been reported.

Table 8.1 Isoforms of cytochrome *c* oxidase. Known sites of expression of tissue-specific isoforms are shown in Tissues column. (P) indicates loss of that gene as pseudogene in human and some ancestral lineages

Subunit	Ubiquitous	Tissue specific	Chr	Tissues	Refs
4	*COX4I1*	*COX4I2*	16, 20	Lung, placenta	Hüttemann et al. (2001, 2007)
6A	*COX6A1*	*COX6A2*	12, 16	Heart, skeletal muscle	Yanamura et al. (1988) and Ewart et al. (1991)
6B	*COX6B1*	*COX6B2*	19, 19	Testis	Hüttemann et al. (2003a)
7A	*COX7A2*	*COX7A1* *COX7A2L*[a]	6, 19, 2	Heart, skeletal muscle	Arnaudo et al. (1992), Van Kuilenburg et al. (1992), and Van Beeumen et al. (1990)
8	*COX8A*	*COX8B(P)* *COX8C*[b]	11, 11, 14	Heart, skeletal muscle	Lomax et al. (1995), Rizzuto et al. (1989), and Hüttemann et al. (2003b)
Cyt *c*	*CYCS*	*CYCT(P)*	7, 2	(Testis)	Hake et al. (1994), Goldberg (2003), and Pierron et al. (2011)

[a]Localizes to Golgi
[b]It is not yet known where COX8C is expressed

hypoxic cells (Fukuda et al. 2007). As originally suggested (Hüttemann et al. 2001), the presence of cysteines that are absent from COX4I1, and that are appropriately located to serve as a potential disulfide redox sensor, supports a role at oxidizing versus reducing oxygen concentrations and/or high reactive oxygen (ROS) environment.

One difficulty in defining a role is in arriving at an understanding of whether the in vitro determined optimum of 4% oxygen is the normal physiological level in some tissues or whether it represents a state of physiological stress we call hypoxia. For example, in the central nervous system in particular, 4% oxygen has been suggested to represent physiological normoxia (Lamanna 2007; Chadwick et al. 2011). In lungs oxygen is about 6% in venous blood and lower within tissues (Kinnula and Crapo 2003). Since no evidence has yet tied COX4I2-containing COX to physiological hypoxia or ischemia, but COX4I2-containing COX has been shown to be a more active enzyme (Hüttemann et al. 2007), it may well be that its main function is to provide a higher turnover to more metabolically active tissues.

Whatever the function carried out by the subunit 4 isoforms, they have arisen more than once: The yeast isoforms COXVa and COXVb have been proposed to be homologous to the mammalian COX4 pair but it is clear from the evolution pattern (Hüttemann et al. 2007) that they arose independently. Furthermore, for the mammalian pair, a recent analysis suggests that *COX4I2* is the ancestral gene and that *COX4I1* arose by duplication (Little et al. 2010).

Little is known about COX6B2 function. COX6B connects the COX monomers in its physiologic (dimeric) form. Its existence as a testis-specific isoform (Hüttemann et al. 2003a) is satisfying since Cyt *c*, which has a testis isoform in mammals, appears to bind to COX by first interacting with COX6B (Sampson and Alleyne 2001; Hüttemann et al. 2003a, b). However, Cyt *c* testis (Cyt *c*t) is lost at the stem of primates (Pierron et al. 2011) whereas *COX6B2* is not. Interestingly, in rodents, which express Cyt *c*t, Cox6b2 is the exclusive testes form; humans, which have lost Cyt *c*t, express both COX6B isoforms in testes.

The remaining isoforms, discussed in Sect. 8.2.1, are contractile muscle specific. However, they are not the exclusive isoform in contractile muscle. Detailed studies are lacking in cases where both isoforms are expressed in the same tissue as to whether there is positional segregation within the cell. It is also unknown whether any hybrid forms occur of the dimer, which under some conditions is the more stable form of COX (Stanicova et al. 2007).

8.3.3 Subunit Silencing in Primates

The presence of a testes-specific Cyt *c* is well known (Hake et al. 1990). It was lost on the primate stem about 65 million years ago via a nonsense mutation that is present in all primates. The loss of the testes isoform produces lower fecundity (Narisawa et al. 2002). A detailed recent analysis of the somatic form of Cyt *c* across primates suggested parallel evolution in both the platyrrhine and catarrhine stems (Pierron et al. 2011). Examination of the sites of evolutionary changes suggests they are focused on the respiratory chain rather than on other Cyt *c* functions, such as apoptosis. In addition to Cyt *c*, a COX subunit was lost as well. The smallest COX subunit, COX8, is present as an L and an H isoform throughout mammals. Where this has been examined—in cow (Yanamura et al. 1988; Lightowlers et al. 1990) and rat (Kadenbach et al. 1990)—subunit 8 is expressed in heart exclusively as the H form. COX8H (COX8B) is also present in strepsirrhine primates (lorises and lemurs) and platyrrhine primates (New World monkeys). Surprisingly, however, COX8B is absent in catarrhines (Old World monkeys, Apes, including human) (Goldberg et al. 2003), where exon 2 has been replaced by repeat elements. Of the three contractile muscle isoforms, perhaps the least is known about the function of COX8. Based on an evolutionary study, *COX8A* (the ubiquitous or L form) evolved under the force of positive selection, including at apparently functionally important positions, in several anthropoid terminal lineages (Goldberg et al. 2003). These changes presumably subsumed the advantages that the H isoform provided to contractile tissues.

8.3.4 Amino Acid Replacement and "Rapid Evolution"

The original discovery that human *COX4I1* evolves more rapidly than expected (Lomax et al. 1992) led to more detailed examination of its evolution (Wu et al. 1997; Wildman et al. 2002) and later that of other subunits of cytochrome oxidase and then of complex III (Grossman et al. 2001; Doan et al. 2005). The COX genes found also to show accelerated evolution in primates are subunit I (Wu et al. 2000), II (Adkins and Honeycutt 1994), 5A (Uddin et al. 2008), 6B (Doan et al. 2004), 6C (Doan et al. 2004), 7A2 (Schmidt et al. 1999), 7C (Doan et al. 2004), and 8A (Goldberg et al. 2003).

For a given subunit, the designation rapid evolution is based on the ratio of a standard measure at each branch of the phylogenetic tree of its nonsynonymous substitution rate (K_A) and its synonymous substitution rate (K_S). The conclusion of positive selection rather than reduction of evolutionary constraints on replacement for these subunits is based on observing a period of increased K_A/K_S followed by a reduction—that is, positive selection followed by purifying selection (Goodman 1982). A K_A/K_S ratio >1 is in accord with the generally accepted criterion of positive selection and is observed explicitly for several subunits (Wu et al. 1997; Wildman et al. 2002). For most subunits, however, K_A/K_S is <1 in anthropoid primates but significantly higher than seen in other mammals. As discussed in Sect. 8.3.4, we have interpreted this as representing positive selection of highly conserved proteins that are undergoing adaptive evolution in the presence of functional constraints.

Although the amino acid replacements can be documented, their functional significance awaits a better understanding of subunit function. The best developed picture stems from the insight that a moderate fraction of the amino acid replacements in mammals are concerned with the binding of Cyt *c* to COX, and that the nature of the interaction between Cyt *c* and COX has changed in anthropoid primates (Schmidt et al. 2005). Both simulated binding (Roberts and Pique 1999) and enzyme kinetic analysis (Osheroff et al. 1983) established that the binding between Cyt *c* and COX is electrostatic. Schmidt et al. (2005), using structural data for Cyt *c* and COX and an interaction model (Roberts and Pique 1999; Zhen et al. 1999), identified 57 of the >1,500 COX residues that take part in binding Cyt *c*. At least 27 of these 57 residues were replaced in the anthropoid primate lineage; what is more, 22 of these 57 residues are charged and 11 of the 22 have been replaced within anthropoids by uncharged residues. The binding of Cyt *c* to COX thus appears to have gone from primarily electrostatic to more hydrophobic. The effect of this remarkable change on the intrinsic reaction parameters has not been investigated so that it remains unclear how this restructuring of the binding site has benefitted anthropoid primates.

8.3.5 *Xenocybrids and Coevolution*

The introduction of transmitochondrial cybrids (Wallace et al. 1975) allowed dissecting nuclear versus mitochondrial contributions to phenotype (e.g., Hayashi et al. 1991). The ability to replace cytoplasms was soon applied also to coevolution. One example was for assessing the ability of human nuclear mitochondrial genes to interact productively with mitochondrial genes from increasingly diverse species (Kenyon and Moraes 1997). Kenyon and Moraes found that they could replace human mtDNA with that from chimpanzee, bonobo, and gorilla with resultant oxygen consumption of the cell lines roughly equal and about 80% the value of the human parental line. However, when human mtDNA was replaced with that from species that diverged as recently as 7–14 million years ago, no oxygen consumption could be demonstrated. This was surprising compared to the expected model of oxygen consumption falling off gradually with increasing evolutionary divergence.

Examination of the mtDNA changes showed about twice as many amino acid replacements between orangutan and human as was found between the chimpanzees or between gorilla and human. With respect to the electron transport chain complexes, the least were found for complex IV. When considered in light of the accelerated evolution reported for many of the nuclear encoded subunits of complexes III and IV during primate descent, it seems likely that the coevolution taking place between subunits of a complex coded by the mitochondrial and nuclear genomes, which presumably acts to optimize protein–protein, protein–RNA, and protein–mtDNA interaction, and therefore function, has become a victim of coevolution when genomes from different species are mixed. The same effect possibly occurs in a more limited way in somatic cell nuclear transfer (Evans et al. 1999).

To investigate the basis of the abrupt drop-off in function after 7–14 million years ago of evolutionary separation, transfer of orangutan chromosomes into human cells lacking mtDNA (ρ°) and repopulated with orangutan mtDNA was investigated (Barrientos et al. 2000). Hybrids that produced some respiratory function were only seen containing many orangutan chromosomes. When examined for activity of individual complexes, they showed a complex IV deficiency. Furthermore, more detailed analysis showed normal synthesis but defective assembly of complex IV components. The most consistent hypothesis was that human components were exerting a dominant negative effect on attempted assembly of an orangutan enzyme so that the low level of activity found resulted from homologous association. Furthermore, it can be speculated that the lack of interference with the other complexes resulted from the higher evolutionary divergence seen for them (Kenyon and Moraes 1997), such that the loose association of interspecies intermediates during the assembly process of the complexes would have been avoided. These studies suggest that as little as 7 million years is sufficient to change key amino acids that can disrupt harmonious assembly and function of these interacting proteins.

The greater evolutionary change seen in complex I components was utilized to study this effect and its role on cell physiology in some detail. Given that there appears to be excess complex I activity, and that reducing it has potential effects on ROS production, mitochondrial membrane potential, and apoptosis, the coevolution of the complex's subunits to optimize function is a clear advantage for long-lived primate species. This was emphasized in detailed studies of complex I deficiencies in transmitochondrial hybrids. In human–chimpanzee hybrids, which suffered a 20–30% decrease in oxygen consumption, only complex I was affected (Barrientos et al. 1998). Furthermore, decreased complex I impairment was correlated with free radical (ROS) production, mitochondrial membrane potential, and apoptosis (Barrientos and Moraes 1999).

The steep effects of evolutionary distance observed for primates were reproduced qualitatively but were quantitatively less significant for rodent xenomitochondrial hybrids. When mtDNA-less cells derived from the common mouse (*Mus musculus domesticus*) were fused to cytoplasts prepared from *Mus musculus*, *Mus spretus*, or rat (*Rattus norvegicus*), a comparable number of respiring clones could be obtained (Dey et al. 2000). However, mouse xenomitochondrial cybrids harboring rat mtDNA had a slower growth rate in medium containing galactose as the carbon source, suggesting a defect in oxidative phosphorylation. Nevertheless, mitochondrial protein synthesis was unaffected. The fact that mild defects could be seen in rodent xenomitochondrial hybrids (Pinkert and Trounce 2002; Trounce et al. 2004), coupled with the difficulties of direct alteration of the mitochondrial genome, led to attempts to use such hybrids as animal models for mitochondrial dysfunction (Cannon et al. 2011; Dunn et al. 2012). The impact thus far has been limited, however, owing to lack of an overt phenotype.

Although the present focus is on primates, it is interesting to consider that mitochondria–nuclear coadaptation has been ongoing probably since the time of the origin of mitochondria. A well-studied and useful system has been the coevolution of nuclear and mitochondrial genes in the copepod *Tigriopus californicus*. When the mitochondria from one population are placed by repeated backcrosses into the nuclear background of a different population, reduced COX activity is observed, suggesting that coadaptation has occurred (Edmands and Burton 1999). This result could be confirmed by showing that Cyt *c* from each population functioned optimally when the COX from the same population was also present (Willett and Burton 2004). In some cases, incompatibilities between populations were present that could be traced to a single amino acid replacement in Cyt *c* (Harrison and Burton 2006). The incompatibilities observed in hybrids led Burton and his collaborators to suggest that the classical reduced fitness of F2 hybrids may stem from reduction of the coevolutionary optimization of mitochondrial function in the parental strains (Burton et al. 2006; Ellison and Burton 2008).

8.4 Evolutionary Mechanism of Primate COX

It is clear that most of the COX subunits and Cyt *c* proteins have evolved rapidly during primate evolution. The reason for this rapid evolution is less clear, however, and the specific function acted upon by evolution is least clear. Indeed, this evolution is necessarily due to the combination of several evolution mechanisms.

8.4.1 Mutation Rate

The accumulation of amino acid changes on the proteins could be due to the imperfect replication of DNA by DNA polymerase. While copying a protein-coding locus, the polymerase can by chance substitute one nucleotide for another and thereby modify a codon. When the codon substitution results in the replacement of the encoded amino acid during protein synthesis the mutation is called "nonsynonymous"; however, due to the redundancy of the genetic code, the substitution sometimes does not result in a change to the encoded amino acid sequence and thus is called "synonymous."

When one mutation event occurs in a germinal stem, the new nucleotide is transmitted to the offspring and so two alleles of this position exist in the same population, the new nucleotide (in the offspring) and the ancestral nucleotide (in the other individuals of the population). Sometimes the new nucleotide spreads over the rest of the population and the ancestral nucleotide disappears. In this chapter, the term substitution event defines this fixation of the new allele in the population. By assuming that the fixation of a synonymous mutation or a mutation on a noncoding position is not subject to selective pressure, it often assumes that the rate of synonymous substitution is proportional to the mutation rate.

Nucleotide mutation is often seen as a random binomial event, with accumulation following Poisson's distribution. However, although the accumulation rate of noncoding and synonymous substitutions is variable, primates appear to have a slower rate of substitution than rodents (Wu and Li 1985; Li et al. 1996) and hominoids have a slower rate than other primates (Wu and Li 1985; Bailey et al. 1991; Steiper et al. 2004). By studying intron (noncoding sequence) and synonymous substitutions on isoform 1 of COX4 (*COX4i1*), Wildman et al. (2002) have shown that neutral substitution slowdown was occurring in parallel with the acceleration of nonsynonymous substitution on COX genes. Therefore, the rapid accumulation of amino acid replacement on COX is not due to an increase of the mutation rate. Surprisingly, in fact, the rapid accumulation of amino acid replacement on COX appears to be opposite to the general slowdown of evolution observed throughout the primate genome compared to the rodent genome.

The primate substitution slowdown was much more drastic for mitochondrial DNA (mtDNA) than for nuclear DNA (nDNA) (Nabholz et al. 2008). Consequently, the mutational ratio has changed between the nuclear and mitochondrial genomes. In Muridae the mitochondrial substitution rate can be 100 times higher than the nuclear one, whereas this ratio is lower than 20 in Hominidae (Nabholz et al. 2008). Nabholz and collaborators have proposed that this could be due to the divergent phenotypic evolution between rodents and primates.

The main hypothesis regarding the nDNA primate slowdown is the increase of generation time (Li et al. 1996; Tsantes and Steiper 2009). Indeed, mutation accumulation occurs during germinal cell duplication. Because the number of germinal cell duplications by generation is almost constant between mammals, the greater the number of generations in a time period, the higher is the number of mutations accumulated during this period. The number of generations on the spermatozoid lineage is much greater than on the egg lineage; thus, most of the mutations occur on the spermatozoid lineage. This "male-driven model" is supported by the fact that the locus on chromosome Y only transmitted by the spermatozoid lineage has a higher mutation rate than other loci (Li et al. 1996).

Interestingly, a study on strepsirrhines has shown a particularly strong correlation between generation time and nDNA neutral replacement accumulation rate but found a lesser correlation with the mtDNA substitution accumulation rate (Tsantes and

Steiper 2009). It seems logical that mtDNA escapes in the male-driven generation time model because of its maternal inheritance. Instead of an impact of spermatozoid stem cell division, it has been proposed that the high mtDNA mutation rate observed in rodent could be directly due to the amount of ROS produced by the high metabolism rate of these species (Pamplona and Barja 2007; Min and Hickey 2008). However, a higher metabolism rate does not necessarily imply higher free radical production and higher free radical production does not directly imply a higher mutation rate (Galtier et al. 2009a). Samuels (2004) has instead proposed that mtDNA mutation rate itself could be subject to specific selective pressures due to life span increase. The neutral mtDNA substitution rate is quite variable across the species of short-lived mammals such as rodent, but constrained to low values in long-lived mammals such as some primates (Nabholz et al. 2008). The low mutation rate of these animals may be an adaptive phenomenon to avoid the accumulation of deleterious mutations on important mtDNA genes, such as mt-COX subunits, on somatic cells across a long life (Galtier et al. 2009a). This type of selective pressure could act on the replication mechanism and it would be interesting to compare the accuracy of gamma polymerase between rodent and primate species but to our knowledge data allowing this kind of comparison are not yet available. Thus, the observed primate mutation slowdown for mtDNA encoded subunits is maternally driven, possibly constrained by the long life span phenotype of primates, whereas the mutation slowdown for nuclear encoded COX subunits is a paternal-driven phenomenon passively due to the increased generation time. What is clear is that the mutation accumulation rate on COX primate genes is not random but is closely linked to the evolution of primate phenotypic life traits.

8.4.2 Mutational Bias

Another major difference between nDNA and mtDNA is that mtDNA presents a strong bias in base composition between the two strands. The transcribed strand (for mt-COX genes) is C-rich and called the light strand based on its buoyant density in a CsCl gradient, whereas the other is G-rich and called the heavy strand. This compositional asymmetry is due to a mutation bias on heavy strand cytosine to thymine (C→T) and adenine to guanine (A→G) as shown on the rat phylogeny of the mt-COII gene (Brown and Simpson 1982) and confirmed by later studies on vertebrates (Reyes et al. 1998; Faith and Pollock 2003). As proposed by Brown and Simpson (1982), the nucleotide bias could be due to mode of mtDNA replication, asymmetric and slow. Indeed, during mtDNA replication the heavy strand remains single stranded for a long time and is exposed to oxidative damage, increasing the deamination of cytosine and adenine, causing transitions to thymine and guanine. This hypothesis is supported by the fact that the proportion of these transitions is related to the time that the heavy strand remains single stranded during replication (Reyes et al. 1998; Faith and Pollock 2003). The mt-COX locus is slightly less sensitive to the phenomenon since it is close to the second replication origin and thus remains single stranded a shorter time.

Anthropoid primates show a stronger compositional shift, with an enrichment of C on the light strand compared to other mammals and nonanthropoid mammals (Schmitz et al. 2002; Gibson et al. 2005). By studying synonymous mutations on the human mtDNA phylogeny, Kivisild et al. (2006) have confirmed a strong mutational bias with 250 transitions of T→C and only 162 C→T despite an already biased composition of 596 codons NNT and 1,444 codons NNC. Interestingly, the transitions A↔G equilibrate with about 230 transitions in each direction, explaining the finding that the percentage of G is constant across mammals (Gibson et al. 2005).

By comparing mtDNA sequences among different orders of mammals, Min and Hickey have shown that the C enrichment is not primate specific but instead is significantly correlated with longevity and generation time (Min and Hickey 2008). No mechanisms have been proposed to explain this correlation yet. Nevertheless, it means that if mutations due to the asymmetric replication are proportionally more important in long-lived species, then some other mutational mechanisms such as inherent misincorporation rate of gamma polymerase have proportionally decreased. This provides a first clue to the selective pressure acting on the global decrease of mutation rate observed in long life span species. Finally, it appears that the evolution of primate life traits has strongly influenced the mutation rate and the composition bias of mt-COX genes.

The strong mutation bias observed in primates should necessarily affect the amino acid composition of mtDNA encoded subunits. Indeed, Gibson et al. have shown a strong correlation between the compositional bias of cytosine (measured by the third position composition) and the amino acid composition. On the most cytosine biased genome, they have highlighted the overrepresentation of amino acids with a C in the codon position (leucine (CTN), proline (CCN), histidine (CAY), threonine (ACN)), and an underrepresentation of those with T (isoleucine, methionine, serine, tyrosine, phenylalanine) (Gibson et al. 2005). Consequently, in primates Schmitz et al. (2002) have shown that nonanthropoid primates such as Tarsius and slow loris exhibit an affinity toward the amino acids isoleucine, lysine, phenylalanine, and tyrosine, which are encoded by T-rich codons, compared to the rest of primates.

It has been shown that mammalian and primate life spans exhibit a positive correlation with the mtDNA frequency of cysteine and negative correlation with the mtDNA frequency of threonine (Kitazoe et al. 2008; Moosmann and Behl 2008). The first hypothesis was that it was a functional adaptive response to reduce free radical production. However, because this correlation has been seen only on the mt-COX genes and not on nuclear encoded COX genes, Jobson et al. (2010) argue that it is a passive phenomenon due to the shift of cytosine mutation pattern linked to increased life span. Nevertheless, the two hypotheses are not exclusive and both link limitation of deleterious mutations on mtDNA encoded genes with long life span. Primate mt-COX gene evolution appears to be strongly constrained by the imperative of limiting deleterious mutations.

8.4.3 *Negative Selection, Recombination, and Effective Population Size*

MtDNA encodes the COX catalytic subunits, which are highly conserved across eukaryotic life. The evolution of these subunits is constrained by strong negative selection—pressure against the spread of new deleterious mutations. This negative selection has to be particularly strong because of the high and biased mutation rate of mtDNA.

Paradoxically, population genetic theory predicts that negative selection should be less efficient on a mtDNA encoded subunit than on a nuclear encoded subunit due to the fact that recombination on primate mtDNA can be considered as marginal (Galtier et al. 2009b). Indeed, maternal inheritance of mtDNA-encoded subunits reduces the effect of negative selection in two ways: (1) by reducing the effective size of the population, so that the fluctuation of mutation frequency is more influenced by random events and less by selective pressures. This larger random effect allows deleterious mutations to become fortuitously fixed despite the negative selection; (2) by excepting reversion, recombination is the only way that offspring do not inherit parental deleterious mutations.

As a consequence, mtDNA of primate lineages should be accumulating deleterious mutations irreversibly, and eventually be condemned to a meltdown. This phenomenon, known as Muller's ratchet, is indeed one of the proposed evolutionary advantages for recombination in the nucleus (Muller 1964; Felsenstein 1974).

Despite this theoretical paradox, the evidence shows that deleterious mutations are actually rapidly removed. Fan et al. (2008) have shown severe mutation (on ND6, complex I) to be selectively eliminated during oogenesis within four generations. Similarly, Stewart et al. (2008), following the mtDNA for a few generations of mutator mice expressing a proofreading-deficient mitochondrial DNA polymerase, have shown a greater accumulation of synonymous than nonsynonymous mutations. This is due to a rapid and strong elimination of nonsynonymous changes in protein-coding genes, a hallmark of negative selection. This result shows that negative selection against mutations in mtDNA protein-coding genes is strong enough to be monitored. Interestingly, they found that the negative selection occurred more strongly on mt-COXI and mt-COXII than on any other mtDNA encoded gene.

These results suggest that deleterious mutations are rapidly removed under efficient negative selection. It is not clear why, but a simple model would be that due the importance of mtDNA-encoded subunits in cell metabolism, mtDNA mutations are directly deleterious for the cell lineage and this lineage disappears by itself or due to competition with other cells. This scenario is strengthened by the fact that mt-COI of cancerous cell lines accumulates much more synonymous than nonsynonymous mutations, suggesting strong purifying selection against mt-COI even for cells less dependent on OxPhos to produce energy (Stafford and Chen-Quin 2010). This result confirms the importance of mt-COI because there was an absence of selection on mt-COIII, NAD3, and NAD4L. Others have proposed a role of the bottleneck phenomenon (drastic reduction of mtDNA copy number during oogenesis) (Wai et al. 2008, 2010); however, its role as a potential promoting or limiting factor for negative selection is still unresolved (Neiman and Taylor 2009).

Whatever the cause, deleterious mutations do not ordinarily spread in the population. The frequency of common mtDNA deleterious mutations such as A3243G is due to mutational hot spots and not to inefficient negative selection (Pierron et al. 2008). However, the fate of mildly deleterious mutations is less clear. For example, deleterious LHON mutations on complex I can spread over several human generations (Carelli et al. 2006). This result was confirmed also on COX genes: the mouse V421A mutation on mt-COI can be transmitted for numerous generations even if the mice are showing mitochondrial pathology symptoms (Fan et al. 2008). Recent results suggest that mildly deleterious mutations can spread across a population to achieve a state of polymorphism before being eventually eliminated. Indeed, based on amino acid properties, young mtDNA polymorphisms are on average more deleterious than old polymorphisms, suggesting a long-term effect of negative selection (Pereira et al. 2011). The long-term effect of negative selection on mtDNA is in agreement with the current explanation of the phylogenetic observation that younger lineages seem to accumulate mutations faster than older stems (Soares et al. 2009).

Nevertheless, deleterious mutations can fortuitously reach fixation. Furthermore, such fixation of deleterious mutations on mtDNA is more frequent in large animals than in small animals (Popadin et al. 2007). The reason proposed is that larger animals have a smaller effective population size. And the smaller is the population, the stronger is the random effect. Therefore, the accumulation of deleterious mutations could have accelerated during the primate lineage along with the mass increase and the decrease of population size. This raises the possibility that the observed rapid replacement of several amino acids of COX is due to a relaxation of negative selection constraints.

8.4.4 *Positive Selection*

The K_A/K_S ratio is often used to test whether an accumulation of amino acid replacements is due to a relaxation of negative selection or due to positive selection, as in the case of COX (see for review Harris 2010). The K_A/K_S ratio is the ratio of nonsynonymous replacement over synonymous replacements. As noted earlier, because synonymous replacements do not impact the protein amino acid sequence, these replacements are considered as minimally affected by any selective force and their accumulation can be considered as a neutral marker of the mutation rate. By contrast, nonsynonymous replacements change the amino acid sequence and potentially impact phenotype and fitness. A K_A/K_S below 1 indicates that nonsynonymous accumulation is slower than synonymous replacement accumulation. This case can be interpreted as nonsynonymous replacements being subject to negative selection and therefore reaching fixation less often than synonymous replacements randomly reaching fixation. A ratio of K_A/K_S over 1 indicates that nonsynonymous replacements reach fixation more often than neutral synonymous replacements, a signal of positive selection.

Although this principle is broadly accepted, it is important to realize the reverse principle is not true. Indeed, a K_A/K_S ratio under 1 does not necessarily reveal the absence of positive selection and conversely a K_A/K_S ratio over 1 does not necessarily mean an absence of negative selection on particular sites.

In agreement with the importance of COX, K_A/K_S on its genes are very low across the mammals, confirming the strong negative selection (Uddin et al. 2008). However, on primates a K_A/K_S higher than 1 has been shown for several subunits (Wildman et al. 2002; Doan et al. 2004). This result clearly shows positive selection on these subunits. For other subunits, K_A/K_S is higher in primates compared to other mammals, but somewhat lower than 1. For these subunits it could be tempting to conclude that K_A/K_S close to 1 means that these proteins are not affected by selective pressure constraints, either positive or negative, but evolve neutrally. However, "neutral" evolution is quite rare and is a short-term state for a protein because it implies the random replacement of amino acids in the sequence. Such random evolution will eventually change the functionality of the protein and allow appearance of a nonsense mutation. Because deleterious mutations on COX genes induce life-threatening symptoms due to mitochondrial pathology, we speculate that a relaxation of negative constraints on these subunits is not likely. Therefore, we propose that the K_A/K_S near 1 observed on primate phylogeny shows the existence of positive selection on these very conserved proteins, such as cytochrome *c*.

Recently, Osada and Akashi (2012) proposed that the positive selection acting on the nuclear subunits results from compensatory evolution due to an accumulation of deleterious mutation on mt-COX genes. Indeed, they argue that only nuclear-encoded components of COX show evidence for positive selection. However, this result is inconsistent with a previous finding (Wu et al. 2000) and may have resulted from the choice of species (five anthropoid primates and two nonprimates but no nonanthropoid primates), missing the evidence of positive selection on mt-COI on the anthropoid stem. Taking all the evidence together, we propose that rapid evolution and positive selection on both nuclear and mitochondrial subunits reinforces the notion of adaptive selection acting on the whole COX complex.

8.5 COX and Primate Phenotypic Evolution

If we allow that the accumulation of amino acid substitutions on COX during the primate lineage is due to positive selection, the next issue is to identify the selective pressure(s) that has acted on COX. It would be tempting to begin by considering that because COX rapid evolution is specific to primates, the selective pressures should also be specific to anthropoid primates. However, for similar pressures it is not unlikely that different taxons can develop different adaptive strategies. For example, amino acid replacement on COX during anthropoid diversification is co-occurring with a change of oxygen concentration in the atmosphere (Falkowski et al. 2005); because oxygen is a substrate of COX, the two events could be linked. But does the fact that other mammals do not present the same COX evolution invalidate this hypothesis? There is no obvious answer to this question and it may well be difficult to move it beyond the speculation level.

Instead we propose here a broader view, to study how COX evolution can be integrated in two major phenotypic primate evolution processes: (1) the history of life evolution, and (2) brain enlargement.

8.5.1 History of Life

As we have seen, COX evolution was closely linked to the evolution of history trait of life in primates. Due to mutation bias, the mutation rate decrease and the life span increase have shaped the COX mtDNA subunits. However, due to OxPhos ROS production, COX has the ability to influence its own mutation rate and mutation bias. Indeed, COX is one of the respiratory chain regulators and thus influences free radical production (Pacelli et al. 2011). It is worth noting that the other OxPhos complexes have also experienced a rapid evolution in primates. Galtier et al. (2009a) have proposed that the decrease of mtDNA mutation rate was positively selected along with the increase of life span. The effect of OxPhos primate evolution on ROS production has to be investigated in order to know if COX evolution could be due to the selective pressure favoring the decrease of mutation rate on somatic cells.

The primate substitution slowdown suggests that this evolution has also led to a decrease of ROS production in the female germ line. In contrast (but not in contradiction), we suggest that evolution of the Cyt *c*-COX couple in primates allowed an increase of ROS production in the male germ line. Indeed, we have shown that a Cyt *c* testis isoform was present in ancestral mammals and it became a pseudogene before primate differentiation (Pierron et al. 2011). Interestingly, when Cyt *c* testis isoform is silenced in knock-out mice, these mice produce functional but less efficient spermatozoids due to reduced free radical scavenging (Narisawa et al. 2002; Liu et al. 2006). Even if mice are not the best model for ancestral primates, this result suggests that one function of the Cyt *c* testis isoform is to maintain efficient spermatozoids. We propose that the loss of testis Cyt *c* in primates could be due to a decrease of selective pressure for efficient spermatozoids on primates compared to rodents. This idea is supported by the fact that primates indeed have a generally smaller litter size and reduced offspring number compared to rodents (De Magalhaes and Costa 2009). It remains to be seen, however, how the ROS scavenging ability of the single primate Cyt *c* compares with the mouse somatic and testis enzymes.

Life span, reproduction age, reproduction rate, and population size are parameters that have drastically changed during primate evolution. Because numerous authors have linked these life history parameters to metabolism evolution, we propose that these parameters should be seen as both potential cause and consequence of OxPhos evolution and particularly the couple COX-Cyt *c* (Fig. 8.1).

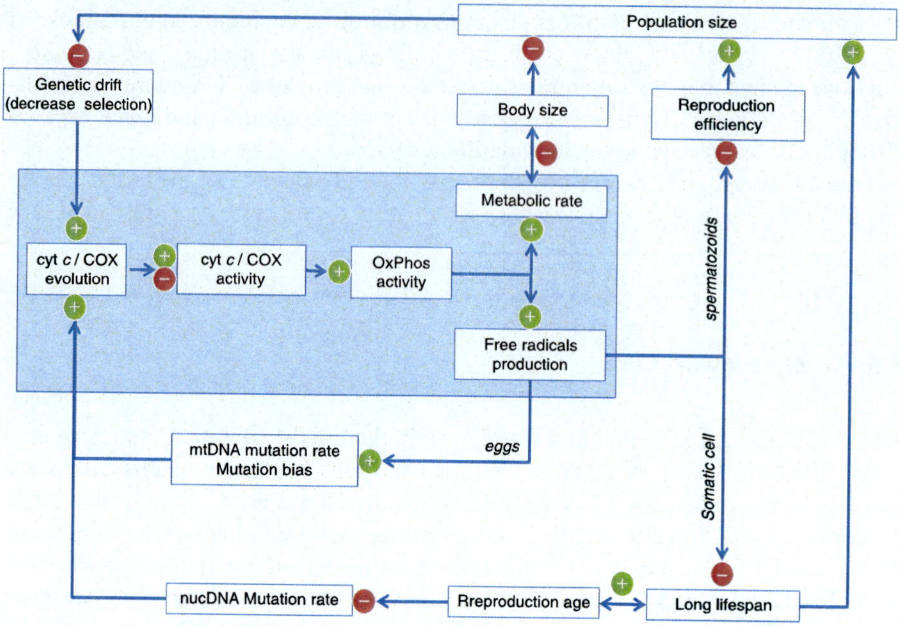

Fig. 8.1 Putative influences of the Cyt *c*/COX couple on their own evolution through nonselective processes. The + and – symbols represent the expected direction of the effect

8.5.2 *Brain Enlargement*

The human brain is approximately six times larger in mass than is expected for a mammal species of its body mass (Jerison 1973). This increase in relative brain mass has a long phylogenetic history. Indeed, all anthropoid primates (i.e., New and Old World monkeys and apes including humans) are at least twice as encephalized as would be expected for creatures of their size, and besides *Homo*, capuchin monkeys (*Cebus* spp.) are the most encephalized mammalian genus (Boddy et al. 2012). Therefore, the process of encephalization must have begun at least 40 million years ago at the time of the last common ancestor of anthropoid primates, and it has continued at variable rates in a wide range of primate taxa such as humans. It should be noted that the encephalization that characterizes humans is due more to expansion of the neocortex than expansion of phylogenetically more ancient brain regions such as the cerebellum (Clark et al. 2001). In general, the human neocortex can be characterized as (a) larger than expected given the species body mass; (b) possessing a significantly more convoluted prefrontal cortex than is seen in other species; and (c) containing a high glia/neuron (i.e., white matter/gray matter) ratio than is found in other anthropoids (Stephan et al. 1988; Rilling and Insel 1999; Sherwood et al. 2006). This increase in the relative amount of white matter in human evolution may serve to enable more connections among the existing neurons. This finding implies that the human neocortex does not have a greater number of neurons associated with its recent expansion, but rather there are more connections among the existing neurons, and it has been proposed that this increase in connectivity enables the associative prowess connected with higher human cognitive abilities (Sultan 2002). It is clear that the expanded neocortex present in humans requires more energy in the form of ATP, but little is known regarding the evolution of energy production in glycolysis and oxidative phosphorylation in neurons vs. glial cells. A promising area of research would involve the quantification and in vivo imaging of mitochondrial activity in the brains of various mammalian species. Studies of gene expression (Caceres et al. 2003; Uddin et al. 2004) have demonstrated that human mRNA expression is upregulated in genes categorized as being involved in aerobic metabolism and neuronal function in humans relative to other primates, but these studies have yet to be confirmed at the protein level and in vivo.

The brain is a metabolically costly organ, and as such it has been proposed that the expansion of the human brain came with the price of reducing the metabolic needs of other organs such as the gut (Aiello and Wheeler 1995). Some have argued that the massive encephalization seen on the human lineage after divergence from chimpanzees was facilitated by the emergence of cooking, and that this innovation both freed up metabolic energy for the brain and enabled early humans to spend more time engaged in social activities instead of chewing food for a significant portion of their waking hours (Wrangham 2009). One recent study compared the weight of visceral organs to brain size in 100 mammal species, but found no negative correlation between brain size and the size of the digestive tract (Navarrete et al. 2011). Instead, in line with Wrangham's proposal, these authors suggested that the evolution of increased brain size in humans was facilitated by the stabilization of energy inputs made possible by the advent of efficient foraging strategies, cooking, and the use of tools. Indeed, the Navarette study indicated that a negative correlation between adiposity and brain size was observed in nonprimate mammals, although that study did not sample the highly encephalized and adipose rich cetaceans. Regardless, humans do possess several means for increasing their energy intake, and it is possible that these innovations played some sort of role in providing the raw material for fueling the energetically costly brain.

The human brain requires energy to function, and much work has suggested that humans possess distinctive aspects of the glucose metabolism pathway. PET studies have shown, on a mass-specific basis, that certain neonatal brain regions consume less glucose than the adult brain (Chugani et al. 1987). Cerebral metabolic rate later exceeds that of the adult by roughly 2–3 years of age and, in at least some regions (e.g., cortex), remains 150–200% higher than the adult brain until late childhood or early adolescence. That human brain metabolic requirements change as a function of age suggests that metabolism of humans has undergone dynamic evolution. In addition to adaptive evolution of the components of the electron transport chain (reviewed in Grossman et al. 2004), the promoters of genes involved in glucose metabolism have been shown to have evolved adaptively during human evolution (Haygood et al. 2007).

Another aspect of molecular evolution that bears directly on energy metabolism in the human brain involves glutamate dehydrogenase, a key mitochondrial enzyme in cellular energy metabolism that plays a role in generating ATP through the Krebs cycle. Humans and other apes possess two genes encoding glutamate dehydrogenase whereas other species of mammals contain only one gene (*GLUD1*) (Burki and Kaessmann 2004). This hominid-specific gene (*GLUD2*) is specifically targeted to the mitochondria due to positive selection in its targeting sequence (Rosso et al. 2008). Conversely, GLUD1 is found in both the mitochondria and cytoplasm. Moreover, it has been posited that the functional role of *GLUD2* is increased in astrocyte metabolism of glutamate (Burki and Kaessmann 2004), and Rosso et al. (2008) further suggest that *GLUD2* has evolved a more positively charged targeting sequence in order to compensate for the relatively low mitochondrial membrane potential in astrocytes compared to tissues where membrane potential is higher (e.g., heart). Taken together, these findings indicate that many aspects of metabolism that are crucial for proper neuronal function have evolved adaptively during primate evolution. It is further interesting, but not thus far connected to brain metabolism, that glutamate dehydrogenase was previously identified as a (bovine) 3′-UTR RNA binding protein for the mRNAs of *COX6A1*, *COX7A2,* and *COX8A* (Preiss et al. 1995; Preiss and Lightowlers 1993).

8.6 Conclusions

The special role of COX and Cyt *c* in terms of isoform expression, allosteric control, and phosphorylation has been emphasized in Chap. 10. In addition to the cited properties, each also shows an evolution pattern that is notable. One of the notable features is the degree to which changes have taken place at the stem of anthropoid primates or within primate lineages. Cyt *c* has undergone several periods of accelerated evolution followed by consolidation (purifying selection) in which replacements were concentrated in regions that participate in OxPhos (Baba et al. 1981; Pierron et al. 2011) and in addition the testes-specific isoform that is present in mammals became inactivated at the stem of anthropoid primates (Pierron et al. 2011). For the 13 subunits of COX, 10 have undergone accelerated evolution in anthropoid primates (Grossman et al. 2004).

The functional nature of the primate evolution changes is not clear. Of the approximately 1,500 amino acids in COX, about 20% or 300 were replaced in anthropoid evolution. Of these replaced residues, nearly 10% or 27 are part of the COX-binding site for Cyt *c*. Furthermore, of the binding site residues, 12 out of 22 were charged and are replaced by noncharged residues. Thus, a significant product of anthropoid evolution for the COX and Cyt *c* couple has been a reconfiguration of the binding interface from an electrostatic to a more hydrophobic surface. How this has modified their functional properties is an ongoing puzzle.

The temporal correlation between the evolution of the electron transport chain and the expansion of the neocortex, which after heart and kidney is the most energy utilizing tissue per gram, has led us to develop a model in which the events are coupled (e.g., Grossman et al. 2004). This correlation is also supported by the increased glia to neuron ratio with brain size (Sherwood et al. 2006). Glia-produced lactate appears to be able to cross-feed adjacent neurons to supply an oxidative energy source (Ames 2000). Anthropoid primates such as humans, in addition to high energy requirements to take account of their size (Isler and Van Schaik 2006; Hasenstaub et al. 2010) and complex social interactions, are also long lived and thus require long-lived neurons (i.e., ones that operate in a low radical environment). At the end of the day, it may be that what has been most refined is the ability to precisely regulate energy expenditure, the brain's most valuable resource, in time and space.

Acknowledgements Supported by NIH GM65580and GM089900, by the National Science Foundation (grants BCS-0550209, BCS0827546 and BCS 9910679), and the Wayne State University Research Excellence fund.

References

Adkins RM, Honeycutt RL (1994) Evolution of the primate cytochrome *c* oxidase subunit II gene. J Mol Evol 38:215–231

Adkins RM, Honeycutt RL, Disotell TR (1996) Evolution of eutherian cytochrome c oxidase subunit II: heterogeneous rates of protein evolution and altered interaction with cytochrome *c*. Mol Biol Evol 13:1393–1404

Aiello LC, Wheeler P (1995) The expensive tissue hypothesis. The brain and digestive system in human evolution. Curr Anthropol 36:199–221

Ames A 3rd (2000) CNS energy metabolism as related to function. Brain Res Brain Res Rev 34:42–68

Andrews TD, Easteal S (2000) Evolutionary rate acceleration of cytochrome *c* oxidase subunit I in simian primates. J Mol Evol 50:562–568

Arnaudo E, Hirano M, Seelan RS, Milatovich A, Hsieh CL, Fabrizi GM, Grossman LI, Francke U, Schon EA (1992) Tissue-specific expression and chromosome assignment of genes specifying 2 isoforms of subunit-VIIa of human cytochrome-*c* oxidase. Gene 119:299–305

Baba ML, Darga LL, Goodman M, Czeluzniak J (1981) Evolution of cytochrome *c* investigated by the maximum parsimony method. J Mol Evol 17:197–213

Bailey WJ, Fitch DH, Tagle DA, Czelusniak J, Slightom JL, Goodman M (1991) Molecular evolution of the psi eta-globin gene locus: gibbon phylogeny and the hominoid slowdown. Mol Biol Evol 8:155–184

Barrientos A, Moraes CT (1999) Titrating the effects of mitochondrial complex I impairment in the cell physiology. J Biol Chem 274:16188–16197

Barrientos A, Kenyon L, Moraes CT (1998) Human xenomitochondrial cybrids – cellular models of mitochondrial complex I deficiency. J Biol Chem 273:14210–14217

Barrientos A, Muller S, Dey R, Wienberg J, Moraes CT (2000) Cytochrome *c* oxidase assembly in primates is sensitive to small evolutionary variations in amino acid sequence. Mol Biol Evol 17:1508–1519

Boddy AM, McGowen MR, Sherwood CC, Grossman LI, Goodman M, Wildman DE (2012) Comparative analysis of encephalization in mammals reveals relaxed constraints on anthropoid primate and cetacean brain scaling. J Evol Biol 25:981–994

Bonne G, Seibel P, Possekel S, Marsac C, Kadenbach B (1993) Expression of human cytochrome-*c* oxidase subunits during fetal development. Eur J Biochem 217:1099–1107

Brown GG, Simpson MV (1982) Novel features of animal mtDNA evolution as shown by sequences of two rat cytochrome oxidase subunit II genes. Proc Natl Acad Sci USA 79:3246–3250

Burki F, Kaessmann H (2004) Birth and adaptive evolution of a hominoid gene that supports high neurotransmitter flux. Nat Genet 36:1061–1063

Burton RS, Ellison CK, Harrison JS (2006) The sorry state of F2 hybrids: consequences of rapid mitochondrial DNA evolution in allopatric populations. Am Nat 168(Suppl 6):S14–S24

Caceres M, Lachuer J, Zapala MA, Redmond JC, Kudo L, Geschwind DH, Lockhart DJ, Preuss TM, Barlow C (2003) Elevated gene expression levels distinguish human from non-human primate brains. Proc Natl Acad Sci USA 100:13030–13035

Cannon MV, Dunn DA, Irwin MH, Brooks AI, Bartol FF, Trounce IA, Pinkert CA (2011) Xenomitochondrial mice: investigation into mitochondrial compensatory mechanisms. Mitochondrion 11:33–39

Carelli V, Achilli A, Valentino ML, Rengo C, Semino O, Pala M, Olivieri A, Mattiazzi M, Pallotti F, Carrara F, Zeviani M, Leuzzi V, Carducci C, Valle G, Simionati B, Mendieta L, Salomao S, Belfort R Jr, Sadun AA, Torroni A (2006) Haplogroup effects and recombination of mitochondrial DNA: novel clues from the analysis of Leber hereditary optic neuropathy pedigrees. Am J Hum Genet 78:564–574

Chadwick W, Boyle JP, Zhou Y, Wang L, Park SS, Martin B, Wang R, Becker KG, Wood WH 3rd, Zhang Y, Peers C, Maudsley S (2011) Multiple oxygen tension environments reveal diverse patterns of transcriptional regulation in primary astrocytes. PLoS One 6:e21638

Chugani HT, Phelps ME, Mazziotta JC (1987) Positron emission tomography study of human brain functional development. Ann Neurol 22:487–497

Clark DA, Mitra PP, Wang SS (2001) Scalable architecture in mammalian brains. Nature 411:189–193

De Magalhaes JP, Costa J (2009) A database of vertebrate longevity records and their relation to other life-history traits. J Evol Biol 22:1770–1774

Dey R, Barrientos A, Moraes CT (2000) Functional constraints of nuclear-mitochondrial DNA interactions in xenomitochondrial rodent cell lines. J Biol Chem 275:31520–31527

Doan JW, Schmidt TR, Wildman DE, Uddin M, Goldberg A, Hüttemann M, Goodman M, Weiss ML, Grossman LI (2004) Coadaptive evolution in cytochrome *c* oxidase: 9 of 13 subunits show accelerated rates of nonsynonymous substitution in anthropoid primates. Mol Phylogenet Evol 33:944–950

Doan JW, Schmidt TR, Wildman DE, Goodman M, Weiss ML, Grossman LI (2005) Rapid nonsynonymous evolution of the iron-sulfur protein in anthropoid primates. J Bioenerg Biomembr 37:35–41

Dunn DA, Cannon MV, Irwin MH, Pinkert CA (2012) Animal models of human mitochondrial DNA mutations. Biochim Biophys Acta 1820(5):601–607.

Edmands S, Burton RS (1999) Cytochrome *c* oxidase activity in interpopulation hybrids of a marine copepod: a test for nuclear-nuclear or nuclear-cytoplasmic coadaptation. Evolution 53:1972–1978

Ellison CK, Burton RS (2008) Interpopulation hybrid breakdown maps to the mitochondrial genome. Evolution 62:631–638

Evans MJ, Gurer C, Loike JD, Wilmut I, Schnieke AE, Schon EA (1999) Mitochondrial DNA genotypes in nuclear transfer-derived cloned sheep. Nat Genet 23:90–93

Ewart GD, Zhang YZ, Capaldi RA (1991) Switching of bovine cytochrome-*c* oxidase subunit VIa isoforms in skeletal muscle during development. FEBS Lett 292:79–84

Faith JJ, Pollock DD (2003) Likelihood analysis of asymmetrical mutation bias gradients in vertebrate mitochondrial genomes. Genetics 165:735–745

Falkowski PG, Katz ME, Milligan AJ, Fennel K, Cramer BS, Aubry MP, Berner RA, Novacek MJ, Zapol WM (2005) The rise of oxygen over the past 205 million years and the evolution of large placental mammals. Science 309:2202–2204

Fan W, Waymire KG, Narula N, Li P, Rocher C, Coskun PE, Vannan MA, Narula J, Macgregor GR, Wallace DC (2008) A mouse model of mitochondrial disease reveals germline selection against severe mtDNA mutations. Science 319:958–962

Felsenstein J (1974) The evolutionary advantage of recombination. Genetics 78:737–756

Follmann K, Arnold S, Ferguson-Miller S, Kadenbach B (1998) Cytochrome *c* oxidase from eucaryotes but not from procaryotes is allosterically inhibited by ATP. Biochem Mol Biol Int 45:1047–1055

Fontanesi F, Soto IC, Horn D, Barrientos A (2006) Assembly of mitochondrial cytochrome *c*-oxidase, a complicated and highly regulated cellular process. Am J Physiol Cell Physiol 291:C1129–C1147
Frank V, Kadenbach B (1996) Regulation of the h+/e(-) stoichiometry of cytochrome *c* oxidase from bovine heart by intramitochondrial ATP/ADP ratios. FEBS Lett 382:121–124
Fukuda R, Zhang H, Kim JW, Shimoda L, Dang CV, Semenza GL (2007) HIF-1 regulates cytochrome oxidase subunits to optimize efficiency of respiration in hypoxic cells. Cell 129:111–122
Galtier N, Jobson RW, Nabholz B, Glemin S, Blier PU (2009a) Mitochondrial whims: metabolic rate, longevity and the rate of molecular evolution. Biol Lett 5:413–416
Galtier N, Nabholz B, Glemin S, Hurst GD (2009b) Mitochondrial DNA as a marker of molecular diversity: a reappraisal. Mol Ecol 18:4541–4550
Gibson A, Gowri-Shankar V, Higgs PG, Rattray M (2005) A comprehensive analysis of mammalian mitochondrial genome base composition and improved phylogenetic methods. Mol Biol Evol 22:251–264
Goldberg A, Wildman DE, Schmidt TR, Hüttemann M, Goodman M, Weiss ML, Grossman LI (2003) Adaptive evolution of cytochrome *c* oxidase subunit VIII in anthropoid primates. Proc Natl Acad Sci USA 100:5873–5878
Goodman M (1982) Positive selection causes purifying selection. Nature 295:630
Goodman M, Sterner KN, Islam M, Uddin M, Sherwood CC, Hof PR, Hou ZC, Lipovich L, Jia H, Grossman LI, Wildman DE (2009) Phylogenomic analyses reveal convergent patterns of adaptive evolution in elephant and human ancestries. Proc Natl Acad Sci USA 106: 20824–20829
Grossman LI, Schmidt TR, Wildman DE, Goodman M (2001) Molecular evolution of aerobic energy metabolism in primates. Mol Phylogenet Evol 18:26–36
Grossman LI, Wildman DE, Schmidt TR, Goodman M (2004) Accelerated evolution of the electron transport chain in anthropoid primates. Trends Genet 20:578–585
Hake LE, Alcivar AA, Hecht NB (1990) Changes in mRNA length accompany translational regulation of the somatic and testis-specific cytochrome *c* genes during spermatogenesis in the mouse. Development 110:249–257
Hake LE, Kuemmerle N, Hecht NB, Kozak CA (1994) The genes encoding the somatic and testis-specific isotypes of the mouse cytochrome *c* genes map to paralogous regions of chromosomes 6 and 2. Genomics 20:503–505
Harris EE (2010) Nonadaptive processes in primate and human evolution. Am J Phys Anthropol 143(Suppl 51):13–45
Harrison JS, Burton RS (2006) Tracing hybrid incompatibilities to single amino acid substitutions. Mol Biol Evol 23:559–564
Hasenstaub A, Otte S, Callaway E, Sejnowski TJ (2010) Metabolic cost as a unifying principle governing neuronal biophysics. Proc Natl Acad Sci USA 107:12329–12334
Hayashi JI, Ohta S, Kikuchi A, Takemitsu M, Goto Y, Nonaka I (1991) Introduction of disease-related mitochondrial DNA deletions into HeLa cells lacking mitochondrial DNA results in mitochondrial dysfunction. Proc Natl Acad Sci USA 88:10614–10618
Haygood R, Fedrigo O, Hanson B, Yokoyama KD, Wray GA (2007) Promoter regions of many neural- and nutrition-related genes have experienced positive selection during human evolution. Nat Genet 39:1140–1144
Horvat S, Beyer C, Arnold S (2006) Effect of hypoxia on the transcription pattern of subunit isoforms and the kinetics of cytochrome *c* oxidase in cortical astrocytes and cerebellar neurons. J Neurochem 99:937–951
Hüttemann M, Kadenbach B, Grossman LI (2001) Mammalian subunit IV isoforms of cytochrome *c* oxidase. Gene 267:111–123
Hüttemann M, Jaradat S, Grossman LI (2003a) Cytochrome *c* oxidase of mammals contains a testes-specific isoform of subunit VIb – the counterpart to testes-specific cytochrome *c*? Mol Reprod Dev 66:8–16
Hüttemann M, Schmidt TR, Grossman LI (2003b) A third isoform of cytochrome *c* oxidase subunit VIII is present in mammals. Gene 312:95–102
Hüttemann M, Lee I, Liu J, Grossman LI (2007) Transcription of mammalian cytochrome *c* oxidase subunit IV-2 is controlled by a novel conserved oxygen responsive element. FEBS J 274:5737–5748

Hüttemann M, Lee I, Pecinova A, Pecina P, Przyklenk K, Doan JW (2008) Regulation of oxidative phosphorylation, the mitochondrial membrane potential, and their role in human disease. J Bioenerg Biomembr 40:445–456

Hüttemann M, Pecina P, Rainbolt M, Sanderson TH, Kagan VE, Samavati L, Doan JW, Lee I (2011) The multiple functions of cytochrome *c* and their regulation in life and death decisions of the mammalian cell: from respiration to apoptosis. Mitochondrion 11:369–381

Isler K, Van Schaik C (2006) Costs of encephalization: the energy trade-off hypothesis tested on birds. J Hum Evol 51:228–243

Isler K, Van Schaik CP (2009) Why are there so few smart mammals (but so many smart birds)? Biol Lett 5:125–129

Jerison HJ (1973) Evolution of the brain and intelligence. Academic, New York

Jobson RW, Dehne-Garcia A, Galtier N (2010) Apparent longevity-related adaptation of mitochondrial amino acid content is due to nucleotide compositional shifts. Mitochondrion 10:540–547

Kadenbach B, Stroh A, Becker A, Eckerskorn C, Lottspeich F (1990) Tissue- and species-specific expression of cytochrome *c* oxidase isozymes in vertebrates. Biochim Biophys Acta 1015:368–372

Kadenbach B, Napiwotzki J, Frank V, Arnold S, Exner S, Hüttemann M (1998) Regulation of energy transduction and electron transfer in cytochrome *c* oxidase by adenine nucleotides. J Bioenerg Biomembr 30:25–33

Kenyon L, Moraes CT (1997) Expanding the functional human mitochondrial DNA database by the establishment of primate xenomitochondrial cybrids. Proc Natl Acad Sci USA 94:9131–9135

Kinnula VL, Crapo JD (2003) Superoxide dismutases in the lung and human lung diseases. Am J Respir Crit Care Med 167:1600–1619

Kitazoe Y, Kishino H, Hasegawa M, Nakajima N, Thorne JL, Tanaka M (2008) Adaptive threonine increase in transmembrane regions of mitochondrial proteins in higher primates. PLoS One 3:e3343

Kivisild T, Shen P, Wall DP, Do B, Sung R, Davis K, Passarino G, Underhill PA, Scharfe C, Torroni A, Scozzari R, Modiano D, Coppa A, De Knijff P, Feldman M, Cavalli-Sforza LL, Oefner PJ (2006) The role of selection in the evolution of human mitochondrial genomes. Genetics 172:373–387

Lamanna JC (2007) Hypoxia in the central nervous system. Essays Biochem 43:139–151

Lane N, Martin W (2010) The energetics of genome complexity. Nature 467:929–934

Lee I, Kadenbach B (2001) Palmitate decreases proton pumping of liver-type cytochrome *c* oxidase. Eur J Biochem 268:6329–6334

Leonard WR, Snodgrass JJ, Robertson ML (2007) Effects of brain evolution on human nutrition and metabolism. Annu Rev Nutr 27:311–327

Li WH, Ellsworth DL, Krushkal J, Chang BH, Hewett-Emmett D (1996) Rates of nucleotide substitution in primates and rodents and the generation-time effect hypothesis. Mol Phylogenet Evol 5:182–187

Li K, Li Y, Shelton JM, Richardson JA, Spencer E, Chen ZJ, Wang X, Williams RS (2000) Cytochrome *c* deficiency causes embryonic lethality and attenuates stress-induced apoptosis. Cell 101:389–399

Lightowlers R, Ewart G, Aggeler R, Zhang YZ, Calavetta L, Capaldi RA (1990) Isolation and characterization of the cDNAs encoding 2 isoforms of subunit-CIX of bovine cytochrome-c oxidase. J Biol Chem 265:2677–2681

Little AG, Kocha KM, Lougheed SC, Moyes CD (2010) Evolution of the nuclear-encoded cytochrome oxidase subunits in vertebrates. Physiol Genomics 42:76–84

Liu Z, Lin H, Ye S, Liu QY, Meng Z, Zhang CM, Xia Y, Margoliash E, Rao Z, Liu XJ (2006) Remarkably high activities of testicular cytochrome *c* in destroying reactive oxygen species and in triggering apoptosis. Proc Natl Acad Sci USA 103:8965–8970

Lomax MI, Hewett-Emmett D, Yang TL, Grossman LI (1992) Rapid evolution of the human gene for cytochrome *c* oxidase subunit-IV. Proc Natl Acad Sci USA 89:5266–5270

Lomax MI, Riggs PK, Womack JE (1995) Structure and chromosomal location of the bovine gene for the heart muscle isoform of cytochrome *c* oxidase subunit VIII. Mamm Genome 6:118–122

Ludwig B, Bender E, Arnold S, Hüttemann M, Lee I, Kadenbach B (2001) Cytochrome *c* oxidase and the regulation of oxidative phosphorylation. Chembiochem 2:392–403

Mcdonald AE, Vanlerberghe GC, Staples JF (2009) Alternative oxidase in animals: unique characteristics and taxonomic distribution. J Exp Biol 212:2627–2634

Mereschkowski C (1905) Natur und ursprung der chromatophoren im pflanzenreiche. Biol Centralbl 25:593–604

Min XJ, Hickey DA (2008) An evolutionary footprint of age-related natural selection in mitochondrial DNA. J Mol Evol 67:412–417

Mishmar D, Ruiz-Pesini E, Golik P, Macaulay V, Clark AG, Hosseini S, Brandon M, Easley K, Chen E, Brown MD, Sukernik RI, Olckers A, Wallace DC (2003) Natural selection shaped regional mtDNA variation in humans. Proc Natl Acad Sci USA 100:171–176

Moosmann B, Behl C (2008) Mitochondrially encoded cysteine predicts animal lifespan. Aging Cell 7:32–46

Muller HJ (1964) The relation of recombination to mutational advance. Mutat Res 106:2–9

Nabholz B, Glemin S, Galtier N (2008) Strong variations of mitochondrial mutation rate across mammals – the longevity hypothesis. Mol Biol Evol 25:120–130

Narisawa S, Hecht NB, Goldberg E, Boatright KM, Reed JC, Millan JL (2002) Testis-specific cytochrome *c*-null mice produce functional sperm but undergo early testicular atrophy. Mol Cell Biol 22:5554–5562

Navarrete A, Van Schaik CP, Isler K (2011) Energetics and the evolution of human brain size. Nature 480:91–93

Neiman M, Taylor DR (2009) The causes of mutation accumulation in mitochondrial genomes. Proc Biol Sci 276:1201–1209

Osada N, Akashi H (2012) Mitochondrial-nuclear interactions and accelerated compensatory evolution: evidence from the primate cytochrome *c* oxidase complex. Mol Biol Evol 29(1): 337–346

Osheroff N, Speck SH, Margoliash E, Veerman ECI, Wilms J, Konig B, Muijsers AO (1983) The reaction of primate cytochromes *c* with cytochrome *c* oxidase. Analysis of the polarographic assay. J Biol Chem 258:5731–5738

Pacelli C, Latorre D, Cocco T, Capuano F, Kukat C, Seibel P, Villani G (2011) Tight control of mitochondrial membrane potential by cytochrome *c* oxidase. Mitochondrion 11:334–341

Pamplona R, Barja G (2007) Highly resistant macromolecular components and low rate of generation of endogenous damage: two key traits of longevity. Ageing Res Rev 6:189–210

Pereira L, Soares P, Radivojac P, Li B, Samuels DC (2011) Comparing phylogeny and the predicted pathogenicity of protein variations reveals equal purifying selection across the global human mtdna diversity. Am J Hum Genet 88:433–439

Pierron D, Rocher C, Amati-Bonneau P, Reynier P, Martin-Negrier ML, Allouche S, Batandier C, Mousson De Camaret B, Godinot C, Rotig A, Feldmann D, Bellanne-Chantelot C, Arveiler B, Pennarun E, Rossignol R, Crouzet M, Murail P, Thoraval D, Letellier T (2008) New evidence of a mitochondrial genetic background paradox: impact of the J haplogroup on the A3243G mutation. BMC Med Genet 9:41

Pierron D, Opazo JC, Heiske M, Papper Z, Uddin M, Chand G, Wildman DE, Romero R, Goodman M, Grossman LI (2011) Silencing, positive selection and parallel evolution: busy history of primate cytochromes *c*. PLoS One 6:e26269

Pierron D, Wildman DE, Hüttemann M, Markondapatnaikuni GC, Aras S, Grossman LI (2012) Cytochrome *c* oxidase: evolution of control via nuclear subunit addition. Biochim Biophys Acta 1817(4):590–597

Pinkert CA, Trounce IA (2002) Production of transmitochondrial mice. Methods 26:348–357

Popadin K, Polishchuk LV, Mamirova L, Knorre D, Gunbin K (2007) Accumulation of slightly deleterious mutations in mitochondrial protein-coding genes of large versus small mammals. Proc Natl Acad Sci USA 104:13390–13395

Preiss T, Lightowlers RN (1993) Post-transcriptional regulation of tissue-specific isoforms – a bovine cytosolic RNA-binding protein, COLBP, associates with messenger RNA encoding the liver-form isopeptides of cytochrome-*c* oxidase. J Biol Chem 268:10659–10667

Preiss T, Sang AE, Chrzanowskalightowlers ZMA, Lightowlers RN (1995) The mRNA-binding protein COLBP is glutamate dehydrogenase. FEBS Lett 367:291–296
Reyes A, Gissi C, Pesole G, Saccone C (1998) Asymmetrical directional mutation pressure in the mitochondrial genome of mammals. Mol Biol Evol 15:957–966
Richards TA, Archibald JM (2011) Cell evolution: gene transfer agents and the origin of mitochondria. Curr Biol 21:R112–R114
Rilling JK, Insel TR (1999) The primate neocortex in comparative perspective using magnetic resonance imaging. J Hum Evol 37:191–223
Rizzuto R, Nakase H, Darras B, Francke U, Fabrizi GM, Mengel T, Walsh F, Kadenbach B, DiMauro S, Schon EA (1989) A gene specifying subunit VIII of human cytochrome *c* oxidase is localized to chromosome 11 and is expressed in both muscle and non- muscle tissues. J Biol Chem 264:10595–10600
Roberts VA, Pique ME (1999) Definition of the interaction domain for cytochrome *c* on cytochrome c oxidase – III. Prediction of the docked complex by a complete, systematic search. J Biol Chem 274:38051–38060
Rosso L, Keller L, Kaessmann H, Hammond RL (2008) Mating system and *AVPR1A* promoter variation in primates. Biol Lett 4:375–378
Ruiz-Pesini E, Mishmar D, Brandon M, Procaccio V, Wallace DC (2004) Effects of purifying and adaptive selection on regional variation in human mtDNA. Science 303:223–226
Sagan L (1967) On the origin of mitosing cells. J Theor Biol 14:255–274
Sampson V, Alleyne T (2001) Cytochrome c/cytochrome c oxidase interaction – direct structural evidence for conformational changes during enzyme turnover. Eur J Biochem 268:6534–6544
Samuels DC (2004) Mitochondrial DNA repeats constrain the life span of mammals. Trends Genet 20:226–229
Schmidt TR, Jaradat SA, Goodman M, Lomax MI, Grossman LI (1997) Molecular evolution of cytochrome *c* oxidase: rate variation among subunit VIa isoforms. Mol Biol Evol 14:595–601
Schmidt TR, Goodman M, Grossman LI (1999) Molecular evolution of the COX7A gene family in primates. Mol Biol Evol 16:619–626
Schmidt TR, Goodman M, Grossman LI (2002) Amino acid replacement is rapid in primates for the mature polypeptides of COX subunits, but not for their targeting presequences. Gene 286:13–19
Schmidt TR, Wildman DE, Uddin M, Opazo JC, Goodman M, Grossman LI (2005) Rapid electrostatic evolution at the binding site for cytochrome *c* on cytochrome *c* oxidase in anthropoid primates. Proc Nat Acad Sci USA 102:6379–6384
Schmitz J, Ohme M, Zischler H (2002) The complete mitochondrial sequence of tarsius bancanus: evidence for an extensive nucleotide compositional plasticity of primate mitochondrial DNA. Mol Biol Evol 19:544–553
Sherwood CC, Stimpson CD, Raghanti MA, Wildman DE, Uddin M, Grossman LI, Goodman M, Redmond JC, Bonar CJ, Erwin JM, Hof PR (2006) Evolution of increased glia-neuron ratios in the human frontal cortex. Proc Natl Acad Sci USA 103:13606–13611
Soares P, Ermini L, Thomson N, Mormina M, Rito T, Rohl A, Salas A, Oppenheimer S, Macaulay V, Richards MB (2009) Correcting for purifying selection: an improved human mitochondrial molecular clock. Am J Hum Genet 84:740–759
Stafford P, Chen-Quin EB (2010) The pattern of natural selection in somatic cancer mutations of human mtDNA. J Hum Genet 55:605–612
Stanicova J, Sedlak E, Musatov A, Robinson NC (2007) Differential stability of dimeric and monomeric cytochrome *c* oxidase exposed to elevated hydrostatic pressure. Biochemistry (Mosc) 46:7146–7152
Steiper ME, Young NM, Sukarna TY (2004) Genomic data support the hominoid slowdown and an early oligocene estimate for the hominoid-cercopithecoid divergence. Proc Natl Acad Sci USA 101:17021–17026
Stephan H, Baron G, Frahm HD (1988) Comparative size of brain and brain components. Comp Primate Biol 4:1–38
Stewart JB, Freyer C, Elson JL, Wredenberg A, Cansu Z, Trifunovic A, Larsson NG (2008) Strong purifying selection in transmission of mammalian mitochondrial DNA. PLoS Biol 6:e10

Sultan F (2002) Analysis of mammalian brain architecture. Nature 415:133–134

Szal B, Jolivet Y, Hasenfratz-Sauder MP, Dizengremel P, Rychter AM (2003) Oxygen concentration regulates alternative oxidase expression in barley roots during hypoxia and post-hypoxia. Physiol Plant 119:494–502

Trounce IA, Mckenzie M, Cassar CA, Ingraham CA, Lerner CA, Dunn DA, Donegan CL, Takeda K, Pogozelski WK, Howell RL, Pinkert CA (2004) Development and initial characterization of xenomitochondrial mice. J Bioenerg Biomembr 36:421–427

Tsantes C, Steiper ME (2009) Age at first reproduction explains rate variation in the strepsirrhine molecular clock. Proc Natl Acad Sci USA 106:18165–18170

Uddin M, Wildman DE, Liu GZ, Xu WB, Johnson RM, Hof PR, Kapatos G, Grossman LI, Goodman M (2004) Sister grouping of chimpanzees and humans as revealed by genome-wide phylogenetic analysis of brain gene expression profiles. Proc Natl Acad Sci USA 101:2957–2962

Uddin M, Opazo JC, Wildman DE, Sherwood CC, Hof PR, Goodman M, Grossman LI (2008) Molecular evolution of the cytochrome *c* oxidase subunit 5Agene in primates. BMC Evol Biol 8:8

Van Beeumen JJ, Van Kuilenburg ABP, Van Bun S, Van Den Bogert C, Tager JM, Muijsers AO (1990) Demonstration of 2 isoforms of subunit-VIIa of cytochrome *c* oxidase from human skeletal muscle – implications for mitochondrial myopathies. FEBS Lett 263:213–216

Van Kuilenburg ABP, Van Beeumen JJ, Van Der Meer NM, Muijsers AO (1992) Subunits-VIIa, b, c of human cytochrome-c oxidase – identification of both heart-type and liver-type isoforms of subunit-VIIa in human heart. Eur J Biochem 203:193–199

Villani G, Attardi G (1997) In vivo control of respiration by cytochrome *c* oxidase in wild-type and mitochondrial DNA mutation-carrying human cells. Proc Natl Acad Sci USA 94:1166–1171

Villani GR, Balzano N, Di Natale P (1998) Two novel mutations of the arylsulfatase b gene in two Italian patients with severe form of mucopolysaccharidosis. Mutations in brief no. 127. Online. Hum Mutat 11:410

Wai T, Teoli D, Shoubridge EA (2008) The mitochondrial DNA genetic bottleneck results from replication of a subpopulation of genomes. Nat Genet 40:1484–1488

Wai T, Ao A, Zhang X, Cyr D, Dufort D, Shoubridge EA (2010) The role of mitochondrial DNA copy number in mammalian fertility. Biol Reprod 83:52–62

Wallace DC, Bunn CL, Eisenstadt JM (1975) Cytoplasmic transfer of chloramphenicol resistance in human tissue culture cells. J Cell Biol 67:174–188

Wildman DE, Wu W, Goodman M, Grossman LI (2002) Episodic positive selection in ape cytochrome *c* oxidase subunit IV. Mol Biol Evol 19:1812–1815

Willett CS, Burton RS (2004) Evolution of interacting proteins in the mitochondrial electron transport system in a marine copepod. Mol Biol Evol 21:443–453

Williams BA, Kay RF, Kirk EC (2010) New perspectives on anthropoid origins. Proc Natl Acad Sci USA 107:4797–4804

Wotton KR, Shimeld SM (2006) Comparative genomics of vertebrate FOX cluster loci. BMC Genomics 7:271

Wrangham RW (2009) Catching fire: how cooking made us human. Basic, New York

Wu CI, Li WH (1985) Evidence for higher rates of nucleotide substitution in rodents than in man. Proc Natl Acad Sci USA 82:1741–1745

Wu W, Goodman M, Lomax MI, Grossman LI (1997) Molecular evolution of cytochrome *c* oxidase subunit IV: evidence for positive selection in simian primates. J Mol Evol 44:477–491

Wu W, Schmidt TR, Goodman M, Grossman LI (2000) Molecular evolution of cytochrome *c* oxidase subunit I in primates: is there co-evolution between mitochondrial and nuclear genomes? Mol Phylogenet Evol 17:294–304

Yanamura W, Zhang YZ, Takamiya S, Capaldi RA (1988) Tissue-specific differences between heart and liver cytochrome *c* oxidases. Biochemistry 27:4909–4914

Zhen YJ, Hoganson CW, Babcock GT, Ferguson-Miller S (1999) Definition of the interaction domain for cytochrome *c* on cytochrome *c* oxidase – I. Biochemical, spectral, and kinetic characterization of surface mutants in subunit II of *Rhodobacter sphaeroides* cytochrome *aa*(3). J Biol Chem 274:38032–38041

Zimmer C (2009) Origins. On the origin of eukaryotes. Science 325:666–668

Chapter 9
Reaction Mechanism of Mammalian Mitochondrial Cytochrome *c* Oxidase

Shinya Yoshikawa, Kazumasa Muramoto, and Kyoko Shinzawa-Itoh

Abstract Cytochrome *c* oxidase (COX) is the terminal oxidase of the mitochondrial respiratory system. This enzyme reduces molecular oxygen (O_2) to water in a reaction coupled with the pumping of protons across the mitochondrial inner membrane. Progress in investigating the reaction mechanism of this enzyme has been limited by the resolution of its X-ray structure. Bovine heart COX has provided the highest resolution (1.8 Å) X-ray structure presently available among the terminal oxidases. The reaction mechanism of the bovine heart enzyme has been the most extensively studied, particularly with respect to (1) the reduction of O_2 to water without release of reactive oxygen species, (2) the mechanism of coupling between the O_2 reduction process and proton pumping, (3) the structural basis for unidirectional proton transfer (proton pumping), and (4) the effective prevention of proton leakage from the proton-pumping pathway to the proton pathway used for generation of water molecules. In this chapter, we will review recent structural studies of bovine heart COX and discuss the mechanisms described earlier in context of the structural data.

9.1 Introduction

Cytochrome *c* oxidase is the terminal oxidase of cell respiration. This enzyme reduces molecular oxygen (O_2) to water at the O_2 reduction site which is composed of two transition metals (Fe_{a3} and Cu_B). This reaction produces the proton motive force across mitochondrial and cell membranes. The electron equivalents used in the reduction of O_2 are received by the second copper site (Cu_A) from cytochrome *c* on the positive side of the membrane and then transferred to the O_2 reduction site via

S. Yoshikawa (✉) • K. Muramoto • K. Shinzawa-Itoh
Department of Life Science, University of Hyogo, Hyogo 678-1297, Japan
e-mail: yoshi@sci.u-hyogo.ac.jp

B. Kadenbach (ed.), *Mitochondrial Oxidative Phosphorylation*, Advances in Experimental Medicine and Biology 748, DOI 10.1007/978-1-4614-3573-0_9,

the second iron site, Fe_a (Yoshikawa et al. 2011). In 1995, the X-ray structures of both bovine and bacterial COXs were solved at 2.8 Å resolution. The structures and the locations of these metal sites and the possible proton transfer pathways in the core subunits (subunits I, II, and III), remarkably, are highly conserved between bovine and bacterial COXs. The subunit composition of the bacterial enzyme (4 subunits) is much simpler than that of bovine enzyme (13 subunits) (Yoshikawa et al. 2011). This suggests that bacterial COX is more suitable for improvement of the resolution of the X-ray structure. Furthermore, bacterial COX is amenable to site-directed mutagenesis. Thus, most functional studies are performed using bacterial (*Paracoccus denitrificans* and *Rhodobacter sphaeroides*) COX (Yoshikawa et al. 2011). However, before the site-directed mutagenesis technique was applied to bacterial COX systems, bovine heart COX was used for most of the structural and functional studies because highly pure and stable preparations were reproducibly available in high yield. Although bovine COX would appear to be unsuitable for improvement of crystallization conditions because of its subunit composition, the improvement of X-ray structural resolution of bovine COX has occurred more rapidly than that of bacterial COX (Yoshikawa et al. 2011). In fact, a simpler subunit composition does not guarantee that high quality crystals will be obtained. Furthermore, a stable expression system for the bovine heart COX gene has been established in HeLa cells and this allows site-directed mutagenesis analyses to be performed (Tsukihara et al. 2003). Therefore, bovine COX remains one of the best materials for structural and functional studies of COX.

Recently, it has been proposed that diverse mechanisms of proton pumping exist among bacterial COXs (Chang et al. 2009). This indicates the possibility that the reaction mechanism of bovine heart COX is different from the mechanisms of bacterial COXs. The implication is that bacterial COX studies will not accurately represent the mechanism of mammalian COX.

In this chapter, we will review recent studies on the mechanism of bovine heart cytochrome *c* oxidase, with a focus on the mechanisms of O_2 reduction and proton pumping, based on X-ray structural findings.

9.2 Composition of Bovine Heart COX

Complete purification of any membrane protein without denaturation and deletion of any of its intrinsic components is quite challenging, especially for membrane proteins which have lipids strongly bound via nonspecific molecular interactions. Crystallization of bovine heart COX as the final step of purification provides fairly large amounts of consistently reproducible COX isolates with respect to lipids and metal content (Yoshikawa et al. 2011). An accurate characterization of the composition of bovine heart COX was recently obtained (Shinzawa-Itoh et al. 2007), 70 years after bovine heart COX was solubilized and isolated for the first time (Yakushiji and Okunuki 1941). However, it remains possible that intrinsic constituents were deleted and that co-crystallization of contaminating materials has occurred.

9.2.1 Protein Moiety

Long before the X-ray structure of bovine heart COX was determined, the presence of 13 different subunits was proposed based on the results of SDS-PAGE analyses (Kadenbach et al. 1983). The presence of 13 different subunits in the X-ray structure of bovine heart enzyme (Tsukihara et al. 1996), consistent with the SDS-PAGE results, provides strong support for the original proposal that bovine COX is composed of 13 different subunits (Kadenbach et al. 1983). However, for complete verification of this proposal, the physiological roles of all subunits must be identified, since co-purified proteins could be specifically bound to COX and co-crystallized. The three largest subunits encoded by mitochondrial genes are surrounded by ten nuclear-coded subunits.

The largest subunit (subunit I) includes three redox active metal sites, a six-coordinated low spin heme (heme *a* including Fe_a), a five-coordinated high spin heme in the reduced state (heme a_3 including Fe_{a3}), and a copper site with trigonal planar coordination by three imidazole groups (Cu_B). The second copper site is located in the third largest subunit (subunit II). This site is a cupredoxin-type dinuclear copper site (Cu_A) in which two cysteine residues bridge the two copper atoms. Each of these metal sites reversibly receives one electron equivalent. COX receives a total of four electron equivalents. This metal content is consistent with the empirical rule that a redox enzyme should have sufficient electron accepting sites to provide electron equivalents for complete reduction of its substrate (in this case, O_2). Fe_{a3} acts as the binding site for O_2 when it is in the reduced state and further acts as the O_2 reduction site in conjunction with Cu_B which is located nearby. Cytochrome *c*, which is located on the positive side of the mitochondrial membrane, donates electron equivalents to the O_2 reduction site via the two other metal sites, Cu_A and heme *a*. Thus, the roles of subunits I and II are obvious. Subunit III, the second largest subunit, contains a possible O_2 pathway, as discussed below (Shinzawa-Itoh et al. 2007). For two of the ten additional subunits, a functional role has been shown: The exchange of bound ADP by ATP on the matrix side of subunit VIa (heart isoform) decreases the H^+/e^- stoichiometry from 1 to 0.5 at high ATP/ADP ratios (Frank and Kadenbach 1996). The exchange of bound ADP by ATP on the matrix side of subunit IV at high ATP/ADP ratios induces an "allosteric ATP-inhibition" of COX activity (Arnold and Kadenbach 1997) of the phosphorylated, but not of the dephosphorylated enzyme (Lee et al. 2001; Napiwotzki et al. 1997). The roles of the other eight subunits are presently unknown. However, as stated earlier, bovine heart COX provides crystals which yield significantly higher resolution in their X-ray structures than those of the bacterial enzymes reported thus far (Yoshikawa et al. 2011). Thus, these ten nuclear-coded subunits are expected to stabilize the conformation and assembly of the three core subunits. Bacterial COXs do not contain these subunits.

9.2.2 Lipids

Crystallization is a powerful method for purification of bovine heart COX as well as many other molecules. In fact, crystallization of bovine heart COX by concentration using a diaflow system, efficiently removes contaminant proteins containing significant amounts of iron. No other method has been found for effectively removing the iron-containing proteins (Mochizuki et al. 1999). The reported molecular extinction coefficients of bovine heart COX determined by the iron content of the COX preparation without purification by crystallization are significantly lower than the values determined for samples purified with crystallization (Mochizuki et al. 1999). Consistent with these results, bovine heart COX purified by crystallization provides reproducible lipid content, suggesting that the crystallization procedure effectively removes nonspecifically bound lipids (Shinzawa-Itoh et al. 2007). Fortunately, the method used for crystallizing bovine heart COX provides sufficient amounts of crystalline COX for lipid structural analyses. However, a method for quantitative determination of the content of each of the lipids by chemical analyses alone has not been established. Thus, the content of each lipid was determined by fitting the lipid structure to the electron density of a high resolution X-ray structure. The complete determination of the intrinsic lipids of bovine heart COX was published in 2007 (Shinzawa-Itoh et al. 2007).

Mass spectrometry (MS) and tandem mass spectrometry (MS-MS) were used for structural determination of each fatty acid tail (including the chain length and the positions of the unsaturated bonds). Phospholipase A_2 treatment was successfully used for determination of the positions of the fatty acyl group in the glycerol backbone of each of the lipids (Shinzawa-Itoh et al. 2007). Lipid fractions extracted from crystalline bovine heart COX showed seven species of phospholipids including cardiolipin (CL), phosphatidylethanolamine (PE), phosphatidylcholine (PC), choline plasmalogen (CP), phosphatidylglycerol (PG), and seven species of triglyceride (TG) as shown in Fig. 9.1 (Shinzawa-Itoh et al. 2007).

The configurations of the unsaturated bonds in vaccenate and oleate were determined by GC analysis of their methyl esters prepared by solvolysis. *Trans*-Vaccenate was detected in a *cis*/*trans* ratio of approximately 13:1 while no *trans* configuration was detected for oleate. The major configuration is expected to be the *cis* isomer in other unsaturated fatty acid tails, although the GC analysis has not been used to investigate the other fatty acids. Four species of choline-containing phospholipids are detectable. Vaccenate (Δ^{11}-octadecenoate) is included in PG in spite of the large abundance of oleate (Δ^{9}-octadecenoate), a closely related fatty acid (Shinzawa-Itoh et al. 2007).

Fitting these lipid structures into the electron density map of the 1.8 Å resolution X-ray structure reveals that 2CLs, 1PC, 3PEs, 4PGs, and 3TGs are detectable (Shinzawa-Itoh et al. 2007). At this resolution, the fatty acid structures of the choline-containing phospholipids cannot be identified in the X-ray structure. However, only one site is detectable for the choline-containing phospholipids. The head group structure could be used to identify the binding site, regardless of the structure of the fatty acid tails. This indicates that these four choline-containing phospholipids species have equal affinity for the site. Alternatively, the structure of the fatty acyl tail portion could determine the site specificity. Thus, only one of the 4 choline-containing phospholipids is selectively bound to the site.

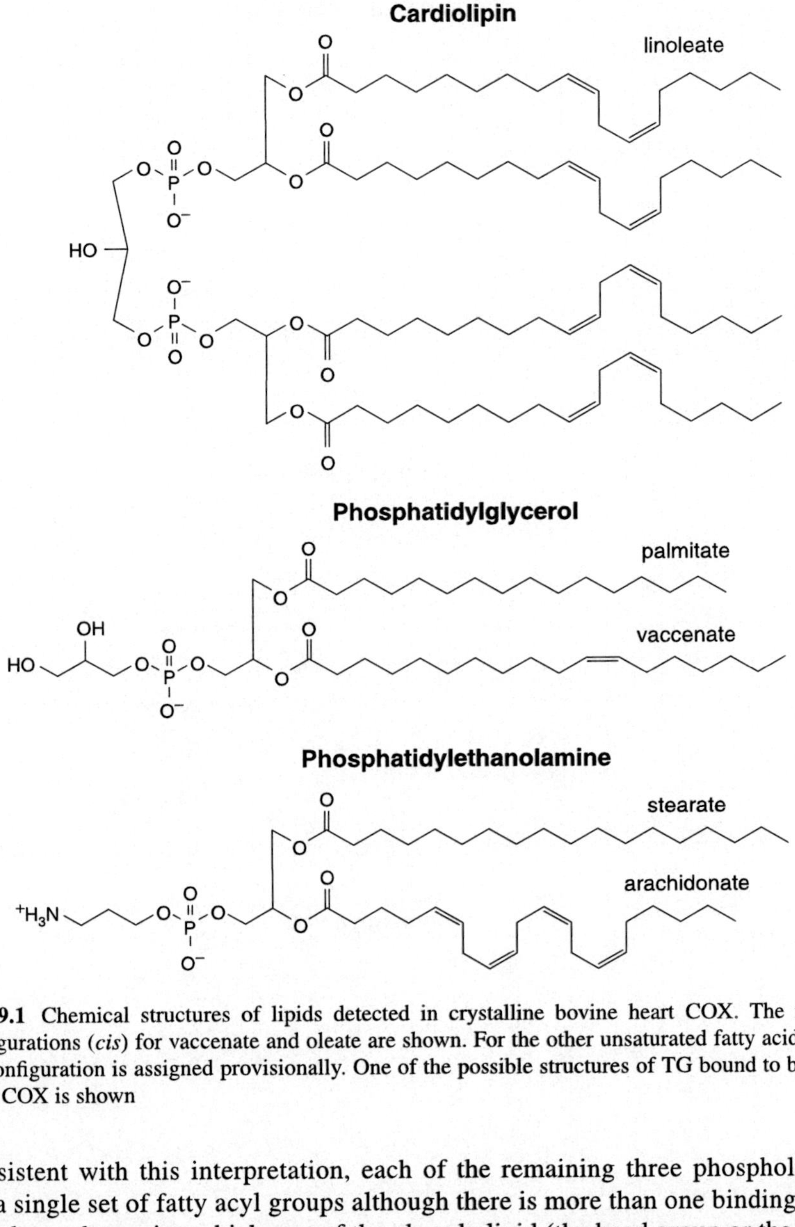

Fig. 9.1 Chemical structures of lipids detected in crystalline bovine heart COX. The major configurations (*cis*) for vaccenate and oleate are shown. For the other unsaturated fatty acids, the *cis*-configuration is assigned provisionally. One of the possible structures of TG bound to bovine heart COX is shown

Consistent with this interpretation, each of the remaining three phospholipids has a single set of fatty acyl groups although there is more than one binding site. In order to determine which part of the phospholipid (the head group or the fatty acyl tail) recognizes its binding site, the overall phospholipid content of the mitochondrial inner membrane was determined using the same method used for the enzyme preparation. The analyses show that the composition of the phospholipids is identical to that of bovine COX. Therefore, each CL, PE, and PG in the X-ray structure has a single set of fatty acyl groups, since the pool (mitochondrial inner membrane) of these phospholipids is composed of only a single species

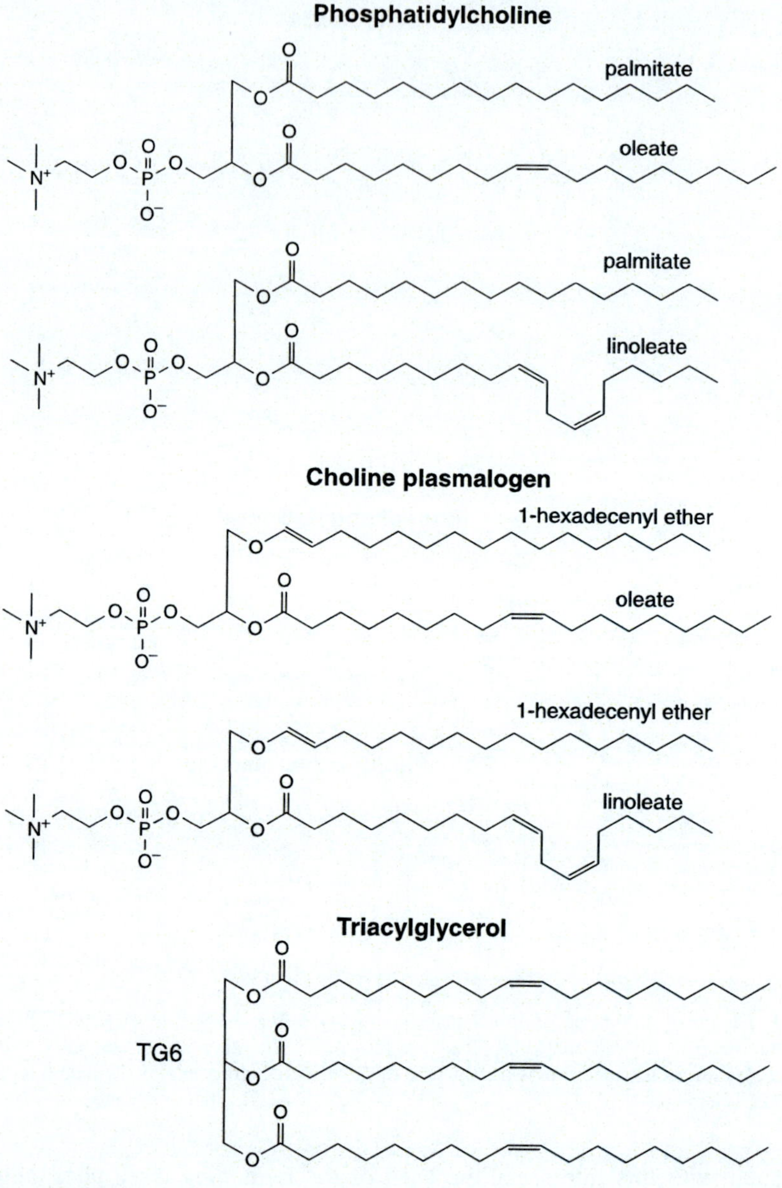

Fig. 9.1 (continued)

for each of the three phospholipids. These results provide strong evidence that the head groups of the phospholipids recognize their binding sites (Shinzawa-Itoh et al. 2007).

As described earlier, seven species of TG were identified by MS and GC analyses of solvolyzed fatty acid methyl esters. Again, the lipid pool only contains seven

TG species. Three binding sites for TG were identified in the X-ray structure. This provides strong evidence that the head group (the fatty acyl group at the terminal carbon of the glycerol backbone) recognizes the TG binding site (Shinzawa-Itoh et al. 2007).

The configuration of vaccenate, characterized in the lipid extract from the mitochondrial inner membrane indicates that the *cis/trans* ratio is 5:1.Thus, the *cis*-configuration is selected by COX as being more favorable than the *trans*-configuration. The electron densities of the two vaccenate molecules in subunit III are clear enough to conclude that the unsaturated bond is 100% in the *cis*-configuration (Shinzawa-Itoh et al. 2007). The X-ray structure indicates that the protein moiety is able to recognize the *cis/trans* configuration. This suggests that these fatty acid tails are involved in specific physiological roles.

Quantitative analysis of the solvolysate of the lipid extract using tripentadecanoic glyceride (an unnatural TG), as an internal standard showed that the total number of fatty acyl groups other than arachidonate is 30.6 (±3.4) per molecule of COX ($n=6$) (Shinzawa-Itoh et al. 2007) (arachidonate is undetectable under these conditions). The number of nonarachidonate fatty acid tails detectable in the X-ray structure is 30. Phosphorous analysis of 40 different batches of the crystalline preparation indicated 12.9 (±1.9) phosphorous atoms per COX molecule. The numbers of fatty acid tails and phosphorous atoms determined experimentally are consistent with the total number of the fatty acid tails detectable in the X-ray structure. This shows that all of the lipids specifically bound to COX which are detectable in the X-ray structure, are extracted quantitatively. Thus, the crystalline COX preparation does not contain nonspecifically bound lipids (Shinzawa-Itoh et al. 2007). The result suggests that all lipids detectable in the X-ray structure are intrinsic constituents of bovine heart COX. Furthermore, they are not expected to be reversibly equilibrated with the lipid pool (mitochondrial inner membrane). Otherwise most of these lipids would be removed during the process of isolating COX.

9.2.3 Complete Determination of the Structure of Heme A

The structure of heme A (the heme moiety of heme *a* and heme a_3) was not determined until 1975 (Caughey et al. 1975). This delay was due to the fact that this cofactor is significantly larger and less stable than protoheme. However, determination of the absolute configuration of the single chiral center of heme A (the carbon with the OH substituent within the hydroxyfarnesylethyl group) using standard methods was not attempted because it is quite challenging to crystallize heme A and to synthesize heme A for determination of the absolute configuration. The absolute configuration was finally identified in 2005 as the *S*-configuration. This was accomplished using a newly developed X-ray structural analysis of heme A in COX crystals (Yamashita et al. 2005). This is the most recent chemical structural determination of a component of bovine heart COX. Structural research activities are presently focused on 3D and vibrational structural aspects.

9.3 Mechanism of O_2 Reduction by Bovine Heart COX

9.3.1 Resonance Raman Analyses of the O_2 Reduction Mechanism of Bovine Heart COX

One-electron reduction of molecular oxygen (O_2) is energetically unfavorable and two-electron reduction is quite favorable. This intrinsic chemical property of O_2 is essential for the stability of oxygenated heme ($Fe^{2+}–O_2$) in hemoglobins and myoglobins. In these globins, the bound O_2 is buried inside the protein and a second electron equivalent for two-electron reduction of the bound O_2 is not readily available (Caughey et al. 1976). On the other hand, the infrared band corresponding to the O–O stretching mode of the bound O_2 in bovine myoglobin indicates that the bound O_2 is in the superoxide (O_2^-) state (Potter et al. 1987). Myoglobin does not release O_2^- under normal physiological conditions. The $Fe^{2+}–O_2$ species is not the O_2^--bound form of Fe^{3+}. The infrared data indicate that the oxygenated form is in the following resonance structure,

$$Fe^{2+} - O_2 \leftrightarrow Fe^{3+} - O_2^-,$$

where the electron density is mostly located on the O_2 molecule. On the other hand, long before the X-ray structures of bovine and bacterial COXs were obtained, the presence of the two transition metals in the O_2 reduction site was well known (Caughey et al. 1976). The structure appears to be suitable for two-electron reduction of O_2. However, it was unexpectedly found in resonance Raman analyses that the initial intermediate is an O_2-bound form known as intermediate A [$Fe_{a3}{}^{2+}–O_2$, $Cu_B{}^{1+}$] (Kitagawa and Ogura 1997). Furthermore, the resonance Raman band of the second intermediate shows that the O–O bond is cleaved in the transition to yield intermediate P [$Fe_{a3}{}^{5+}=O^{2-}$, $Cu_B{}^{2+}–OH^-$] in which $Fe_{a3}{}^{5+}$ denotes the formal oxidation state of Fe_{a3} (Kitagawa and Ogura 1997). The one-oxidation equivalent is likely to be located on the OH group of Tyr244 which is covalently bound to one of the three imidazole groups coordinated to Cu_B. Thus, Tyr244OH, which is located very close to the heme-copper system, is expected to donate one electron equivalent (and one proton) to the bound O_2. Then, during normal enzymatic turnover, the following intermediates appear sequentially: intermediate F [$Fe_{a3}{}^{4+}=O^{2-}$, $Cu_B{}^{2+}–OH^-$], intermediate O [$Fe_{a3}{}^{3+}–OH^-$, $Cu_B{}^{2+}–OH^-$], intermediate E [$Fe_{a3}{}^{3+}–OH^-$, $Cu_B{}^{1+}$], and intermediate R [$Fe_{a3}{}^{2+}$, $Cu_B{}^{1+}$]. During each transition, one electron equivalent is received from cytochrome *c* via Cu_A and heme *a* (Yoshikawa et al. 2011). The one-step transition from intermediate A to intermediate P suggests that this strategy enables cleavage of the O–O bond (involving complete reduction), without release of any reactive oxygen species.

9.3.2 The Structural Basis for the O_2 Reduction Mechanism of Bovine COX

The reaction cycle described in the previous section was established almost 20 years ago on the basis of resonance Raman analyses conducted by three different research groups (Kitagawa and Ogura 1997). However, no structural basis for the reaction cycle has been identified. For example, it is not known why the oxygenated form (or intermediate A) is stable enough to be identified as one of the intermediates. If the second possible electron donor, Cu_B, donates electrons to the bound O_2, the rate of formation of intermediate A is definitely slower than the rate of electron transfer from Cu_B^{1+} to the O_2 at Fe_{a3}. This causes intermediate A to be undetectable under normal enzymatic turnover. For elucidation of the mechanism responsible for the stability of the intermediate, it is necessary to elucidate the structure of Cu_B site at high resolution. However, resonance Raman analyses do not provide structural information with respect to the O_2 reduction site. X-ray structural analyses of these intermediate species are desirable for elucidation of the O_2 reduction mechanism. However, most of these intermediate species are unstable and unsuitable for crystallization. Thus, the functions of the O_2 reduction site have been probed by high resolution X-ray structural analysis of COX derivatives with various stable O_2 analogues (Muramoto et al. 2010).

It has been shown that the most accurate O_2 analogue, NO, binds to Fe_{a3}^{2+} in a bent, end-on fashion in the X-ray structure of NO-bound, fully reduced COX at 1.8 Å resolution (Fig. 9.2). The distance between Cu_B^{1+} and the bound ligand is 2.5 Å, suggesting that the interaction between these entities is weak. Cu_B^{1+} is in a trigonal planar coordination arrangement with three histidine imidazole groups. In general, trigonal planar cuprous compounds are very stable and thus are poor electron donors as well as poor ligand acceptors. These X-ray structural findings indicate that Cu_B is unlikely to participate in the catalytic turnover as the second electron donor (Muramoto et al. 2010). On the other hand, Tyr244, which is covalently linked to His240, one of the imidazole groups coordinated to Cu_B (Muramoto et al. 2010), is located near the external ligand bound to Fe_{a3}^{2+}. Although a deprotonated Tyrosine-OH group is expected to donate electrons reversibly in a manner similar to that of a transition metal, the X-ray structure of the NO-bound form of COX indicates that the imidazole ring of His240 effectively blocks access of the OH group to the bound ligand. The X-ray structure of the CO-bound form also indicates structural characteristics which are essentially the same as those of the NO-bound form (Muramoto et al. 2010). These X-ray structures of respiratory inhibitor derivatives of COX provide clear evidence of the stability of the O_2-bound Fe_{a3}^{2+} species.

When CO is bound to Fe_{a3}^{2+}, it is readily photolyzed and trapped at Cu_B (Fiamingo et al. 1982). The process of rebinding to Fe_{a3}^{2+} is blocked at low temperature. Therefore, the CO-bound Fe_{a3}^{2+} derivative cannot be stabilized at 100 K, at which X-ray diffraction experiments for the bovine heart COX have been conducted

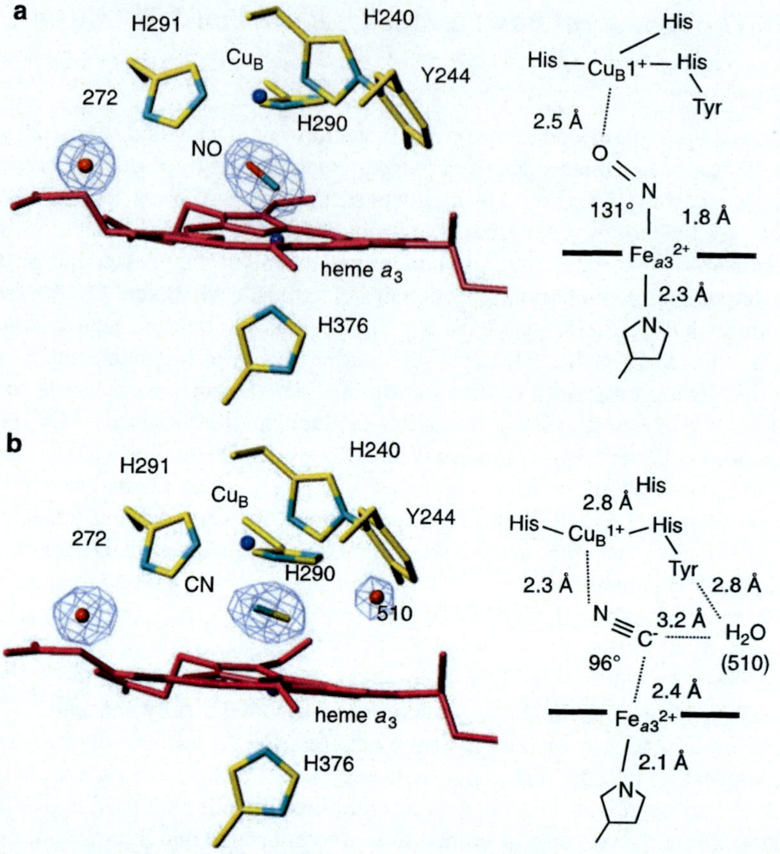

Fig. 9.2 The X-ray structures of the O_2 reduction sites of (**a**) the NO- and (**b**) CN^--bound forms of fully reduced COX at 1.8 and 2.05 Å resolutions, respectively. The F_O–F_C maps are contoured at (**a**) 6.4 σ and (**b**) 5.8 σ levels. The ligand-binding structures are shown in F_O–F_C maps. The digits without letters in the F_O–F_C maps indicate the numbering of fixed water molecules. The structural characteristics are indicated in the schematic representations

recently. The X-ray structure determined at 100 K shows that CO is bound to Cu_B in a side-on fashion (Muramoto et al. 2010). The binding of CO to Cu_B at low temperatures has been examined in infrared analyses (Fiamingo et al. 1982). However, the binding geometry is visible in a 100 K X-ray structure at 1.8 Å resolution. The structure suggests that O_2 is trapped at Cu_B before it binds to Fe_{a3}^{2+}, as suggested by kinetic analyses (Oliveberg and Malmstrom 1992). It appears that Cu_B must control the supply of O_2 to Fe_{a3}^{2+} without forming a bridging O_2 structure between Cu_B^{1+} and Fe_{a3}^{2+} which could produce reactive oxygen species spontaneously.

Subunit III, the second largest subunit, has an O_2 supply control system. The dicyclohexylurea (DCU)-bound form of COX has no O_2 reduction activity (Shinzawa-Itoh et al. 2007). In the X-ray structure of this derivative, the DCU unit bound to

Glu90, which is located within a putative O_2 pathway in subunit III, induces significant conformational changes in the two palmitate tails of PGs to effectively block the pathway without producing a significant structural change in the protein moiety (Shinzawa-Itoh et al. 2007). This is consistent with the observation of complete blockage of O_2 reduction activity with binding of DCU (Shinzawa-Itoh et al. 2007). The conformational changes in the palmitate tails which are detectable in the X-ray structure are independent of the protein conformation. This suggests that the palmitate tails are able to actively control the O_2 transfer process in response to the flow of O_2 through the pathway. Interestingly, a bacterial COX preparation depleted of subunit III undergoes suicide inhibition. This process is characterized by decreased enzymatic activity induced by repeated enzymatic turnover (Gilderson et al. 2003). This suggests that subunit III controls the rate of supply of O_2 to the O_2 reduction site to prevent the formation of reactive oxygen species (Shinzawa-Itoh et al. 2007).

It is well known that cyanide (CN^-) is a strong ligand to ferric hemes (Fe^{3+}) but a weak ligand to ferrous hemes (Fe^{2+}). Only COX and peroxidases appreciably bind CN^- in the ferrous form with Kd values in the millimolar range (Yoshikawa et al. 1985, 1995). A fairly large conformational change is detectable in the O_2 reduction site of the CN^- derivative of fully reduced COX, as shown in Fig. 9.2 (Muramoto et al. 2010). A fixed water molecule is introduced to bridge the Tyr244 OH group and the bound ligand, concomitantly with a translational shift of the plane of heme a_3. Furthermore, CN^- bridges between Fe_{a3}^{2+} and Cu_B^{1+} to form a new trigonal planar structure which includes the two imidazole groups of His240 and His291 and the bound CN^-. This causes one of the three histidine imidazole groups to be released from Cu_B. The plane of the new trigonal planar arrangement is perpendicular to the plane of heme a_3 in fully reduced COX. The incorporation of the water molecule concomitantly with the translational shift of the heme a_3 plane is expected to limit the rate of the conformational changes induced by cyanide, since the fixed water molecule is expected to be transferred from a water storage site located 7 Å away from the O_2 reduction site (Muramoto et al. 2010).

As stated earlier, the O_2-bound form of heme a_3 has a resonance structure between $Fe^{2+}–O_2$ and $Fe^{3+}–O_2^-$ and the electronic structure is essentially the same as that of $Fe^{3+}–O_2^-$ which is not completely identical to the combination of ferric heme and O_2^- (Potter et al. 1987). Thus, O_2 bound to Fe_{a3}^{2+} also forms the above resonance structure which is mimicked by the cyanide derivative of fully reduced COX. Therefore, once Fe_{a3}^{2+} has received O_2, it is expected to induce the structural change which is essentially identical to the structural change induced by binding of CN^-. The conformational change induced by binding of CN^- which is shown in Fig. 9.2 provides three possible electron transfer pathways from Fe_{a3}^{3+}, Cu_B^{1+}, and Tyr244-OH via the fixed water molecule. The overall conformational change is likely to be rate limited by the introduction of H_2O. Thus, O_2^- at Fe_{a3} is expected to receive three electron equivalents nonsequentially through the three pathways. This completely reduces O_2^- to the oxide level, without release of any reactive oxygen species (Muramoto et al. 2010).

These results provide strong evidence that the O_2 reduction mechanism is the mechanism represented by the scheme shown in Fig. 9.3 (Muramoto et al. 2010). The supply of O_2 to Fe_{a3}^{2+} is controlled in two steps by phospholipids in the O_2

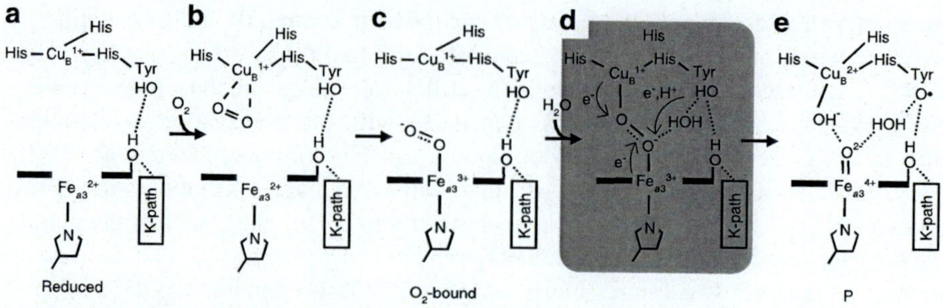

Fig. 9.3 A schematic representation of O_2 reduction in COX. The proposed intermediate is shown in the *shadowed area*

pathway in subunit III and by Cu_B^{1+}. When both metals in the O_2 reduction site are in the reduced state, Cu_B^{1+} receives O_2 from the O_2 transfer pathway in subunit III (Shinzawa-Itoh et al. 2007) and donates it to Fe_{a3}^{2+} with appropriate timing and without forming the bridging intermediate (Fe_{a3}^{2+}–O_2–Cu_B^{1+}) which could produce reactive oxygen species. The O_2-bound form adopts the resonance structure described earlier in which O_2 is essentially in the one-electron reduced form (O_2^-). However, prior to introduction of the fixed water molecule, the conformation of the O_2 reduction site is similar to that of NO-bound form (Muramoto et al. 2010). The slow introduction of the fixed water molecule increases the lifetime of the O_2 (or O_2^-) bound form so that it is detectable by resonance Raman analysis (Kitagawa and Ogura 1997). The nonsequential three-electron reduction of O_2^- is triggered by formation of the transient structure which is shaded in Fig. 9.3. The rate of this process is limited by introduction of the fixed water molecule (Muramoto et al. 2010). This mechanism, proposed on the basis of X-ray structural analyses, could be examined by infrared spectroscopy to characterize the Cu_B-bound imidazoles and the Tyr244 OH group.

9.3.3 *The Structure and Possible Role of the Resting Oxidized Form*

The fully oxidized form, in which both metals are in the oxidized state under turnover conditions, reacts strongly and rapidly with cyanide (Thornstrom et al. 1988). The resonance Raman band of the O form at 450 cm^{-1} indicates that the ligand of Fe_{a3}^{3+} is OH^- (Kitagawa and Ogura 1997). It has been established that both O to E and E to R transitions, each induced by one-electron reduction from cytochrome *c*, are coupled to the process of proton pumping (Yoshikawa et al. 2011). However, under nonturnover aerobic conditions, another type of the oxidized form (the resting oxidized form) appears. This form binds CN^- slowly and lacks the proton-pumping function for each of the two initial electron transfers. Furthermore, six electron equivalents are required for complete reduction of the resting oxidized form (Mochizuki et al. 1999). On the other hand, oxidative titration of the fully reduced form of the bovine heart COX preparation with O_2 yields a COX species which requires four electron equivalents for complete reduction. The O_2-oxidized COX preparation, once exposed to air for 30 min, requires six electrons for complete reduction (Mochizuki et al. 1999). These results provide strong support for the proposal that a peroxide bridge exists between Fe_{a3}^{3+} and Cu_B^{2+}.

Comparison of the concrete structure of the O_2 reduction site of the resting oxidized form with the structure of the O form would provide a number of important insights with regard to the proton-pumping mechanism. X-ray structural analysis of the resting oxidized form was conducted by limited X-ray exposure using about 400 single crystals in order to minimize the effect of the hydrated electrons generated by the strong X-ray beams at SPring-8. The strength of these beams significantly reduces the electron density of the bridging molecule between Fe_{a3} and Cu_B (Aoyama et al. 2009). The structure, determined at 1.95 Å resolution, shows a peroxide bridge between the two metals in the O_2 reduction site. However, the O–O bond distance (1.7 Å) is significantly longer than the typical distance of a peroxide bridge between two metals (1.44 Å) (Aoyama et al. 2009), although the presence of a covalent bond between the two oxygen atoms is supported by the reductive titration data (Mochizuki et al. 1999). Furthermore, a resonance Raman band assignable to the O–O stretch of peroxide in the resting oxidized form was identified using 647.1 nm line excitation (Sakaguchi et al. 2010). All examples of laser excitation at the shorter wavelengths appear to reduce the peroxide. Many attempts to identify the O–O stretch using excitation at shorter wavelengths were unsuccessful. The isotopic shift experiments using unevenly labeled O_2 would provide important insights into the function of the O_2 reduction site.

The size of the chloride atom indicates that a single chloride ion with electron density evenly distributed between the two sites is indistinguishable from the electron density of peroxide, at the present resolution of X-ray structure. In fact, it has long been proposed that a chloride ion is located between the two metals in the O_2 reduction site (Fabian et al. 2001). Thus, the possibility of the presence of chloride was examined by an anomalous dispersion analysis (Suga et al. 2011). For direct elemental analysis of the O_2-reduction site, the difference anomalous electron density between Fe_{a3} and Cu_B was examined for the resting-oxidized form of bovine heart COX using 1.7470 Å wavelength X-ray beams which provide a slightly larger anomalous scattering factor for Cl relative to Fe and S. An anomalous peak for Cl was not detected between Cu_B and Fe_{a3} while anomalous peaks for Cu_B and Fe_{a3} were clearly detected in the same map. These results support the conclusion that the resting oxidized form of bovine heart COX has a peroxide bridge between Fe_{a3}^{3+} and Cu_B^{2+}. In the reductive titration of the resting oxidized form, a clear lag phase in the titration curve is detectable during input of the initial two electron equivalents. The results provide strong evidence that the initial two electron equivalents are used for reducing the bridging peroxide to the oxide level without affecting the oxidation state of both metals in the O_2 reduction site (Mochizuki et al. 1999). Thus, the peroxide decouples the proton pump by trapping the electron equivalents to avoid reduction of the two metals. Reduction of heme a_3 or Cu_B triggers the proton pump. The O_2 reduction site is not only a simple electron sink but also contributes to the proton-pumping process.

O_2 is highly hydrophobic and can readily diffuse into the interior of the protein where it is likely to produce reactive oxygen species when it encounters metal ions fixed in the O_2 reduction site. Therefore, the bound peroxide in resting oxidized COX would effectively prevent spontaneous interactions of O_2 with these metal ions in the O_2 reduction site, under conditions of limited supply of electrons in the respiratory system. The resting oxidized form is very stable. In the crystalline state, no significant absorption spectral change is detectable even after this form of the enzyme is stored for more than 1 year at 4°C. Presently, no visible structural features appear to account for this unusual stability.

9.4 Mechanism of Proton Pumping by Bovine Heart COX

9.4.1 Requirement of Conformational Changes for Proton Pumping

In general, a redox-driven proton-pump system must be facilitated by conformational changes which induce redox-coupled pKa and accessibility changes in the proton-loading site. It is impossible to evaluate the magnitude of the conformational changes required for proton pumping. Thus, improvement of the resolution of the X-ray structure in both oxidation states is necessary in order to elucidate the proton-pumping mechanism. Of course, it is ideal to solve the X-ray structure at the hydrogen atom level to identify the proton transfer directly. However, the microenvironment of acidic or basic amino acid side chains clearly indicates the protonation state, since the pKa of acidic and basic amino acid residues is very sensitive to the polarity of the solvent. In fact, the pKa of acetic acid increases to 9 when water is exchanged for methanol as a solvent (Isaacs 1995). Thus, resolutions allowing the detection of fixed water molecules, the protonation state of acidic and basic amino acid residues can be assigned fairly accurately according to the features of the microenvironment. For example, a redox-coupled conformational change of an acidic residue of bovine heart COX, Asp51, was discovered at 2.3 Å resolution (Yoshikawa et al. 1998). The carboxyl group of Asp51 in the oxidized state was found to be hydrogen bonded to two peptide NH groups and two serine OH groups. The hydrogen-bonding structure suggests that the effective dielectric constant of the microenvironment of the carboxyl group is quite low. On the other hand, in the reduced state, the carboxyl group is hydrogen bonded to three fixed water molecules and one serine OH group. This suggests that the carboxyl group has a fairly high effective dielectric constant. The conformational changes strongly suggest that the carboxyl group undergoes almost 100% reversible protonation/deprotonation upon oxidation/reduction of the enzyme. The protonation state change has been confirmed by an infrared analysis which identified bands at 1,585 cm^{-1} and 1,738 cm^{-1}. These bands were assigned to COO^- and COOH, respectively (Tsukihara et al. 2003).

9.4.2 Structure of the Proton Pump System of Bovine Heart COX

The conformational change of Asp51 described earlier provides strong evidence that this residue is involved in the proton-pumping system of bovine heart COX. The X-ray structure at 2.3 Å resolution indicates that Asp51 is located near the molecular surface facing the positive side, where it makes tandem connections with the molecular surface facing the negative side via a hydrogen-bond network and a water channel. Asp51 is located near the positive side end of the hydrogen-bond network. The negative side end of the hydrogen-bond network is connected to the water channel which extends to the molecular surface on the negative side. Thus, the water molecules in the negative side phase are accessible to the negative side end of the hydrogen-bond network. Arg38 is located at this position, as shown in Fig. 9.4 (Yoshikawa et al. 2011; Tsukihara et al. 2003; Yoshikawa et al. 1998). The water channel, identified by a molecular surface calculation, is composed of water pathways and cavities. The water pathway is too narrow to permit stable accumulation of water molecules. The water cavity is able to trap at least one water molecule.

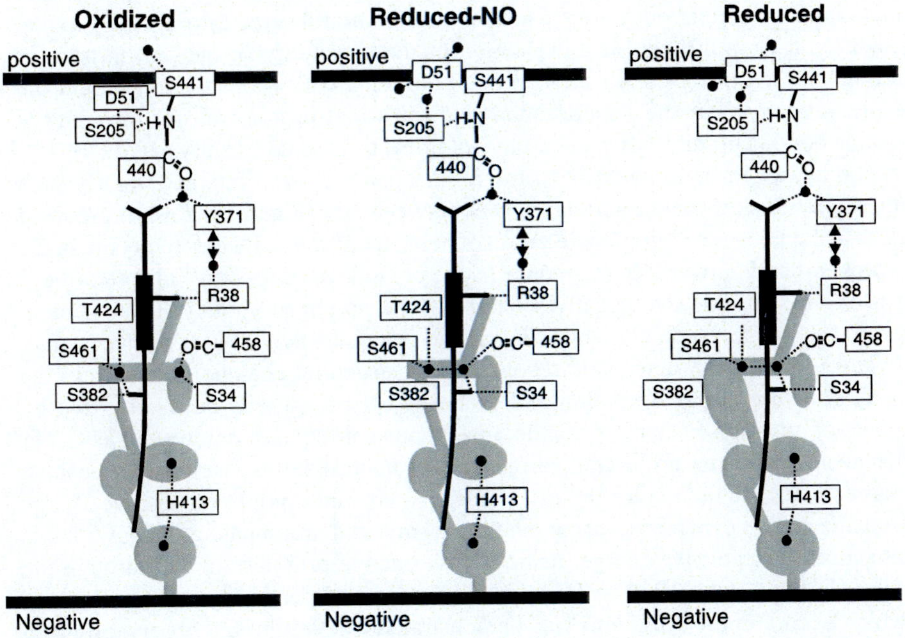

Fig. 9.4 Schematic representation of the structure of the H-pathway and the conformational changes induced by the oxidation state and the ligand-binding state

Thus, Arg38, which is located at the negative side end of the hydrogen-bond network, is protonically equilibrated with the negative side aqueous phase (Tsukihara et al. 2003; Muramoto et al. 2010). The putative proton-conducting system which extends across the enzyme is known as the H-pathway.

In the oxidized state, Asp51 is hydrogen bonded to the peptide NH of Ser441. The peptide bond between Ser441 and Tyr440 is included in the hydrogen-bond network, as shown in Fig. 9.4. Protons are transferred through the peptide bond when an imidic acid intermediate is formed as follows: upon protonation of the peptide C=O group, an imidic acid intermediate (–C(OH)=N^{+}H–) is formed, followed by extraction of a proton to give the enol form of the peptide (–C(OH)=N–). Since the enol form of the peptide is much less stable than the keto form (–CO–NH–), the enol form spontaneously transforms to the keto form. Thus, proton transfer through the peptide bond has a unidirectional character which prevents back leakage of protons from the positive side. The peptide bond is expected to provide the hydrogen-bond network with unidirectional character instead of acting as an accessibility switch system (Tsukihara et al. 2003).

It should be noted that in the enol to keto transition which occurs during proton transfer through the peptide bond, the proton initially added to the peptide C=O to form the imidic acid intermediate must be transferred to the nitrogen atom of the enol form (=N–) to provide the peptide NH. Otherwise, net proton transfer through the peptide bond cannot be obtained. In the interior of the protein, direct proton transfer through the peptide bond is likely to be restricted so that the enol nitrogen (=N–) could receive a proton from the enol nitrogen side to form the imidic acid intermediate. At that point, reverse proton transfer would occur through the peptide. However, a careful theoretical evaluation of the possibility of net proton transfer through the peptide bond in the hydrogen-bond network (taking into account the X-ray structure near the peptide bond) indicates that proton transfer through the peptide bond is possible within the physiological time scale (Kamiya et al. 2007).

The hydrogen-bond network in the proton-pump system interacts tightly with heme *a* by forming two hydrogen bonds between Arg38 and the formyl group of heme *a* and between a propionate group of heme *a* and a fixed water molecule in the hydrogen-bond network. In the reduced state of heme *a*, the positive charge of Fe_a^{2+} is neutralized by the two negative charges of the porphyrin moiety. Thus, heme *a* has no net charge. Upon oxidation, Fe_a creates one equivalent of net positive charge. However, there is no detectable redox-coupled structural change in heme *a* in the X-ray structure at 1.8 Å resolution. The net positive charge is not neutralized by uptake of an anion. If heme *a* was in an aqueous solution, an anion such as Cl^- or OH^- would be taken up to compensate for (or neutralize) the increase in positive charge. The X-ray structure indicates that a significant amount of positive charge delocalization is expected to occur over the formyl and propionate groups. With this delocalized net positive charge, heme *a* drives active proton transport through the hydrogen-bond network. The large positive charge increase in the formyl group (which occurs upon oxidation) was demonstrated in 1989 by the observation of a resonance Raman C–O stretch shift from 1,610 cm^{-1} to 1,650 cm^{-1} (Sasaroli et al. 1989). These X-ray structures of the hydrogen-bond network as well as the resonance

Raman results provide strong evidence that the features described earlier provide the coupling site between O_2 reduction and proton pumping.

As stated earlier, the hydrogen-bond network in the proton-pumping system is protonically equilibrated with the aqueous negative side phase because the water channel bridges one end of the hydrogen-bond network with the molecular surface of the negative side. The X-ray structural analyses of various oxidation and ligand binding states indicate that the largest water cavity located near the junction point to the hydrogen-bond network is detectable only when both Fe_{a3} and Cu_B are in the reduced state. A long water pathway appears after elimination of the largest water channel as shown in Fig. 9.4 (Muramoto et al. 2010). The process of water exchange within the water channel depends upon the thermal motion of the protein moiety. Thus, in the absence of the biggest cavity, the efficiency of the water exchange (and thus proton equilibration with the hydrogen-bond network) is expected to be quite low, at least within the physiological time scale. The conformational states of the water channel with and without the largest water cavity are designated as open and closed states, respectively. In the closed state, the water channel would block the backward leakage of protons from the hydrogen-bond network. As stated earlier, the peptide bond located near the positive side end of the hydrogen-bond network blocks leakage of protons from the positive side. In the normal catalytic cycle, as described earlier, four electron equivalents are transferred to the P state of the O_2 reduction site from cytochrome *c* one at a time. This produces intermediate F, intermediate O, and intermediate E with regeneration of the R state. Each electron transfer is coupled to the proton-pumping process (Yoshikawa et al. 2011). The water channel is in the closed conformation during the catalytic progression from intermediate A to intermediate R. Thus, all of the proton-pumping processes proceed when the channel is closed. The backward leakage of protons is blocked at both ends of the hydrogen-bond network (Muramoto et al. 2010).

9.4.3 Functional Examination of the Bovine Heart Proton-Pumping Pathway (H-Pathway)

The proton-pumping mechanism described earlier has been deduced from X-ray structures in various oxidation states and ligand binding states. The X-ray structure, which shows the 3D locations of the atoms in the functional site, provides crucial information with regard to the mechanism by which the protein drives its physiological reaction at the functional site. However, the reaction mechanism deduced by X-ray structural analyses must be confirmed functionally. Any proposed function must be shown experimentally. For this purpose, site-directed mutagenesis approaches are quite common for bacterial protein systems. This is one of the reasons why bacterial COX have been extensively studied. A bacterial COX system would provide an excellent model of mammalian COX if the physiological function of the bacterial system is essentially identical to that of the mammalian system. However, one of the key residues of the H-pathway in mammalian

COXs, Asp51, is absent in bacterial and plant COXs. Therefore, in order to apply site-directed mutagenesis to bovine heart COX, a stable HeLa cell-based expression system of the gene of bovine heart subunit I, which contains all of the H-pathway amino acid residues, has been constructed (Tsukihara et al. 2003). Subunit I is encoded by a mitochondrial gene. In the expression system, subunit I is expressed in the cytosol with a mitochondrial targeting peptide for transfer to the mitochondrial inner membrane where it forms a hybrid COX with the other 12 subunits produced by the human mitochondrial genes of the HeLa cells (Tsukihara et al. 2003).

Using this hybrid enzyme, three mutant enzymes have been constructed and characterized. Asp51Asn was constructed to abolish the redox-coupled proton transfer mechanism operating near the molecular surface of the positive side (Tsukihara et al. 2003). Tyr440Pro was prepared in order to block proton transfer through the peptide bond. The nitrogen atom of the peptide cannot be protonated in this mutant. Although a proline mutation will often induce large conformational changes which complicate the straightforward interpretation of the mutation results, the dihedral angle of the peptide bond between Tyr440 and Ser441 suggests that the conformational change induced by the proline mutation is minimal. The double mutant Val386Leu/Met390Trp was generated to block water exchange in the water channel by introducing bulkier residues (Shimokata et al. 2007).

All three of the mutant enzymes provide a strictly identical phenotype, involving complete abolishment of proton pumping without a decrease in the O_2 reduction activity. These results provide functional confirmation of the proton-pumping proposal deduced from X-ray structural analyses. The system which includes the hydrogen-bond network and heme *a* interacting with each other via two hydrogen bonds couples the electron transfer process (and thus O_2 reduction) and proton active transport (proton pumping). The phenotype suggests that the coupling is not completely tight and this is consistent with the X-ray structure. The driving force for proton active transport is the electrostatic repulsion between the delocalized positive charge and protons in the hydrogen-bond network. The electrostatic repulsion itself does not determine the direction of proton transfer. The unidirectionality of the proton transfer is provided by the two structures which block reverse proton transfer. These structures are located on both sides of the hydrogen-bond network. Thus, the reverse process (reverse proton transfer through the hydrogen-bond network) cannot completely block the electron transfer from heme *a* to the O_2 reduction site.

It should be noted that another important aspect of the mechanism of proton pumping through the H-pathway is the complete isolation of the proton-pumping pathway from the pathways used to donate protons to the O_2 reduction site for production of water molecules. If the proton-pumping pathway is connected to the proton pathway taken by protons used for water formation or if it is accessible to the O_2 reduction site, protons which were to be used in the pumping process would be sent to the O_2 reduction site to neutralize oxide moieties and generate water molecules. The process would dissipate the free energy change generated by O_2 reduction, as suggested long ago (Williams 1995).

9.4.4 *Diversity in the Proton-Pumping Mechanism of COX*

9.4.4.1 Proton-Pumping Mechanism for Bacterial COX Proposed by Mutagenesis Data

Eukaryotic COXs and many bacterial COXs have two possible proton transfer pathways which make connections to the molecular surface of the negative side from the O_2 reduction site (the D- and K-pathways) in addition to the H-pathway. The structures are quite well conserved, and the structure of the H-pathway is not well conserved. For example, although COXs of most animals have Asp51 as one of the key residues in the H-pathway, plant and bacterial COXs do not have this residue (Yoshikawa et al. 2011). This lack of Asp51 is apparently not consistent with the proposal that Asp51 functions as the proton-loading site of proton-pumping system of mammalian COX, since proton pumping is one of the basic functions of COX.

On the other hand, the D-pathway is quite well conserved. A proton-pumping mechanism operating through this pathway has been proposed based on the results of site-directed mutagenesis experiments. Both of the two known D-pathway mutants, Glu242Gln and Asp91Asn (with reference to the bovine COX numbering system), abolish both proton pumping and electron transfer activities (Konstantinov et al. 1997). Furthermore, the Asn98Asp mutant abolishes the process of proton pumping without decreasing O_2 reduction activity (Pawate et al. 2002). From these experimental results, it has been proposed that the D-pathway transfers the protons for pumping and for generating water molecules. However, electron transfer to the O_2 reduction site is likely to be tightly coupled with the transfer of protons used for generation of water molecules. Thus, blockage of proton transfer through the D-pathway by the above mutation would abolish O_2 reduction at the O_2 reduction site. Pumping of protons would then be prevented, since no driving force is available. Thus, it cannot be concluded that the D-pathway transfers protons for both proton pumping and generation of water molecules, based on the data obtained from Glu242Gln and Asp91Asn. These data do not identify the location of the proton-pumping system. Another important point of the proposal involving the D-pathway is that a structure responsible for sorting protons to be pumped from the protons to be used for generation of water molecules (which would prevent leakage of the pumping protons to the O_2 reduction site) has not been identified in the X-ray structure. Furthermore, a clear pathway from the D-pathway to the positive side surface of COX has not been identified.

Interpretation of the data obtained for the Asn98Asp mutant is not straightforward (Pawate et al. 2002). First of all, it is unlikely that Asp98 would block proton transfer through the D-pathway. The simplest interpretation for the observed phenotype is that the Asn/Asp exchange influences the integrity of the proton-pump system and abolishes the proton-pumping function without impairing the process of generating water molecules with proton transfer through D-pathway. Anyhow, the data obtained from this mutant does not identify the location of the proton-pumping site. It should be noted that conclusive experimental evidence for proton pumping via the D-pathway is needed.

9.4.4.2 Diversity in Proton Transfer Pathways

It has been recently shown that the D-pathway is not completely conserved among all of the heme-copper oxygen reductases (Chang et al. 2009; Hemp et al. 2007). The common elements across all families of the heme-copper oxygen reductases are the heme a_3/Cu_B dinuclear site, the four histidine imidazole groups that coordinate the metals, and the covalently linked His–Tyr moiety. Thus, it has been suggested that these elements reduce O_2 via a mechanism coupled to the proton-pumping process and that these proton-conducting pathways are simply used for proton delivery. However, there is an alternative interpretation which would explain this diversity. A process responsible for reduction of O_2 without release of any reactive oxygen species must be a well-organized complex reaction process. No alternative system with efficiency higher than that of the Fe_{a3}/Cu_B system has been developed during the course of evolution of aerobic organisms. On the other hand, since proton pumping is a chemically simple process, various amino acid residues can facilitate the pumping function to provide diversity.

Acknowledgments This work is supported in part by the Grant-in-Aid for Scientific Research 2247012 (S.Y.), the Targeted Protein Research Program, and the Global Center of Excellence Program, each provided by the Japanese Ministry of Education, Culture, Sports, Science and Technology. S.Y. is a Senior Visiting Scientist in the RIKEN Harima Institute.

References

Aoyama H, Muramoto K, Shinzawa-Itoh K, Hirata K et al (2009) A peroxide bridge between Fe and Cu ions in the O_2 reduction site of fully oxidized cytochrome *c* oxidase could suppress the proton pump. Proc Natl Acad Sci USA 106:2165–2169

Arnold S, Kadenbach B (1997) Cell respiration is controlled by ATP, an allosteric inhibitor of cytochrome-*c* oxidase. Eur J Biochem 249:350–354

Caughey WS, Smythe GA, O'Keefe DH, Maskasky JE, Smith ML (1975) Heme A of cytochrome *c* oxidase. J Biol Chem 250:7602–7622

Caughey WS, Wallace WJ, Volpe JA, Yoshikawa S (1976) Cytochrome *c* oxidase. In: Boyer PD (ed) The enzymes, volume XIII. Oxidation-reduction, Part C, 3rd edn. Academic, New York

Chang HY, Hemp J, Chen Y, Fee JA, Gennis RB (2009) The cytochrome ba_3 oxygen reductase from *Thermus thermophilus* uses a single input channel for proton delivery to the active site and for proton pumping. Proc Natl Acad Sci USA 106:16169–16173

Fabian M, Skultety L, Brunel C, Palmer G et al (2001) Cyanide stimulated dissociation of chloride from the catalytic center of oxidized cytochrome *c* oxidase. Biochemistry 40:6061–6069

Fiamingo FG, Altshuld RA, Moh PP, Alben JO (1982) Dynamic interactions of CO with a_3Fe and Cu_B in cytochrome *c* oxidase in beef heart mitochondria studied by Fourier transform infrared spectroscopy at low temperatures. J Biol Chem 257:1639–1650

Frank V, Kadenbach B (1996) Regulation of the H^+/e^- stoichiometry of cytochrome *c* oxidase from bovine heart by intramitochondrial ATP/ADP ratios. FEBS Lett 382:121–124

Gilderson G, Salomonson L, Aagaard A, Gray J et al (2003) Subunit III of cytochrome *c* oxidase of *Rhodobacter sphaeroides* is required to maintain rapid proton uptake through the D pathway at physiologic pH. Biochemistry 42:7400–7409

Hemp J, Han H, Roh JH, Kaplan S et al (2007) Comparative genomics and site-directed mutagenesis support the existence of only one input channel for protons in the C-family (*cbb*$_3$ oxidase) of heme-copper oxygen reductases. Biochemistry 46:9963–9972

Isaacs NS (1995) Physical organic chemistry, 2nd edn. Longman, Essex

Kadenbach B, Ungibauer M, Jarausch J, Buge U, Kuhn-Nentwig L (1983) The complexity of respiratory complexes. Trends Biochem Sci 8:398–400

Kamiya K, Boero M, Tateno M, Shiraishi K, Oshiyama A (2007) First-principles molecular dynamics study of proton transfer mechanism in bovine cytochrome *c* oxidase. J Am Chem Soc 129:9663–9673

Kitagawa T, Ogura T (1997) Oxygen activation mechanism at binuclear site of heme-copper oxidase superfamily as revealed by time-resolved resonance Raman spectroscopy. Prog Inorg Chem 45:431–479

Konstantinov A, Siletsky S, Mitchell D, Kaulen A, Gennis RB et al (1997) The roles of the two proton input channels in cytochrome *c* oxidase from *Rhodobacter sphaeroides* probed by the effects of site-directed mutations on time-resolved electrogenic intraprotein proton transfer. Proc Natl Acad Sci USA 94:9085–9090

Lee I, Bender E, Arnold S, Kadenbach B (2001) New control of mitochondrial membrane potential and ROS formation–a hypothesis. Biol Chem 382:1629–1636

Mochizuki M, Aoyama H, Shinzawa-Itoh K, Usui T et al (1999) Quantitative reevaluation of the redox active sites of crystalline bovine heart cytochrome *c* oxidase. J Biol Chem 274:33403–33411

Muramoto K, Ohta K, Shinzawa-Itoh K, Kanda K et al (2010) Bovine cytochrome *c* oxidase structures enable O_2 reduction with minimization of reactive oxygens and provide a proton-pumping gate. Proc Natl Acad Sci USA 107:7740–7745

Napiwotzki J, Shinzawa-Itoh K, Yoshikawa S, Kadenbach B (1997) ATP and ADP bind to cytochrome *c* oxidase and regulate its activity. Biol Chem 378:1013–1021

Oliveberg M, Malmstrom BG (1992) Reaction of dioxygen with cytochrome *c* oxidase reduced to different degrees: indications of a transient dioxygen complex with copper-B. Biochemistry 31:3560–3563

Pawate AS, Morgan J, Namslauer A, Mills D et al (2002) A mutation in subunit I of cytochrome oxidase from *Rhodobacter sphaeroides* results in an increase in steady-state activity but completely eliminates proton pumping. Biochemistry 41:13417–13423

Potter WT, Tucker MP, Houtchens RA, Caughey WS et al (1987) Oxygen infrared spectra of oxyhemoglobins and oxymyoglobins. Evidence of two major liganded O_2 structures. Biochemistry 26:4699–4707

Sakaguchi M, Shinzawa-Itoh K, Yoshikawa S, Ogura T (2010) A resonance Raman band assignable to the O-O stretching mode in the resting oxidized state of bovine heart cytochrome *c* oxidase. J Bioenerg Biomembr 42:241–243

Sasaroli M, Ching Y-C, Dasgupta S, Rousseau DL (1989) Cytochrome *c* oxidase: evidence for interaction of water molecules with cytochrome *a*. Biochemistry 28:3128–3132

Shimokata K, Katayama Y, Murayama H, Suematsu M et al (2007) The proton pumping pathway of bovine heart cytochrome *c* oxidase. Proc Natl Acad Sci USA 104:4200–4205

Shinzawa-Itoh K, Aoyama H, Muramoto K, Terada H et al (2007) Structures and physiological roles of 13 integral lipids of bovine heart cytochrome *c* oxidase. EMBO J 26:1713–1725

Suga M, Yano N, Muramoto K, Shinzawa-Itoh K et al (2011) Distinguishing between Cl^- and O_2^{2-} as the bridging element between Fe^{3+} and Cu^{2+} in resting-oxidized cytochrome *c* oxidase. Acta Crystallogr D Biol Crystallogr D67:742–744

Thornstrom PE, Nilsson T, Malmstrom BG (1988) The possible role of the closed-open transition in proton pumping by cytochrome *c* oxidase: the pH dependence of cyanide inhibition. Biochim Biophys Acta 935:103–108

Tsukihara T, Aoyama H, Yamashita E, Tomizaki T et al (1996) The whole structure of the 13-subunit oxidized cytochrome *c* oxidase at 2.8 Å. Science 272:1136–1144

Tsukihara T, Shimokata K, Katayama Y, Shimada H et al (2003) The low-spin heme of cytochrome *c* oxidase as the driving element of the proton-pumping process. Proc Natl Acad Sci USA 100:15304–15309

Williams RJP (1995) Bioenergetics. Purpose of proton pathways. Nature 376:643

Yakushiji E, Okunuki K (1941) Isolierung der a-Komponents des Cytochroms und ihre Eigenschaften. Proc Imp Acad Jpn 17:38–40

Yamashita E, Aoyama H, Yao M, Muramoto K et al (2005) Absolute configuration of the hydroxyfarnesylethyl group of haem A, determined by X-ray structural analysis of bovine heart cytochrome *c* oxidase using methods applicable at 2.8 Angstrom resolution. Acta Crystallogr D Biol Crystallogr D61:1373–1377

Yoshikawa S, O'Keeffe DH, Caughey WS (1985) Investigations of cyanide as an infrared probe of hemeprotein ligand binding sites. J Biol Chem 260:3518–3528

Yoshikawa S, Mochizuki M, Zhao XJ, Caughey WS (1995) Effects of overall oxidation state on infrared spectra of heme a_3 cyanide in bovine heart cytochrome *c* oxidase. Evidence of novel mechanistic roles for Cu_B. J Biol Chem 270:4270–4279

Yoshikawa S, Shinzawa-Itoh K, Nakashima R, Yaono R et al (1998) Redox-coupled crystal structural changes in bovine heart cytochrome *c* oxidase. Science 280:1723–1729

Yoshikawa S, Muramoto K, Shinzawa-Itoh K (2011) Proton-pumping mechanism of cytochrome *c* oxidase. Annu Rev Biophys 40:205–223

Chapter 10
Phosphorylation of Mammalian Cytochrome *c* and Cytochrome *c* Oxidase in the Regulation of Cell Destiny: Respiration, Apoptosis, and Human Disease

Maik Hüttemann, Icksoo Lee, Lawrence I. Grossman, Jeffrey W. Doan, and Thomas H. Sanderson

Abstract The mitochondrial oxidative phosphorylation (OxPhos) system not only generates the vast majority of cellular energy, but is also involved in the generation of reactive oxygen species (ROS), and apoptosis. Cytochrome *c* (Cytc) and cytochrome *c* oxidase (COX) represent the terminal step of the electron transport chain (ETC), the proposed rate-limiting reaction in mammals. Cytc and COX show unique regulatory features including allosteric regulation, isoform expression, and regulation through cell signaling pathways. This chapter focuses on the latter and discusses all mapped phosphorylation sites based on the crystal structures of COX and Cytc. Several signaling pathways have been identified that target COX including protein kinase A and C, receptor tyrosine kinase, and inflammatory signaling. In addition, four phosphorylation sites have been mapped on Cytc with potentially

M. Hüttemann (✉) • L.I. Grossman
Center for Molecular Medicine and Genetics, and Cardiovascular Research Institute, Wayne State University School of Medicine, Detroit, MI, USA

Karmanos Cancer Institute, Detroit, MI, USA
e-mail: mhuttema@med.wayne.edu; l.grossman@wayne.edu

I. Lee • J.W. Doan
Center for Molecular Medicine and Genetics, Wayne State University School of Medicine, Detroit, MI 48201, USA
e-mail: ile@med.wayne.edu

T.H. Sanderson
Cardiovascular Research Institute, Wayne State University School of Medicine, Detroit, MI 48201, USA

Department of Emergency Medicine, Wayne State University School of Medicine, Detroit, MI 48201, USA

B. Kadenbach (ed.), *Mitochondrial Oxidative Phosphorylation*, Advances in Experimental Medicine and Biology 748, DOI 10.1007/978-1-4614-3573-0_10,

large implications due to its multiple functions including apoptosis, a pathway that is overactive in stressed cells but inactive in cancer. The role of COX and Cytc phosphorylation is reviewed in a human disease context, including cancer, inflammation, sepsis, asthma, and ischemia/reperfusion injury as seen in myocardial infarction and ischemic stroke.

10.1 Introduction

The mitochondrial oxidative phosphorylation machinery (OxPhos) is essential for cell function, maintenance, and survival. OxPhos in mammals provides more than 90% of cellular energy. OxPhos consists of the electron transport chain (ETC), which generates the mitochondrial proton motive force by pumping protons across the inner mitochondrial membrane, and ATP synthase (complex V), which couples the backflow of protons into the matrix with the synthesis of ATP from ADP and phosphate. The ETC consists of three proton-pumping complexes, NADH dehydrogenase (complex I), bc_1-complex (complex III), and cytochrome *c* oxidase (COX; complex IV), and nonproton pumping succinate dehydrogenase (complex II), as well as the nonprotein two-electron carrier ubiquinone and the small one-electron carrier cytochrome *c* (Cytc). Electrons enter the ETC mainly through complex I from NADH. In addition, complex II feeds electrons derived from succinate directly into the ubiquinone/ubiquinol pool, linking the Krebs cycle with OxPhos.

OxPhos dysfunction is devastating as can be seen in patients with "traditional" mitochondrial diseases, caused, for example, by mutations in the mitochondrial DNA or nuclear encoded assembly factors of OxPhos complexes. The tissues that are most prominently affected are those that rely most heavily on aerobic energy production including skeletal muscle, brain, and the visual system. More recently, mitochondrial dysfunction has been implicated in an increasing number of human diseases, including the most common pathologies such as cardiovascular disease, diabetes, cancer, and ischemia/reperfusion injury as seen in myocardial infarction and stroke. These "nontraditional" mitochondrial diseases can be better understood when viewed in light of cellular signaling pathways and regulatory control mechanisms, which are often dysregulated in those pathologies. Recent studies have begun to establish a connection between cell signaling and OxPhos, and more than 20 phosphorylation sites have been mapped on the OxPhos complexes (Hüttemann et al. 2007). There is strong evidence that additional sites are phosphorylated, and given the hydrophobic nature of the many membrane-spanning subunits of the OxPhos complexes, several additional phosphorylation sites will likely be revealed in the future due to technological improvements, specifically in mass spectrometry (MS) and related methodologies such as enrichment of phospho-peptides (Thingholm et al. 2006; Walther and Mann 2010).

10.2 Composition and Function of Cytochrome *c* Oxidase and Cytochrome *c*

This chapter focuses on the regulation of cytochrome *c* oxidase (COX) and cytochrome *c* (Cytc) by cell signaling pathways. To gain a better insight into this topic, we will first briefly discuss the basic properties of both enzymes.

Mammalian COX contains 13 subunits per monomer, and cow heart COX has been crystallized as a dimer suggesting that this is the functional form in mammals (Tsukihara et al. 1996). Each COX monomer contains two heme and two copper redox centers, which are located in catalytic subunits I and II. Of the 13 subunits, three are encoded by mitochondrial DNA and the remaining ten by nuclear DNA (Fig. 10.1).

Mammalian Cytc is a one-electron carrier that shuttles electrons from bc_1-complex to COX. It contains 104 amino acids and a heme group, which is covalently attached to cysteines 14 and 17 (Fig. 10.1, top left). Cytc is highly positively charged with a p*I* of 9.6 and is located in the mitochondrial intermembrane space,

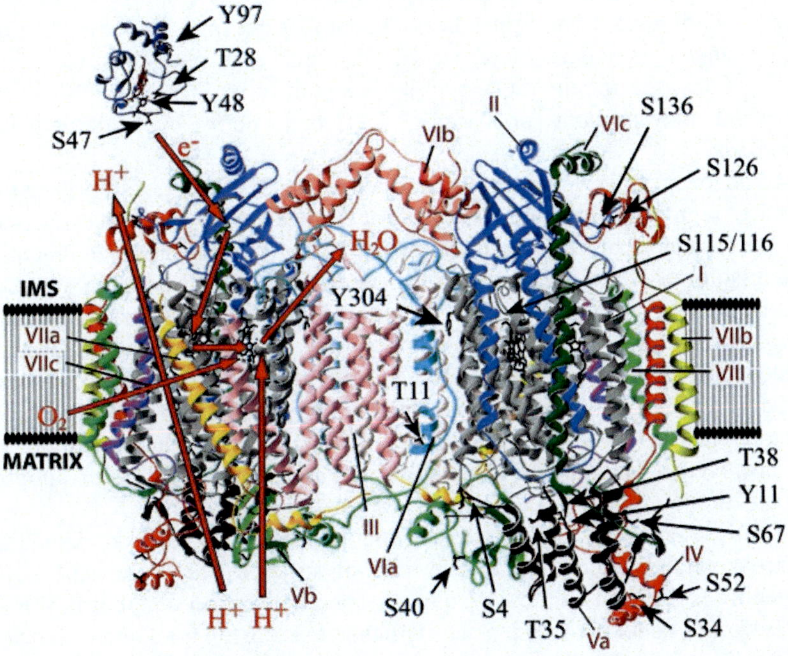

Fig. 10.1 Identified phosphorylation sites on cytochrome *c* and cytochrome *c* oxidase. Crystal structure data of horse heart cytochrome *c* (Sanishvili et al. 1995) and cow heart COX (Tsukihara et al. 1996) were processed with the program Swiss-PDBViewer 3.7. Identified phosphorylated amino acids in mammals are shown in sticks. See Table 10.1 for a detailed description of the sites including phospho-epitopes and references. Note that subunit IV residue Thr52 in rabbit corresponds to Ser52 in cow

where it is associated with negatively charged phospholipids of the inner membrane, in particular cardiolipin.

Cytc and COX represent the last step of the ETC, and catalyze the successive transfer of four electrons to molecular oxygen, which is reduced to water. At the same time, COX pumps protons across the inner mitochondrial membrane, with a stoichiometry of $1H^+/1e^-$ under normal conditions. The precise mechanism of how electron transfer is coupled to proton pumping and the location of the proton exit pathways remain unclear and are a matter of heated debate (von Ballmoos et al. 2011; von der Hocht et al. 2011; Yoshikawa et al. 2011). In addition to the pumped protons, the consumption of "chemical" protons from the matrix side for the water formation reaction contributes to the generation of the proton motive force.

The reaction catalyzed by COX with a free energy of $\Delta G^{\circ\prime} = -100$ kJ/mol (Hinkle et al. 1991) is essentially irreversible. In mammals, it is the proposed rate-limiting step of the ETC in intact cells (Villani and Attardi 1997; Villani et al. 1998; Acin-Perez et al. 2003; Piccoli et al. 2006; Dalmonte et al. 2009; Pacelli et al. 2011), but not in isolated mitochondria. The latter phenomenon may be explained by unsuitable mitochondria isolation methods, which disrupt cellular structures and signaling networks, leading to a loss of regulatory properties of COX and Cytc, likely via dephosphorylation reactions. It has been shown that not only mitochondrial morphology changes dramatically after isolation but also that, in contrast to the other ETC complexes, specifically COX activity is significantly increased (Picard et al. 2011), which may account for a loss of ETC flux control by COX in isolated mitochondria.

Based on the central role of the terminal step of the ETC, it is not surprising that a multitude of regulatory mechanisms are in place in addition to reversible phosphorylation. One other mechanism is the expression of tissue-specific and developmentally-regulated isoforms. Cytc occurs as somatic and testis-specific isoforms in rodents (Goldberg et al. 1977), but not in humans, where the genomic region syntenic to the testes-isoform in rodents now contains a nontranscribed pseudogene (Hüttemann et al. 2003; Zhang and Gerstein 2003). The slight differences in amino acid composition between the isoforms affect its function. In comparison with the somatic isoform, testes Cytc shows a threefold increased activity to reduce hydrogen peroxide; however, it also shows a fourfold increased ability to trigger apoptosis (Liu et al. 2006).

For COX, six subunit isoforms have been identified in mammals to date. Those are heart/skeletal-muscle-specific isoforms of subunit VIa, VIIa, and VIII, a lung-specific isoform of subunit IV, a testes-specific isoform of subunit VIb, and a third isoform of subunit VIII [see Hüttemann et al. (2011) for a current review]. All isoforms are encoded by separate genes. Expression of tissue-specific isoforms may not only directly affect the activities of COX and Cytc but it can also provide a platform for tissue-specific cell signaling. For example, a phosphorylation site was mapped in the heart/skeletal muscle-specific subunit VIa of COX (Table 10.1) (Tsukihara et al. 2003). The corresponding epitope in the liver-type isoform is distinct suggesting that cell signaling is adapted to tissue-specific signals and needs.

Table 10.1 Identified phospho-epitopes in mammalian cytochrome *c* and cytochrome *c* oxidase[a]

Enzyme	Species and tissue	Phosphorylated amino acid[b]	Phospho-epitope[c]	Reference	Method
Cytc	Human skeletal muscle	Thr28	EKGGKHKTGPNLHGL	Zhao et al. (2010)	HT-MS[d]
Cytc	Human skeletal muscle	Ser47	TGQAPGYSYTAANKN	Zhao et al. (2010)	HT-MS
Cytc	Cow liver	Tyr48	GQAPGFSYTDANKKN	Yu et al. (2008a)	MS
Cytc	Cow heart	Tyr97	EREDLIAYLKKATNE	Lee et al. (2006)	MS
COX	Cow liver	Tyr304, SU I	MDVDTRAYFTSATMI	Lee et al. (2005)	MS, cAMP-dependent
COX	Cow liver	Tyr304, SU I	MDVDTRAYFTSATMI	Samavati et al. (2008)	Phospho-epitope-specific antibody; TNFα-dependent
COX	Rabbit heart	Ser115 and Ser116, SU I	SLHLAGVSSILGAINF	Fang et al. (2007)	MS; after I/R[e]
COX	Cow heart	Ser126, SU II	DSYMIPTSELKPGEL	Hüttemann et al. (2012)	MS
COX	Cow liver	Tyr11, SU IV-1	SVVKSEDYALPSYVD	Lee et al. (2006)	MS
COX	Cow heart	Ser34, SU IV-1	VAHVKNLSASQKALK	Helling et al. (2008)	MS
COX	HeLa cells	Ser67, SU IV-1	YRIKFKESFAEMNRG	Olsen et al. (2010)	HT-MS
COX	Rabbit heart	Thr52, SU IV-1	KAPWGSLTRDEKVEL	Fang et al. (2007)	MS; after I/R
COX	HeLa cells	Ser136, SU IV-1	NPIQGLASKWDYEKN	Olsen et al. (2010)	HT-MS
COX	Cow heart	Ser4, SU Va	SHGSHETDEEF	Helling et al. (2008)	MS
COX	Cow heart	Thr35, SU Va	ELRKGMNTLVGYDLV	Helling et al. (2008)	MS
COX	HeLa cells	Thr35 and Thr38, SU Va	LRKGINTLVTYDMVPE	Olsen et al. (2010)	HT-MS
COX	Rabbit heart	Ser40, SU Vb	MLPPKAASGTKEDPN	Fang et al. (2007)	MS; after I/R
COX	Cow heart	Thr11, SU VIa	AKGDHGGTGARTWRF	Tsukihara et al. (2003)	Crystal structure

[a]Identified phosphorylation sites are presented in an ascending amino acid and subunit (SU) order
[b]Numbering according to mature peptide based on cow
[c]Sequence is based on the species where the phosphorylation site was identified
[d]High-throughput mass spectrometry sequencing
[e]Ischemia/reperfusion

Another mechanism is allosteric regulation via binding of ATP and ADP, a built-in energy sensor in COX that adapts COX activity to energy demand (see Chap. 11). Interestingly, ATP also binds to Cytc, contributing to the inhibition of the reaction between Cytc and COX (Ferguson-Miller et al. 1976) but no such regulation has yet been reported for the other OxPhos complexes. Additional regulatory mechanisms acting on COX include (1) competitive binding of nitric oxide, making COX one of the most important targets for NO signaling within the cell (Martinez-Ruiz et al. 2011); (2) binding of small molecules such as thyroid hormone T2, leading to an activation of COX even in the presence of allosteric inhibitor ATP (Arnold et al. 1998) and the fatty acid palmitate, which reduces the H^+/e^- stoichiometry in liver COX, making the COX reaction less efficient (Lee and Kadenbach 2001); and (3) direct protein–protein interactions with nitric oxide synthase (Persichini et al. 2005), the androgen receptor (Beauchemin et al. 2001), the epidermal growth factor receptor (EGFR) (Boerner et al. 2004), and Smad4, a downstream executer of TGF-β signaling, which binds to COX during apoptosis (Pang et al. 2011).

10.3 Regulation of Mitochondrial OxPhos via Phosphorylation of Cytochrome *c* Oxidase and Cytochrome *c*

Single-celled and particularly multicellular organisms employ mechanisms that allow the exchange of information within and between cells and with the cellular environment. Posttranslational modifications such as phosphorylations allow rapid modification of protein function, adapting it to cellular changes. Since the mitochondrial OxPhos system is a central functional unit primarily linked to energy production it is a logical target of cell signaling. Other functions such as ROS production and the participation of Cytc in apoptosis also require tight regulation.

Phosphorylation sites have been identified on all mammalian OxPhos complexes. For the vast majority, signaling pathways, kinases and phosphatases mediating these phosphorylations remain unknown or uncertain (Hüttemann et al. 2007). Future work should focus on the functional consequences of individual phosphorylations, their dynamics, i.e., regarding both extent and change over time, and the identification of kinases and phosphatases acting on OxPhos. Among OxPhos complexes, phosphorylation of COX and Cytc have been studied more deeply and will be discussed here, including their potential role in pathological conditions.

10.3.1 Phosphorylation of Cytochrome c

Cytc is a pivotal component of both apoptosis and electron transfer. There are good reasons to believe that such an important molecule would be the target of cell signaling pathways, but such pathways had never been discovered, even though Cytc had been studied for more than a century. To discover phosphorylations of Cytc, if they existed, it would be necessary to isolate Cytc from animal tissue under stringent conditions that would preserve the physiological phosphorylation state. Carrying out isolation of Cytc from cow heart tissue using such stringent conditions, we found by mass spectrometry (MS) that it was phosphorylated on Tyr97 (Fig. 10.1) (Lee et al. 2006). We then investigated the effects of that phosphorylation. Previous investigators had found that the heme iron-Met80 absorption band (normally at 695 nm) is an important indicator of functional intactness of Cytc. We found that this band was shifted to 687 nm upon phosphorylation of Tyr97. Phosphorylated Cytc showed functional differences: a shifted K_m of COX for Cytc from 2.5 μM compared to 5.5 μM and enhanced sigmoidal kinetics indicating an inhibition in the reaction with COX.

Cytc isolated from cow liver tissue, we later found, is phosphorylated too, but on a different residue, Tyr48 (Fig. 10.1) (Yu et al. 2008a). This finding surprised us at first, but is consistent with what is known about differential regulation of COX from tissue type to tissue type. The effects of this phosphorylation were distinct from those found in Tyr97 phosphorylation in cow heart in that in liver Cytc there were no spectral changes, and Tyr48 phosphorylation produced a hyperbolic response, similar to the unphosphorylated Cytc. However, at maximal turnover COX activity was more than 50% reduced with Tyr48-phosphorylated Cytc. Thus, phosphorylation in both cases (Tyr97 in heart, Tyr48 in liver) causes partial, but not full, inhibition of the reaction between Cytc and COX.

The functional explanation for inhibition of mitochondrial respiration by Cytc phosphorylation may be that it provides ample mitochondrial membrane potential $\Delta\Psi_m$ for the production of ATP without the production of excessive free radicals that are concomitant with high $\Delta\Psi_m$ levels. However, excess capacity is needed in certain conditions as a trigger of type II apoptosis, which is initiated by a transient hyperpolarization of $\Delta\Psi_m$, followed by a burst of ROS as a signal to commit to cell death, a model that we will revisit below.

Experiments with phosphomimetic mutant Cytc indicate that Cytc phosphorylation at Tyr48 may have important functional effects for apoptosis, the second key role of Cytc. We replaced Tyr48 with Glu, which mimics the negative charge of the phosphate group (Pecina et al. 2010). The mutant Cytc showed a 45 mV reduction of its midpoint redox potential compared to wild-type unphosphorylated Cytc. The reaction kinetics of phosphomimetic Cytc with COX were similar to that of the Tyr48 Cytc with COX, suggesting that the phosphomimetic form is a good model for Tyr48-phosphorylated Cytc. Strikingly, the ability of Cytc to trigger downstream caspase activation, a requirement of apoptosis, was completely lost in the phosphomimetic mutant. The possibility that phosphorylation of Cytc regulates programmed cell death has potentially important therapeutic implications for diseases like cancer, in which apoptosis is inhibited. Once the kinases and phosphatases are identified that act on Cytc, these enzymes could be specifically targeted to promote apoptosis by dephosphorylation of Cytc in conditions such as cancer. In addition, phosphorylation of Cytc could be induced in conditions of stress such as ischemia/reperfusion injury where cell survival strategies would be beneficial.

A fraction of Cytc is normally bound to the mitochondria-specific lipid cardiolipin, tethering Cytc to the inner mitochondrial membrane. It has been suggested that the release of Cytc from cardiolipin is one of the first steps of the participation of Cytc in apoptosis, and that this is mediated via cardiolipin peroxidase activity of Cytc (Kagan et al. 2009). Upon oxidation of cardiolipin by Cytc in the presence of ROS or lipid peroxides that serve as substrates, Cytc binding affinity is reduced, leading to a dissociation of Cytc from cardiolipin. Interestingly, peroxidase activity of Tyr48Glu phosphomimetic Cytc was inducible only at high cardiolipin concentrations, unlike controls, suggesting that Cytc phosphorylation may suppress the cardiolipin oxidation reaction (Pecina et al. 2010). This could be a second safeguard mechanism to ensure that apoptosis is well regulated, through modulation of Cytc attachment to the inner mitochondrial membrane and thus its release.

Two more phosphorylation sites (Thr28 and Ser47; Fig. 10.1, Table 10.1) have recently been mapped on Cytc in skeletal muscle tissue by high-throughput phosphoproteomic MS analysis (Zhao et al. 2010). Although their function is unknown, these phosphorylations suggest regulation by a different pathway than those operating in heart and liver.

It should be kept in mind that there are other functions of Cytc than the well-known roles of Cytc in electron transport and apoptosis. Cytc also scavenges ROS under healthy conditions (Korshunov et al. 1999; Wang et al. 2003), and generates ROS through the $p66^{shc}$ pathway (Giorgio et al. 2005).

10.3.2 Phosphorylation of Cytochrome c Oxidase

The first indication of phosphorylation of COX was presented by Steenaart and Shore (1997), who found that COX was phosphorylated on subunit IV-1 by labeling mitochondrial proteins with radioactive ATP. Since then 14 phosphorylation sites have been mapped on COX (Table 10.1; Fig. 10.1) and several signaling pathways have been studied in some depth. Although many pieces of the puzzle are still missing it is clear that phosphorylation of COX can decisively regulate its activity.

10.3.2.1 The cAMP-Dependent Pathway

In comparison to other signaling pathways PKA signaling has received the most attention. In vitro work first suggested that in cow heart COX incubated with PKA, cAMP, and γ-^{32}ATP, subunit Vb, perhaps subunit II or III, and subunit I are all phosphorylated (Bender and Kadenbach 2000; Lee et al. 2002). However, in vitro work does not usually include various auxiliary components such as scaffolding proteins (e.g., A-kinase anchoring proteins for PKA), and therefore phosphorylation may lack specificity (Welch et al. 2010). Based on our experience, from observation of phosphorylation patterns of COX after in vitro incubation with several kinases, which did not match results obtained in vivo after stimulation or inhibition of the corresponding signaling pathway, we caution the reader against pursuing an in vitro approach.

Our laboratory pursued an in vivo approach to elucidating the issue of cAMP-dependent phosphorylation of COX in liver tissue, where cAMP is a starvation signal triggered by the hormone glucagon. High cAMP levels were generated by treatment with phosphodiesterase inhibitor theophylline. Vanadate and fluoride, two unspecific tyrosine and serine/threonine phosphatase inhibitors, respectively, were included during the purification of mitochondria and subsequently of COX. The treated enzyme, but not the untreated control, showed phosphorylation on Tyr304 of subunit I (Lee et al. 2005). Tyr304 is located adjacent to the oxygen-binding center on COX, and structural modifications near that site can be expected to exhibit functional consequences. Experiments revealed that Tyr304-phosphorylated COX was inhibited strongly to completely at up to 10 μM Cytc substrate concentrations, even in the presence of allosteric activator ADP (Lee et al. 2005). A similar inhibitory effect was seen with other agents that increase cAMP levels, i.e., the adenylyl cyclase activator forskolin and the physiological starvation hormone glucagon. Functionally, this finding makes sense because in starved conditions it would be adaptive for energy production via the electron transport chain to be decreased. Mechanistically, since PKA does not phosphorylate tyrosine residues it appears that a tyrosine kinase downstream of PKA phosphorylates COX.

A similar inhibitory effect along with lowered ATP levels resulted when cow lung tissue was treated with theophylline (Lee et al. 2005). The concentrations of theophylline were those used in asthma therapy, for which theophylline has long been used effectively. The efficacy of theophylline, then, may be in part related to its effect on cAMP, which would then lead to COX phosphorylation. The general mechanism of action would be the following. Theophylline treatment increases cAMP levels, which causes COX phosphorylation, thus reducing COX activity and decreasing ATP production. Decreased ATP production would lower the ability of the airway to constrict, which is an energy-intensive process, and thus would counter the airway constriction that is a hallmark of asthma. If the energy hypothesis is correct, mitochondrial OxPhos could be specifically targeted for asthma treatment, preferably by agents that act only on lung.

It is known that the response of COX to cAMP signaling is tissue-specific. In neuronal tissue, theophylline treatment leads to COX activation (Hüttemann et al. 2010), whereas in heart theophylline and also 3-isobutyl-1-methylxanthine (IBMX, a heart-specific cAMP inducer) had no significant effects on COX activity in our hands. Other phosphorylations have been found in ischemic rabbit heart, and the possibility was raised that these could be a result of cAMP/PKA signaling (Prabu et al. 2006) because the phosphorylations were not found in the presence of kinase inhibitor H89. However, H89 does not affect only PKA and instead inhibits multiple kinases (Bain et al. 2007). The sites were later identified by MS and are Ser115 and Ser116 of subunit I, Thr52 of subunit IV, and Ser40 of subunit Vb (Fang et al. 2007). Since the above sites are not PKA consensus sequences PKA is probably not directly involved in these phosphorylations.

Although COX subunit I Tyr304 phosphorylation, located toward the intermembrane side, inhibits the enzyme in mammalian liver, a distinct effect was found by the Manfredi group in their investigation of a carbon dioxide/bicarbonate-regulated adenylyl cyclase, which localizes to the mitochondrial matrix (Acin-Perez et al. 2009). This enzyme links nutrient availability to OxPhos activity through sensing of CO_2 generated by the citric acid cycle. The study was performed in HeLa cells, and COX subunits I and IV were phosphorylated. The authors proposed that this phosphorylation is mediated by a matrix-localized PKA, although the phosphorylation sites need to be mapped to further confirm PKA involvement. The effect of such signaling is activation of COX (Acin-Perez et al. 2009). Another example of matrix-localized PKA signaling to COX had been reported earlier: the RIα regulatory subunit of PKA can bind to matrix-localized subunit Vb of COX (Yang et al. 1998).

10.3.2.2 Other Signaling Pathways

Various other studies suggest that COX is a target of signaling pathways, without identification of the phosphorylation sites. For example, in addition to its primary cytosolic localization, nonreceptor tyrosine kinase Src has been shown to localize to the mitochondrial intermembrane space (Salvi et al. 2002). In osteoblasts Src targets COX subunit II for phosphorylation, leading to increased COX activity (Miyazaki et al. 2003).

10.3.2.3 Signaling Pathways Unknown

Additional phosphorylation sites have been mapped on COX for which the signaling pathways involved and the functional consequences are unknown. Those sites are Tyr11 of subunit IV-1 in isolated cow liver (Lee et al. 2006), Ser67 and Ser136 of subunit IV-1, Thr35 and Thr38 of subunit Va in human HeLa cells by high-throughput MS (Olsen et al. 2010), Ser34 of subunit IV-1 and Ser4 and Thr35 of subunit Va in cow heart (Helling et al. 2008), Thr11 of subunit VIa heart isoform, which was identified in the cow heart crystal structure (Tsukihara et al. 2003), and Ser126 on catalytic subunit II, which was identified in three independent COX isolations (Hüttemann et al. 2012) (Table 10.1; Fig. 10.1).

10.3.3 Calcium Signaling

Finally, a promising avenue of research would be to investigate the direct or indirect action of calcium on COX and Cytc phosphorylation. Calcium has been proposed to be the strongest signal for mitochondrial activation (Robb-Gaspers et al. 1998). Calcium also plays a key role during conditions of cellular stress, where it leads to hyperactive ETC complexes and increased $\Delta\Psi_m$ levels, leading to excessive ROS production as discussed in Sect. 10.4.1.1. Mammalian COX contains a calcium–sodium exchange site in subunit I (Kirichenko et al. 1998, 2005). In addition to a possible direct functional effect on COX, which remains to be studied, calcium may indirectly affect COX and Cytc through changes in the phosphorylation state. It was shown in pig heart mitochondria that calcium leads to dephosphorylation of most mitochondrial proteins (Hopper et al. 2006). Since dephosphorylation of COX at subunit I Tyr304 and of Cytc at Tyr48 and Tyr97 results in increased ETC activity, this may explain some of the activating effects of calcium signaling.

10.3.4 Mitochondrial Tyrosine Phosphatase Shp-2 and Noonan Syndrome

Protein tyrosine phosphatase Shp-2 was the first identified tyrosine phosphatase with mitochondrial co-localization. It was identified in the intermembrane space and the outer mitochondrial membrane in addition to the cytoplasm in rat brain (Salvi et al. 2004). Shp-2 is mostly associated with Ras/mitogen-activated signaling, where it acts as a positive modulator. Mutations in the gene encoding Shp-2 (*PTPN11*) account for about half of the cases of Noonan syndrome (Tartaglia et al. 2001), a relatively common autosomal dominant disorder characterized by congenital heart defects, dysmorphic facial features, webbed neck, short stature, chest deformity, and variable cognitive deficits (Tartaglia et al. 2011). In these Noonan patients with *PTPN11* mutations, Shp-2 shows increased basal phosphatase activity (Neel et al. 2003). Because COX and Cytc are currently the only OxPhos components with mapped tyrosine phosphorylation sites, three of which localize to the IMS (Fig. 10.1) as does Shp-2, they might be targets of Shp-2. We thus analyzed patient and mouse cell lines with Noonan syndrome mutations and observed significantly increased COX activity (Lee et al. 2010). COX and Cytc protein levels were downregulated in the mutant cells, suggesting a compensatory mechanism to counterbalance increased COX activity. In addition, mutant cells showed 30% decreased ATP and increased ROS levels, both of which may interfere with organ development. Currently it is unknown if any of the identified tyrosine phosphorylation sites on COX or Cytc are targets of Shp-2, but the observation above, i.e., the combination of significantly increased COX activity with reduced COX and Cytc protein levels, suggests changes in protein phosphorylation as a mechanistic explanation.

10.4 Role of Cytochrome *c* Oxidase and Cytochrome *c* in Human Disease

COX and Cytc have been implicated in numerous diseases, some of which we briefly discussed above. This final section focuses on three distinct pathological conditions, ischemia/reperfusion injury, cancer, and inflammation, all of which involve COX and/or Cytc in a highly specific and distinct manner.

10.4.1 Cytochrome c Oxidase and Cytochrome c in Ischemia/Reperfusion Injury and the Role of Protein Kinase C Signaling

Loss of blood flow and thus delivery of oxygen and nutrients to tissues causes extensive damage. Different cell types have varying sensitivities or resistance to ischemic damage, depending on the extent to which the tissue relies on OxPhos for ATP production. While tissue damage caused by ischemia can be extensive, of greater clinical interest is the additional damage induced by restoration of blood flow, or reperfusion injury (Oliver et al. 1990). Currently, the only treatment for ischemia is prompt restoration of blood flow. However, paradoxically, the act of restoring oxygen supply to tissue that has undergone ischemic stress causes profound damage (Campbell et al. 1986; Zweier 1988; Aronowski et al. 1997; Fellman and Raivio 1997; Tilney and Guttmann 1997). This, in turn, results in worsened morbidity and mortality in a host of clinical disorders, including stroke, cardiac arrest/resuscitation, myocardial infarction, acute tubular necrosis, and neonatal hypoxic/ischemic encephalopathy (Zweier 1988; Schumer et al. 1992; Aronowski et al. 1997; Fellman and Raivio 1997; Roger et al. 2011). Importantly, reperfusion injury can, in theory, be treated therapeutically.

10.4.1.1 The Connection Between Respiratory Activity, the Mitochondrial Membrane Potential, and the Production of Reactive Oxygen Species

Mitochondria play a central role in reperfusion injury, primarily as a major source of ROS. As we discussed in detail previously (Hüttemann et al. 2008, 2011), mitochondrial ROS are generated by OxPhos, specifically at high $\Delta\Psi_m$. One of the proposed mechanisms of increased ROS at high $\Delta\Psi_m$ is an overall reduction of electron flux in the ETC because high $\Delta\Psi_m$ levels inhibit the proton pumping activity of complexes I, III, and COX from further proton pumping. This causes an increased half-life of the ubisemiquinone radical intermediates, which now have more time to transfer their unpaired electron to oxygen (Liu 1999). As the proposed rate-limiting step of the ETC in intact cells (see Sect. 10.1), Cytc and COX can be considered primary regulators of mitochondrial ROS generation (Piantadosi and Zhang 1996). Here we put forth a model of ROS generation during reperfusion in multiple tissues, and present studies suggesting the validity of this model. Specifically, we propose that altered phosphorylation of COX and Cytc, caused by ischemia, results in high "uncontrolled" respiratory rates, high $\Delta\Psi_m$ levels, and thus a reperfusion-induced ROS burst upon restoration of blood flow (Fig. 10.2a).

Our model proposes that cells modify the activity of OxPhos by posttranslational modifications on COX and/or Cytc under normal physiologic conditions. Under conditions of energy depletion, cells attempt to augment their energy production by

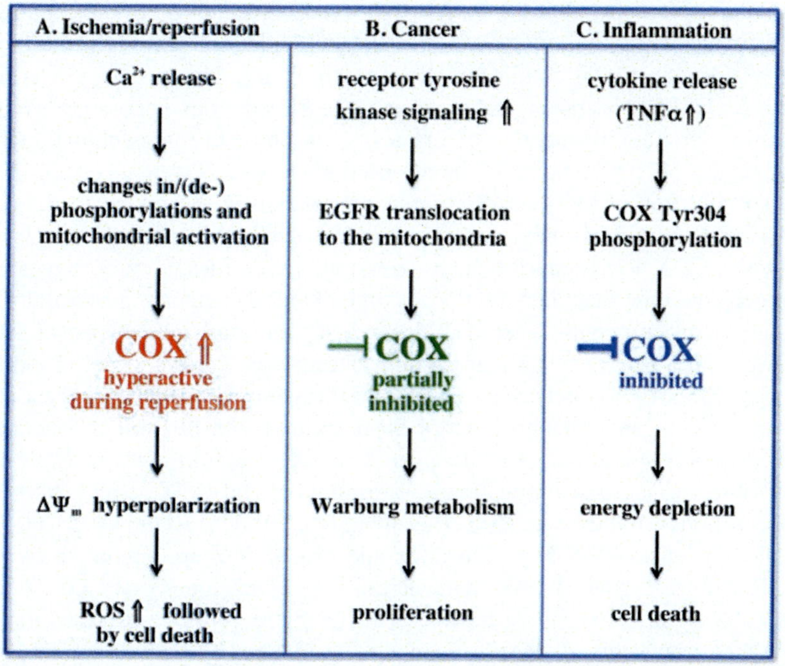

Fig. 10.2 Role of COX in human disease. In our proposed model, the activities of COX and Cytc are regulated by cell signaling pathways. Under healthy conditions both proteins are phosphorylated in vivo leading to controlled respiration. This generates healthy mitochondrial membrane potentials ($\Delta\Psi_m$), allowing efficient energy production but preventing the production of ROS, which are only generated at pathologically high $\Delta\Psi_m$ levels. During pathological conditions the phosphorylation patterns change, leading to an imbalance of ROS and ATP production. (**a**) Ischemic stroke and myocardial infarction are common human pathologies and the only current treatment is rapid resumption of blood flow, i.e., reperfusion. Reperfusion causes the major part of ischemia/reperfusion-related injury through the following sequence of events. During episodes of ischemia, nutrients and oxygen become depleted causing excessive calcium release and alterations of PKC signaling. This results in changes of phosphorylation and/or dephosphorylation of COX and/or Cytc (Cytc is not shown). In the ischemic phase COX does not turn over due to lack of substrates. In the reperfusion phase, oxygen and nutrients are reintroduced leading to a rapid reestablishment of $\Delta\Psi_m$. Because COX is still in a hyperactive state, $\Delta\Psi_m$ increases further, leading to a $\Delta\Psi_m$ hyperpolarization and the production of excessive ROS. ROS in turn serve as a signal for triggering apoptosis leading to cell death. (**b**) Many cancers are characterized by hyperactive receptor tyrosine kinase signaling including EGFR signaling. Here, activation of the pathway leads to an internalization of the receptor and translocation to the mitochondria where EGFR directly interacts with and phosphorylates COX subunit II, leading to a partial inhibition of COX activity. This step allows shifting of aerobic energy metabolism to Warburg metabolism: increased glycolytic and pentose phosphate pathway activity now provide essential building blocks for the cell, enabling rapid proliferation. (**c**) Acute inflammation as in sepsis leads to the release of cytokines including TNFα resulting in phosphorylation of Tyr304 on catalytic subunit I. COX is strongly inhibited leading to decreased $\Delta\Psi_m$ levels, and eventually depletion of ATP. In septic patients, this model would explain organ failure and death through energetic failure

the OxPhos system by altering the phosphorylation status of these complexes. However, under conditions of pathologic energy depletion, such as the ischemic state, energy exhaustion is coupled to oxygen deprivation. In this state, the terminal substrate for OxPhos is absent, and no aerobic ATP production is possible. We propose that the apparent dysfunction of this normal response to energy depletion is the proximal cause of ROS generation upon reperfusion.

The specific role of COX and Cytc phosphorylation in the setting of ischemia/reperfusion remains largely unknown at present. However, many studies have identified alterations in COX activity during the progression of reperfusion injury suggesting a role for cell signaling that alters OxPhos activity. During ischemia, intracellular calcium concentrations increase when ATP-dependent pumps fail (Rosenthal et al. 1987), and mitochondria actively sequester calcium during early restoration of blood flow (Zaidan and Sims 1994). Increased mitochondrial calcium is a potent signal for phosphatase activation, and calcium induces dephosphorylation of most mitochondrial proteins (Robb-Gaspers et al. 1998; Hopper et al. 2006). Interestingly, multiple studies have observed increased mitochondrial respiration, and COX activity, upon calcium sequestration by mitochondria (Rosenthal et al. 1987; Fiskum et al. 2004). In vitro, calcium causes COX hyperactivation and loss of allosteric inhibition by ATP, and the Kadenbach group showed that calcium did not act directly on COX (Bender and Kadenbach 2000). In contrast, they observed an indirect effect, most likely through changes in posttranslational modifications of COX induced by calcium. Additionally, mitochondria treated with calcium had increased state 4 respiration (Vlessis et al. 1990), which may be explained by COX dephosphorylation. Dephosphorylation of COX in vitro results in loss of the ability of ATP to allosterically inhibit COX (Hüttemann et al. 2008). This effect, when occurring in the setting of ischemia/reperfusion injury would result in a condition where COX could generate high $\Delta\Psi_m$ that subsequently leads to the generation of ROS, triggering death processes (Fig. 10.2a).

As discussed above, all functionally studied phosphorylations of Cytc inhibit respiration. Therefore, if calcium-induced dephosphorylation of Cytc were to occur, this would contribute to increased OxPhos flux and thus the hyperpolarization of $\Delta\Psi_m$. Recent studies from our group found that Cytc phosphorylation is lost when brain is rendered ischemic (unpublished). This dephosphoryated Cytc would also have the full capability to induce apoptosis (Pecina et al. 2010).

10.4.1.2 Protein Kinase C

In the context of ischemia/reperfusion injury, PKC signaling has been studied in some detail. PKC is a stress-activated kinase involved in modulating calcium uptake by mitochondria, production of free radicals, and the induction of apoptosis. Importantly, several PKC subtypes have been implicated in regulating these deleterious events following ischemia/reperfusion injury in multiple tissues. It is generally believed that PKC signaling serves to protect the cells during episodes of ischemia/reperfusion but the intricate connection between calcium and PKC signaling may complicate the picture.

There is clear evidence of a direct action of PKC on mitochondria. For example, following ischemia/reperfusion in the heart, PKCε translocates to the mitochondrial inner membrane (Budas et al. 2010). In a model of global brain ischemia, PKCβ translocates to the mitochondria in neurons resistant to cell death, while in brain regions that go on to die (the CA1 hippocampus), no translocation of PKCβ was seen (Kowalczyk et al. 2012). Interestingly, PKCβ was found in the mitochondria, associated with components of the electron transport chain. Indeed, inhibition of PKC family kinases leads to increased $\Delta\Psi_m$ and ROS production (Lu et al. 2011). An important protective effect of PKCε in ischemia/reperfusion injury may be due to its interaction with the calcium-sensing receptor, leading to a reduction of calcium release (Dong et al. 2010). This would prevent calcium-activated dephosphorylation and hyperactivation of mitochondrial proteins. Interestingly, administration of compounds that induce PKC activation provides protection from ischemia/reperfusion injury of the heart (Sivaraman et al. 2009) and brain (Della-Morte et al. 2011). The compound tribulosin protects the heart from ischemia/reperfusion injury through activation of PKCε and parallel activation of superoxide dismutase (Zhang et al. 2010). Another selective PKCε activator, psivarepsilon-RACK, demonstrated significant neuroprotection from reperfusion-induced neuronal cell death (Della-Morte et al. 2011). Translocation of PKCδ to the mitochondria does not occur during ischemia, but was observed within 5 min of reperfusion after 30 min ischemia in rat hearts, followed by a decrease in mitochondrial respiration and an increase in superoxide radical production (Churchill and Szweda 2005).

The above studies underscore the potential contribution of PKC signaling in reversible phosphorylation of OxPhos complexes, and the central role PKC may play in the pathophysiology of ischemia/reperfusion injury. As discussed in Sect. 10.3.2.1, ischemia directly affects COX and leads to phosphorylation of several subunits (Prabu et al. 2006), and it is possible that this phosphorylation is mediated by a yet-to-be-identified PKC isozyme. In rat heart, ischemia/reperfusion resulted in changes of the immunoreactivity of several COX subunits, especially subunit I, and the authors proposed that those subunits are lost from the holoenzyme (Yu et al. 2008b). An alternative and perhaps more likely explanation might be masking of the epitope recognized by the antibody by phosphorylation.

The ischemia-triggered phosphorylations lead to a partial inhibition of COX activity (Prabu et al. 2006), which would be protective during reperfusion. However, other studies have shown that the opposite effect, i.e., activation of COX, is possible. In rat neonatal cardiac myocytes, activation of PKC with diacylglycerol or 4β-PMA caused phosphorylation of COX subunit IV in vitro and resulted in about two- to fourfold increased COX activity (Ogbi et al. 2004; Ogbi and Johnson 2006). PKCε is a possible candidate for this phosphorylation because the authors showed that it co-immunoprecipitated with COX (Guo et al. 2007).

Based on the extent of the stress impact, calcium signaling may prevail leading to an overall activation of the ETC proton pumps. Based on our model (Fig. 10.2a), upon restoration of blood flow, if oxygen reaches a COX enzyme that has been posttranslationally modified to increase its activity, these alterations would initially aid in the restoration of $\Delta\Psi_m$, and the reestablishment of cellular energy levels (Ekholm et al. 1993). However, in this hyperactive state, the proton pumps would not be inhibited by physiological $\Delta\Psi_m$ levels <140 mV and would continue to generate pathologically high $\Delta\Psi_m$ levels >140 mV, which would lead to ROS generation (Liu 1999). The majority of mitochondrial ROS are created during this early reperfusion interval (Fabian et al. 1995), and the ETC is a primary source of ROS during reperfusion (Piantadosi and Zhang 1996). Reperfusion of ischemic brain results in a rapid restoration of $\Delta\Psi_m$, followed by a transient hyperpolarization of $\Delta\Psi_m$ and substantial ROS generation (Liu and Murphy 2009). These findings position COX and Cytc as potential regulatory sites that can indirectly control ROS generation by regulating overall ETC flux, thereby controlling $\Delta\Psi_m$ and ROS. It is thus possible that therapeutic interventions targeting COX and/or Cytc phosphorylations may be neuroprotective in the context of ischemia/reperfusion injury.

In summary, we propose a model of reperfusion-induced ROS generation and cell death where posttranslational modifications of Cytc and/or COX play a critical role in controlling the eventual fate of the cell. Specifically, we propose that cell signaling systems, most notably PKC signaling, retain Cytc and/or COX in a phosphorylated "controlled" state. However, ischemic stress can result in excessive calcium release and subsequent dephosphorylation of OxPhos complexes, thus tipping the balance from controlled respiration to $\Delta\Psi_m$ hyperpolarization, subsequent ROS generation, and cell death (Fig. 10.2a). Therefore, regulation of OxPhos phosphorylation may represent a novel method to minimize reperfusion injury following ischemic events in multiple tissues.

10.4.2 Cancer and Inflammation: Cytochrome c Oxidase and Cytochrome c as Functional Targets

In this section, we will discuss mechanisms underlying cancer and inflammation and discuss the emerging link between the two (Fig. 10.2b, c). We will start at the macroscopic level by identifying similarities between cancer and inflammation and eventually focus on COX and Cytc at the molecular level.

10.4.2.1 Inflammation as a Promoter of Cancer

Metabolism changes during carcinogenesis, and most solid tumors show a 25–60% reduction of mitochondrial mass compared to healthy differentiated tissue (Pedersen 1978). During carcinogenesis, cells shift their metabolism from aerobic energy production to glycolysis. The shift takes place even in the presence of oxygen and is therefore referred to as aerobic glycolysis, and it is known as the Warburg effect (Warburg et al. 1924; Warburg 1956). Since Warburg's discovery, considerable work has been done on the role of mitochondria in cancer. Generally, two primary mitochondrial cancer-promoting factors have been tied to cancer, via metabolic switching to provide building blocks for the growing cells (Weinberg and Chandel 2009), and/or via increased mitochondrial ROS production resulting in the emergence of some cells with oncogenic mutations (Ralph et al. 2010).

Inflammation is an immune response initiated by the vascular system to fight various compounds including pathogens, irritants, and even cells of the organism itself. It is involved acutely and chronically in numerous pathological conditions, such as sepsis, asthma, and rheumatoid arthritis. More recently, inflammation has also become a widely accepted component in different stages of tumor development (Rakoff-Nahoum 2006; Mantovani et al. 2008). Grivennikov et al. (2010) proposed a two-step model in which reactive oxygen and nitrogen species (RONS) produced by inflammatory cells first cause mutations in neighboring cells. Tumor initiation is further amplified by cytokine-mediated increased RONS production in premalignant cells. The second step, tumor promotion, is accompanied by immune cell-mediated cytokine production, which activates key transcription factors in premalignant cells, including NF-κB and STAT3. This induces pro-tumorigenic processes, including survival, proliferation, growth, angiogenesis, and invasion. Since it has turned out that key factors such as ROS are found in *both* inflammation and cancer and are connected to mitochondria, this has stimulated examination of the links that may tie them together (Kamp et al. 2011).

The observations that tumors often arise at a site of chronic inflammation, and that they contain inflammatory cells, are more than 100 years old. Recent interest in this topic was promoted by several types of observations, including substantial epidemiological evidence, such as the beneficial effect on cancer prevalence of chronic use of nonsteroidal anti-inflammatory drugs, and the unraveling and manipulating in animal models of the molecular pathways. The considerable epidemiological evidence includes a wide array of chronic infections, such as by *Helicobacter pylori*, exposure to a wide range of irritants that trigger inflammation, such as tobacco smoke, and autoimmune conditions. Strong evidence also comes from studies of inflammatory bowel disease. Patients with ulcerative colitis, an inflammatory bowel disease, have a five- to sevenfold increased risk of colorectal cancer; this risk is reduced by 80% by administration of cyclooxygenase-2 inhibitors (Kamp et al. 2011).

Germ-line mutations are rare causes of cancer; about 90% of cancers results from a combination of environmental factors and somatic mutations. Detailed studies have uncovered two pathways of inflammatory connection to cancer. An extrinsic pathway, which raises cancer risk, is provided by a chronic site-specific inflammatory condition such as pancreatitis or inflammatory bowel disease (Mantovani et al. 2008). In the intrinsic pathway oncogenes are activated, which is the driving force in an environment provided by the extrinsic pathway. Studies of *RAS* family mutations and *MYC* show the early induction of chemokines and inflammatory cytokines as part of the remodeling of the tissue microenvironment.

Inflammation has been shown to participate in all of the recognized stages in tumorigenesis—initiation, promotion, and metastasis. Initiation usually consists of accumulation of multiple mutations in the same cell. This is more likely in an inflammatory microenvironment, either via RONS produced by activated inflammatory cells or via cytokines produced by inflammatory cells that stimulate RONS production in neighboring epithelial cells. In tumor promotion, an initiated cell transforms into a tumor. Initiated cells are aided by inflammation both to proliferate and to survive, such as via the ROS-stimulated production of HIF-1α (Hamanaka and Chandel 2010). In addition, inflammatory mediators such as STAT3 and NF-κB promote the HIF-1α stimulation of angiogenesis that is needed by growing tumors to provide adequate blood supply (Grivennikov and Karin 2010). Lastly, since more than 90% of cancer mortality results from metastasis, this component is of the greatest clinical importance. Metastasis is clearly linked to inflammation. Many initiated cells express chemokine receptors on their surface (Balkwill 2004). During metastasis, such cells utilize chemokines to aid their migration to distant sites, and to aid their survival upon arriving (Kim et al. 2005, 2009). Furthermore, both autocrine and paracrine signaling by cytokines like TNFα upregulate receptor expression and thereby increase invasive capacity and facilitate metastasis (Kulbe et al. 2005).

The epidemiological connection between inflammation and cancer stimulated numerous studies designed to elicit the mechanistic basis. As signaling pathways were unearthed, it became clear that central mediators of inflammation-associated cancer are RONS. Among those, ROS are the most studied reactive species and arise primarily from NADPH oxidase in phagocytes but elsewhere largely (>90%) from metabolism via the mitochondrial electron transport chain.

Work being pursued from another direction has turned out to be germane. The Warburg hypothesis connected metabolism and cancer in the original version by noticing that tumors are more glycolytic than normal tissues and that, therefore, metabolism was a component of cancer. This observation has also stimulated numerous studies, which have led to a more nuanced picture than the original hypothesis that takes into account a number of metabolic adaptations made by tumors to promote their growth. Except for the situation where hypoxia is found centrally in solid tumors, glycolysis appears not to be utilized for ATP production but for producing intermediates through the pentose phosphate pathway for nucleotide and phospholipid synthesis (Hamanaka and Chandel 2010; Weinberg et al. 2010). Mitochondrial metabolism per se in normoxic conditions may be dispensable, as suggested by a study that utilized a mutation in a complex III gene to block electron transport. Cells containing this mutation reduced but did not abolish a proxy for tumor growth, whereas ρ° cells did not show anchorage-independent growth in a Kras tumor model (Weinberg et al. 2010). An important function of mitochondria appears to be ROS production as a signal for cell proliferation.

A provisional consolidated picture that emerges from considering these dual connections is that inflammation-stimulated ROS, produced in mitochondria under stimulation by cytokines and hypoxia, can act as a tumor promoter. Furthermore, such ROS would be amplified by ROS already produced as growth stimulators by emerging tumors. The increased ROS level could therefore act also as a mutagen to initiate feed-forward cycles of cellular decline, thereby playing a role in both the extrinsic and the intrinsic pathways (Mantovani et al. 2008).

10.4.2.2 Cancer Signaling Targets Cytochrome *c* Oxidase

In addition to metabolic changes, cancers manage to evade apoptosis, and both changes are likely caused, at least in part, by receptor tyrosine kinase signaling that is upregulated in many cancers. Increased epidermal growth factor receptor (EGFR) signaling is implicated in numerous cancers including breast, colon, and lung cancers, and it is the first example of a tyrosine kinase receptor with a direct effect on COX. After stimulation with EGF it was shown in breast cancer cell lines that EGFR translocates to the mitochondria where it physically interacts with COX subunit II (Boerner et al. 2004). Only the activated, Tyr845-phosphorylated EGFR receptor binds to COX together with Src kinase leading to an increase in COX subunit II phosphorylation in vitro as was shown after incubation with [γ-^{32}P]ATP (Demory et al. 2009). Although the phosphorylation site remains to be identified, those findings are in line with the Warburg effect because COX activity was decreased by 60% after cells were treated with EGF (Fig. 10.2b). It is possible that other receptor tyrosine kinases may follow a similar mechanism and that they target other OxPhos components. The only other OxPhos complex where tyrosine phosphorylation has been shown to date is ATP synthase (Ko et al. 2002): NIH3T3 and kidney cells treated with platelet-derived growth factor (PDGF) displayed tyrosine phosphorylation of the δ-subunit of ATP synthase.

Future studies of receptor tyrosine kinase signaling on multiple OxPhos components might reveal a concerted mode of action, i.e., the parallel targeting of several enzymes for phosphorylation to adapt OxPhos activity to cancer-specific energy metabolism. Specifically, Cytc might be targeted for phosphorylation for two reasons: since the functionally studied Cytc phosphorylations on tyrosines 48 and 97 both lead to an inhibition of respiration as discussed above, increased phosphorylation would contribute to the Warburg effect. In addition, Cytc phosphorylation may interfere with apoptosis as suggested by studies with phosphomimetic Cytc, which was not able to trigger any measureable caspase activation (Pecina et al. 2010). Since cancers manage to evade apoptosis, increased phosphorylation of Cytc via cancer signaling might provide a mechanism for the suppression of apoptosis.

10.4.2.3 Inflammatory Signaling Targets Cytochrome *c* Oxidase

The effect of acute inflammation on mitochondrial function has been studied in some detail. Acute inflammation as seen in sepsis is a major medical problem and leading cause of mortality in intensive care units with 210,000 deaths annually in the USA alone (Hotchkiss and Karl 2003). It can be caused by pathogenic infections of the blood and is therefore often referred to as blood poisoning. Sepsis can affect various organs such as the brain, heart, and liver, and it can result in multiple organ dysfunction syndrome (MODS) (Ruggieri et al. 2010).

Genetic evidence strongly suggests that mitochondria play a key role in sepsis. The mitochondrial DNA composition is a genetic predictor for survival after sepsis. Mitochondrial DNA can be grouped into evolutionarily related DNA families, i.e., mitochondrial DNA haplogroups. Interestingly, septic patients belonging to haplogroup H, which is common in Europeans, have a more than twofold higher chance of survival compared to patients with other haplogroups (Baudouin et al. 2005).

Somewhat similarly to cancer, acute inflammation is accompanied by metabolic changes and a suppression of mitochondrial respiration. Thus septic patients show increased rates of lactate production and blood lactate levels (Revelly et al. 2005). Increased systemic delivery of oxygen during the course of sepsis does not improve outcome (Hayes et al. 1994), suggesting that oxygen consumption rather than uptake and delivery is impaired, a condition referred to as cytopathic hypoxia (Fink 2002). Therefore, the ETC and specifically COX seem to be a logical target of inflammatory signaling. Indeed, in endotoxin-treated rats, a commonly used animal model for sepsis, ETC complexes I, II, and COX were downregulated both at the transcript and protein levels within 24 h after treatment (Callahan and Supinski 2005). A recent study with 96 septic patients analyzed COX in platelets. The authors demonstrated a highly significant positive correlation between survival and COX activity and amount (Lorente et al. 2011). The Levy group demonstrated in a cecal ligation sepsis animal model that oxidation of Cytc by COX was competitively and reversibly inhibited, whereas in later stages it became irreversible and noncompetitive (Levy et al. 2004). Others have further shown that cellular energy levels are significantly reduced in septic animals (Astiz et al. 1988), which can be explained with the suppression of mitochondrial respiration. In septic rats, a 70% reduction of tissue ATP appears to be a critical threshold for cellular survival, since a further reduction appears to be incompatible with sustaining cellular functions, resulting in death (Duvigneau et al. 2008).

The above reports point to alterations in COX function during the course of sepsis. To gain a better molecular understanding, we tested the effect of tumor necrosis factor α (TNFα) on COX. TNFα is a proinflammatory cytokine that is strongly induced during sepsis and is a promoter of the septic state (Duvigneau et al. 2008). TNFα alters cellular metabolism by inducing lactate production in vitro and in vivo (Lee et al. 1987; Tracey et al. 1987), indicating a metabolic switch from respiration to glycolysis, as seen in septic patients.

We therefore analyzed the effect of TNFα on cow and mouse liver tissue as well as mouse hepatocytes in culture. TNFα treatment of liver homogenates caused a 60% reduction of COX activity within 5 min after treatment (Samavati et al. 2008). To identify the molecular mechanism explaining this effect, we isolated COX from cow liver with and without TNFα treatment. Further analysis revealed phosphorylation of subunit I tyrosine 304 after TNFα treatment, the same site that was targeted for phosphorylation by the cAMP-dependent pathway in liver as discussed above (Fig. 10.1). TNFα treatment resulted in a reduction of $\Delta\Psi_m$ and a 35 and 64% decrease of cellular ATP levels in mouse liver tissue and H2.35 cells, respectively (Samavati et al. 2008).

The reader may ask, why does such an inflammatory mechanism with potentially disastrous consequences exist in humans? We recently proposed the following scenario

(Hüttemann et al. 2012): The septic state is an extreme inflammatory condition that affects major parts of or an entire organism, and thus is not localized to a restricted small area, a much more common inflammatory situation. Sepsis is very rare compared to conditions of localized inflammation, such as small wounds that may occur on a daily basis. Here shutdown of cellular processes and specifically OxPhos is understandable because several pathogens take over the host infrastructure and energy production system. For example, Chlamydiae bacteria express several nucleotide transporters that facilitate the uptake of molecules such as ATP (Knab et al. 2011). As a result cutting off essential metabolites locally at the infected area will help the organism fight the pathogen. If inflammation gets out of control and becomes a systemic reaction, MODS and death can occur due to energy failure of entire organs (Fig. 10.2c).

10.5 Conclusion

In contrast to the other ETC complexes, COX and Cytc show all three main regulatory mechanisms found in key metabolic enzymes: isoform expression, allosteric control, and phosphorylation. This supports the suggested rate-limiting role of this step in the ETC and thus makes COX and Cytc prime target candidates for therapeutic interventions in the future in the numerous pathological conditions where mitochondrial energy and ROS production are dysregulated. Identification of kinases and phosphatases that act on COX and Cytc will be a central step in this endeavor and allow specific manipulation of signaling pathways.

Functional consequences of cell signaling also have to be carefully analyzed in order to gain a better understanding of the structure–function relationships and the effect at the physiological or organismal level. For example, from what is known about inflammatory and cancer signaling to date, both pathways affect COX at the molecular level and both pathways lead to an inhibition of COX activity (Fig. 10.2b, c). COX is a target via phosphorylation on subunits I or II triggered by TNFα and EGFR, respectively. The resultant changes in COX kinetics are different, however, since TNFα causes a shift from hyperbolic to sigmoidal kinetics with very low COX activities at low Cytc substrate concentrations (Samavati et al. 2008), whereas EGFR signaling does not change the hyperbolic kinetics but decreases maximal turnover by 60% (Demory et al. 2009). Thus, inflammatory signaling can function as an off-switch whereas growth factor signaling leads to a general partial inhibition of COX, in support of the Warburg hypothesis.

Acknowledgements This work was supported by grant GM089900 from the National Institutes of Health, a Department of Defence USAMRAA National Oncogenomic and Molecular Imaging Center contract through the Karmanos Cancer Institute, Detroit, the Center for Molecular Medicine and Genetics, and the Cardiovascular Research Institute, Wayne State University School of Medicine, Detroit.

References

Acin-Perez R, Bayona-Bafaluy MP, Bueno M, Machicado C, Fernandez-Silva P, Perez-Martos A, Montoya J, Lopez-Perez MJ, Sancho J, Enriquez JA (2003) An intragenic suppressor in the cytochrome c oxidase I gene of mouse mitochondrial DNA. Hum Mol Genet 12:329–339

Acin-Perez R, Salazar E, Kamenetsky M, Buck J, Levin LR, Manfredi G (2009) Cyclic AMP produced inside mitochondria regulates oxidative phosphorylation. Cell Metabolism 9:265–276

Arnold S, Goglia F, Kadenbach B (1998) 3,5-Diiodothyronine binds to subunit Va of cytochrome-c oxidase and abolishes the allosteric inhibition of respiration by ATP. Eur J Biochem 252:325–330

Aronowski J, Strong R, Grotta JC (1997) Reperfusion injury: demonstration of brain damage produced by reperfusion after transient focal ischemia in rats. J Cereb Blood Flow Metab 17:1048–1056

Astiz M, Rackow EC, Weil MH, Schumer W (1988) Early impairment of oxidative metabolism and energy production in severe sepsis. Circ Shock 26:311–320

Bain J, Plater L, Elliott M, Shpiro N, Hastie CJ, McLauchlan H, Klevernic I, Arthur JS, Alessi DR, Cohen P (2007) The selectivity of protein kinase inhibitors: a further update. Biochem J 408:297–315

Balkwill F (2004) Cancer and the chemokine network. Nat Rev Cancer 4:540–550

Baudouin SV, Saunders D, Tiangyou W, Elson JL, Poynter J, Pyle A, Keers S, Turnbull DM, Howell N, Chinnery PF (2005) Mitochondrial DNA and survival after sepsis: a prospective study. Lancet 366:2118–2121

Beauchemin AM, Gottlieb B, Beitel LK, Elhaji YA, Pinsky L, Trifiro MA (2001) Cytochrome c oxidase subunit Vb interacts with human androgen receptor: a potential mechanism for neurotoxicity in spinobulbar muscular atrophy. Brain Res Bull 56:285–297

Bender E, Kadenbach B (2000) The allosteric ATP-inhibition of cytochrome c oxidase activity is reversibly switched on by cAMP-dependent phosphorylation. FEBS Lett 466:130–134

Boerner JL, Demory ML, Silva C, Parsons SJ (2004) Phosphorylation of Y845 on the epidermal growth factor receptor mediates binding to the mitochondrial protein cytochrome c oxidase subunit II. Mol Cell Biol 24:7059–7071

Budas GR, Churchill EN, Disatnik MH, Sun L, Mochly-Rosen D (2010) Mitochondrial import of PKCepsilon is mediated by HSP90: a role in cardioprotection from ischaemia and reperfusion injury. Cardiovasc Res 88:83–92

Callahan LA, Supinski GS (2005) Downregulation of diaphragm electron transport chain and glycolytic enzyme gene expression in sepsis. J Appl Physiol 99:1120–1126

Campbell CA, Przyklenk K, Kloner RA (1986) Infarct size reduction: a review of the clinical trials. J Clin Pharmacol 26:317–329

Churchill EN, Szweda LI (2005) Translocation of deltaPKC to mitochondria during cardiac reperfusion enhances superoxide anion production and induces loss in mitochondrial function. Arch Biochem Biophys 439:194–199

Dalmonte ME, Forte E, Genova ML, Giuffre A, Sarti P, Lenaz G (2009) Control of respiration by cytochrome c oxidase in intact cells: role of the membrane potential. J Biol Chem 284:32331–32335

Della-Morte D, Raval AP, Dave KR, Lin HW, Perez-Pinzon MA (2011) Post-ischemic activation of protein kinase C epsilon protects the hippocampus from cerebral ischemic injury via alterations in cerebral blood flow. Neurosci Lett 487:158–162

Demory ML, Boerner JL, Davidson R, Faust W, Miyake T, Lee I, Hüttemann M, Douglas R, Haddad G, Parsons SJ (2009) Epidermal growth factor receptor translocation to the mitochondria: regulation and effect. J Biol Chem 284:36592–36604

Dong S, Teng Z, Lu FH, Zhao YJ, Li H, Ren H, Chen H, Pan ZW, Lv YJ, Yang BF, Tian Y, Xu CQ, Zhang WH (2010) Post-conditioning protects cardiomyocytes from apoptosis via PKC(epsilon)-interacting with calcium-sensing receptors to inhibit endo(sarco)plasmic reticulum-mitochondria crosstalk. Mol Cell Biochem 341:195–206

Duvigneau JC, Piskernik C, Haindl S, Kloesch B, Hartl RT, Hüttemann M, Lee I, Ebel T, Moldzio R, Gemeiner M, Redl H, Kozlov AV (2008) A novel endotoxin-induced pathway: upregulation of heme oxygenase 1, accumulation of free iron, and free iron-mediated mitochondrial dysfunction. Lab Invest 88:70–77

Ekholm A, Katsura K, Kristian T, Liu M, Folbergrova J, Siesjo BK (1993) Coupling of cellular energy state and ion homeostasis during recovery following brain ischemia. Brain Res 604:185–191

Fabian RH, DeWitt DS, Kent TA (1995) In vivo detection of superoxide anion production by the brain using a cytochrome c electrode. J Cereb Blood Flow Metab 15:242–247

Fang JK, Prabu SK, Sepuri NB, Raza H, Anandatheerthavarada HK, Galati D, Spear J, Avadhani NG (2007) Site specific phosphorylation of cytochrome c oxidase subunits I, IVi1 and Vb in rabbit hearts subjected to ischemia/reperfusion. FEBS Lett 581:1302–1310

Fellman V, Raivio KO (1997) Reperfusion injury as the mechanism of brain damage after perinatal asphyxia. Pediatr Res 41:599–606

Ferguson-Miller S, Brautigan DL, Margoliash E (1976) Correlation of the kinetics of electron transfer activity of various eukaryotic cytochromes c with binding to mitochondrial cytochrome c oxidase. J Biol Chem 251:1104–1115

Fink MP (2002) Bench-to-bedside review: cytopathic hypoxia. Crit Care 6:491–499

Fiskum G, Rosenthal RE, Vereczki V, Martin E, Hoffman GE, Chinopoulos C, Kowaltowski A (2004) Protection against ischemic brain injury by inhibition of mitochondrial oxidative stress. J Bioenerg Biomembr 36:347–352

Giorgio M, Migliaccio E, Orsini F, Paolucci D, Moroni M, Contursi C, Pelliccia G, Luzi L, Minucci S, Marcaccio M, Pinton P, Rizzuto R, Bernardi P, Paolucci F, Pelicci PG (2005) Electron transfer between cytochrome c and p66Shc generates reactive oxygen species that trigger mitochondrial apoptosis. Cell 122:221–233

Goldberg E, Sberna D, Wheat TE, Urbanski GJ, Margoliash E (1977) Cytochrome c: immunofluorescent localization of the testis-specific form. Science 196:1010–1012

Grivennikov SI, Karin M (2010) Dangerous liaisons: STAT3 and NF-kappaB collaboration and crosstalk in cancer. Cytokine Growth Factor Rev 21:11–19

Grivennikov SI, Greten FR, Karin M (2010) Immunity, inflammation, and cancer. Cell 140:883–899

Guo D, Nguyen T, Ogbi M, Tawfik H, Ma G, Yu Q, Caldwell RW, Johnson JA (2007) Protein kinase C-epsilon coimmunoprecipitates with cytochrome oxidase subunit IV and is associated with improved cytochrome-c oxidase activity and cardioprotection. Am J Physiol Heart Circ Physiol 293:H2219–H2230

Hamanaka RB, Chandel NS (2010) Mitochondrial reactive oxygen species regulate cellular signaling and dictate biological outcomes. Trends Biochem Sci 35:505–513

Hayes MA, Timmins AC, Yau EH, Palazzo M, Hinds CJ, Watson D (1994) Elevation of systemic oxygen delivery in the treatment of critically ill patients. N Engl J Med 330:1717–1722

Helling S, Vogt S, Rhiel A, Ramzan R, Wen L, Marcus K, Kadenbach B (2008) Phosphorylation and kinetics of mammalian cytochrome c oxidase. Mol Cell Proteomics 7:1714–1724

Hinkle PC, Kumar MA, Resetar A, Harris DL (1991) Mechanistic stoichiometry of mitochondrial oxidative phosphorylation. Biochemistry 30:3576–3582

Hopper RK, Carroll S, Aponte AM, Johnson DT, French S, Shen RF, Witzmann FA, Harris RA, Balaban RS (2006) Mitochondrial matrix phosphoproteome: effect of extra mitochondrial calcium. Biochemistry 45:2524–2536

Hotchkiss RS, Karl IE (2003) The pathophysiology and treatment of sepsis. N Engl J Med 348:138–150

Hüttemann M, Jaradat S, Grossman LI (2003) Cytochrome *c* oxidase of mammals contains a testes-specific isoform of subunit VIb – the counterpart to testes-specific cytochrome *c*? Mol Reprod Dev 66:8–16

Hüttemann M, Lee I, Samavati L, Yu H, Doan JW (2007) Regulation of mitochondrial oxidative phosphorylation through cell signaling. Biochim Biophys Acta 1773:1701–1720

Hüttemann M, Lee I, Pecinova A, Pecina P, Przyklenk K, Doan JW (2008) Regulation of oxidative phosphorylation, the mitochondrial membrane potential, and their role in human disease. J Bioenerg Biomembr 40:445–456

Hüttemann M, Nantwi KD, Lee I, Liu J, Mohiuddin S, Petrov T (2010) Theophylline treatment improves mitochondrial function after upper cervical spinal cord hemisection. Exp Neurol 223:523–528

Hüttemann M, Helling S, Sanderson TH, Sinkler C, Samavati L, Mahapatra G, Varughese A, Lu G, Liu J, Ramzan R, Vogt S, Grossman LI, Doan JW, Marcus K, Lee I (2012) Regulation of mitochondrial respiration and apoptosis through cell signaling: cytochrome c oxidase and cytochrome c in ischemia/reperfusion injury and inflammation. Biochim Biophys Acta 1817:598–609

Hüttemann M, Pecina P, Rainbolt M, Sanderson TH, Kagan VE, Samavati L, Doan JW, Lee I (2011) The multiple functions of cytochrome c and their regulation in life and death decisions of the mammalian cell: from respiration to apoptosis. Mitochondrion 11:369–381

Kagan VE, Bayir HA, Belikova NA, Kapralov O, Tyurina YY, Tyurin VA, Jiang J, Stoyanovsky DA, Wipf P, Kochanek PM, Greenberger JS, Pitt B, Shvedova AA, Borisenko G (2009) Cytochrome c/cardiolipin relations in mitochondria: a kiss of death. Free Radic Biol Med 46:1439–1453

Kamp DW, Shacter E, Weitzman SA (2011) Chronic inflammation and cancer: the role of the mitochondria. Oncology 25:400–410, 413

Kim J, Takeuchi H, Lam ST, Turner RR, Wang HJ, Kuo C, Foshag L, Bilchik AJ, Hoon DS (2005) Chemokine receptor CXCR4 expression in colorectal cancer patients increases the risk for recurrence and for poor survival. J Clin Oncol 23:2744–2753

Kim S, Takahashi H, Lin WW, Descargues P, Grivennikov S, Kim Y, Luo JL, Karin M (2009) Carcinoma-produced factors activate myeloid cells through TLR2 to stimulate metastasis. Nature 457:102–106

Kirichenko A, Vygodina T, Mkrtchyan HM, Konstantinov A (1998) Specific cation binding site in mammalian cytochrome oxidase. FEBS Lett 423:329–333

Kirichenko AV, Pfitzner U, Ludwig B, Soares CM, Vygodina TV, Konstantinov AA (2005) Cytochrome c oxidase as a calcium binding protein. Studies on the role of a conserved aspartate in helices XI-XII cytoplasmic loop in cation binding. Biochemistry 44:12391–12401

Knab S, Mushak TM, Schmitz-Esser S, Horn M, Haferkamp I (2011) Nucleotide parasitism by Simkania negevensis (Chlamydiae). J Bacteriol 193:225–235

Ko YH, Pan W, Inoue C, Pedersen PL (2002) Signal transduction to mitochondrial ATP synthase: evidence that PDGF-dependent phosphorylation of the delta-subunit occurs in several cell lines, involves tyrosine, and is modulated by lysophosphatidic acid. Mitochondrion 1:339–348

Korshunov SS, Krasnikov BF, Pereverzev MO, Skulachev VP (1999) The antioxidant functions of cytochrome c. FEBS Lett 462:192–198

Kowalczyk JE, Kawalec M, Beresewicz M, Debski J, Dadlez M, Zablocka B (2012) Protein kinase C beta in postischemic brain mitochondria. Mitochondrion 12(1):138–143

Kulbe H, Hagemann T, Szlosarek PW, Balkwill FR, Wilson JL (2005) The inflammatory cytokine tumor necrosis factor-alpha regulates chemokine receptor expression on ovarian cancer cells. Cancer Res 65:10355–10362

Lee I, Kadenbach B (2001) Palmitate decreases proton pumping of liver-type cytochrome c oxidase. Eur J Biochem 268:6329–6334

Lee MD, Zentella A, Vine W, Pekala PH, Cerami A (1987) Effect of endotoxin-induced monokines on glucose metabolism in the muscle cell line L6. Proc Natl Acad Sci USA 84:2590–2594

Lee I, Bender E, Kadenbach B (2002) Control of mitochondrial membrane potential and ROS formation by reversible phosphorylation of cytochrome c oxidase. Mol Cell Biochem 234–235:63–70

Lee I, Salomon AR, Ficarro S, Mathes I, Lottspeich F, Grossman LI, Hüttemann M (2005) cAMP-dependent tyrosine phosphorylation of subunit I inhibits cytochrome *c* oxidase activity. J Biol Chem 280:6094–6100

Lee I, Salomon AR, Yu K, Doan JW, Grossman LI, Hüttemann M (2006) New prospects for an old enzyme: mammalian cytochrome *c* is tyrosine-phosphorylated in vivo. Biochemistry 45:9121–9128

Lee I, Pecinova A, Pecina P, Neel BG, Araki T, Kucherlapati R, Roberts AE, Hüttemann M (2010) A suggested role for mitochondria in Noonan syndrome. Biochim Biophys Acta 1802:275–283
Levy RJ, Vijayasarathy C, Raj NR, Avadhani NG, Deutschman CS (2004) Competitive and non-competitive inhibition of myocardial cytochrome c oxidase in sepsis. Shock 21:110–114
Liu SS (1999) Cooperation of a "reactive oxygen cycle" with the Q cycle and the proton cycle in the respiratory chain–superoxide generating and cycling mechanisms in mitochondria. J Bioenerg Biomembr 31:367–376
Liu RR, Murphy TH (2009) Reversible cyclosporin A-sensitive mitochondrial depolarization occurs within minutes of stroke onset in mouse somatosensory cortex in vivo: a two-photon imaging study. J Biol Chem 284:36109–36117
Liu Z, Lin H, Ye S, Liu QY, Meng Z, Zhang CM, Xia Y, Margoliash E, Rao Z, Liu XJ (2006) Remarkably high activities of testicular cytochrome c in destroying reactive oxygen species and in triggering apoptosis. Proc Natl Acad Sci USA 103:8965–8970
Lorente L, Martin MM, Lopez-Gallardo E, Iceta R, Sole-Violan J, Blanquer J, Labarta L, Diaz C, Jimenez A, Lafuente N, Hernandez M, Mendez F, Medina N, Ferrer-Aguero JM, Ferreres J, Llimiñana MC, Mora ML, Lubillo S, Sanchez-Palacios M, Montoya J, Ruiz-Pesini E (2011) Platelet cytochrome c oxidase activity and quantity in septic patients. Crit Care Med 39:1289–1294
Lu N, Wang W, Liu J, Wong CW (2011) Protein kinase C epsilon affects mitochondrial function through estrogen-related receptor alpha. Cell Signal 23(9):1473–1478
Mantovani A, Allavena P, Sica A, Balkwill F (2008) Cancer-related inflammation. Nature 454:436–444
Martinez-Ruiz A, Cadenas S, Lamas S (2011) Nitric oxide signaling: classical, less classical, and nonclassical mechanisms. Free Radic Biol Med 51:17–29
Miyazaki T, Neff L, Tanaka S, Horne WC, Baron R (2003) Regulation of cytochrome c oxidase activity by c-Src in osteoclasts. J Cell Biol 160:709–718
Neel BG, Gu H, Pao L (2003) The 'Shp'ing news: SH2 domain-containing tyrosine phosphatases in cell signaling. Trends Biochem Sci 28:284–293
Ogbi M, Johnson JA (2006) Protein kinase Cε interacts with cytochrome c oxidase subunit IV and enhances cytochrome c oxidase activity in neonatal cardiac myocyte preconditioning. Biochem J 393:191–199
Ogbi M, Chew CS, Pohl J, Stuchlik O, Ogbi S, Johnson JA (2004) Cytochrome c oxidase subunit IV as a marker of protein kinase Cε function in neonatal cardiac myocytes: implications for cytochrome c oxidase activity. Biochem J 382:923–932
Oliver CN, Starke-Reed PE, Stadtman ER, Liu GJ, Carney JM, Floyd RA (1990) Oxidative damage to brain proteins, loss of glutamine synthetase activity, and production of free radicals during ischemia/reperfusion-induced injury to gerbil brain. Proc Natl Acad Sci USA 87:5144–5147
Olsen JV, Vermeulen M, Santamaria A, Kumar C, Miller ML, Jensen LJ, Gnad F, Cox J, Jensen TS, Nigg EA, Brunak S, Mann M (2010) Quantitative phosphoproteomics reveals widespread full phosphorylation site occupancy during mitosis. Sci Signal 3:ra3
Pacelli C, Latorre D, Cocco T, Capuano F, Kukat C, Seibel P, Villani G (2011) Tight control of mitochondrial membrane potential by cytochrome c oxidase. Mitochondrion 11:334–341
Pang L, Qiu T, Cao X, Wan M (2011) Apoptotic role of TGF-beta mediated by Smad4 mitochondria translocation and cytochrome c oxidase subunit II interaction. Exp Cell Res 317:1608–1620
Pecina P, Borisenko GG, Belikova NA, Tyurina YY, Pecinova A, Lee I, Samhan-Arias AK, Przyklenk K, Kagan VE, Hüttemann M (2010) Phosphomimetic substitution of cytochrome c tyrosine 48 decreases respiration and binding to cardiolipin and abolishes ability to trigger downstream caspase activation. Biochemistry 49:6705–6714
Pedersen PL (1978) Tumor mitochondria and the bioenergetics of cancer cells. Prog Exp Tumor Res 22:190–274
Persichini T, Mazzone V, Polticelli F, Moreno S, Venturini G, Clementi E, Colasanti M (2005) Mitochondrial type I nitric oxide synthase physically interacts with cytochrome c oxidase. Neurosci Lett 384:254–259

Piantadosi CA, Zhang J (1996) Mitochondrial generation of reactive oxygen species after brain ischemia in the rat. Stroke 27:327–332

Picard M, Taivassalo T, Ritchie D, Wright KJ, Thomas MM, Romestaing C, Hepple RT (2011) Mitochondrial structure and function are disrupted by standard isolation methods. PLoS One 6:e18317

Piccoli C, Scrima R, Boffoli D, Capitanio N (2006) Control by cytochrome c oxidase of the cellular oxidative phosphorylation system depends on the mitochondrial energy state. Biochem J 396:573–583

Prabu SK, Anandatheerthavarada HK, Raza H, Srinivasan S, Spear JF, Avadhani NG (2006) Protein kinase A-mediated phosphorylation modulates cytochrome c oxidase function and augments hypoxia and myocardial ischemia-related injury. J Biol Chem 281:2061–2070

Rakoff-Nahoum S (2006) Why cancer and inflammation? Yale J Biol Med 79:123–130

Ralph SJ, Rodriguez-Enriquez S, Neuzil J, Saavedra E, Moreno-Sanchez R (2010) The causes of cancer revisited: "mitochondrial malignancy" and ROS-induced oncogenic transformation – why mitochondria are targets for cancer therapy. Mol Aspects Med 31:145–170

Revelly JP, Tappy L, Martinez A, Bollmann M, Cayeux MC, Berger MM, Chiolero RL (2005) Lactate and glucose metabolism in severe sepsis and cardiogenic shock. Crit Care Med 33:2235–2240

Robb-Gaspers LD, Burnett P, Rutter GA, Denton RM, Rizzuto R, Thomas AP (1998) Integrating cytosolic calcium signals into mitochondrial metabolic responses. EMBO J 17:4987–5000

Roger VL, Go AS, Lloyd-Jones DM, Adams RJ, Berry JD, Brown TM, Carnethon MR, Dai S, de Simone G, Ford ES, Fox CS, Fullerton HJ, Gillespie C, Greenlund KJ, Hailpern SM, Heit JA, Ho PM, Howard VJ, Kissela BM, Kittner SJ, Lackland DT, Lichtman JH, Lisabeth LD, Makuc DM, Marcus GM, Marelli A, Matchar DB, McDermott MM, Meigs JB, Moy CS, Mozaffarian D, Mussolino ME, Nichol G, Paynter NP, Rosamond WD, Sorlie PD, Stafford RS, Turan TN, Turner MB, Wong ND, Wylie-Rosett J (2011) Heart disease and stroke statistics – 2011 update: a report from the American Heart Association. Circulation 123:e18–e209

Rosenthal RE, Hamud F, Fiskum G, Varghese PJ, Sharpe S (1987) Cerebral ischemia and reperfusion: prevention of brain mitochondrial injury by lidoflazine. J Cereb Blood Flow Metab 7:752–758

Ruggieri AJ, Levy RJ, Deutschman CS (2010) Mitochondrial dysfunction and resuscitation in sepsis. Crit Care Clin 26:567–575, x–xi

Salvi M, Brunati AM, Bordin L, La Rocca N, Clari G, Toninello A (2002) Characterization and location of Src-dependent tyrosine phosphorylation in rat brain mitochondria. Biochim Biophys Acta 1589:181–195

Salvi M, Stringaro A, Brunati AM, Agostinelli E, Arancia G, Clari G, Toninello A (2004) Tyrosine phosphatase activity in mitochondria: presence of Shp-2 phosphatase in mitochondria. Cell Mol Life Sci 61:2393–2404

Samavati L, Lee I, Mathes I, Lottspeich F, Hüttemann M (2008) Tumor necrosis factor α inhibits oxidative phosphorylation through tyrosine phosphorylation at subunit I of cytochrome *c* oxidase. J Biol Chem 283:21134–21144

Sanishvili R, Volz KW, Westbrook EM, Margoliash E (1995) The low ionic strength crystal structure of horse cytochrome c at 2.1 Å resolution and comparison with its high ionic strength counterpart. Structure 3:707–716

Sivaraman V, Hausenloy DJ, Kolvekar S, Hayward M, Yap J, Lawrence D, Di Salvo C, Yellon DM (2009) The divergent roles of protein kinase C epsilon and delta in simulated ischaemia-reperfusion injury in human myocardium. J Mol Cell Cardiol 46:758–764

Steenaart NA, Shore GC (1997) Mitochondrial cytochrome c oxidase subunit IV is phosphorylated by an endogenous kinase. FEBS Lett 415:294–298

Tartaglia M, Mehler EL, Goldberg R, Zampino G, Brunner HG, Kremer H, van der Burgt I, Crosby AH, Ion A, Jeffery S, Kalidas K, Patton MA, Kucherlapati RS, Gelb BD (2001) Mutations in PTPN11, encoding the protein tyrosine phosphatase SHP-2, cause Noonan syndrome. Nat Genet 29:465–468

Tartaglia M, Gelb BD, Zenker M (2011) Noonan syndrome and clinically related disorders. Best practice & research. J Clin Endocrinol Metab 25:161–179

Thingholm TE, Jorgensen TJ, Jensen ON, Larsen MR (2006) Highly selective enrichment of phosphorylated peptides using titanium dioxide. Nat Protoc 1:1929–1935
Tilney NL, Guttmann RD (1997) Effects of initial ischemia/reperfusion injury on the transplanted kidney. Transplantation 64:945–947
Tracey KJ, Lowry SF, Fahey TJ, Albert JD, Fong Y, Hesse D, Beutler B, Manogue KR, Calvano S, Wei H, Cerami A, Shires GT (1987) Cachectin/tumor necrosis factor induces lethal shock and stress hormone responses in the dog. Surg Gynecol Obstet 164:415–422
Tsukihara T, Aoyama H, Yamashita E, Tomizaki T, Yamaguchi H, Shinzawa-Itoh K, Nakashima R, Yaono R, Yoshikawa S (1996) The whole structure of the 13-subunit oxidized cytochrome c oxidase at 2.8 Å. Science 272:1136–1144
Tsukihara T, Shimokata K, Katayama Y, Shimada H, Muramoto K, Aoyama H, Mochizuki M, Shinzawa-Itoh K, Yamashita E, Yao M, Ishimura Y, Yoshikawa S (2003) The low-spin heme of cytochrome c oxidase as the driving element of the proton-pumping process. Proc Natl Acad Sci USA 100:15304–15309
Villani G, Attardi G (1997) In vivo control of respiration by cytochrome c oxidase in wild-type and mitochondrial DNA mutation-carrying human cells. Proc Natl Acad Sci USA 94:1166–1171
Villani G, Greco M, Papa S, Attardi G (1998) Low reserve of cytochrome c oxidase capacity in vivo in the respiratory chain of a variety of human cell types. J Biol Chem 273:31829–31836
Vlessis AA, Widener LL, Bartos D (1990) Effect of peroxide, sodium, and calcium on brain mitochondrial respiration in vitro: potential role in cerebral ischemia and reperfusion. J Neurochem 54:1412–1418
von Ballmoos C, Gennis RB, Adelroth P, Brzezinski P (2011) Kinetic design of the respiratory oxidases. Proc Natl Acad Sci USA 108:11057–11062
von der Hocht I, van Wonderen JH, Hilbers F, Angerer H, MacMillan F, Michel H (2011) Interconversions of P and F intermediates of cytochrome c oxidase from Paracoccus denitrificans. Proc Natl Acad Sci USA 108:3964–3969
Walther TC, Mann M (2010) Mass spectrometry-based proteomics in cell biology. J Cell Biol 190:491–500
Wang ZB, Li M, Zhao Y, Xu JX (2003) Cytochrome c is a hydrogen peroxide scavenger in mitochondria. Protein Pept Lett 10:247–253
Warburg O (1956) On the origin of cancer cells. Science 123:309–314
Warburg O, Posener K, Negelein E (1924) Über den Stoffwechsel der Carcinomzelle. Biochem Z 152:309–344
Weinberg F, Chandel NS (2009) Mitochondrial metabolism and cancer. Ann N Y Acad Sci 1177:66–73
Weinberg F, Hamanaka R, Wheaton WW, Weinberg S, Joseph J, Lopez M, Kalyanaraman B, Mutlu GM, Budinger GR, Chandel NS (2010) Mitochondrial metabolism and ROS generation are essential for Kras-mediated tumorigenicity. Proc Natl Acad Sci USA 107:8788–8793
Welch EJ, Jones BW, Scott JD (2010) Networking with AKAPs: context-dependent regulation of anchored enzymes. Mol Interv 10:86–97
Yang WL, Iacono L, Tang WM, Chin KV (1998) Novel function of the regulatory subunit of protein kinase A: regulation of cytochrome c oxidase activity and cytochrome c release. Biochemistry 37:14175–14180
Yoshikawa S, Muramoto K, Shinzawa-Itoh K (2011) The O_2 reduction and proton pumping gate mechanism of bovine heart cytochrome c oxidase. Biochim Biophys Acta 1807:1279–1286
Yu H, Lee I, Salomon AR, Yu K, Hüttemann M (2008a) Mammalian liver cytochrome *c* is tyrosine-48 phosphorylated in vivo, inhibiting mitochondrial respiration. Biochim Biophys Acta 1777:1066–1071
Yu Q, Nguyen T, Ogbi M, Caldwell RW, Johnson JA (2008b) Differential loss of cytochrome-c oxidase subunits in ischemia-reperfusion injury: exacerbation of COI subunit loss by PKC-epsilon inhibition. Am J Physiol 294:H2637–H2645
Zaidan E, Sims NR (1994) The calcium content of mitochondria from brain subregions following short-term forebrain ischemia and recirculation in the rat. J Neurochem 63:1812–1819

Zhang Z, Gerstein M (2003) The human genome has 49 cytochrome c pseudogenes, including a relic of a primordial gene that still functions in mouse. Gene 312:61–72

Zhang S, Li H, Yang SJ (2010) Tribulosin protects rat hearts from ischemia/reperfusion injury. Acta Pharmacol Sin 31:671–678

Zhao X, Leon IR, Bak S, Mogensen M, Wrzesinski K, Hojlund K, Jensen ON (2010) Phosphoproteome analysis of functional mitochondria isolated from resting human muscle reveals extensive phosphorylation of inner membrane protein complexes and enzymes. Mol Cell Proteomics 10(1):M110.000299

Zweier JL (1988) Measurement of superoxide-derived free radicals in the reperfused heart. Evidence for a free radical mechanism of reperfusion injury. J Biol Chem 263:1353–1357

Chapter 11
Individual Biochemical Behaviour Versus Biological Robustness: Spotlight on the Regulation of Cytochrome *c* Oxidase

Rabia Ramzan, Petra Weber, Bernhard Kadenbach, and Sebastian Vogt

Abstract During evolution from prokaryotes to eukaryotes, the main function of cytochrome *c* oxidase (COX), i.e., the coupling of oxygen reduction to proton translocation without the production of ROS (reactive oxygen species) remained unchanged demonstrating its robustness. A new regulation of respiration by the ATP/ADP ratio was introduced in eukaryotes based on nucleotide interaction with the added COX subunit IV. This *allosteric ATP-inhibition* was proposed to keep the mitochondrial membrane potential ($\Delta\Psi_m$) at low *healthy* values and thus prevents the formation of ROS at complexes I and III. ROS have been implicated in various degenerative diseases. The *allosteric ATP-inhibition* of COX is reversibly switched on and off by phosphorylation of COX at a serine or threonine. In more than 100 individual preparations of rat heart and liver mitochondria, prepared under identical conditions, the extent of *allosteric ATP-inhibition* varied. This variability correlates with the variable inhibition of uncoupled respiration in intact isolated mitochondria by ATP. It is concluded that in higher organisms the *allosteric ATP-inhibition* is continually switched on and off by neuronal signalling in order to change oxidative phosphorylation from optimal efficiency with lower rate of ATP synthesis under resting conditions (low $\Delta\Psi_m$ and ROS production) to maximal rate of ATP synthesis under active (working, stress) conditions (elevated $\Delta\Psi_m$ and ROS production).

R. Ramzan • P. Weber • S. Vogt
Biomedical Research Center, Cardiovascular Laboratory, Philipps-University, Marburg 35032, Germany
e-mail: rabiaramzan23@yahoo.com; vogts@med.uni-marburg.de

B. Kadenbach (✉)
Fachbereich Chemie, Philipps-University Marburg, Marburg 35032, Germany
e-mail: kadenbach@staff.uni-marburg.de

B. Kadenbach (ed.), *Mitochondrial Oxidative Phosphorylation*, Advances in Experimental Medicine and Biology 748, DOI 10.1007/978-1-4614-3573-0_11,

11.1 Introduction

Biological systems are characterized by robustness, which is the ability to maintain performance in the face of perturbations (Stelling et al. 2004; Kitano 2004). The robustness holds for whole organisms as well as for metabolic pathways and enzyme complexes. Biological organisms, in addition, are characterized by homeostasis of their metabolites, e.g., at variable work loads (Balaban et al. 1986). To assure the homeostasis, a number of feedback mechanisms are involved which maintain the steady state (Freeman 2000). Cytochrome *c* oxidase (COX) is suggested to represent the rate-limiting enzyme for respiration of aerobic organisms in vivo (Villani and Attardi 1997, 2001; Piccoli et al. 2006; Dalmonte et al. 2009; Pacelli et al. 2011) but not in isolated mitochondria (Groen et al. 1982). It is characterized in particular by evolutionary robustness. From the bacterial enzyme with three subunits up to the mammalian enzyme with 13 subunits, the basic catalytic activities remained unchanged. Both enzymes use molecular oxygen to oxidize reducing equivalents from food via cytochrome *c* forming water accompanied by translocation of protons and electrons across the membrane without formation of reactive oxygen species (ROS) (Ludwig et al. 2001; Yu et al. 2011). In fact, by comparing COX from *Paracoccus denitrificans* and bovine heart, no difference in the catalytic properties including the development of $\Delta\Psi$ and ΔpH and the pumping of protons was measured (Pardhasaradhi et al. 1991; Hendler et al. 1991).

In animals, respiration and the synthesis of energy (ATP and heat) are strongly regulated by the variable demands of tissues/organs as well as by endogenous and environmental signals. From the 13 subunits of mammalian COX only subunits I–III (encoded in mitochondrial DNA and homologous to the three bacterial COX subunits) are essential for the catalytic activity. The ten additional nuclear-encoded subunits occur in tissue-specific (Anthony et al. 1990; Hüttemann et al. 2001; 2003), developmental-specific (Bonne et al. 1993), and species-specific isoforms (Linder et al. 1995), suggesting their regulatory function. Isoforms have not been found in other complexes of oxidative phosphorylation (OxPhos), although they also contain multiple subunits with mostly unknown functions: 45 subunits in complex I (NADH dehydrogenase), 4 subunits in complex II (succinate dehydrogenase), 11 subunits in complex III (cytochrome bc_1), and 16 subunits in complex V (ATP synthase) (McKenzie et al. 2009).

In addition to the expression of isoforms, COX activity is also regulated by allosteric effectors like the ATP/ADP ratio (Frank and Kadenbach 1996; Arnold and Kadenbach 1999; Beauvoit and Rigoulet 2001), 3,5-diiodothyronine (Arnold et al. 1998), and palmitate (Lee and Kadenbach 2001). This review describes, in particular, the *allosteric ATP-inhibition* of COX, which is suggested to keep the mitochondrial membrane potential ($\Delta\Psi_m$) at low healthy values, and thus prevents the formation of ROS at complexes I and III (Dröse and Brandt 2008; Kussmaul and Hirst 2006; see also Chap. 6) at high $\Delta\Psi_m$ (Liu 1997; Korshunov et al. 1997; Murphy 2009; Rottenberg et al. 2009). ROS have been implicated in various degenerative diseases (Dalle-Donne et al. 2006; Valko et al. 2007; Trachootham et al. 2008).

11.2 The ATP/ADP Ratio, an Allosteric Effector of COX

The inhibition of COX activity by ATP, first described by Ferguson-Miller et al. (1976), was confirmed by Lin et al. (1995) with purified COX photoaffinity labelled with 8-azido-ATP. The interaction of COX with adenine nucleotides is based on tight noncatalytic binding sites for ADP/ATP with dissociation constants K_d in the micromolar range. Noncatalytic binding sites for ADP/ATP have also been identified in many other enzymes/proteins (Yegutkin and Burnstock 1999; Rajagopalan et al. 1999; Bjornson and Modrich 2003; Warnock and Raines 2004; Lamb et al. 2006; Inoue and Shingyoji 2007; Malyan 2010a, b).

In isolated bovine heart COX, 10 binding sites for ADP were determined by equilibrium dialysis from which seven are exchanged by ATP at high ATP/ADP ratios (Rieger et al. 1995; Napiwotzki et al. 1997; Napiwotzki and Kadenbach 1998). This result demonstrates that not ATP but the ATP/ADP ratio represents the allosteric effector of COX and probably of many other enzymes. In living cells, where the concentration of ATP+ADP is in the millimolar range (Schwenke et al. 1981), noncatalytic binding sites for ADP are always occupied and are exchanged by ATP only at high ATP/ADP ratios. In fact, it has been calculated that only 5% of total cellular ADP represents *free ADP* (Veech et al. 1979). Therefore, the chemically determined ATP/ADP ratios of 6–7 in the cytosol and 1 in the mitochondrial matrix (Schwenke et al. 1981) do not represent the free ATP/ADP ratios. From ^{31}P-NMR data of the perfused rat heart, cytosolic ATP/ADP ratios are in the range between 100 and 400, but also values up to 1,000 have been measured by From et al. (1990; see also Wiseman and Kushmerick 1997; and for a review Balaban 1990).

Binding of ADP to purified COX from bovine heart, which was isolated in the presence of cholate, is a slow process and is accompanied by a spectral change. The spectral change is specific for adenine nucleotides, since GDP, CDP, and UDP had no effect and IDP caused only a small spectral change (Napiwotzki et al. 1997). The ten binding sites for ADP in bovine heart COX have been corroborated in the crystal structure of the enzyme by 10 mol of tightly bound cholate molecules per mole of COX monomer, measured with [carboxyl-^{14}C]cholate in decylmaltoside-washed COX crystals (Napiwotzki et al. 1997). Cholate is structurally very similar to the ADP molecule, but different from ATP. Up to now, four cholate molecules have been visualized in the crystal structure of bovine heart COX at 1.8 Å resolution. The other six molecules of cholate may be seen at a resolution of 1.4 Å (in preparation), since many unidentified electron densities at the electron density maps have been found at higher resolution (S. Yoshikawa, personal communication).

Specific binding sites for ATP/ADP have been identified in COX from bovine heart in subunits IV and VIa-H (heart isoform). On the cytosolic side of the transmembrane subunit IV high ATP/ADP ratios were shown to increase the K_M for cytochrome *c* (Napiwotzki and Kadenbach 1998). On the matrix side of subunit IV high ATP/ADP ratios induce an *allosteric ATP-inhibition*, indicated by a sigmoidal inhibition curve. Half-maximal induction is found at ATP/ADP=28 (Arnold and Kadenbach 1999). On the matrix side of subunit VIa-H (Anthony et al. 1993) the exchange of bound ADP by ATP at high ATP/ADP ratios is accompanied by a decrease of the proton pumping stoichiometry (H^+/e^-) from 1.0 to 0.5 (Frank and Kadenbach 1996). Half-maximal decrease of H^+/e^- occurs at ATP/ADP = 100. A similar decrease of H^+/e^- stoichiometry is obtained in COX from bovine liver with low concentrations of free palmitate (Lee and Kadenbach 2001).

11.3 The *Allosteric ATP-Inhibition* of COX Activity

The *allosteric ATP-inhibition* of COX is based on the exchange of bound ADP by ATP at the matrix domain of the nuclear-encoded COX subunit IV at high ATP/ADP ratios, as demonstrated with monoclonal antibodies against subunit IV (Arnold and Kadenbach 1997). At high ATP/ADP ratios and low ferrocytochrome *c* concentrations oxygen consumption is completely inhibited. It is found with COX from eukaryotic cells including yeast, but is absent in prokaryotic COX lacking subunit IV (Follmann et al. 1998).

11.3.1 Kinetic Measurement of the Allosteric ATP-Inhibition of COX

The *allosteric ATP-inhibition* of COX can only be visualized by kinetic analysis. In mitochondria ascorbate respiration is measured polarographically in the presence of a weak nonionic detergent like Tween-20 with increasing concentrations of cytochrome *c*. No detergent is required with COX reconstituted in asolectin-liposomes or in case of mitoplasts. The measurement has to be done in the presence of both ADP and ATP with an ATP-regenerating system (phosphoenolpyruvate + pyruvate kinase) (Arnold and Kadenbach 1997). The ATP regenerating system is required to maintain the high ATP/ADP ratio. The oxygen consumption with ascorbate is plotted against increasing concentrations of cytochrome *c* (see Fig. 11.1a, right side).

Many assay protocols for the measurement of COX activity use the strong nonionic detergent dodecylmaltoside (Miró et al. 1998; Baden et al. 2007; Murray et al. 2007), which results in partial loss of subunits (Thompson and Ferguson-Miller 1983; Weishaupt and Kadenbach 1992) and disturbs the nucleotide interaction (Arnold 1997). In addition, dodecylmaltoside was shown to dissociate the dimeric enzyme into monomers (Suarez et al. 1984; Bolli et al. 1985; Hakvoort et al. 1987). The *allosteric ATP-inhibition* of COX, however, is suggested to involve the cooperativity of the two cytochrome *c* binding sites in the dimeric structure (Arnold and Kadenbach 1997).

Other assay protocols include tetramethyl-*p*-phenylenediamine (TMPD) (Capaldi et al. 1995; Bylund-Fellenius et al. 1982; Kuznetsov and Gnaiger 2010) resulting in higher rates but abolition of the regulatory effects of ADP and ATP from the matrix side on reconstituted COX from bovine heart (Rohdich and Kadenbach 1993).

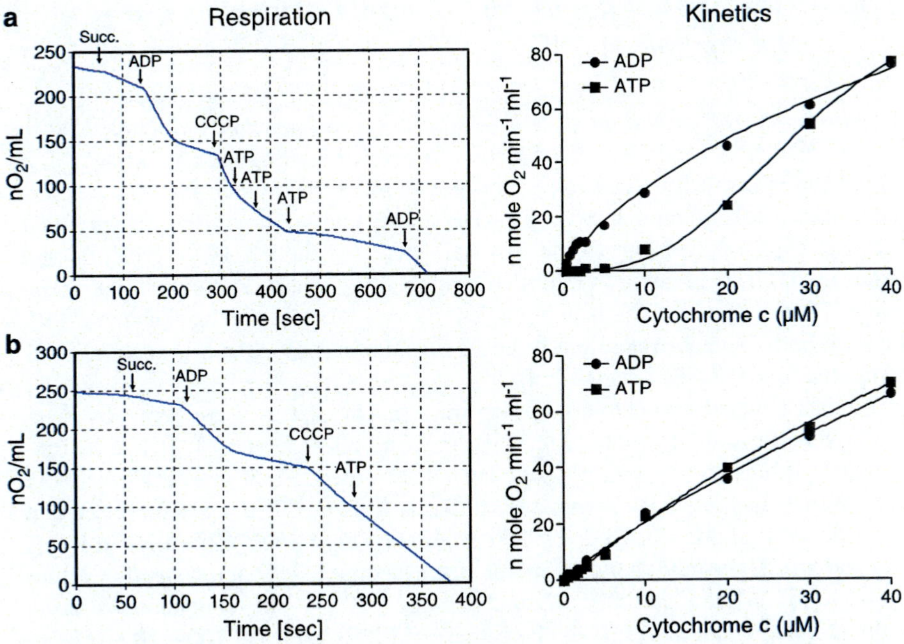

Fig. 11.1 Respiration and kinetics of two preparations of rat liver mitochondria (**a** and **b**). Mitochondria were isolated and respiration and kinetic measurements were done as previously described (Ramzan et al. 2010). *Respiration (left side)*: the $\Delta\Psi_m$-independent control of respiration by the ATP/ADP ratio varies in the two mitochondrial preparations. The decrease of oxygen concentration (nmol O_2/ml) was recorded polarographically. Preparation (**a**): To the mitochondria (0.8 mg protein/ml) in respiratory medium (250 mM sucrose, 10 mM Hepes (pH 7.2), 5 mM KPi, 5 mM $MgSO_4$, 0.2 mM EDTA, 0.5% bovine serum albumin) was added sequentially: 5 mM succinate (Succ), 0.2 mM ADP, 3 μM CCCP (carbonyl cyanide *m*-chlorophenylhydrazone), 1 mM ATP, 1 mM ATP, 8 mM ATP, and 1 mM ADP. RCR = 5.9. Preparation (**b**): To the mitochondria (0.7 mg protein/ml) in kinetics medium was added sequentially: 5 mM succinate, 0.2 mM ADP, 3 μM CCCP, and 10 mM ATP. RCR = 4.2. *Kinetics (right side)*: The polarographic measuring medium at room temperature (250 mM sucrose, 20 mM Hepes (pH 7.4), 1 mM EDTA, 2 mM EGTA, 25 mM NaF, 10 nM okadaic acid, 5 mM $MgSO_4$, and 1% Tween-20) contained in addition 16 mM ascorbate and either 5 mM ADP or 5 mM ATP and the ATP-regenerating system (10 mM phosphoenolpyruvate, 160 U/ml pyruvate kinase) as indicated. To the mitochondrial preparations (**a**) and (**b**) (0.6 mg protein/ml) in measuring medium cytochrome *c* was added subsequently from 0.2 to 60 μM and the rates of oxygen consumption (nmol O_2 min^{-1} ml^{-1}) were calculated at each concentration of cytochrome *c* and presented graphically using the Microsoft Excel program

In the presence of TMPD, the modulating effects of nucleotides on COX activity are not measured (Hüther and Kadenbach 1987), because the rate-limiting step of COX activity, the dissociation of cytochrome *c* from the enzyme, is bypassed by TMPD (Ferguson-Miller et al. 1978). Also DAB (3,3′ diaminobenzidine tetrachloride), which is frequently used for histochemical assays (Chrzanowska-Lightowlers et al. 1993), excludes the measurement of the *allosteric ATP-inhibition* of COX.

11.3.2 Two Types of Respiratory Control in Mitochondria and the Role of $\Delta\Psi_m$

The *allosteric ATP-inhibition* of COX—or the *second mechanism of respiratory control* (Kadenbach and Arnold 1999)—differs from the generally known *first mechanism of respiratory control*, which was defined as stimulation of the respiration of intact isolated mitochondria by ADP (Lardy and Wellman 1952; Chance and Williams 1955). According to the Mitchell theory (Mitchell 1966; Nicholls and Ferguson 2002), respiration is inhibited at high $\Delta\Psi_m$ values. In contrast, the *allosteric ATP-inhibition* of COX is independent of $\Delta\Psi_m$ (Arnold and Kadenbach 1999). But with both mechanisms in place the ATP-inhibited mitochondrial respiration is stimulated by ADP (see Sect. 11.2.4).

In intact isolated mitochondria $\Delta\Psi_m$ values of 180–220 mV are generally measured (Nicholls and Ferguson 2002; O'Brien et al. 2008; Lim et al. 2010). In contrast, $\Delta\Psi_m$ values of 100–140 mV have been determined in living cells (Zhang et al. 2001; Wan et al. 1993; for a review see Hüttemann et al. 2008). It was proposed that the feedback inhibition of COX by ATP maintains high ATP/ADP ratios in living cells and arrests the mitochondrial membrane potential ($\Delta\Psi_m$) at low *healthy* values (100–140 mV) (Lee et al. 2001; Kadenbach et al. 2010). This conclusion is based on the $\Delta\Psi_m$ dependence of F_0F_1-ATP synthase. At $\Delta\Psi_m$ values of 100–120 mV F_0F_1-ATP synthase is saturated and the rate of ATP synthesis is maximal (Kaim and Dimroth 1999). At maximal rates of ATP synthesis the ATP/ADP ratio will be high and the inhibition of COX is maximal, thus preventing further increase of $\Delta\Psi_m$. In fact, the high $\Delta\Psi_m$ values of isolated rat liver mitochondria (230 mV) could be decreased to low healthy values (120 mV) by increasing the ATP/ADP ratio with phosphoenolpyruvate and pyruvate kinase (Ramzan et al. 2010). By keeping $\Delta\Psi_m$ at low values the *allosteric ATP-inhibition* prevents the formation of ROS in mitochondria since its production increases exponentially at $\Delta\Psi_m$ values above 140 mV (Liu 1997; Korshunov et al. 1997; Murphy 2009; Rottenberg et al. 2009).

In many studies with intact cells an increase of $\Delta\Psi_m$ (*hyperpolarization*) by multiple stress factors was measured. Publications including more than 20 stress factors have been listed in Kadenbach et al. (2010). In some cases, the *hyperpolarization* was associated with increased ROS formation and subsequent apoptosis. The *hyperpolarization* of $\Delta\Psi_m$ by stress factors was suggested to be based on the loss of *allosteric ATP-inhibition* by dephosphorylation of COX (Lee et al. 2001; Kadenbach et al. 2004, 2010).

11.3.3 Switching On and Off the Allosteric ATP-Inhibition by Reversible Phosphorylation

The *allosteric ATP-inhibition* of COX is only effective with the phosphorylated enzyme and is switched off by dephosphorylation (Bender and Kadenbach 2000; Lee et al. 2001, 2002). In the isolated bovine heart enzyme (dissolved in 1% Tween-20) the *allosteric ATP-inhibition* can be switched on in vitro by incubation with protein kinase A (PKA)+cAMP+ATP, and switched off by subsequent incubation with protein phosphatase 1 (PP1) (Bender and Kadenbach 2000; Lee et al. 2001). The responsible phosphorylation site (serine or threonine) is located on the cytosolic side of COX, because in the reconstituted enzyme (in asolectin liposomes) the *allosteric ATP-inhibition* is only switched on by phosphorylation from the outer side (Lee et al. 2001, 2002). The responsible protein kinase is located at the intermembrane space, because incubation of bovine liver mitochondria only with cAMP + ATP switched it on (cAMP cannot penetrate the inner mitochondrial membrane), whereas subsequent incubation with calcium switched it off (Lee et al. 2002). Phosphorylation of isolated COX with [γ-^{32}P]ATP labelled mainly subunit I (Bender and Kadenbach 2000), and a Western blot of COX, incubated with PKA+cAMP+ATP, indicated mainly phosphorylation of serine at subunit I (Helling et al. 2008). In the crystal structure of bovine heart COX (Tsukihara et al. 1996), only one consensus sequence for PKA-dependent phosphorylations (Pearson and Kemp 1991) occurs on the cytosolic side of subunit I. Therefore, the conserved Ser-441 at subunit I (Lee et al. 2002) was suggested to represent the phosphorylation site responsible for the *allosteric ATP-inhibition* (Lee et al. 2001). A fragment of bovine heart COX subunit I, containing Ser-441 (PRRYSDYPDAYTM) was isolated and identified by mass spectrometry but the peptide was not phosphorylated, independent of the presence or absence of *allosteric ATP-inhibition* in the mitochondria or isolated COX from which it was prepared (Helling et al. 2012). It cannot be excluded, however, that it became dephosphorylated during preparation of COX from mitochondria by BN-PAGE or of the peptide for mass spectrometric identification.

11.3.4 Measurement of the Allosteric ATP-Inhibition in Intact Isolated Mitochondria

The second mechanism of respiratory control (*allosteric ATP-inhibition* of COX) can be distinguished from the generally known respiratory control (see Sect. 11.3.2) by the ATP-inhibition of uncoupled respiration, since in contrast to the first mechanism it is independent of $\Delta\Psi_m$ (Arnold and Kadenbach 1999). As previously shown for heart mitochondria (Ramzan et al. 2010), also respiration of liver mitochondria is inhibited by ATP in the presence of an uncoupler. In Fig. 11.1a (left side), succinate respiration of intact isolated rat liver mitochondria is stimulated by ADP (respiratory control ratio = 5.9) and the following decreased state 4 respiration was again stimulated by the uncoupler CCCP to the rate of state 3 respiration. Addition of small amounts of ATP inhibited the respiration rate successively and resulted in almost full inhibition at 10 mM ATP. Further addition of ADP released the ATP-inhibition as expected from the dependence of the allosteric ATP-inhibition on the ATP/ADP ratio (>20/1) (Arnold and Kadenbach 1999).

However, it was found that different preparations of intact rat liver mitochondria exhibited different extents of ATP-inhibition of uncoupled respiration. In Fig. 11.1b (left side), another preparation of rat liver mitochondria with a respiratory control ratio of 4.2 did not show an ATP-inhibition of the uncoupled respiration. This variation of ATP-inhibition parallels the variation of the kinetically measured allosteric ATP-inhibition (Fig. 11.1, right side). In more than 50 rat liver and rat heart mitochondrial preparations, a large individual variation of the extent of ATP-inhibition of uncoupled respiration was measured using either succinate, ketoglutarate, or glutamate + malate as substrates. We conclude that the inhibition of uncoupled mitochondrial respiration by ATP is based on the *allosteric ATP-inhibition* of COX, since both vary to the same extent, depending on the individual preparation of mitochondria.

11.3.5 Variation of the Allosteric ATP-Inhibition of COX in Individual Preparations of Isolated Mitochondria

The kinetics of ATP-inhibition of COX was measured in more than hundred preparations of rat heart mitochondria. The extent of *allosteric ATP-inhibition*, however, varied strongly as did the kinetic properties in each individual mitochondrial preparation, although the conditions of sacrificing the rat (guillotine) and isolation of mitochondria were the same. Four types of kinetic properties can be distinguished as shown in Fig. 11.2. In Fig. 11.2a, the typical *allosteric ATP-inhibition* is shown. In Fig. 11.2b, very little inhibition by ATP is seen as compared to the activity in the presence of ADP so that the kinetics in the presence of ADP and ATP are almost the same. In Fig. 11.2c, sigmoidal inhibition curves are observed at increasing cytochrome *c* concentrations in the presence of both ATP and ADP. The degree of ATP-inhibition as compared to the curve in the presence of ADP, however, was found to vary. In Fig. 11.2d, no sigmoidal curve is found in the presence of ATP so that thew-percentage inhibition compared to the activity in the presence of ADP was almost the same at all cytochrome *c* concentrations.

In order to show the large variability of the *allosteric ATP-inhibition*, we have calculated the percentage inhibition of activity by ATP as related to the activity in the presence of ADP = 100%. In Fig. 11.3, the ATP-inhibition of 42 individual preparations of rat heart mitochondria at four concentrations of cytochrome *c* is shown. Clearly, a large variability of COX inhibition by ATP is observed ranging from 0% to almost 100% (complete inhibition of activity).

Large variations in the extent of *allosteric ATP-inhibition* have also been described for bovine COX in freshly prepared mitochondria from heart, liver, and kidney of three different cows. Mitochondria of the three cows exhibited either a clear, a small, or no *allosteric ATP-inhibition* of COX in the three tissues from the same cow (Ramzan et al. 2010), suggesting a systemic switching on and off of the *allosteric ATP-inhibition* in the animal.

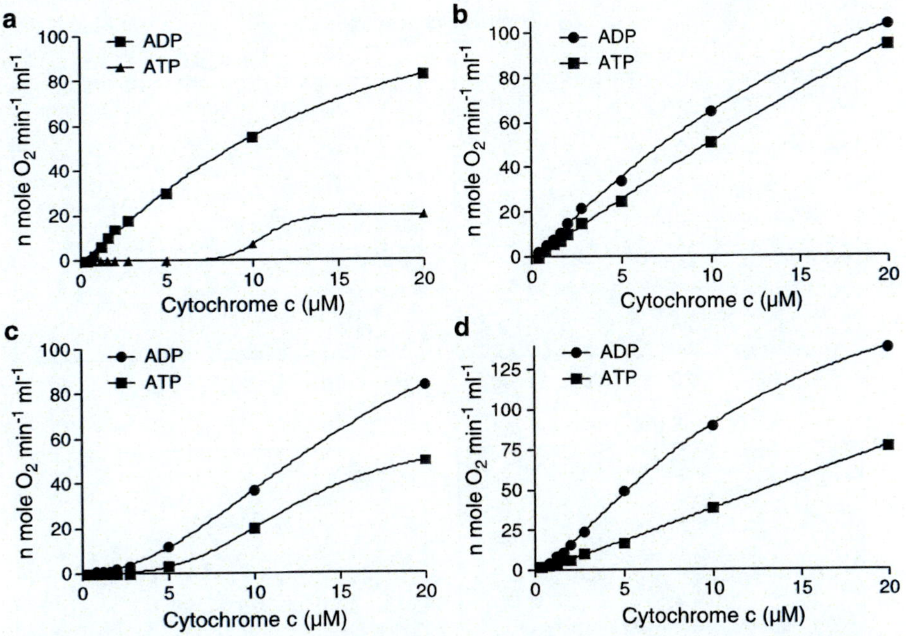

Fig. 11.2 From polarographic measurements of COX activity of more than hundred individual preparations of rat heart mitochondria (Ramzan et al. 2010) four types could be distinguished kinetically (**a–d**). All measurements were performed at room temperature in the polarographic measuring medium as described in the legend to Fig. 11.1. The mitochondrial protein content in the kinetics medium was between 0.3 and 0.5 mg/ml

11.4 Reversible Phosphorylations of COX and Their Functions

In a previous study, no protein phosphorylation was detected in preparations of isolated bovine heart COX by matrix-assisted laser desorption/ionization mass spectrometry (Marx et al. 1998). In other COX preparations, phosphorylated subunits were identified by Western blots and specific phosphorylated amino acids were identified by mass spectrometry (Helling et al. 2008). Up to now, 18 phosphorylation sites have been identified in mammalian COX as presented in Table 11.1. But more phosphorylation sites are expected to occur based on consensus sequences for protein kinase A (Pearson and Kemp 1991). In fact, 53 potential phosphorylation sites occur only for serine or threonine in the bovine heart enzyme (11 in subunit I, 10 in II, 3 in III, 6 in IV, 3 in Va, 3 in Vb, 4 in VIa, 3 in VIb, 2 in VIc, 2 in VIIa, 1 in VIIb, 1 in VIIc, and 4 in VIII). We suggest that the variable and complex COX kinetics, measured in isolated mitochondria (see Fig. 11.2), are based on multiple and reversible phosphorylations of the enzyme.

The kinetic behaviour of COX in Fig. 11.2c characterized by sigmoidal curves in the presence of ATP and of ADP, indicating COX inhibition independent of the ATP/ADP ratio, has been related to the phosphorylation of Tyr-304 of subunit I

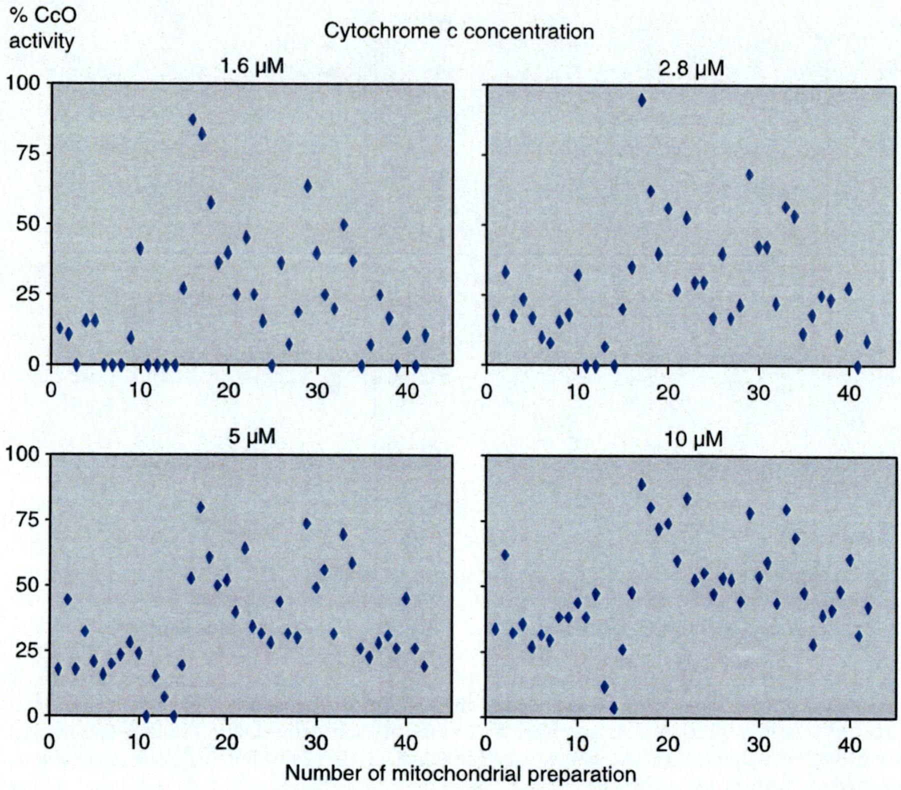

Fig. 11.3 Percentage inhibition of COX activity by ATP related to the activity in the presence of ADP = 100% of heart mitochondrial preparations from 42 different rats. The kinetics of isolated mitochondria were measured polarographically in the presence of ADP and of ATP+PEP+PK. The percentage inhibition by ATP related to ADP = 100% was calculated at 4 different cytochrome *c* concentrations (1.6, 2.8, 5, and 10 µM)

(Lee et al. 2005). This phosphorylation is turned on by glucagon via cAMP and leads to the inhibition of cell respiration in liver tissue. It was also shown to be triggered by inflammatory signalling resulting in sepsis and in many cases leading to human death (Samavati et al. 2008; see also Chap. 10). This phosphorylation, however, does not switch on the *allosteric ATP-inhibition* (Fig. 11.2a).

A new phosphorylation site was proposed to occur in mouse COX subunit IV at Ser-58 by Acin-Perez et al. (2011), using protein mutagenesis, molecular dynamics simulations, and induced fit docking. Since this site is close to the postulated binding site for ATP at the matrix side of subunit IV (Hüttemann et al. 2001), its phosphorylation was concluded to turn on the *allosteric ATP-inhibition* of COX. The presented kinetics, however, exclude this possibility because no sigmoidal kinetics were obtained in the presence of ATP. Instead, phosphorylation of Ser-58 of subunit IV induces the kinetic type of ATP-inhibition shown in Fig. 11.2d. Therefore, the phosphorylation site responsible for the *allosteric ATP-inhibition* remains to be identified.

Table 11.1 Identified phospho-epitopes in mammalian cytochrome *c* oxidase

Subunit	Site	Peptide	Tissue	Conditions	Reference
I	Ser-115	SLHLAGVSSILGAINF	Rabbit heart	After ischemia	Fang et al. (2007)
I	Ser-116	SLHLAGVSSILGAINF	Rabbit heart	After ischemia	Fang et al. (2007)
I	Tyr-304	MDVDTRAYFTSATMI	Bovine liver	Glucagon treated	Lee et al. (2005)
II	Ser-126	IPTSELKPGELR	Bovine heart	Isolated enzyme	Hüttemann et al. (2012)
II	Tyr-218	PIVLELVPLKYFEK	Bovine heart	Isolated enzyme	Helling et al. (2012)
IV-1	Tyr-11	SVVKSEDYALPSYVD	Bovine liver	Isolated enzyme	Lee et al. (2009)
IV-1	Ser-34	NLSASQKALKEKEK	Bovine heart	Isolated enzyme	Helling et al. (2008)
IV-1	Thr-52	KAPWGSLTRDEKVEL	Rabbit heart	After ischemia	Fang et al. (2007)
IV-1	Ser-67	YRIKFKESFAEMNRG	HeLa cells	Cell culture	Olsen et al. (2010)
IV-1	Ser-136	NPIQGLASKWDYEKN	HeLa cells	Cell culture	Olsen et al. (2010)
Va	Ser-1	SHGSHETDEEFDAR	Bovine heart	Isolated enzyme	Helling et al. (2012)
Va	Ser-4	SHGSHETDEEFDAR	Bovine heart	Isolated enzyme	Helling et al. (2008)
Va	Thr-35	GMNPTLVGYDLVPEPK	Bovine heart	Isolated enzyme	Helling et al. (2008)
Va	Thr-38	LRKGINTLVTYDMVPE	HeLa cells	Cell culture	Olsen et al. (2010)
Vb	Ser-2	ASGGGVPTDEEQATGLER	Bovine heart	Isolated enzyme	Helling et al. (2012)
Vb	Ser-40	MLPPKAASGTKEDPN	Rabbit heart	After ischemia	Fang et al. (2007)
VIa-H	Thr-11	AKGDHGGTGARTWRF	Bovine heart	Crystal structure	Tsukihara et al. (2003)
VIIc	Ser-1	SHYEEGPGK	Bovine heart	Isolated enzyme	Helling et al. (2012)

In bovine liver COX phosphorylation of Tyr-11 in subunit IV was identified (Lee et al. 2009). This phosphorylation site is also located on the matrix side close to the postulated binding site for ATP (Hüttemann et al. 2001) and could modify the allosteric ATP-inhibition of COX.

In the crystal structure of bovine heart COX, the phosphorylation of Thr-11 in the membrane region of subunit VIa-H was identified (Tsukihara et al. 2003) and is suggested to participate in the dimeric structure of the 13 subunit enzyme complex. Subunit VIa-H is one subunit that bridges the two monomers in the dimeric structure.

Four phosphorylation sites were mapped on rabbit heart COX after ischemia: Ser-115 and Ser-116 of subunit I, Thr-52 of subunit IV, and Ser-40 of subunit Vb (Prabu et al. 2006; Fang et al. 2007). Since the above sites are not PKA consensus sequences PKA is probably not directly involved in these phosphorylations.

11.5 Concluding Remarks

The robustness of COX activity, i.e., the coupling of oxygen reduction to proton translocation without the production of ROS, is evident throughout the evolution from aerobic bacteria to mammals. The diversification of organisms during evolution was accompanied by regulatory specialization. COX became the main regulator of OxPhos mainly by three mechanisms (1) by tissue-, developmental-, and species-specific expression of nuclear-encoded subunits; (2) by feedback interaction with its product ATP (i.e. the ATP/ADP ratio); and (3) by reversible phosphorylation. The multiplicity of interactions became possible during evolution by adding up to 10 subunits, tightly associated with the enzyme complex. The multi-subunit COX allows complex conformational changes modulating electron transport as well as the stoichiometry of proton translocation.

An essential feature of eukaryotic COX is the *allosteric ATP-inhibition*, since it allows the adaptation of OxPhos to large variations of ATP demands at resting and active conditions. According to the Mitchell theory, the control of respiration (inhibition) occurs only at high $\Delta\Psi_m$ (180–220 mV), a condition under which ROS at complex I–complex III are formed, which have been implicated in various degenerative diseases. The control of respiration by the *allosteric ATP-inhibition* is independent of $\Delta\Psi_m$ and suggested to maintain low *healthy* $\Delta\Psi_m$ values (100–140 mV), thus preventing the formation of deleterious ROS.

The *allosteric ATP-inhibition* is suggested to be continuously switched on and off by reversible phosphorylation of COX. Under resting conditions it maintains maximal efficiency of OxPhos at lower rates of ATP synthesis and low ROS production. At work (or activity, stress) it is switched off allowing maximal rates of ATP synthesis at lower efficiency and increased $\Delta\Psi_m$ and ROS production. We suggest that mainly neuronal signals switch off the *allosteric ATP-inhibition*, while intracellular signals switch it on.

The signal pathways and the protein kinase(s) and phosphatase(s) interacting with COX, as well as the phosphorylation site(s) which switch on and off the *allosteric*

ATP-inhibition are unknown. Their identification will be a great challenge for future biochemical research in order to understand numerous degenerative diseases.

Acknowledgements Annika Rhiel is gratefully acknowledged for excellent technical assistance. Dominic Böth is acknowledged for help in respiratory measurements. This paper was supported by the Deutsche Forschungsgemeinschaft (DFG Ka 192/40-1).

References

Acin-Perez R, Gatti DL, Bai Y, Manfredi G (2011) Protein phosphorylation and prevention of cytochrome oxidase inhibition by ATP: coupled mechanisms of energy metabolism regulation. Cell Metab 13:712–719

Anthony G, Stroh A, Lottspeich F, Kadenbach B (1990) Different isozymes of cytochrome c oxidase are expressed in bovine smooth muscle and skeletal or heart muscle. FEBS Lett 277:97–100

Anthony G, Reimann A, Kadenbach B (1993) Tissue-specific regulation of bovine heart cytochrome c oxidase by ADP via interaction with subunit VIa. Proc Natl Acad Sci USA 90:1652–1656

Arnold S (1997) The structure of the cytochrome c oxidase of various vertebrates and its regulation by thyreoid hormone and the energy metabolism of cells. Dissertation, Fachbereich Chemie, Philipps-Universität, Marburg

Arnold S, Kadenbach B (1997) Cell respiration is controlled by ATP, an allosteric inhibitor of cytochrome c oxidase. Eur J Biochem 249:350–354

Arnold S, Kadenbach B (1999) The intramitochondrial ATP/ADP-ratio controls cytochrome c oxidase activity allosterically. FEBS Lett 443:105–108

Arnold S, Goglia F, Kadenbach B (1998) 3,5-diiodothyronine binds to subunit Va of cytochrome c oxidase and abolishes the allosteric inhibition of respiration by ATP. Eur J Biochem 252:325–330

Baden KN, Murray J, Capaldi RA, Guillemin K (2007) Early developmental pathology due to cytochrome *c* oxidase deficiency is revealed by a new zebrafish model. J Biol Chem 282:34839–34849

Balaban RS (1990) Regulation of oxidative phosphorylation in the mammalian cell. Am J Physiol 258:C377–C389

Balaban RS, Kantor HL, Katz LA, Briggs RW (1986) Relation between work and phosphate metabolite in the in vivo paced mammalian heart. Science 232(4754):1121–1123

Beauvoit B, Rigoulet M (2001) Regulation of cytochrome c oxidase by adenylic nucleotides. Is oxidative phosphorylation feedback regulated by its end-products? IUBMB Life 52:143–152

Bender E, Kadenbach B (2000) The allosteric ATP-inhibition of cytochrome c oxidase is reversibly switched on by cAMP-dependent phosphorylation. FEBS Lett 466:130–134

Bjornson KP, Modrich P (2003) Differential and simultaneous adenosine di- and triphosphate binding by MutS. J Biol Chem 278:18557–18562

Bolli R, Nalecz KA, Azzi A (1985) The interconversion between monomeric and dimeric bovine heart cytochrome c oxidase. Biochimie 67:119–128

Bonne G, Seibel P, Possekel S, Marsac C, Kadenbach B (1993) Expression of humen cytochrome c oxidase subunits during fetal development. Eur J Biochem 217:1099–1107

Bylund-Fellenius AC, Davidsson M, Arvidsson A, Elander A, Scherstén T (1982) Optimal conditions for assay of cytochrome-c-oxidase activity in human skeletal muscle tissue. Clin Physiol 2:71–79

Capaldi RA, Marusich MF, Taanman JW (1995) Mammalian cytochrome-c oxidase: characterization of enzyme and immunological detection of subunits in tissue extracts and whole cells. Methods Enzymol 260:117–132

Chance B, Williams GR (1955) Respiratory enzymes in oxidative phosphorylation. III. The steady state. J Biol Chem 217:409–427

Chrzanowska-Lightowlers ZM, Turnbull DM, Lightowlers RN (1993) A microtiter plate assay for cytochrome c oxidase in permeabilized whole cells. Anal Biochem 214:45–49

Dalle-Donne I, Rossi R, Colombo R, Giustarini D, Milzani A (2006) Biomarkers of oxidative damage in human disease. Clin Chem 52:601–623

Dalmonte ME, Forte E, Genova ML, Giuffrè A, Sarti P, Lenaz G (2009) Control of respiration by cytochrome c oxidase in intact cells: role of the membrane potential. J Biol Chem 284:32331–32335

Dröse S, Brandt U (2008) The mechanism of mitochondrial superoxide production by the cytochrome bc1 complex. J Biol Chem 283:21649–21654

Fang J-K, Prabu SK, Sepuri NB, Raza H, Anandatheerthavarada HK, Galati D, Spear J, Avadhani NG (2007) Site specific phosphorylation of cytochrome c oxidase subunits I, IVi1 and Vb in rabbit hearts subjected to ischemia/reperfusion. FEBS Lett 581:1302–1310

Ferguson-Miller S, Brautigan DL, Margoliash E (1976) Correlation of the kinetics of electron transfer activity of various eukaryotic cytochromes c with binding to mitochondrial cytochrome c oxidase. J Biol Chem 251:1104–1115

Ferguson-Miller S, Brautigan DL, Margoliash E (1978) Definition of cytochrome c binding domains by chemical modification. III. Kinetics of reaction of carboxydinitrophenyl cytochromes c with cytochrome c oxidase. J Biol Chem 253:149–159

Follmann K, Arnold S, Ferguson-Miller S, Kadenbach B (1998) Cytochrome c oxidase activity from eukaryotes but not from prokaryotes is allosterically inhibited by ATP. Biochem Mol Biol Intern 45:1047–1055

Frank V, Kadenbach B (1996) Regulation of the H+/e- stoichiometry of cytochrome c oxidase from bovine heart by intraliposomal ATP/ADP ratios. FEBS Lett 382:121–124

Freeman M (2000) Feedback control of intercellular signalling in development. Nature 408(6810):313–319

From AHL, Zimmer SD, Michurski SP, Mohanakrishnan P, Ulstad VK, Thoma WJ, Ugurbil K (1990) Regulation of the oxidative phosphorylation rate in the intact cell. Biochemistry 29:3731–3743

Groen AK, Wanders RJA, Westerhoff HV, van der Meer R, Tager JM (1982) Quantification of the contribution of various steps to the control of mitochondrial respiration. J Biol Chem 257:2754–2757

Hakvoort TBM, Moolenaar K, Lankvelt AHM, Singorgo KMC, Dekker HL, Muijsers OA (1987) Separation, stability and kinetics of monomeric and dimeric bovine heart cytochrome c oxidase. Biochim Biophys Acta 894:347–354

Helling S, Vogt S, Rhiel A, Ramzan R, Wen L, Marcus K, Kadenbach B (2008) Phosphorylation and kinetics of mammalian cytochrome c oxidase. Mol Cell Proteomics 7:1714–1724

Helling S, Hüttemann H, Ramzan R, Kim SH, Lee I, Müller T, Langenfeld E, Meyer HE, Kadenbach B, Vogt S, Marcus K (2012) Multiple phosphorylations of cytochrome c oxidase and their functions. Proteomics 12:950–959

Hendler RW, Pardhasaradhi K, Reynafarje B, Ludwig B (1991) Comparison of energy-transducing capabilities of the two- and three-subunit cytochromes aa3 from *Paracoccus denitrificans* and the 13-subunit beef heart enzyme. Biophys J 60:415–423

Hüther F-J, Kadenbach B (1987) ADP increases the affinity for cytochrome c by interaction with the matrix side of bovine heart cytochrome c oxidase. Biochem Biophys Res Commun 147:1268–1275

Hüttemann M, Kadenbach B, Grossman LI (2001) Mammalian subunit IV isoforms of cytochrome c oxidase. Gene 267:111–123

Hüttemann M, Jaradat S, Grossman LI (2003) Cytochrome c oxidase of mammals contains a testes-specific isoform of subunit VIb – the counterpart to testes-specific cytochrome c? Mol Reprod Dev 66:8–16

Hüttemann M, Lee I, Pecinova A, Pecina P, Przyklenk K, Doan JW (2008) Regulation of oxidative phosphorylation, the mitochondrial membrane potential, and their role in human disease. J Bioenerg Biomembr 40:445–456

Hüttemann M, Helling S, Sanderson TH, Sinkler C, Samavati L, Mahapatra G, Varughese A, Lu G, Liu J, Ramzan R, Vogt S, Grossman LI, Doan JW, Marcus K, Lee I (2012) Regulation of mitochondrial respiration and apoptosis through cell signaling: cytochrome *c* oxidase and cytochrome *c* in ischemia/reperfusion injury and inflammation. Biochim Biophys Acta 1817:598–609

Inoue Y, Shingyoji C (2007) The roles of noncatalytic ATP binding and ADP binding in the regulation of dynein motile activity in flagella. Cell Motil Cytoskeleton 64:690–704

Kadenbach B, Arnold S (1999) A second mechanism of respiratory control. FEBS Lett 447:131–134

Kadenbach B, Arnold S, Lee I, Hüttemann M (2004) The possible role of cytochrome c oxidase in stress-induced apoptosis and degenerative diseases. Biochim Biophys Acta 1655:400–408

Kadenbach B, Ramzan R, Wen L, Vogt S (2010) New extension of the Mitchell Theory for oxidative phosphorylation in mitochondria of living organisms. Biochim Biophys Acta 1800:205–212

Kaim G, Dimroth P (1999) ATP synthesis by F-type ATP synthase is obligatorily dependent on the transmembrane voltage. EMBO J 18:4118–4127

Kitano H (2004) Biological robustness. Nat Rev Genet 5:826–837

Korshunov SS, Skulachev VP, Starkov AA (1997) High protonic potential actuates a mechanism of production of reactive oxygen species in mitochondria. FEBS Lett 416:15–18

Kussmaul L, Hirst J (2006) The mechanism of superoxide production by NADH:ubiquinone oxidoreductase (complex I) from bovine heart mitochondria. Proc Natl Acad Sci USA 103:7607–7612

Kuznetsov AV, Gnaiger E (2010) Oxygraph assay of cytochrome c oxidase activity: chemical background correction. Mitochondr Physiol Netw 06(06):1–4

Lamb HK, Mee C, Xu W, Liu L, Blond S, Cooper A, Charles IG, Hawkins AR (2006) The affinity of a major Ca2+ binding site on GRP78 is differentially enhanced by ADP and ATP. J Biol Chem 281:8796–8805

Lardy HA, Wellman H (1952) Oxidative phosphorylations; role of inorganic phosphate and acceptor systems in control of metabolic rates. J Biol Chem 195:215–224

Lee I, Kadenbach B (2001) Palmitate decreases proton pumping of liver-type cytochrome c oxidase. Eur J Biochem 268:6329–6334

Lee I, Bender E, Arnold S, Kadenbach B (2001) New control of mitochondrial membrane potential and ROS-formation. Biol Chem 382:1629–1633

Lee I, Bender E, Kadenbach B (2002) Control of mitochondrial membrane potential and ROS formation by reversible phosphorylation of cytochrome c oxidase. Mol Cell Biochem 234(235):63–70

Lee I, Salomon AR, Ficarro S, Mathes I, Lottspeich F, Grossman JI, Hüttemann M (2005) cAMP-dependent tyrosine phosphorylation of subunit I inhibits cytochrome c oxidase activity. J Biol Chem 280:6094–6100

Lee I, Salomon AR, Yu K, Samavati L, Pecina P, Pecinova A, Hüttemann M (2009) Isolation of regulatory-competent, phosphorylated cytochrome c oxidase. Methods Enzymol 457:193–210

Lim TS, Dávila A, Wallace DC, Burke P (2010) Assessment of mitochondrial membrane potential using an on-chip microelectrode in a microfluidic device. Lab Chip 10:1683–1688

Lin J, Wu S, Chan SI (1995) Electron transfer from cytochrome c to 8-azido-ATP-modified cytochrome c oxidase. Biochemistry 34:6335–6343

Linder D, Freund R, Kadenbach B (1995) Species-specific expression of cytochrome c oxidase isozymes. Comp Biochem Physiol 112B:461–469

Liu SS (1997) Generating, partitioning, targeting and functioning of superoxide in mitochondria. Biosci Rep 17:259–272

Ludwig B, Bender E, Arnold S, Hüttemann M, Lee I, Kadenbach B (2001) Cytochrome c oxidase and the regulation of oxidative phosphorylation. Chembiochem 2:392–403

Malyan AN (2010a) Nucleotide binding to noncatalytic sites is essential for ATP-dependent stimulation and ADP-dependent inactivation of the chloroplast ATP synthase. Photosynth Res 105:243–248

Malyan AN (2010b) Role of short conserved segments of alpha- and beta-subunits that link F(1)-ATPase catalytic and noncatalytic sites. Biochemistry (Mosc) 75:81–84
Marx MK, Mayer-Posner F, Souliname T, Buse G (1998) Matrix-assisted laser desorption/ionization mass spectrometry analysis and thiol-group determination of isoforms of bovine cytochrome c oxidase, a hydrophobic multisubunit membrane protein. Anal Biochem 256:192–199
McKenzie M, Lazarou M, Ryan MT (2009) Analysis of respiratory chain complex assembly with radiolabeled nuclear- and mitochondrial-encoded subunits. Methods Enzymol 456:321–339
Miró O, Cardellach F, Barrientos A, Casademont J, Rötig A, Rustin P (1998) Cytochrome c oxidase assay in minute amounts of human skeletal muscle using single wavelength spectrophotometers. J Neurosci Methods 80(1):107–111
Mitchell P (1966) Chemiosmotic coupling in oxidative and photosynthetic phosphorylation. Biol Rev 41:445–502
Murphy MP (2009) How mitochondria produce reactive oxygen species. Biochem J 417:1–13
Murray J, Schilling B, Row RH, Yoo CB, Gibson BW, Marusich MF, Capaldi RA (2007) Small-scale immunopurification of cytochrome *c* oxidase for a high-throughput multiplexing analysis of enzyme activity and amount. Biotechnol Appl Biochem 48:167–178
Napiwotzki J, Shinzawa-Itoh K, Yoshikawa S, Kadenbach B (1997) ATP and ADP bind to cytochrome c oxidase and regulate its activity. Biol Chem 378:1013–1021
Napiwotzki J, Kadenbach B (1998) Extramitochondrial ATP/ADP-ratios regulate cytochrome c oxidase activity via binding to the cytosolic domain of subunit IV. Biol Chem 379:335–339
Nicholls DG, Ferguson SJ (2002) Bioenergetics 3. Academic, San Diego, pp 63–66
O'Brien TM, Oliveira PJ, Wallace KB (2008) Inhibition of the adenine nucleotide translocator by N-acetyl perfluorooctane sulfonamides in vitro. Toxicol Appl Pharmacol 227:184–195
Olsen JV, Vermeulen M, Santamaria A, Kumar C, Miller ML, Jensen LJ, Gnad F, Cox J, Jensen TS, Nigg EA, Brunak S, Mann M (2010) Quantitative phosphoproteomics reveals widespread full phosphorylation site occupancy during mitosis. Sci Signal 3:ra3
Pacelli C, Latorre D, Cocco T, Capuano F, Kukat C, Seibel P, Villani G (2011) Tight control of mitochondrial membrane potential by cytochrome c oxidase. Mitochondrion 11(2):334–341
Pardhasaradhi K, Ludwig B, Hendler RW (1991) Potentiometric and spectral studies with the two-subunit cytochrome aa3 from Paracoccus denitrificans. Comparison with the 13-subunit beef heart enzyme. Biophys J 60:408–414
Pearson RB, Kemp BE (1991) Protein kinase phosphorylation site sequences and consensus specificity motifs: tabulations. Methods Enzymol 200:62–81
Piccoli C, Scrima R, Boffoli D, Capitanio N (2006) Control by cytochrome c oxidase of the cellular oxidative phosphorylation system depends on the mitochondrial energy state. Biochem J 396:573–583
Prabu SK, Anandatheerthavarada HK, Raza H, Srinivasan S, Spear JF, Avadhani NG (2006) Protein kinase A-mediated phosphorylation modulates cytochrome c oxidase function and augments hypoxia and myocardial ischemia-related injury. J Biol Chem 281:2061–2070
Rajagopalan K, Watt DS, Haley BE (1999) Orientation of GTP and ADP within their respective binding sites in glutamate dehydrogenase. Eur J Biochem 265:564–571
Ramzan R, Staniek K, Kadenbach B, Vogt S (2010) Mitochondrial respiration and membrane potential are regulated by the allosteric ATP-inhibition of cytochrome c oxidase. Biochim Biophys Acta 1797:1672–1680
Rieger T, Napiwotzki J, Kadenbach B (1995) On the number of nucleotide binding sites in cytochrome c oxidase. Biochem Biophys Res Commun 217:34–40
Rohdich F, Kadenbach B (1993) Tissue-specific regulation of cytochrome c oxidase efficiency by nucleotides. Biochemistry 32:8499–8503
Rottenberg H, Covian R, Trumpower BL (2009) Membrane potential greatly enhances superoxide generation by the cytochrome bc_1 complex reconstituted into phospholipid vesicles. J Biol Chem 284:19203–19210
Samavati L, Lee I, Mathes I, Lottspeich F, Hüttemann M (2008) Tumor necrosis factor alpha inhibits its oxidative phosphorylation through tyrosine phosphorylation at subunit I of cytochrome c oxidase. J Biol Chem 283:21134–21144

Schwenke W-D, Soboll S, Seitz HJ, Sies H (1981) Mitochondrial and cytosolic ATP/ADP ratios in rat liver in vivo. Biochem J 200:405–408

Stelling J, Sauer U, Szallasi Z, Doyle FJ 3rd, Doyle J (2004) Robustness of cellular functions. Cell 118:675–685

Suarez MD, Revzin A, Natlock R, Keper ES, Thompson DA, Ferguson-Miller S (1984) The functional and physical form of mammalian cytochrome c oxidase determined by gel filtration, radiation inactivation, and sedimentation equilibrium analysis. J Biol Chem 259:13791–13799

Thompson DA, Ferguson-Miller S (1983) Lipid and subunit III depleted cytochrome c oxidase purified by horse cytochrome c affinity chromatography in lauryl maltoside. Biochemistry 22(13):3178–3187

Trachootham D, Lu W, Ogasawara MA, Nilsa RD, Huang P (2008) (2008) Redox regulation of cell survival. Antioxid Redox Signal 10:1343–1374

Tsukihara T, Aoyama H, Yamashita E, Tomizaki T, Yamaguchi H, Shinzawa-Itoh K, Nakashima R, Yaono R, Yoshikawa S (1996) The whole structure of the 13-subunit oxidized cytochrome c oxidase at 2.8 A. Science 272:1136–1144

Tsukihara T, Shimokata K, Katayama Y, Shimada H, Muramoto K, Aoyama H, Mochizuki M, Shinzawa-Itoh K, Yamashita E, Yao M, Ishimura Y, Yoshikawa S (2003) The low-spin heme of cytochrome c oxidase as the driving element of the proton-pumping process. Proc Natl Acad Sci USA 100:15304–15309

Valko M, Leibfritz D, Moncola J, Cronin MTD, Mazura M, Telser J (2007) Free radicals and antioxidants in normal physiological functions and human disease. Int J Biochem Cell Biol 39:44–84

Veech RL, Lawson JWR, Cornell NW, Krebs HA (1979) Cytosolic phosphorylation potential. J Biol Chem 254:6538–6547

Villani G, Attardi G (1997) In vivo control of respiration by cytochrome c oxidase in wild-type and mitochondrial DNA mutation-carrying human cells. Proc Natl Acad Sci USA 94:1166–1171

Villani G, Attardi G (2001) In vivo measurements of respiration control by cytochrome c oxidase and in situ analysis of oxidative phosphorylation. Methods Cell Biol 65:119–131

Wan B, Doumen C, Duszynski J, Salama G, Vary TC, LaNoue KF (1993) Effects of cardiac work on electrical potential gradient across mitochondrial membrane in perfused rat hearts. Am J Physiol 265:H453–H460

Warnock LJ, Raines SA (2004) Restoration of wild-type conformation to full-length and truncated p53 proteins: specific effects of ATP and ADP. Cancer Biol Ther 3:634–637

Weishaupt A, Kadenbach B (1992) Selective removal of subunit VIb increases the activity of cytochrome c oxidase. Biochemistry 31:11477–11481

Wiseman RW, Kushmerick MJ (1997) Phosphorus metabolite distribution in skeletal muscle: quantitative bioenergetics using creatine analogs. Mol Cell Biochem 174:23–28

Yegutkin GG, Burnstock G (1999) Steady-state binding of adenine nucleotides ATP, ADP and AMP to rat liver and adipose plasma membranes. J Recept Signal Transduct Res 19:437–448

Yu MA, Egawa T, Shinzawa-Itoh K, Yoshikawa S, Yeh SR, Rousseau DL, Gerfen GJ (2011) Radical formation in cytochrome c oxidase. Biochim Biophys Acta 1807(10):1295–1304

Zhang H, Huang HM, Carson RC, Mahmood J, Thomas HM, Gibson GE (2001) Assessment of membrane potentials of mitochondrial populations in living cells. Anal Biochem 298:170–180

Chapter 12
Bigenomic Regulation of Cytochrome *c* Oxidase in Neurons and the Tight Coupling Between Neuronal Activity and Energy Metabolism

Margaret T.T. Wong-Riley

Abstract Cytochrome *c* oxidase is the terminal enzyme of the mitochondrial electron transport chain, without which oxidative metabolism cannot be carried to completion. It is one of only four unique, bigenomic proteins in mammalian cells. The holoenzyme is made up of three mitochondrial-encoded and ten nuclear-encoded subunits in a 1:1 stoichiometry. The ten nuclear subunit genes are located in nine different chromosomes. The coordinated regulation of such a multisubunit, multichromosomal, bigenomic enzyme poses a challenge. It is especially so for neurons, whose mitochondria are widely distributed in extensive dendritic and axonal processes, resulting in the separation of the mitochondrial from the nuclear genome by great distances. Neuronal activity dictates COX activity that reflects protein amount, which, in turn, is regulated at the transcriptional level. All 13 *COX* transcripts are up- and downregulated by neuronal activity. The ten nuclear *COX* transcripts and those for *Tfam* and *Tfbm*s important for mitochondrial *COX* transcripts are transcribed in the same transcription factory. Bigenomic regulation of all 13 transcripts is mediated by nuclear respiratory factors 1 and 2 (NRF-1 and NRF-2). NRF-1, in addition, also regulates critical neurochemicals of glutamatergic synaptic transmission, thereby ensuring the tight coupling of energy metabolism and neuronal activity at the molecular level in neurons.

12.1 Cytochrome *c* Oxidase

Cytochrome *c* oxidase (COX, cytochrome *aa*3, ferrocytochrome *c* oxygen oxidoreductase, complex IV, E.C. 1.9.3.1) is the terminal enzyme of the mitochondrial electron transport chain. It catalyzes the oxidation of its substrate, cytochrome *c*,

M.T.T. Wong-Riley (✉)
Department of Cell Biology, Neurobiology and Anatomy, Medical College of Wisconsin, Milwaukee, WI 53226, USA
e-mail: mwr@mcw.edu

B. Kadenbach (ed.), *Mitochondrial Oxidative Phosphorylation*, Advances in Experimental Medicine and Biology 748, DOI 10.1007/978-1-4614-3573-0_12,

and the reduction of molecular oxygen to water. It assists in the pumping of protons from the matrical to the cytosolic side of the inner mitochondrial membrane, setting up the electrochemical proton gradient that drives the synthesis of ATP from ADP and phosphate by ATP synthase (complex V). Inactivation of COX by cyanide, azide, or carbon monoxide is incompatible with life, as oxidative metabolism cannot be carried to completion, and no ATP can be generated from mitochondria. Thus, highly oxidative organs such as the heart, liver, kidney, skeletal muscles, and especially the brain, are critically dependent on COX for their normal functioning and survival.

COX is one of the most ancient enzymes known, parts of it evolved more than a billion years ago. It is one of only four bigenomic proteins in mammalian cells: complexes I, III, IV, and V of the electron transport chain, each of which has subunits from either the nuclear or the mitochondrial genome, and none of which is encoded entirely by a single genome. This implies that (a) the mitochondrial genome retains its control of key subunits of the electron transport chain through evolution; and (b) the two genomes have to work closely together to ensure proper functioning of the oxidative phosphorylation machinery. COX holoenzyme has 13 subunits with 1:1 stoichiometry (Kadenbach et al. 1983). The largest three subunits (COX I, II, and III) are encoded in the maternally inherited mitochondrial genome, and the remaining ten (COX IV, Va,b, VIa,b,c, VIIa,b,c, and VIII) are nuclear-encoded in nine different chromosomes. To form a functional holoenzyme, precise coordination between the two genomes is necessary. The mechanism of regulating such a complex, multisubunit, bigenomic enzyme appears daunting and poorly understood until recent years, when the regulatory machinery was beginning to be revealed.

12.2 Cytochrome *c* Oxidase as a Metabolic Marker for Neurons

Neurons are ideal cells for investigating the regulation of COX. Unlike glial cells that can fare quite well under anaerobic condition, neurons are dependent almost entirely on oxidative metabolism for their function and survival. Being postmitotic, they do not undergo constant turnover and rebirth, and hence their metabolic activity reflects primarily their constitutive functional demands. Of all the ATP-demanding functions of neurons, such as the synthesis of proteins and other molecules, turnover of transmitters and receptors, active anterograde and retrograde transport of proteins and organelles, and active transport of ions against their concentration and electrical gradients, the first two consume relatively little energy, the third one accounts for only a minor fraction of the energy, but the last one is by far the most energy-demanding function of neurons, especially with regard to repolarizing the membrane after excitatory depolarizing activity (Wong-Riley 1989, 2010; Attwell and Laughlin 2001). Indeed, it is the neuronal activity that controls energy expenditure, and not vice versa (Lowry 1975; Wong-Riley 1989). The control is precise, such that energy is not generated until energy is spent. As different compartments of a single neuron require varying amounts of energy, the entire neuron is not metabolically homogeneous. Dendrites, being the major receptive sites of incoming depolarizing input and whose membranes have to be constantly repolarized, consume the bulk of energy, whereas axonal trunks, especially myelinated axons, consume very little energy (Wong-Riley 1989). Energy demand of cell bodies is largely dependent on the magnitude and frequency of excitatory input they receive, and that of axon terminals reflects how tonically active they are (Wong-Riley 1989).

Can cytochrome *c* oxidase serve as a sensitive and reliable metabolic marker for neurons? At the biochemical, histochemical, immunohistochemical, cytochemical, and molecular levels, this proves to be the case, and they correlate well with the neuron's functional activity (Wong-Riley 1979; Wong-Riley et al. 1989a, b; Hevner et al. 1995; Hevner and Wong-Riley 1989, 1991). In altering the functional demands of neurons, such as with tetrodotoxin (TTX)-induced impulse blockade or with KCl-mediated depolarizing stimulation, the levels of cytochrome *c* oxidase in affected neurons are down- or upregulated accordingly (Wong-Riley and Carroll 1984; Hevner and Wong-Riley 1990; DeYoe et al. 1995; Zhang and Wong-Riley 1999). Such alterations exist not only in the activity of the enzyme, but also in its protein and mRNA amount, indicating that the activity reflects mainly the protein amount, which, in turn, is regulated mainly at the transcriptional level (Wong-Riley et al. 1998a).

12.3 The Challenge of Bigenomic Coordination in Neurons

As COX subunit transcripts and proteins are of two genomic sources, this poses a unique challenge for neurons. For, unlike most cells, neurons have mitochondria-laden dendrites as well as axons that can extend far from the cell bodies, hence the nuclear and the mitochondrial genomes can be separated by great distances. Do the nuclear transcripts migrate from the cell body to distal dendrites for local translation, or do they stay within the cell body? To answer this question, in situ hybridization was done at both the light and electron microscopic (EM) levels. It was found that, indeed, the mitochondrial mRNAs are located within the mitochondria that are distributed throughout the cell bodies, dendrites, and axons, but the nuclear transcripts are restricted only to the cell bodies, indicating that translation of nuclear transcripts occurs only in the cell bodies (Hevner and Wong-Riley 1991; Wong-Riley et al. 1997). How, then, can the nuclear-encoded subunit proteins get to distal dendrites, where they are most needed? Do they take the intra- or the extra-mitochondrial route in their transit from the cell body to distal processes?

The answer for at least one of the nuclear-encoded subunits, COXIV, is that the precursor protein, which contains the mitochondrial-targeting presequence, is translated in the cell body, incorporated into the mitochondria within the cell body, and is translocated intra-mitochondrially to distal processes (Liu and Wong-Riley 1994). There, it can remain as precursor protein, or be processed into the mature form to be incorporated into the holoenzyme with the other nuclear- and mitochondrial subunit polypeptides. Thus, neurons have devised a mechanism by which the precursor proteins are not immediately processed upon entry into the mitochondria, as in the case of yeast and rat hepatocytes (Mori et al. 1981; Reid et al. 1982), but rather, can form a precursor pool in dendrites and axons until such time when additional energy demand triggers further processing into their mature forms. This mechanism ensures conservation of energy and bypasses the need for constant shuttling of individual mitochondrion back to the cell body for a fresh supply of precursor proteins each time a new stock of holoenzyme is called for.

12.4 Bigenomic Coordination of Cytochrome *c* Oxidase in Response to Changing Neuronal Activity

Are all 13 subunits of cytochrome *c* oxidase coordinately or disparately regulated by neuronal activity? Both in vivo and in vitro approaches have been used to probe this question. In vitro, all 13 *COX* subunit transcripts are significantly upregulated after 5 h of depolarizing stimulation, and they are all downregulated by TTX blockade (Liang et al. 2006). However, the levels of the three mitochondrial-encoded transcripts fall earlier than those of the ten nuclear-encoded ones (2 versus 4 days). By the 6th day after inactivation, all 13 transcripts are downregulated to about the same extent (to ~20% of controls). Likewise, in vivo sensory deprivation with retinal impulse blockade or enucleation induces an earlier and more severe downregulation of the mitochondrial- than the nuclear-encoded subunit transcripts (Hevner and Wong-Riley 1993; Liang et al. 2006). This implies that the mitochondrial genome exerts a greater control over the activity and amount of the enzyme in neurons. The merit of such a mechanism includes (a) mitochondria in distal dendrites and axon terminals are strategically located at the "business" ends of neurons, where they can sense local energy demand and adjust the supply of holoenzymes accordingly; (b) the mitochondrial genome is responsible for the largest three subunits that form the catalytic core of the enzyme, although the ten nuclear-encoded subunits also play important roles in energy metabolism (Kadenbach et al. 2000); and (c) as stated above, there is a reservoir of nuclear-encoded subunit proteins in distal neuronal processes, so the downregulation of these subunits may be delayed. Ultimately, however, all 13 subunits are up- or downregulated by neuronal activity (Liang et al. 2006).

12.4.1 Synthesis Versus Degradation of Bigenomic Transcripts

Are activity-induced changes in *COX* transcripts due to RNA synthesis rate or stability, or both? The answer came from an experiment in which primary neurons in culture were stimulated with 20-mM KCl for 5 h (Zhang and Wong-Riley 2000a). It was found that the synthesis rate of both the mitochondrial-encoded *Cox2* and the nuclear-encoded *Cox4* transcripts is increased significantly after 3 h of depolarizing stimulation. The rate of *Cox2* remains higher than that of controls at 4 and 5 h of stimulation, but that of *Cox4* returns to control levels after 3 h. The degradation rate was monitored by ^{3}H-uridine pulse-chase labeling, and it revealed a half-life of 84 min for *Cox2* and 50 min for *Cox4* mRNA. With KCl stimulation, the half life of *Cox2* transcripts remains relatively constant, whereas that of *Cox4* increases to 102 min. These data indicate that the mitochondrial transcripts are regulated mainly at the transcriptional level, but that the nuclear transcripts are regulated at both the synthetic and degradative levels, and that both are tightly governed by neuronal activity (Zhang and Wong-Riley 2000a).

12.5 Transcription Factors as Bigenomic Coordinators

The bigenomic nature of COX imposes a special need for transcriptional coordination between the two genomes. Is there a transcription factor or factors that may serve such a role? Two factors have been proposed to mediate nuclear-mitochondrial interactions. They are nuclear respiratory factors 1 and 2 (NRF-1 and NRF-2) (Scarpulla 2008).

12.5.1 Role of Nuclear Respiratory Factor 2

NRF-2 is the human homologue of the murine GA-binding protein (GABP) (Evans and Scarpulla 1990; Thompson et al. 1991; Virbasius et al. 1993a). It belongs to the *Ets* (E26 transformation-specific) family of transcription factors, recognizing the consensus sequence (C/A)GGA(A/T)(A/G) (LaMarco et al. 1991; Thompson et al. 1991; Virbasius and Scarpulla 1991). NRF-2 is a heteromeric protein made up of mainly α and β subunits (β1, β2), and γ1 and γ2 are splice variants of the β subunit. The α subunit has the Ets domain and is required for DNA binding. The β subunit contains four Notch-ankyrin repeats that mediate dimerization with the α subunit, but is incapable of binding DNA alone. The β subunit also has the transactivating domain and the nuclear localizing signal. Heterodimerization of α and β is required for stabilization and specificity of α-DNA binding (De la Brousse et al. 1994; Batchelor et al. 1998). The homodimerization domain of the β subunit enables the formation of $\alpha_2\beta_2$ heterotetramer that binds to tandem repeats of NRF-2 in target gene promoters (Scarpulla 2002). DNA binding and in vitro studies have implicated the transactivational activity of NRF-2 and its regulatory role in the expression of a number of subunits of respiratory chain enzymes, especially some of the nuclear-encoded COX subunits (rat and mouse subunits IV and Vb, human subunit Vb, bovine subunit VIIaL (Virbasius and Scarpulla 1991; Carter et al. 1992; Virbasius et al. 1993a; Carter and Avadhani 1994; Scarpulla 1997), and human COX VIaL (Ongwijitwat and Wong-Riley 2004). In addition, NRF-2 also regulates genes that encode mitochondrial transcription factors A and B (TFAM, TFB1M, and TFB2M) (Virbasius and Scarpulla 1994; Gleyzer et al. 2005), which are nuclear-derived and function inside the mitochondria as regulators of mtDNA transcription and replication (Fisher and Clayton 1988; Falkenberg et al. 2002). NRF-2 also regulates three of the four human succinic dehydrogenase subunit genes, as well as genes for human TOMM20, mitochondrial transcription termination factor (mTERF), RNA polymerase POLRMT, and the B subunit of the DNA Polγ, among others (Au and Scheffler 1998; Blesa et al. 2007; Scarpulla 2008; Bruni et al. 2010). Knockout of NRF-2/GABP is embryonically lethal before implantation, attesting to the essential role of this transcription factor in embryogenesis; whereas the heterozygous nulls appear normal (Ristevski et al. 2004). Thus, NRF-2 potentially links the nucleus and mitochondria by regulating COX-related gene expression in the two genomes (Scarpulla 2008).

12.5.1.1 NRF-2 Itself Responds to Changes in Neuronal Activity

What is known about the significance of NRF-2 in neurons? Remarkably, the pattern of NRF-2's distribution in the primate visual cortex is virtually identical to that of COX (Nie and Wong-Riley 1999), a unique feature not shared by any other transcription factors studied thus far. NRF-2 is also more strongly expressed in cell types that have higher COX activity than those with lower activity (Wong-Riley et al. 2005). Moreover, NRF-2 itself responds to monocular impulse blockade by downregulating its protein and message levels in deprived cortical columns and neurons in which COX activity is suppressed (Nie and Wong-Riley 1999; Guo et al. 2000; Wong-Riley et al. 2005). In response to KCl depolarization in cultured primary neurons, NRF-2 protein is upregulated prior to the upregulation of COX subunit message and activity (Zhang and Wong-Riley 2000b), and both α and β subunits of NRF-2 respond to increased neuronal activity by translocating from the cytoplasm to the nucleus, where they associate primarily with euchromatin to activate their target genes (Yang et al. 2004).

12.5.1.2 NRF-2 Regulates All 13 Cytochrome *c* Oxidase Subunit Genes in Neurons

To determine if NRF-2 regulates all COX subunit genes in neurons, in vitro electrophoretic mobility shift (EMSA) and supershift assays, in vivo chromatin immunoprecipitation (ChIP) assays, and promoter mutational analysis were performed. It was found that, indeed, NRF-2 functionally regulates all ten nuclear-encoded subunit genes of *COX* (Ongwijitwat and Wong-Riley 2005). Moreover, functional silencing of NRF-2 with small hairpin interference RNA (shRNA) significantly reduces the expression of all ten nuclear-encoded *COX* subunit genes, as well as of

Tfam and *Tfb1m*, which regulate the expression of the three mitochondrial-encoded *COX* subunit genes (Ongwijitwat et al. 2006). As discussed above, the role of NRF-2 in directly regulating *Tfam* and *Tfbms* has been established (Virbasius and Scarpulla 1994; Gleyzer et al. 2005). These findings, then, are consistent with NRF-2's proposed role as a transcriptional activator of *COX* and that its own expression is regulated by neuronal activity (reviewed in Wong-Riley et al. 2008).

12.5.2 Role of Nuclear Respiratory Factor 1

NRF-1 was first discovered as a transcriptional regulator of the somatic cytochrome *c*, the substrate for COX (Evans and Scarpulla 1989). NRF-1 also activates other genes whose products function within the mitochondria, such as a few of the nuclear-encoded COX subunits (Vb and VIa in humans, Vb and VIc in rats, and VIIaL in cows), specific nuclear-encoded subunits of complexes I, II, III, and V, mitochondrial RNA processing (MRP) RNA, as well as 5-aminolevulinate synthase, which is important for regulating the supply of heme to the cytochromes and other hemoproteins (reviewed in Kelly and Scarpulla 2004). Together with NRF-2, NRF-1 also activates *Tfam* and the *Tfbms* (Virbasius and Scarpulla 1994; Gleyzer et al. 2005). Thus, NRF-1 is another potential coordinator of mitochondrial- and nuclear-encoded subunits of COX.

NRF-1, unlike NRF-2, is a single-gene product whose gene is mapped to human chromosome 7 (7q31) (Gopalakrishnan and Scarpulla 1995) and is ~104-kb long (Huo and Scarpulla 1999). The DNA-binding domain is at the amino terminus and is highly conserved, whereas the transactivation domain is at the carboxy terminus, which is quite divergent among the species (Virbasius et al. 1993b; Gugneja et al. 1996). NRF-1 binds the palindromic consensus sequence (T/C)GCGCA(T/C) GCGC(A/G) (Evans and Scarpulla 1990; Virbasius et al. 1993b; Scarpulla 1997). However, the GCA core is found to be invariant, whereas the flanking GC-rich sequences can be somewhat variable (Dhar et al. 2008). Phosphorylation of NRF-1 greatly enhances its DNA-binding and transactivational activity (Gugneja and Scarpulla 1997). Homozygous NRF-1 knockout mice are embryonically lethal at E3.5 to E6.5, and the blastocysts have greatly reduced mtDNA levels (Huo and Scarpulla 2001). This is consistent with the key role of NRF-1 in the maintenance of mtDNA and respiratory chain function during early embryogenesis. On the other hand, heterozygous mice developed normally, and no apparent deficits have been detected. Interestingly, mutations in a homologue of NRF-1 in the zebrafish, known as *Not really finished*, cause a progressive degeneration of photoreceptors and other cells in the retina, optic tectum, and the brain (Becker et al. 1998).

12.5.2.1 NRF-1 Itself Responds to Changes in Neuronal Activity

What is the role of NRF-1 in neurons? Does it respond to changes in neuronal activity? By means of light and EM immunohistochemistry, western blotting, and real-time quantitative PCR, it was found that both NRF-1 protein and mRNA are present in mammalian visual cortical neurons, and that both are regulated by neuronal activity (Liang and Wong-Riley 2006; Yang et al. 2006). In vitro impulse blockade with TTX and in vivo monocular enucleation lead to a significant down-regulation of NRF-1 mRNA and protein in deprived neurons after 6 or 7 days of deprivation (Liang and Wong-Riley 2006). On the other hand, depolarizing stimulation with KCl progressively upregulates both NRF-1 message and protein in a time-dependent manner, increasing above controls after 1 h and remaining high at 3, 5, and 7 h (Yang et al. 2006). NRF-1 message increases in both the nucleus and the cytoplasm of stimulated neurons, and EM quantification of immunogold particles is consistent with an activity-induced cytoplasmic-to-nuclear translocation of NRF-1 protein. Levels of NRF-1 mRNA and protein progressively decline when the stimulation is withdrawn, with the former reaching basal levels by 5 h and the latter by 7 h (Yang et al. 2006). Thus, NRF-1 upregulates swiftly to functional stimulation but declines more slowly with functional impulse blockade in neurons. These findings are consistent with an activity dependency of the synthesis, distribution, and possibly stability of NRF-1 mRNA and protein in neurons, and that the regulation is primarily at the transcriptional level.

12.5.2.2 NRF-1 Regulates All 13 Cytochrome *c* Oxidase Subunit Genes in Neurons

Does NRF-1 regulate all 13 COX subunit genes? Its indirect activation of the three mitochondrial-encoded subunit genes via TFAM and TFBMs is already known (Virbasius and Scarpulla 1994; Gleyzer et al. 2005). In silico analysis of the rat genome revealed the typical NRF-1 binding motif in only three of the ten nuclear-encoded *COX* subunit promoters: *COX5b*, *6a1*, and *6c* (Ongwijitwat and Wong-Riley 2005). These three have been reported previously in humans and rats (Bachman et al. 1996; Ongwijitwat and Wong-Riley 2004; Evans and Scarpulla 1990). In silico analysis has revealed another NRF-1 binding site on the bovine *Cox7a2* promoter (Seelan et al. 1996). However, after using multiple approaches, such as EMSA, supershift, ChIP, and promoter mutational analysis, it was found that NRF-1 functionally regulates all ten nuclear subunits of COX in neurons (Dhar et al. 2008). The reason that in silico analysis failed to detect those sites is that the classical NRF-1 *cis* motif can actually vary slightly with respect to the sequence of GCs, as long as the GCA core remains intact (Dhar et al. 2008). Silencing of NRF-1 with shRNA significantly downregulates all ten nuclear *COX* mRNAs, as well as messages for *TFAM*, *TFB1M*, *TFB2M*, *SURF1* (surfeit 1), *VDAC* (voltage-dependent anion channel), and *TOM20* (transporter of outer mitochondrial membrane) (Dhar et al. 2008), the last five are known target genes of NRF-1 (Scarpulla 2002; Kelly and Scarpulla 2004; Gleyzer et al. 2005). The extent of reduction ranges from ~35% to 70% ($P<0.05$–0.01). Thus, both NRF-1 and NRF-2 prove to be key bigenomic coordinators for transcriptional regulation of all COX subunit genes in neurons.

12.6 Transcriptional Coactivators: Peroxisome Proliferator-Activated Receptor Gamma Coactivator 1α (PGC-1α)

In recent years, an important transcriptional coactivator of NRF-1 and NRF-2 has been identified as the peroxisome proliferator-activated receptor gamma coactivator 1 (PGC-1 or PGC-1α) (Wu et al. 1999). It was first cloned from a brown fat cDNA library to be induced by cold exposure in both the brown fat and skeletal muscles of mice (Puigserver et al. 1998). Now it is known that it belongs to a family of regulated coactivators, which include PGC-1α, PGC-1β, and PGC-1α-related coactivator (PRC) (reviewed in Scarpulla 2008). As a coactivator, PGC-1α does not bind DNA directly, but rather, in response to appropriate signals in a tissue-specific manner, such as cold exposure in brown adipose tissue and muscle, fasting in the heart, and prolonged physical exercise in skeletal muscles, it interacts with nuclear receptors and transcription factors to activate genes involved in energy and nutrient homeostasis (Puigserver et al. 1998; Lehman et al. 2000; Goto et al. 2000; Baar et al. 2002). These factors include peroxisome proliferator-activated receptor gamma (PPARγ) and alpha (PPARα), thyroid hormone receptor, estrogen-related receptor (ERRα), glucocorticoid receptor, mineralocorticoid receptor, myocyte enhancer factor 2C, Ying Yang 1 (YY1), as well as NRF-1 and NRF-2 (Puigserver et al. 1998; Wu et al. 1999; Knutti and Kralli 2001; Scarpulla 2011). The target genes of these factors include those that encode for the uncoupling proteins (UCPs), subunits of mitochondrial electron transport chain complexes, TFAM, and TFBMs, among others (Puigserver et al. 1998; Knutti and Kralli 2001; Scarpulla 2008, 2011). Thus, PGC-1α plays a key role in adaptive thermogenesis, glucose and fatty acid metabolism, skeletal muscle fiber type switching, heart development, and mitochondrial biogenesis (Puigserver et al. 1998; Knutti and Kralli 2001; Scarpulla 2011). Surprisingly, PGC-1α knockout mice are viable, but exhibit multisystem abnormalities, decreased mitochondrial function, defective thermogenic response, and lesions in the striatum and cerebral cortex (Lin et al. 2004; Leone et al. 2005). Over-expression of PGC-1α increases mitochondrial content, induces the expression of genes involved in energy production and transduction pathways, and protects cultured cells from oxidative stress-induced death (Lehman et al. 2000; St-Pierre et al. 2006; Scarpulla 2011). However, cardiac-specific over-expression of PGC-1α in transgenic mice can lead to uncontrolled mitochondrial proliferation and dilated cardiomyopathy (Lehman et al. 2000).

12.6.1 PGC-1α Responds to Changes in Neuronal Activity

In neurons, PGC-1α is localized mainly to nuclear euchromatin and cytoplasmic free ribosomes (Meng et al. 2007). Depolarizing stimulation for 0.5 h significantly increases PGC-1α in both the nucleus and the cytoplasm (Meng et al. 2007). The level is sustained up to 3 h of stimulation, but decreases from 5 h and returns to baseline level by 10 h. Thus, PGC-1α responds very early to increased neuronal activity (earlier than either NRF-1 or NRF-2) by upregulating its own synthesis in the cytoplasm and having the protein being translocated to the nucleus for gene activation. When neuronal activity is reduced by impulse blockade in vitro or sensory deprivation in vivo, the levels of PGC-1α mRNA and proteins are significantly downregulated earlier than those of NRF-1 and NRF-2 (Liang and Wong-Riley 2006). Neuronal activity, therefore, directly regulates PGC-1α, and PGC-1α is likely to be a critical sensor of activity-dependent energy demand in neurons.

12.6.2 Regulation of PGC-1α in Neurons

Depolarizing activation of PGC-1α in neurons is found to be mediated by p38 mitogen-activated protein kinase (MAPK) and calcium channels (Liang et al. 2010). Stimulation upregulates PGC-1α mRNA and protein levels in 0.5 and 1 h, respectively, but both p38 MAPK and phosphorylated p38 MAPK levels are increased after only 15 min. Such upregulation is suppressed by 30 min of pretreatment with SB203580 (a blocker of p38 MAPK that also blocks the upregulation of PGC-1α by KCl) or with nifedipine (a Ca^{2+} channel blocker). Furthermore, a knockdown of p38 MAPK with shRNA significantly suppresses both PGC-1α mRNAs and proteins (Liang et al. 2010). Thus, both p38 MAPK and Ca^{2+} are critical in mediating signaling in depolarization-induced activation of PGC-1α in neurons.

Taken together, PGC-1α is an early sensor of changes in neuronal activity, and it recruits NRF-1 and NRF-2 (among other factors) to regulate the expression of target genes, such as *COX*, important in energy metabolism that is tightly coupled to neuronal activity. Such a chain of events regulates not only the three mitochondrial-encoded *COX* subunits via TFAM and TFBMs, but also all ten of the nuclear-encoded *COX* subunit genes in neurons (Fig. 12.1).

12.7 Is There a Transcription Factory for the 13 Genomic Loci Involved in the Bigenomic Transcription of *COX* in Neurons?

Transcription factories have been described as dynamic but discrete loci in the nucleus that actively transcribe related genes. These sites are thought to contain several RNA polymerase II molecules, relevant transcription factors, and loops of chromatin-containing genes to be transcribed together (Jackson et al. 1998; Osborne et al. 2004; Zhou et al. 2006). To demonstrate such long-range interactions among related genes, chromosome conformation capture (3C) has been developed (Dekker et al. 2002; Miele and Dekker 2009). This technique converts chromatin conformation and physical interactions in vivo into specific ligation products demonstrable with polymerase chain reaction. Interactions between loci from the same chromosome or from two different chromosomes have been described (Spilianakis and Flavell 2004;

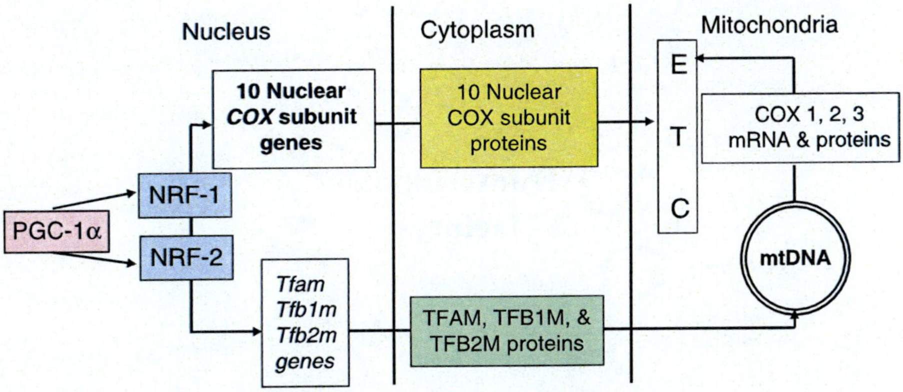

Fig. 12.1 Schematic diagram depicting that PGC-1α coordinates the induction of NRF-1 and NRF-2 in regulating the transcription of all ten nuclear-encoded *COX* subunit genes, as well as the genes for *Tfam*, *Tfb1m*, and *Tfb2m* in the nucleus. Translation in the cytoplasm leads to the generation of the respective proteins, all of which enter into the mitochondria. Within the mitochondrion, the three mitochondrial-encoded *COX* genes are transcribed aided by the TFs and translated into polypeptides. Together, the nuclear- and mitochondrial-encoded COX subunits form the holoenzyme that is complex IV of the electron transport chain (ETC)

Ling et al. 2006; Lomvardas et al. 2006; Schoenfelder et al. 2010). However, to demonstrate interactions among ten genomic loci of the ten nuclear-encoded *COX* subunit genes located in nine different chromosomes poses a distinct challenge.

By means of 3C, it was found that not only do these ten genomic loci interact in the same transcription factory, but that genes from three chromosomes encoding *Tfam*, *Tfb1m*, and *Tfb2m* that are critical for the transcription of the three mitochondrial-encoded *COX* subunit genes all occupy common intranuclear sites in the murine neuronal nuclei (Dhar et al. 2009a). Moreover, interactions between *COX* subunit and *Tf* genes are upregulated by depolarizing stimulation and downregulated by impulse blockade in primary neurons in culture (Dhar et al. 2009a). No doubt, such "transcription factories" are dynamic entities regulated by the energy demand of neurons. Taken together, there is indeed an exquisite mechanism in place for a coordinated and synchronized transcriptional regulation of the multisubunit, multichromosomal, bigenomic COX enzyme in neurons (Fig. 12.2).

12.8 Tight Coupling Between Neuronal Activity and Energy Metabolism at the Transcriptional Level

The tight coupling between neuronal activity and energy metabolism has been well established at the cellular level (Wong-Riley 1989; Wong-Riley et al. 1998a, 2008). As discussed above, repolarization of membrane potentials after depolarizing stimulation consumes the bulk of energy in neurons (Wong-Riley 1989). The more excitatory input a neuron receives, the greater its energy demand.

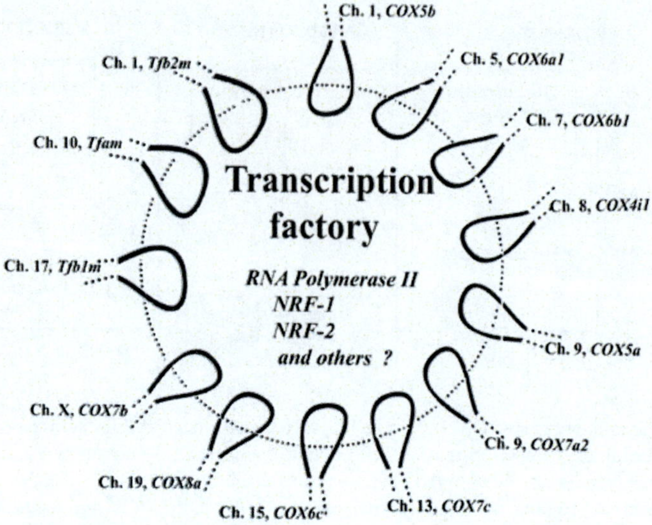

Fig. 12.2 Schematic rendition of a dynamic transcription factory in which the loops of 13 genomic loci for the ten nuclear-encoded *COX* subunit genes and genes for *Tfam*, *Tfb1m*, and *Tfb2m* are cotranscribed, with the aid of RNA polymerase II, NRF-1, NRF-2, and possibly other transcription factors and coactivators (Reproduced with permission from Dhar et al. 2009a)

12.8.1 Glutamatergic System in Neurons

The main depolarizing agent in the brain is glutamate, a major and the most prevalent excitatory neurotransmitter (Fonnum 1984; Streit 1984). Its action is mediated by two major types of receptors, *N*-methyl-D-aspartate (NMDA) and non-NMDA (reviewed in Nakanishi 1992). NMDA receptor is an ionotropic, ligand-gated calcium channel with voltage-dependent magnesium block, and it is made up of the ubiquitous and obligatory NR1 (GluN1) subunit in a heterotetrameric complex with one or more of NR2 (NR2A-D or GluN2A-D) and/or NR3 (NR3A-B or GluN3A-B) subunits (Orrego and Villanueva 1993; Mori and Mishina 1995; Dingledine et al. 1999; Salussolia et al. 2011). Among the NR subunits, NR2B is important in synaptic signaling, long-term potentiation, learning and memory, as well as involvement in a number of human neurological disorders (Loftis and Janowsky 2003; Babb et al. 2005). Within the non-NMDA receptor category, the AMPA (α-amino-3-hydroxyl-5-methyl-4-isoxazolpropionic acid) type is ionotropic, ligand-gated, mediates fast synaptic transmission, and is made up of GluR1-4 (GluR-A-D or GluA1-4) subunits in various combinations (Keinänen et al. 1990). These subunits each undergoes RNA editing and alternative splicing, yielding either the flip or flop variants (Sommer et al. 1990; Lomeli et al. 1994). AMPA receptors play a critical role in synaptogenesis, neural circuitry formation, and synaptic plasticity (Tanaka et al. 2000; Palmer et al. 2005). Subunit 2 (GluR2, GluR-B, or GluA2) is of special interest because the glutamine residue in its transmembrane segment 2 is mRNA-edited to the positively charged arginine residue, and it is the only subunit that impedes Ca^{2+} entry into neurons (Verdoorn et al. 1991). AMPA receptors that contain the GluR2 subunit are impermeable to Ca^{2+}, show simple outward rectification, and are insensitive to blockage by external polyamines (Hollmann et al. 1991; Burnashev et al. 1992; Washburn et al. 1997). Thus, GluR2 is dominant in determining the functional properties of heteromeric AMPA receptors (Jonas et al. 1994; Tanaka et al. 2000). The downregulation of GluR2 mRNA has been implicated in enhanced neurotoxicity with increased Ca^{2+} permeability in the affected neurons (reviewed in Pellegrini-Giampietro et al. 1997).

In the CNS, regions rich in COX also have higher levels of glutamatergic and NMDA receptor-mediated synapses. When these regions are deprived of their excitatory input, the levels of both COX and NMDA receptor NR1 are downregulated (Wong-Riley et al. 1998b, c). The expressions of GluR2 mRNA and proteins are also governed by neuronal activity (Wong-Riley and Jacobs 2002; Bai and Wong-Riley 2003). If excitatory glutamatergic neurotransmission goes hand in hand with COX expression, the question naturally arises as to whether the coupling between these two activities exists at the molecular level? That is, can the same transcription factor or factors regulate COX as well as neurochemicals of glutamatergic neurotransmission?

12.8.2 Does NRF-1 Coordinate the Transcriptional Regulation of Neurochemicals and Cytochrome c Oxidase in Neurons?

NRF-1 is a natural candidate for such an inquiry. Its role in regulating all 13 *COX* subunit genes has been well defined (see above). A consensus recognition sequence for NRF-1 has been reported for the GC-rich proximal promoter of the rat *Gria2* (for GluR2) gene, but it has not been rigorously tested (Myers et al. 1998). Whether NRF-1 regulates the other AMPA (*Gria*) and any of the NMDA receptor (*Grin*) subunit genes was entirely unknown. By means of in silico analyses, in vitro EMSA and supershift assays, in vivo ChIP, promoter mutational analyses, and shRNA, NRF-1 was found to functionally regulate *Grin1* and *Grin2b*, but not the other, subunits of the NMDA receptor genes, and *Gria2*, but not the other, subunits of the AMPA receptor genes (Dhar and Wong-Riley 2009; Dhar et al. 2009b). The transcripts are upregulated by KCl depolarizing stimulation and downregulated by TTX impulse blockade in cultured primary neurons. However, silencing of NRF-1 blocks the upregulation, and over-expression of NRF-1 rescues the downregulation, of *Grin1*, *Grin2b*, *Gria2*, as well as *COX* subunit transcripts in neurons (Dhar and Wong-Riley 2009; Dhar et al. 2009b). NRF-1-binding sites on these genes are also highly conserved among rats, mice, and humans. As discussed above, NR1 is an essential subunit, NR2B is critical for a number of basic structural and functional attributes of the NMDA receptors, and GluR2 is an important regulatory subunit of AMPA receptors.

NRF-1 also functionally regulates neuronal nitric oxide synthase (*Nos1*) (Dhar et al. 2009c), which links NMDA receptor transmission to the cGMP second messenger cascade (Garthwaite 1991). This regulation is specific, as NRF-1 does not control the expressions of either inducible NOS (iNOS or *Nos2*) or endothelial NOS (eNOS or *Nos3*) (Dhar et al. 2009c). Silencing NRF-1 not only downregulates *Nos1* mRNA and proteins, but also the transcripts of guanylyl cyclase, a downstream target of the nitric oxide pathway (Dhar et al. 2009c).

12.8.3 Is There a Transcription Factory for Cytochrome c Oxidase and Genes of Glutamatergic Neurotransmission?

To verify that there is a coordinated transcription of COX and those of neurochemicals coregulated by NRF-1, chromosome confirmation capture was utilized. Indeed, interactions were found among genomic loci for *COX*, *Grin1*, *Grin2b*, *Gria2*, and *Nos1* in neurons, but not in C2C12 muscle cells, indicating that such a "factory" is neuron-specific (Dhar and Wong-Riley 2010). *COX* subunit genes also do not interact with *Grin3a*, *Gria4*, or *Nos3*, genes that are not regulated by NRF-1, nor with genes for calreticulin, a non-mitochondrial protein (Dhar and Wong-Riley 2010). Depolarizing stimulation increases the interaction frequencies between *COX* and neurochemical genes, whereas TTX impulse blockade or KCN inhibition of COX downregulates such interactions in neurons (Dhar and Wong-Riley 2010). Hence, these data are consistent with coordinated transcription of *COX* and specific glutamatergic neurochemical genes in the same transcription factory in neurons.

12.8.4 NRF-1 Coregulates NR2B and Its Transport Motor KIF17 in Neurons

More recently, NRF-1 was found to also regulate the expression of the kinesin superfamily protein KIF17 (Dhar and Wong-Riley 2011), which transports NR2B along microtubules specifically from the cell body to the dendrites, where it forms part of the NMDA receptor complex (Setou et al. 2000). Interestingly, *Kif17* is not regulated by NRF-2, and NRF-1 does not regulate other *Kif* transcripts, such as *Kif1a* (Dhar and Wong-Riley 2011). This is a clear example of how the same transcription factor regulates the expression of both the motor and its specific synaptic cargo in neurons.

12.8.5 Molecular Coupler(s) of Energy Metabolism and Neuronal Activity

NRF-1 plays the heretofore unrecognized and unappreciated role of dually coordinating the expressions of neurochemicals of glutamatergic neurotransmission and agents of energy metabolism (COX). This coordinated expression ensures that

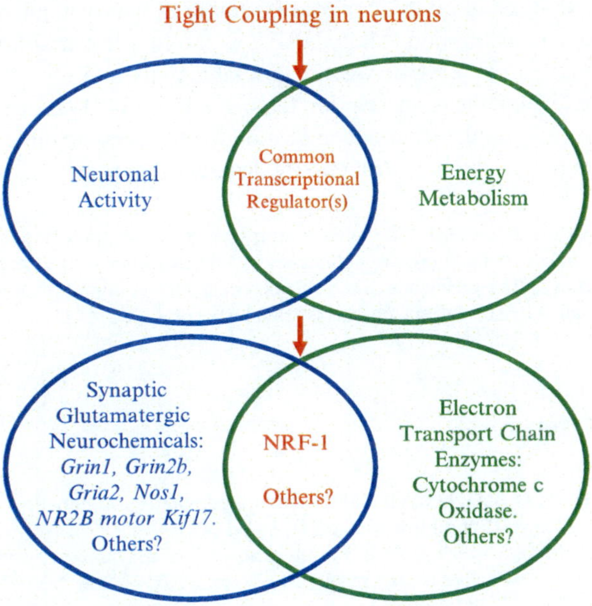

Fig. 12.3 A Venn diagram illustrating the tight coupling between neuronal activity and energy metabolism at the molecular level by having the same transcription factor, NRF-1, coregulating genes for critical glutamatergic neurochemicals (*Grin1*, *Grin2b*, *Gria2*, *Nos1*, and NR2B motor *Kif17*) as well as all 13 subunits of *COX*. Such coupling ensures that energy production exquisitely matches energy demand of neuronal activity

energy production precisely matches energy utilization and thereby mediates the tight coupling between neuronal activity and energy metabolism at the molecular level (Fig. 12.3). Whether NRF-2 and/or other transcription factors also participate in this coupling remains to be explored. For example, it is unknown if NRF-2 coregulates glutamatergic neurochemicals together with NRF-1 in a complementary, concurrent, or a combination of complementary and concurrent manner. Whether NRF-1 and NRF-2 interact as they activate their common target genes is also not known at this time. However, silencing each of them with shRNA does not affect the expression of the other (Ongwijitwat et al. 2006; Dhar et al. 2008), suggesting that the two may function independently of each other in neurons.

12.9 Conclusions

Cytochrome *c* oxidase is one of the most ancient enzymes known. Its critical roles in the complete oxidation of carbohydrates, amino acids, and fatty acids and in the generation of a proton gradient necessary for ATP synthesis within the mitochondria are well recognized. The absolute dependence of neurons on COX for their proper functioning and survival is without question. However, only in recent years has the transcriptional regulation of this multisubunit, multichromosomal, bigenomic enzyme been extensively explored in neurons. NRF-1 and NRF-2 are proven bigenomic transcriptional coordinators of all 13 *COX* subunit transcripts from the two genomes, and both of them are under strict regulation of neuronal activity.

NRF-1, in addition, regulates a number of neurochemicals crucial for glutamatergic neurotransmission. Thus, NRF-1 is the first transcription factor known to mediate the tight coupling between neuronal activity and energy metabolism at the molecular level. Other transcription factors, such as NRF-2, and coactivators, such as PGC-1α, may well participate in the coupling process to ensure the exquisite matching of energy production with energy demand of synaptic transmission in neurons.

Acknowledgments The author expresses her deep appreciation for the valuable contributions from her past and present colleagues in her laboratory as well as in the relevant fields of bioenergetics and neuroscience. She also appreciates the continuous support from the National Institutes of Health. The current support is from NIH grant EY018441.

References

Attwell D, Laughlin SB (2001) An energy budget for signaling in the grey matter of the brain. J Cereb Blood Flow Metab 21:1133–1145

Au HC, Scheffler IE (1998) Promoter analysis of the human succinate dehydrogenase iron-protein gene – both nuclear respiratory factors NRF-1 and NRF-2 are required. Eur J Biochem 251:164–174

Baar K, Wende AR, Jones TE, Marison M, Nolte LA, Chen M, Kelly DP, Holloszy JO (2002) Adaptations of skeletal muscle to exercise: rapid increase in the transcriptional coactivator PGC-1. FASEB J 16:1879–1886

Babb TL, Mikuni N, Najm I, Wylie C, Olive M, Dollar C, MacLennan H (2005) Pre- and postnatal expression of NMDA receptors 1 and 2B subunit proteins in the normal rat cortex. Epilepsy Res 64:23–30

Bachman NJ, Yang TL, Dasen JS, Ernst RE, Lomax MI (1996) Phylogenetic footprinting of the human cytochrome c oxidase subunit VB promoter. Arch Biochem Biophys 333:152–162

Bai X, Wong-Riley MTT (2003) Neuronal activity regulates protein and gene expressions of GluR2 in postnatal rat visual cortical neurons in culture. J Neurocytol 32:71–78

Batchelor AH, Piper DE, De la Brousse FC, McKnight SL, Wolberger C (1998) The structure of GABPalpha/beta: an ETS domain-ankyrin repeat heterodimer bound to DNA. Science 279:1037–1041

Becker TS, Burgess SM, Amsterdam AH, Allende ML, Hopkins N (1998) *Not really finished* is crucial for development of the zebrafish outer retina and encodes a transcription factor highly homologous to human nuclear respiratory factor-1 and avian initiation binding repressor. Development 125:4369–4378

Blesa JR, Prieto-Ruiz JA, Hernández JM, Hernández-Yago J (2007) NRF-2 transcription factor is required for human *TOMM20* gene expression. Gene 391:198–208

Bruni F, Polosa PL, Gadaleta MN, Cantatore P, Roberti M (2010) Nuclear respiratory factor 2 induces the expression of many but not all human proteins acting in mitochondrial DNA transcription and replication. J Biol Chem 285:3939–3948

Burnashev N, Monyer H, Seeburg PH, Sakmann B (1992) Divalent ion permeability of AMPA receptor channels is dominated by the edited form of a single subunit. Neuron 8:189–198

Carter RS, Avadhani NG (1994) Cooperative binding of GA-binding protein transcription factors to duplicated transcription initiation region repeats of cytochrome c oxidase subunit IV gene. J Biol Chem 269:4381–4387

Carter RS, Bhat NK, Basu A, Avadhani NG (1992) The basal promoter elements of murine cytochrome c oxidase subunit IV gene consist of tandemly duplicated ETS motifs that bind to GABP-related transcription factors. J Biol Chem 267:23418–23426

De la Brousse FC, Birkenmeier EH, King DS, Rowe LB, McKnight SL (1994) Molecular and genetic characterization of GABP beta. Genes Dev 8:1853–1865
Dekker J, Rippe K, Dekker M, Kleckner N (2002) Capturing chromosome conformation. Science 295:1306–1311
DeYoe EA, Trusk TC, Wong-Riley MTT (1995) Activity correlates of cytochrome oxidase-defined compartments in granular and supragranular layers of primary visual cortex of the macaque monkey. Vis Neurosci 12:629–639
Dhar SS, Wong-Riley MTT (2009) Coupling of energy metabolism and synaptic transmission at the transcriptional level: role of nuclear respiratory factor 1 in regulating both cytochrome c oxidase and NMDA glutamate receptor subunit genes. J Neurosci 29:483–492
Dhar SS, Wong-Riley MTT (2010) Chromosome conformation capture of transcriptional interactions between cytochrome c oxidase genes and genes of glutamatergic synaptic transmission in neurons. J Neurochem 115:676–683
Dhar SS, Wong-Riley MTT (2011) The kinesin superfamily protein KIF17 is regulated by the same transcription factor (NRF-1) as its cargo NR2B in neurons. BBA – Mol. Cell Res 1813:403–411
Dhar SS, Ongwijitwat S, Wong-Riley MTT (2008) Nuclear respiratory factor 1 regulates all ten nuclear-encoded subunits of cytochrome c oxidase in neurons. J Biol Chem 283:3120–3129
Dhar SS, Ongwijitwat S, Wong-Riley MTT (2009a) Chromosome conformation capture of all 13 genomic loci in the transcriptional regulation of the multi-subunit bigenomic cytochrome c oxidase in neurons. J Biol Chem 284:18644–18650
Dhar SS, Liang HL, Wong-Riley MTT (2009b) Nuclear respiratory factor 1 co-regulates AMPA glutamate receptor subunit 2 and cytochrome c oxidase: tight coupling of glutamatergic transmission and energy metabolism in neurons. J Neurochem 108:1595–1606
Dhar SS, Liang HL, Wong-Riley MTT (2009c) Transcriptional coupling of synaptic transmission and energy metabolism: role of nuclear respiratory factor 1 in co-regulating neuronal nitric oxide synthase and cytochrome c oxidase genes in neurons. BBA – Mol. Cell Res 1793:1604–1613
Dingledine R, Borges K, Bowie D, Traynelis SF (1999) The glutamate receptor ion channels. Pharmacol Rev 51:7–61
Evans MJ, Scarpulla RC (1989) Interaction of nuclear factors with multiple sites in the somatic cytochrome c promoter. Characterization of upstream NRF-1, ATF, and intron Sp1 recognition sequences. J Biol Chem 264:14361–14368
Evans MJ, Scarpulla RC (1990) NRF-1: a trans-activator of nuclear-encoded respiratory genes in animal cells. Genes Dev 4:1023–1034
Falkenberg M, Gaspari M, Rantanen A, Trifunovic A, Larsson NG, Gustafsson CM (2002) Mitochondrial transcription factors B1 and B2 activate transcription of human mtDNA. Nat Genet 31:289–294
Fisher RP, Clayton DA (1988) Purification and characterization of human mitochondrial transcription factor 1. Mol Cell Biol 8:3496–3509
Fonnum F (1984) Glutamate: a neurotransmitter in mammalian brain. J Neurochem 42:1–11
Garthwaite J (1991) Glutamate, nitric oxide and cell-cell signaling in the nervous system. Trends Neurosci 14:60–67
Gleyzer N, Vercauteren K, Scarpulla RC (2005) Control of mitochondrial transcription specificity factors (TFB1M and TFB2M) by nuclear respiratory factors (NRF-1 and NRF-2) and PGC-1 family coactivators. Mol Cell Biol 25:1354–1366
Gopalakrishnan L, Scarpulla RC (1995) Structure, expression, and chromosomal assignment of the human gene encoding nuclear respiratory factor 1. J Biol Chem 270:18019–18025
Goto M, Terada S, Kato M, Katoh M, Yokozeki T, Tabata I, Shimokawa T (2000) cDNA cloning and mRNA analysis of PGC-1 in epitrochlearis muscle in swimming-exercised rats. Biochem Biophys Res Commun 274:350–354
Gugneja S, Scarpulla RC (1997) Serine phosphorylation within a concise amino-terminal domain in nuclear respiratory factor 1 enhances DNA binding. J Biol Chem 272:18732–18739

Gugneja S, Virbasius CM, Scarpulla RC (1996) Nuclear respiratory factors 1 and 2 utilize similar glutamine-containing clusters of hydrophobic residues to activate transcription. Mol Cell Biol 16:5708–5716

Guo AL, Nie F, Wong-Riley MTT (2000) Human brain nuclear respiratory factor (NRF) 2α cDNA: isolation, subcloning, sequencing and *in situ* hybridization of transcripts in normal and visually deprived macaque visual system. J Comp Neurol 417:221–232

Hevner RF, Wong-Riley MTT (1989) Brain cytochrome oxidase: purification, antibody production, and immunohistochemical/histochemical correlations in the CNS. J Neurosci 9:3884–3898

Hevner RF, Wong-Riley MTT (1990) Regulation of cytochrome oxidase protein levels by functional activity in the macaque monkey visual system. J Neurosci 10:1331–1340

Hevner RF, Wong-Riley MTT (1991) Neuronal expression of nuclear and mitochondrial genes for cytochrome oxidase (CO) subunits analyzed by *in situ* hybridization; comparison with CO activity and protein. J Neurosci 11:1942–1958

Hevner RF, Wong-Riley MTT (1993) Mitochondrial and nuclear gene expression for cytochrome oxidase subunits are disproportionately regulated by functional activity in neurons. J Neurosci 13:1805–1819

Hevner RF, Liu S, Wong-Riley MTT (1995) A metabolic map of cytochrome oxidase in the rat brain: histochemical, densitometric and biochemical studies. Neuroscience 65:313–342

Hollmann M, Hartley M, Heinemann S (1991) Ca^{2+} permeability of KA-AMPA-gated glutamate receptor channels depends on subunit composition. Science 252:851–853

Huo L, Scarpulla RC (1999) Multiple 5′-untranslated exons in the nuclear respiratory factor 1 gene span 47 kb and contribute to transcript heterogeneity and translational efficiency. Gene 233:213–224

Huo L, Scarpulla RC (2001) Mitochondrial DNA instability and peri-implantation lethality associated with targeted disruption of nuclear respiratory factor 1 in mice. Mol Cell Biol 21:644–654

Jackson DA, Iborra FJ, Manders EM, Cook PR (1998) Numbers and organization of RNA polymerases, nascent transcripts, and transcription units in HaLa nuclei. Mol Biol Cell 9:1523–1536

Jonas P, Racca C, Sakmann B, Seeburg PH, Monyer H (1994) Differences in Ca^{2+} permeability of AMPA-type glutamate receptor channels in neocortical neurons caused by differential GluR-B subunit expression. Neuron 12:1281–1289

Kadenbach B, Jaraush S, Hartmann R, Merle P (1983) Separation of mammalian cytochrome c oxidase into 13 polypeptides by a sodium dodecyl sulfate- gel electrophoresis procedure. Anal Biochem 129:517–521

Kadenbach B, Huttermann M, Arnold S, Lee I, Bender E (2000) Mitochondrial energy metabolism is regulated via nuclear-coded subunits of cytochrome c oxidase. Free Radic Biol Med 29:211–221

Keinänen K, Wisden W, Sommer B, Werner P, Herb A, Verdoorn TA, Sakmann B, Seeburg PH (1990) A family of AMPA-selective glutamate receptors. Science 249:556–560

Kelly DP, Scarpulla RC (2004) Transcriptional regulatory circuits controlling mitochondrial biogenesis and function. Genes Dev 18:357–368

Knutti D, Kralli A (2001) PGC-1, a versatile coactivator. Trends Endocrinol Metab 12:360–365

LaMarco K, Thompson CC, Byers BP, Walton EM, McKnight SL (1991) Identification of Ets- and Notch-related subunits in GA-binding protein. Science 253:789–792

Lehman JJ, Barger PM, Kovacs A, Saffitz JE, Medeiros DM, Kelly DP (2000) Peroxisome proliferator-activated receptor gamma coactivator-1 promotes cardiac mitochondrial biogenesis. J Clin Invest 106:847–856

Leone TC, Lehman JJ, Finck BN, Schaeffer PJ, Wende AR, Boudina S, Courtois M, Wozniak DF, Sambandam N, Bernal-Mizrachi C, Chen Z, Holloszy JO, Medeiros DM, Schmidt RE, Saffitz JE, Abel ED, Semenkovich CF, Kelly DP (2005) PGC-1alpha deficiency causes multi-system energy metabolic derangements: muscle dysfunction, abnormal weight control and hepatic steatosis. PLoS Biol 3:e101

Liang HL, Wong-Riley MTT (2006) Activity-dependent regulation of nuclear respiratory factor-1, nuclear respiratory factor-2, and peroxisome proliferators-activated receptor gamma coactivator-1 in neurons. Neuroreport 17:401–405

Liang HL, Ongwijitwat S, Wong-Riley MTT (2006) Bigenomic functional regulation of all 13 cytochrome c oxidase subunit transcripts in rat neurons *in vitro* and *in vivo*. Neuroscience 140:177–190

Liang HL, Dhar SS, Wong-Riley MTT (2010) p38 Mitogen-activated protein kinase and calcium channels mediate signaling in depolarization-induced activation of peroxisome proliferator-activated receptor gamma coactivator-1α in neurons. J Neurosci Res 88:640–649

Lin J, Wu PH, Tarr PT, Lindenberg KS, St-Pierre J, Zhang CY, Mootha VK, Jager S, Vianna CR, Reznick RM, Cui L, Manieri M, Donovan MX, Wu Z, Cooper MP, Fan MC, Rohas LM, Zavacki AM, Cinti S, Shulman GI, Lowell BB, Krainc D, Spiegelman BM (2004) Defects in adaptive energy metabolism with CNS-linked hyperactivity in PGC-1alpha null mice. Cell 119:121–135

Ling JQ, Li T, Hu JF, Vu TH, Chen HL, Qiu XW, Cherry AM, Hoffman AR (2006) CTCF mediates interchromosomal colocalization between Igf2/H19 and Wsb1/Nf1. Science 312:269–272

Liu S, Wong-Riley M (1994) Nuclear-encoded mitochondrial precursor protein: intramitochondrial delivery to dendrites and axon terminals of neurons and regulation by neuronal activity. J Neurosci 14:5338–5351

Loftis JM, Janowsky A (2003) The N-methyl-D-aspartate receptor subunit NR2B: localization, functional properties, regulation, and clinical implications. Pharmacol Ther 97:55–85

Lomeli H, Mosbacher J, Melcher T, Höger T, Geiger JRP, Kuner T, Monyer H, Higuchi M, Bach A, Seeburg PH (1994) Control of kinetic properties of AMPA receptor channels by nuclear RNA editing. Science 266:1709–1713

Lomvardas S, Barnea G, Pisapia DJ, Mendelsohn M, Kirkland J, Axel R (2006) Interchromosomal interactions and olfactory receptor choice. Cell 126:403–413

Lowry OH (1975) Energy metabolism in brain and its control. In: Ingvar DH, Lassen NA (eds) Brain work: the coupling of function, metabolism, and blood flow in the brain. Alfred Benzon symposium VIII. Academic, New York, pp 48–64

Meng H, Liang HL, Wong-Riley M (2007) Quantitative immune-electron microscopic analysis of depolarization-induced expression of PGC-1alpha in cultured rat visual cortical neurons. Brain Res 1175:10–16

Miele A, Dekker J (2009) Mapping cis- and trans-chromatin interaction networks using chromosome conformation capture (3C). Methods Mol Biol 464:105–121

Mori H, Mishina M (1995) Structure and function of the NMDA receptor channel. Neuropharmacology 34:1219–1237

Mori M, Morita T, Ikeda F, Amaya Y, Tatibana M, Cohen PP (1981) Synthesis, intracellular transport, and processing of the precursor for mitochondrial ornithine transcarbamylase and carbamoylphosphate synthetase I in isolated hepatocytes. Proc Natl Acad Sci USA 78:6056–6060

Myers SJ, Peters J, Huang Y, Comer MB, Barthel F, Dingledine R (1998) Transcriptional regulation of the GluR2 gene: neural-specific expression, multiple promoters, and regulatory elements. J Neurosci 18:6723–6739

Nakanishi S (1992) Molecular diversity of glutamate receptors and implications for brain function. Science 258:597–603

Nie F, Wong-Riley M (1999) Nuclear respiratory factor-2 subunit protein: correlation with cytochrome oxidase and regulation by functional activity in the monkey primary visual cortex. J Comp Neurol 404:310–320

Ongwijitwat S, Wong-Riley MTT (2004) Functional analysis of the rat cytochrome c oxidase subunit 6A1 promoter in primary neurons. Gene 337:163–171

Ongwijitwat S, Wong-Riley MTT (2005) Is nuclear respiratory factor 2 a master transcriptional coordinator for all ten nuclear-encoded cytochrome c oxidase subunits in neurons? Gene 360:65–77

Ongwijitwat S, Liang HL, Graboyes EM, Wong-Riley MTT (2006) Nuclear respiratory factor 2 senses changing cellular energy demands and its silencing down-regulates cytochrome oxidase and other target gene mRNAs. Gene 374:39–49

Orrego F, Villanueva S (1993) The chemical nature of the main central excitatory transmitter: a critical appraisal based upon release studies and synaptic vesicle localization. Neuroscience 56:539–555

Osborne CS, Chakalova L, Brown KE, Carter D, Horton A, Debrand E, Goyenechea B, Mitchell JA, Lopes S, Reik W, Fraser P (2004) Active genes dynamically colocalize to shared sites of ongoing transcription. Nat Genet 36:1065–1071

Palmer CL, Lin W, Hastie PGR, Toward M, Korolchuk VI, Burbidge SA, Banting G, Collingridge GL, Isaac JTR, Henley JM (2005) Hippocalcin functions as a calcium sensor in hippocampal LTD. Neuron 47:487–494

Pellegrini-Giampietro DE, Gorter JA, Bennett MV, Zukin RS (1997) The GluR2 (GluR-B) hypothesis: Ca^{2+}-permeable AMPA receptors in neurologic disorders. Trends Neurosci 20:464–470

Puigserver P, Wu Z, Park CW, Graves R, Wright M, Spiegelman BM (1998) A cold-inducible coactivator of nuclear receptors linked to adaptive thermogenesis. Cell 92:829–839

Reid GA, Yonetani T, Schatz G (1982) Import of proteins into mitochondria: import and maturation of the mitochondrial intermembrane space enzyme cytochemistry *b2* and cytochrome c peroxidase in intact yeast cells. J Biol Chem 257:13068–13074

Ristevski S, O'Leary DA, Thornell AP, Owen MJ, Kola I, Hertzog PJ (2004) The ETS transcription factor GABPalpha is essential for early embryogenesis. Mol Cell Biol 24:5844–5849

Salussolia CL, Prodromou ML, Borker P, Wollmuth LP (2011) Arrangement of subunits in functional NMDA receptors. J Neurosci 31:11295–11304

Scarpulla RC (1997) Nuclear control of respiratory chain expression in mammalian cells. J Bioenerg Biomembr 29:109–119

Scarpulla RC (2002) Nuclear activators and coactivators in mammalian mitochondrial biogenesis. Biochim Biophys Acta 1576:1–14

Scarpulla RC (2008) Nuclear control of respiratory chain expression by nuclear respiratory factors and PGC-1 related coactivator. Ann N Y Acad Sci 1147:321–334

Scarpulla RC (2011) Metabolic control of mitochondrial biogenesis through the PGC-1 family regulatory network. Biochim Biophys Acta 1813:1269–1278

Schoenfelder S, Sexton T, Chakalova L, Cope NF, Horton A, Andrews S, Kurukuti S, Mitchell JA, Umlauf D, Dimitrova DS, Eskiw CH, Luo Y, Wei CL, Ruan Y, Bieker JJ, Fraser P (2010) Preferential associations between co-regulated genes reveal a transcriptional interactome in erythroid cells. Nat Genet 42:53–61

Seelan RS, Gopalakrishnan L, Scarpulla RC, Grossman LI (1996) Cytochrome c oxidase subunit VIIa liver isoform. Characterization and identification of promoter elements in the bovine gene. J Biol Chem 271:2112–2120

Setou M, Nakagawa T, Seog DH, Hirokawa N (2000) Kinesin superfamily motor protein KIF17 and mLin-10 in NMDA receptor-containing vesicle transport. Science 288:1796–1802

Sommer B, Keinänen K, Verdoorn TA, Wisden W, Burnashev N, Herb A, Köhler M, Takagi T, Sakmann B, Seeburg PH (1990) Flip and flop: a cell-specific functional switch in glutamate-operated channels of the CNS. Science 249:1580–1585

Spilianakis CG, Flavell RA (2004) Long-range intrachromosomal interactions in the T helper type 2 cytokine locus. Nat Immunol 5:1017–1027

St-Pierre J, Drori S, Uldry M, Silvaggi JM, Rhee J, Jager S, Handschin C, Zheng K, Lin J, Yang W, Simon DK, Bachoo R, Spiegelman BM (2006) Suppression of reactive oxygen species and neurodegeneration by the PGC-1 transcriptional coactivators. Cell 127:397–408

Streit P (1984) Glutamate and aspartate as transmitter candidate for systems of the cerebral cortex. In: Jones EG, Peters A (eds) Cerebral cortex, vol 2, Functional properties of cortical cells. Plenum, New York, pp 119–143

Tanaka H, Grooms SY, Bennett MVL, Zukin RS (2000) The AMPAR subunit GluR2: still front and center-stage. Brain Res 886:190–207

Thompson CC, Brown TA, McKnight SL (1991) Convergence of Ets- and notch-related structural motifs in a heteromeric DNA binding complex. Science 253:762–768

Verdoorn TA, Burnashev N, Monyer H, Seeburg PH, Sakmann B (1991) Structural determinants of ion flow through recombinant glutamate receptor channels. Science 252:1715–1718

Virbasius JV, Scarpulla RC (1991) Transcriptional activation through ETS domain binding sites in the cytochrome c oxidase subunit IV gene. Mol Cell Biol 11:5631–5638

Virbasius JV, Scarpulla RC (1994) Activation of the human mitochondrial transcription factor A gene by nuclear respiratory factors: a potential regulatory link between nuclear and mitochondrial gene expression in organelle biogenesis. Proc Natl Acad Sci USA 91:1309–1313

Virbasius JV, Virbasius CA, Scarpulla RC (1993a) Identity of GABP with NRF-2: a multisubunit activator of cytochrome oxidase expression, reveals a cellular role for an ETS domain activator of viral promoters. Genes Dev 7:380–392

Virbasius CA, Virbasius JV, Scarpulla RC (1993b) NRF-1, an activator involved in nuclear-mitochondrial interaction, utilizes a new DNA-binding domain conserved in a family of developmental regulators. Genes Dev 7:2431–2445

Washburn MS, Numberger M, Zhang S, Dingledine R (1997) Differential dependence on GluR2 expression of three characteristic features of AMPA receptors. J Neurosci 17:9393–9406

Wong-Riley M (1979) Changes in the visual system of monocularly sutured or enucleated cats demonstrable with cytochrome oxidase histochemistry. Brain Res 171:11–28

Wong-Riley MTT (1989) Cytochrome oxidase: an endogenous metabolic marker for neuronal activity. Trends Neurosci 12:94–101

Wong-Riley MTT (2010) Energy metabolism of the visual system. Eye Brain 2:99–116

Wong-Riley M, Carroll EW (1984) Effect of impulse blockage on cytochrome oxidase activity in monkey visual system. Nature 307:262–264

Wong-Riley MTT, Jacobs P (2002) AMPA glutamate receptor subunit 2 in normal and visually deprived macaque visual cortex. Vis Neurosci 19:563–573

Wong-Riley MTT, Tripathi SC, Trusk TC, Hoppe DA (1989a) Effect of retinal impulse blockage on cytochrome oxidase-rich zones in the macaque striate cortex: I. Quantitative electron-microscopic (EM) analysis of neurons. Vis Neurosci 2:483–497

Wong-Riley MTT, Trusk TC, Tripathi SC, Hoppe DA (1989b) Effect of retinal impulse blockage on cytochrome oxidase-rich zones in the macaque striate cortex: II. Quantitative electron-microscopic (EM) analysis of neuropil. Vis Neurosci 2:499–514

Wong-Riley MTT, Mullen MA, Huang Z, Guyer C (1997) Brain cytochrome oxidase subunit complementary DNAs: isolation, subcloning, sequencing, light and electron microscopic *in situ* hybridization of transcripts, and regulation by neuronal activity. Neuroscience 76:1035–1055

Wong-Riley MTT, Nie F, Hevner RF, Liu S (1998a) Brain cytochrome oxidase. In: Gonzalez-Lima F (ed) Cytochrome oxidase in neuronal metabolism and Alzheimer's disease. Plenum, New York, pp 1–53

Wong-Riley M, Anderson B, Liebl W, Huang Z (1998b) Neurochemical organization of the macaque striate cortex: correlation of cytochrome oxidase with Na^+K^+ATPase, NADPH-diaphorase, nitric oxide synthase, and N-methyl-D-aspartate receptor subunit 1. Neuroscience 83:1025–1045

Wong-Riley MT, Huang Z, Liebl W, Nie F, Xu H, Zhang C (1998c) Neurochemical organization of the macaque retina: effect of TTX on levels and gene expression of cytochrome oxidase and nitric oxide synthase and on the immunoreactivity of Na^+K^+ATPase and NMDA receptor subunit 1. Vision Res 38:1455–1477

Wong-Riley MTT, Yang SJ, Liang HL, Ning G, Jacobs P (2005) Quantitative immuno-electron microscopic analysis of nuclear respiratory factor 2 alpha and beta subunits: normal distribution and activity-dependent regulation in mammalian visual cortex. Vis Neurosci 22:1–18

Wong-Riley MTT, Liang HL, Ongwijitwat S (2008) Activity-dependent bigenomic transcriptional regulation of cytochrome c oxidase in neurons, Chapter 11. In: Dudek SM (ed) Transcriptional regulation by neuronal activity: to the nucleus and back. Springer, New York, pp 209–228

Wu Z, Puigserver P, Andersson U, Zhang C, Adelmant G, Mootha V, Troy A, Cinti S, Lowell B, Scarpulla RC, Spiegelman BM (1999) Mechanisms controlling mitochondrial biogenesis and respiration through the thermogenic coactivator PGC-1. Cell 98:115–124

Yang SJ, Liang HL, Ning G, Wong-Riley MTT (2004) Ultrasturctural study of depolarization-induced translocation of NRF-2 transcription factor in cultured rat visual cortical neurons. Eur J Neurosci 19:1153–1162

Yang SJ, Liang HL, Wong-Riley MTT (2006) Activity-dependent transcriptional regulation of nuclear respiratory factor-1 in cultured rat visual cortical neurons. Neuroscience 141:1181–1192

Zhang C, Wong-Riley M (1999) Expression and regulation of NMDA receptor subunit R1 and neuronal nitric oxide synthase in cortical neuronal cultures: correlation with cytochrome oxidase. J Neurocytol 28:525–539

Zhang C, Wong-Riley MTT (2000a) Synthesis and degradation of cytochrome oxidase subunit mRNA in neurons: differential bigenomic regulation by neuronal activity. J Neurosci Res 60:338–344

Zhang C, Wong-Riley M (2000b) Depolarization stimulation upregulates GA-binding protein in neurons: a transcription factor involved in the bigenomic expression of cytochrome oxidase subunits. Eur J Neurosci 12:1013–1023

Zhou GL, Xin L, Song W, Di LJ, Liu G, Wu XS, Liu DP, Liang CC (2006) Active chromatin hub of the mouse alpha-globin locus forms in a transcription factory of clustered house-keeping genes. Mol Cell Biol 26:5096–5105

Chapter 13
Cytochrome *c* Oxidase and Its Role in Neurodegeneration and Neuroprotection

Susanne Arnold

Abstract A hallmark of neurodegenerative diseases, such as Alzheimer's and Parkinson's diseases, and stroke is a malfunction of mitochondria including cytochrome *c* oxidase (COX), the terminal enzyme complex of the respiratory chain. COX is ascribed a key role based on mainly two regulatory mechanisms. These are the expression of isoforms and the binding of specific allosteric factors to nucleus-encoded subunits. These characteristics represent a unique feature of COX compared with the other respiratory chain complexes. Additional regulatory mechanisms, such as post-translational modification, substrate availability, and allosteric feedback inhibition by products of the COX reaction, control the enzyme activity in a complex way. In many tissues and cell types, COX represents the rate-limiting enzyme of the respiratory chain which further emphasizes the impact of the regulation of COX as a central site for regulating energy metabolism and oxidative stress. Two of the best-analyzed regulatory mechanisms of COX to date are the allosteric feedback inhibition of the enzyme by its indirect product ATP and the expression of COX subunit IV isoforms. This ATP feedback inhibition of COX requires the expression of COX isoform IV-1. At high ATP/ADP ratios, ADP is exchanged for ATP at the matrix side of COX IV-1 leading to an inhibition of COX activity, thus enabling COX to sense the energy level and to adjust ATP synthesis to energy demand. However, under hypoxic, toxic, and degenerative conditions, COX isoform IV-2 expression is up-regulated and exchanged for COX IV-1 in the enzyme complex. This COX IV isoform switch causes an abolition of the allosteric ATP feedback inhibition of COX and consequently the loss of sensing the energy level. Thus, COX activity is increased leading to higher levels of ATP in neural cells independently of the cellular energy level. Concomitantly, ROS production is increased. Thus, under pathological conditions, neural cells are provided with ATP to meet the energy demand, but at the expense of elevated oxidative stress.

S. Arnold (✉)
Institute for Neuroanatomy, Faculty of Medicine, RWTH Aachen University, Wendlingweg 2, Aachen 52074, Germany
e-mail: sarnold2012@googlemail.com

B. Kadenbach (ed.), *Mitochondrial Oxidative Phosphorylation*, Advances in Experimental Medicine and Biology 748, DOI 10.1007/978-1-4614-3573-0_13,

This mechanism explains the functional relevance of COX subunit IV isoform expression for cellular energy sensing, ATP production, and oxidative stress levels. This, in turn, affects neural cell function, signaling, and survival. Thus, COX is a crucial factor in etiology, progression, and prevalence of numerous human neurodegenerative diseases and represents an important target for developing diagnostic and therapeutic tools against those diseases.

13.1 Introduction

The central nervous system (CNS) depends on oxidative metabolism of glucose more than any other organ in the mammalian organism. Thus, oxygen and glucose are the two essential energy substrates to support the function of brain and spinal cord. It points to the importance of aerobic energy metabolism by mitochondria. One of the key players in aerobic mitochondrial energy metabolism is the terminal and highly regulated enzyme complex IV of the respiratory chain, cytochrome *c* oxidase (COX). COX is responsible for reduction of up to 95% of the oxygen taken up by mammalian organisms which are characterized by a mainly aerobic mitochondrial energy metabolism. The bigenomic COX complex is regulated mainly by isoform expression and allosteric effector binding, two features unique for the COX within the respiratory chain. Together with its central role in energy production, signaling and oxidative stress formation in neural cells, COX takes center stage in regulating neural cell function and survival under physiological, pathological, and protective conditions in the CNS.

13.2 Correlation of Structure and Function of the Mammalian Cytochrome *c* Oxidase

The mammalian COX (EC 1.9.3.1.) represents a dimeric enzyme complex each monomer of which is composed of 13 subunits with a molecular weight of 205 kDa (Tsukihara et al. 1996). In mammals, the biogenesis of COX involves the coordinated assembly of three subunits encoded by mitochondrial DNA (mtDNA) and ten by nuclear DNA (ncDNA, Figs. 13.1 and 13.2, nomenclature after Kadenbach et al. 1983). The mitochondria-encoded COX subunits contain the three catalytic centers of the enzyme and are essential and sufficient for the catalytic COX activity in eukaryotes and prokaryotes and show high evolutionary conservation (Babcock and Wikström 1992; Ferguson-Miller and Babcock 1996; Ostermeier et al. 1997; Yoshikawa et al. 1998). Subunit II contains the two-copper center Cu_A which represents the binding site for cytochrome *c*. Heme a and the oxygen-binding heme a_3/Cu_B centers are located in subunit I. The catalytic functions imply the transfer of electrons from ferrocytochrome *c* to oxygen, accompanied by the vectorial uptake of protons for the formation of water and the outward translocation of protons building

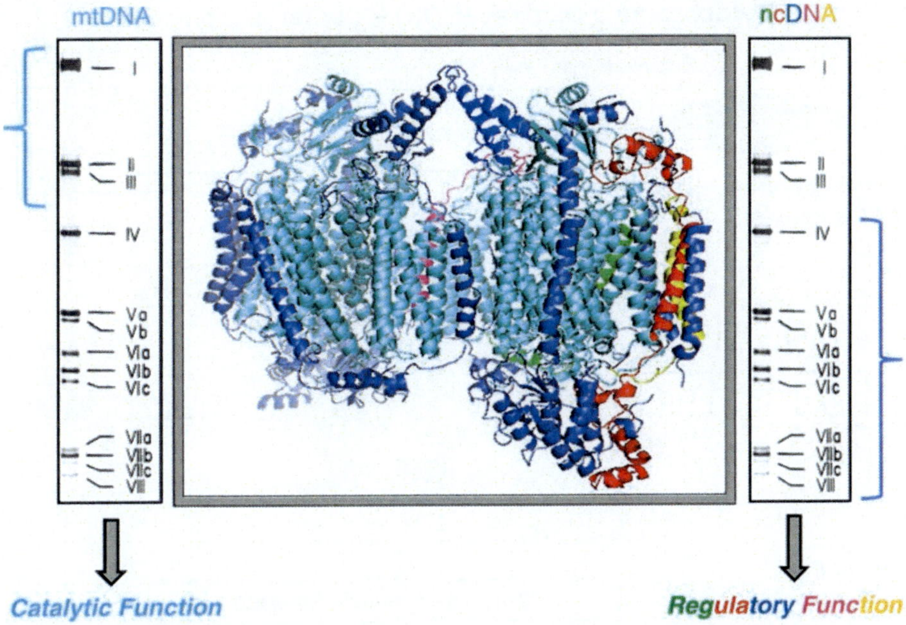

Fig. 13.1 Schematic representation of cytochrome *c* oxidase structure and function. Dimeric cytochrome *c* oxidase (COX) is positioned at the center of the scheme. Crystallographic data of dimeric bovine heart COX (Tsukihara et al. 1996) were taken from PDB entry 1OCC and processed with the software program RASMOL 2.7. The three mitochondria-encoded subunits in each monomer are represented as cyan-colored helices, the ten nuclear-encoded subunits are depicted in various colors except *cyan*. The *left panel* points to three mtDNA-encoded and catalytic subunits I–III, the *right panel* to ten ncDNA-encoded and regulatory subunits IV–VIII visualized after sodium dodecylsulfate polyacrylamide gel electrophoresis

up a proton gradient across the inner mitochondrial membrane as part of the mitochondrial membrane potential ($\Delta\Psi m$). The proton gradient serves the phosphorylation of ADP and inorganic phosphate to ATP through the F0F1-ATP synthase. Both the proton gradient and ATP can, therefore, be considered as products of the COX reaction. The catalytic activities of the mammalian and bacterial enzymes are similar when studied as isolated enzymes under standard conditions (Hendler et al. 1991). However, large differences in the catalytic activities occur between the two enzymes under more physiological conditions indicating a significant role of nucleus-encoded subunits of the eukaryotic COX complex in electron transport and proton pumping (Kadenbach 1986; Kadenbach et al. 2000; Ludwig et al. 2001).

Whereas the catalytic core consisting of the three mtDNA-encoded COX subunits are synthesized within the mitochondria, ten peripheral COX subunits are synthesized in the cytoplasm and are subsequently imported into the mitochondria and incorporated into the COX complex. These so-called "additional" mammalian COX subunits are required for the stability of the enzyme core and the regulation of its activity in response to particular cellular energy demand and signaling processes of neural cells to support their function and survival.

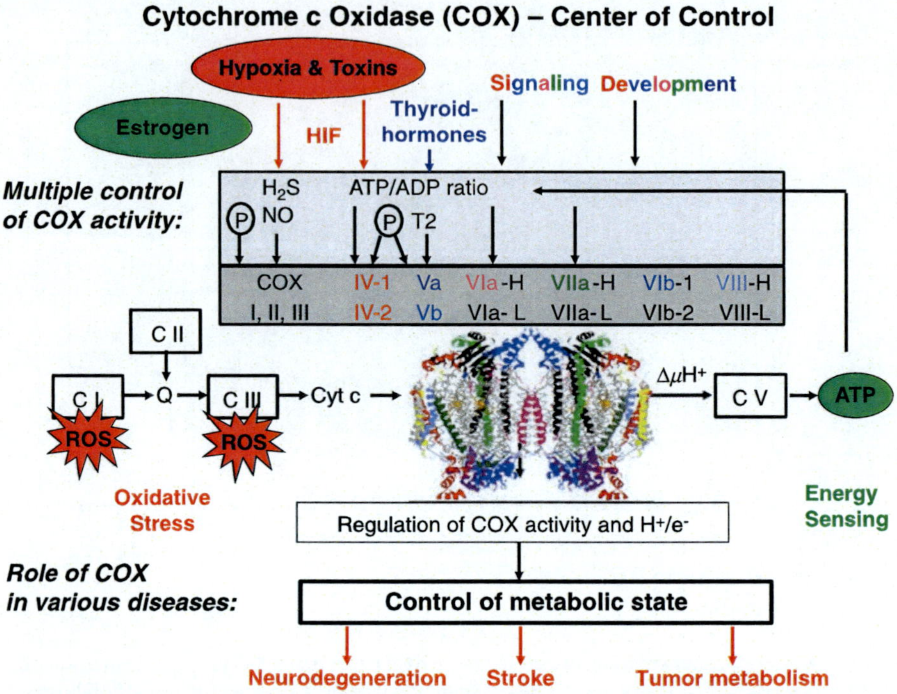

Fig. 13.2 Schematic representation of cytochrome *c* oxidase regulation in neurodegeneration and neuroprotection. Cytochrome *c* oxidase (COX) is positioned at the center of the scheme (mitochondrial matrix is located at the *bottom* of the COX crystal structure, the intermembrane space at the *top*). Crystallographic data of dimeric bovine heart COX (Tsukihara et al. 1996) were taken from PDB entry 1OCC and processed with the software program RASMOL 2.7.Three mitochondria-encoded catalytic subunits in each monomer are represented as peptide backbone traces (*gray*) with their redox centers highlighted in *blue* (copper atoms of Cu_A, Cu_B centers) and *orange* (heme a, a3). The helices of the ten nuclear-encoded, regulatory subunits are depicted in color (IV-*red*, Va-*purple*, VIa-*magenta*, VIIa-*green*, VIb-*blue*, VIII-*light-blue*). *Black arrows* pointing from respiratory chain complexes I and II to complex III (C I–C III) via ubiquinone (Q) and further to cytochrome *c* (Cyt c) and COX indicate the electron transfer, which serves to build up a proton gradient ($\Delta\mu H^+$) across the inner mitochondrial membrane, and which in turn is used to drive the ATP synthase (C V). The produced ATP can be considered as an indirect product of COX reaction functioning as an allosteric inhibitor of COX activity in a negative feedback reaction (*arrow* directing to ATP/ADP ratio in the *upper panel*).
Upper panel: Multiple effectors (phosphorylation (P), H_2S, NO, ATP/ADP ratio, 3,5-diiodothyronine (T2)) resulting from developmental, signaling or pathophysiological processes (protection by estrogen, hypoxia via HIF, thyroid hormone action, mitochondrial neurotoxin exposure, such as azide, cyanide, cobalt, NPA, 6-OHDA, MPP^+) interact with nucleus-encoded regulatory COX subunits (IV–VIII, indicated by *black arrows*). COX subunit isoforms (IV, VIa, VIIa, VIb, VIII) are listed and depicted in subunit-specific color.
Lower panel: The interaction of effector molecules with COX subunits induces regulatory mechanisms affecting COX activity and H^+/e^- stoichiometry which in turn affect the respiratory chain (ATP and ROS production) with respect to COX being the key and rate-limiting respiratory chain complex. The metabolic control by COX leads to physiological and/or pathological signaling, protecting from, promoting or inducing diseases, such as neurodegeneration. For further explanations and references, see text

13.3 Mechanisms of Regulation of Cytochrome *c* Oxidase

Mitochondria are the main producers of ATP in eukaryotic cells and as such they fulfill the cellular energy demand. The control of respiration, named "respiratory control," is explained according to the chemiosmotic hypothesis. Peter Mitchell (1961) identified a universal principle of the proton motive force Δp_m as the intermediate energy storage in all organisms. The transmembrane electrochemical potential $\Delta\mu H^+$ consists mainly of the transmembrane electrical potential $\Delta\Psi m$ and the transmembrane proton gradient ΔpH (Mitchell 1961). Thus, "respiratory control" is reflected as inhibition of the mitochondrial proton pumping respiratory chain complexes, I, III, and IV by high $\Delta\Psi m$. Although the mechanisms of respiration and oxidative phosphorylation in mitochondria have been principally clarified (Nicholls and Ferguson 2002), their regulation in living eukaryotic organisms under physiological and pathological conditions remains largely unknown.

"Metabolic flux control analyses" revealed little rate-limiting control of the terminal enzyme of the respiratory chain in isolated mitochondria (Groen et al. 1982; Letellier et al. 1993). In contrast, COX represents the rate-limiting step of the mitochondrial electron transport chain with a small excess capacity and tight in vivo control of respiration in cultured cells and saponin-permeabilized muscle fibers (Dalmonte et al. 2009; Kunz et al. 2000; Li et al. 2006; Villani and Attardi 1997, 2000; Villani et al. 1998). Thus, COX takes center stage for metabolic and cellular signaling control mechanisms (Fig. 13.2). This rate-limiting feature of COX for mitochondrial respiration becomes even more crucial with respect to the regulatory mechanisms known so far only for the nuclear-encoded COX subunits, such as allosteric effector binding (adenine nucleotides and diiodothyronine), posttranslational modifications, e.g., phosphorylation, and COX subunit isoform expression (Arnold 2011).

13.3.1 Cytochrome c Oxidase Substrates and Products

The most obvious mechanism for regulating the catalysis of an enzyme is by the availability of substrates which includes oxygen as a direct COX substrate and NADH, ADP, and inorganic phosphate as mitochondrial respiratory chain substrates (Lardy and Wellman 1952). Oxygen is the most important substrate to fulfill energy requirements of organs with high energy demand, such as the brain. Therefore, hypoxic conditions cause detrimental effects, especially to the high energy-demanding neurons.

The proton motive force Δp_m can be considered as a product of the respiratory chain activity and thereby of COX affecting directly its activity. ATP, however, represents an indirect product of COX activity. At increasing Δp_m and at high matrix ATP/ADP ratios the reconstituted COX enzyme shows a decrease in catalytic activity, thereby causing a diminished ATP production (Fig. 13.2, Arnold and Kadenbach 1997; Murphy and Brand 1987; Papa et al. 1991). Furthermore, if the availability of

ADP in the mitochondrial matrix becomes limiting, Δp_m rises and, in turn, inhibits the proton pumps at high values by a mechanism known as "respiratory control" and described by the chemiosmotic hypothesis (Chance and Williams 1955; Mitchell 1961). Yet another negative feedback mechanism is based on the allosteric inhibition of COX by ATP, the indirect product of the COX reaction, at high intramitochondrial ATP/ADP ratios. It was discovered by us (Arnold and Kadenbach 1997, 1999) and we named it "second mechanism of respiratory control" (Kadenbach and Arnold 1999). This "allosteric inhibition of COX by ATP" was verified by high-affinity binding of ATP to the matrix domain of COX subunit IV accompanied by sigmoidal enzyme kinetics with a Hill coefficient of 2 and occurring independently of $\Delta\Psi m$ (Fig. 13.2, Arnold and Kadenbach 1997, 1999). The regulation of COX by adenine nucleotides was also considered by Beauvoit and Rigoulet (2001) as being a feedback regulation of oxidative phosphorylation by its end-products.

13.3.2 Cytochrome c Oxidase Effector Molecule Binding

With respect to the complexity of the COX structure, especially the presence of "additional" ten nuclear-encoded COX subunits, a function for these subunits in the context with regulation of enzyme catalysis becomes obvious. So far, regulatory functions have been ascribed to mammalian (not to bacterial) COX subunit IV isoforms, subunit Va, and subunit VIa isoforms (Fig. 13.2, Arnold et al. 1998; Boyalla et al. 2011; Horvat et al. 2006; Ludwig et al. 2001; Singh et al. 2009). The regulation of respiration and ATP synthesis in higher organisms is mainly accomplished by allosteric effectors, such as metabolites, ions, hormones, and their binding to specific sites at nuclear-encoded subunits of COX. This has been postulated 25 years ago by Kadenbach (1986) and such regulatory mechanisms were suggested to change the activity and coupling degree of proton/electron transfer of COX.

13.3.3 Adenine Nucleotides: Signaling an Allosteric Feedback Inhibition of COX

Electron transfer and energy transduction in COX are regulated by multiple binding sites for adenine nucleotides at the ncDNA-encoded COX subunits (Antonini et al. 1988; Bisson et al. 1987; Kadenbach et al. 1998; Malatesta et al. 1987; Reimann et al. 1988). The bovine heart COX contains ten high affinity ADP-binding sites, seven of which are exchanged for ATP at high ATP/ADP ratios (Napiwotzki and Kadenbach 1998; Napiwotzki et al. 1997). Interaction of adenine nucleotides with its multiple binding sites at the COX subunits confers changes in energy transduction efficiency, i.e., H^+/e^- stoichiometry (Frank and Kadenbach 1996), and an allosteric feedback inhibition causing modifications in cytochrome *c* affinity, enzyme cooperativity, and catalytic activity (Arnold 2011; Arnold and Kadenbach 1997, 1999).

With respect to the regulation of COX function by adenine nucleotides, subunit IV has been shown to be a key regulatory subunit. COX subunit IV revealed two binding sites for ATP or ADP at the intermembrane domain and at the matrix domain. The binding site at the intermembrane domain causes decreased affinity of the enzyme for cytochrome *c*, when ATP instead of ADP is bound (Napiwotzki and Kadenbach 1998). The adenine nucleotide binding site located at the matrix domain of COX subunit IV confers the allosteric inhibition of COX activity at high ATP/ADP levels by binding of ATP, thereby regulating the catalytic activity, substrate affinity and cooperativity of the substrate binding sites of the mammalian dimeric COX complex (Fig. 13.1, Arnold and Kadenbach 1997). Thus, at high cellular energy levels (ATP/ADP ratios) ATP functions as an allosteric inhibitor of COX activity showing a sigmoidal titration curve of COX activity in dependence on the substrate cytochrome *c* concentration and a Hill coefficient of 2 (Arnold and Kadenbach 1999). Both kinetic parameters point to a cooperativity of two cytochrome *c*-binding sites in the dimeric enzyme complex in tuna and bovine heart (Arnold and Kadenbach 1997, 1999) and in astrocytes from different mouse brain regions (Horvat et al. 2006; Singh et al. 2009). Furthermore, they indicate an inhibition of enzyme activity at physiological concentrations of cytochrome *c* and high ATP/ADP ratios. The physiological relevance of this mechanism lies in enabling COX to sense the ATP/ADP ratio, thereby playing an important role in adjusting energy production to cellular energy demand (Arnold and Kadenbach 1997, 1999; Boyalla et al. 2011; Horvat et al. 2006; Misiak et al. 2010a, b; Roemgens et al. 2010; Singh et al. 2009, 2010).

This allosteric COX inhibition by ATP and enzyme cooperativity depend on the presence of cardiolipin and are abolished when COX activity is measured in the presence of TMPD (*N*,*N*,*N*′,*N*′-tetramethyl-p-phenylenediamine) as an electron donor which remains bound to cytochrome *c* throughout the electron transfer from ascorbate onto COX (Arnold and Kadenbach 1997, 1999). Furthermore, solubilization of the enzyme causing a monomerization of COX also leads to a loss of the cooperativity in the dimeric enzyme complex (Arnold and Kadenbach 1997, 1999). The allosteric inhibition of COX by ATP is independent of the mitochondrial membrane potential (Arnold and Kadenbach 1999) and can, therefore, be considered as the "second mechanism of respiratory control" (Kadenbach and Arnold 1999) contrasting the "classical" or "first" respiratory control mechanism by Mitchell which is based on the inhibition of mitochondrial respiration by a high membrane potential (Nicholls and Ferguson 2002).

Furthermore, this allosteric feedback inhibition of COX by ATP was proposed to keep the $\Delta\Psi m$ in living cells and tissues at low values (100–140 mV), when the matrix ATP/ADP ratios are high, thus preventing the generation of reactive oxygen species (ROS) (Kadenbach et al. 2004, 2009, 2010). Supportively, Kadenbach and colleagues (Ramzan et al. 2010) observed a reversible decrease of $\Delta\Psi m$ from 233 to 123 mV in isolated rat liver mitochondria with glutamate plus malate as substrates after addition of phosphoenolpyruvate and pyruvate kinase which were added to keep the ATP/ADP ratio maximal. The allosteric inhibition of COX by ATP occurs in living cells and requires a phosphorylation of COX subunit I as shown by correlating COX subunit I phosphorylation with sigmoidal inhibition kinetics in the presence of ATP (Helling et al. 2008). Tyr-304 phosphorylation was described for COX subunit I in liver tissue after activation with glucagon or forskolin via the cAMP/PKA signaling pathway and was also paralleled by enhanced allosteric COX inhibition kinetics (Lee et al. 2005).

13.3.4 Cytochrome c Oxidase Isoform Expression

Expression of respiratory chain complex subunits as isoforms is a feature unique to COX and absent from the other respiratory chain complexes. COX isoforms are described for five out of the ten nuclear-encoded COX subunits. In mammals, isoforms for subunits IV, VIa, VIb, VIIa, and VIII are expressed in a tissue-specific and/or developmentally regulated way (Fig. 13.2; Capaldi 1990; Grossman and Lomax 1997; Hüttemann et al. 2001, 2003a, b; Kadenbach 1986; Kadenbach and Reimann 1992). As we found recently, COX subunit IV isoforms are additionally expressed in a cell type-, hypoxia-, and toxin-dependent manner (Boyalla et al. 2011; Horvat et al. 2006; Misiak et al. 2010a, b; Roemgens et al. 2010; Singh et al. 2009, 2010).

13.3.4.1 COX Subunit IV isoform Expression: A Role in Oxygen and Toxin Sensing

With respect to the function of COX complex, subunit IV has been shown to be a key regulatory subunit. Any isoform expression of this subunit is, therefore, of particular interest. In yeast, tuna fish, and mammals, two isoforms of COX subunit IV (IV-1 and IV-2) were detected (Hüttemann et al. 2001; Poyton et al. 1995). While COX IV-1 is ubiquitously expressed in all mammalian tissues, COX IV-2 showed high expression levels in fetal tissue (lung and muscle), in adult lung tissue (Hüttemann et al. 2001), and in neurons (Horvat et al. 2006; Misiak et al. 2010a; Singh et al. 2010).

For the first time, we discovered a hypoxia-mediated and toxin-dependent up-regulation of COX isoform IV-2 expression in mammalian cells (Fig. 13.2; Boyalla et al. 2011; Horvat et al. 2006; Misiak et al. 2010a, b; Roemgens et al. 2010; Singh et al. 2009, 2010). We demonstrated that astrocytes, the major glial cell type in the CNS, express COX IV-1 under normoxic conditions and that COX IV-2 is present in these cells to an only marginal degree. However, hypoxia induced a significant up-regulation of COX IV-2 transcript and protein levels (Horvat et al. 2006). A similar observation of increased COX IV-2 transcript and protein levels was made after treatment of astrocytes with mitochondrial (neuro-)toxins, such as chemical inducers of hypoxia (azide, cyanide, cobalt), NPA (3-nitropropionic acid), 6-OHDA (6-hydroxydopamine), or MPP^+ (1-methyl-4-phenylpyridinium) (Boyalla et al. 2011; Misiak et al. 2010a, b; Roemgens et al. 2010; Singh et al. 2009, 2010). In our experiments, we were also able to demonstrate the effect of an increased COX IV-2 expression on kinetic properties of COX correlating it with intracellular ATP levels, mitochondrial peroxide production, and cell survival (Singh et al. 2009).

Hypoxia caused decreased intracellular ATP levels due to an insufficient oxygen supply for aerobic mitochondrial energy production (Horvat et al. 2006). At physiologically relevant cytochrome *c* concentrations, COX activity was increased at high ATP/ADP ratios, thereby abolishing the allosteric inhibition of COX by ATP. Interestingly, at low ATP/ADP levels, COX activity was decreased. These two observations indicate a dysregulation of COX sensitivity and activity in dependence on the cellular energy level under hypoxic conditions as less energy is produced when necessary (at low ATP/ADP ratios) and more energy when cells are provided with high ATP/ADP levels (Horvat et al. 2006).

After toxin exposure, elevated expression of COX isoform IV-2 confers a higher catalytic activity irrespective of the cellular/mitochondrial energy status. Thus, COX expressing isoform IV-2 lost its sensitivity toward the cellular energy level. Toxin-treated neural cells showed increased intracellular ATP levels accompanied by an elevated mitochondrial peroxide production (Boyalla et al. 2011; Misiak et al. 2010a, b; Roemgens et al. 2010; Singh et al. 2009, 2010). This indicates increased ATP production at the expense of elevated oxidative stress in toxin-treated neural cells (Arnold 2011).

The first indications for a possible mammalian COX subunit IV isoform expression being regulated by oxygen concentration originated from studies by Poyton and colleagues in yeast. The authors analyzed the expression of nuclear-encoded yeast COX subunit V isoforms (Va and Vb), which are homologous to mammalian COX subunit IV isoforms (Poyton et al. 1995). Under normoxic conditions, the isoform Va is expressed, whereas at low oxygen concentration (below 0.5 μM O_2), the isoform Vb is transcribed (Kwast et al. 1998; Burke and Poyton 1998). The yeast COX complex comprising the isoform Vb shows an accelerated internal electron transfer step from heme a to heme a3, resulting in a higher turnover rate (Allen et al. 1995). Although these results in yeast (Kwast et al. 1999; Wilson et al. 1994) suggested an important role of the 2001 published mammalian COX subunit IV isoforms (Hüttemann et al. 2001) in oxygen sensing, no evidence was reported until the effect of hypoxia on COX IV isoform expression in astrocytes was observed (Horvat et al. 2006).

Our observation of a hypoxia-mediated regulation of COX isoform IV-2 expression in astrocytes and neurons (Horvat et al. 2006) was later supported by Fukuda et al. (2007) demonstrating an hypoxia-mediated up-regulation of COX IV-2 in cell lines, mouse embryo fibroblasts, and pulmonary artery smooth muscle cells. The hypoxia-inducible factor (HIF) is involved in the reciprocal regulation of COX subunit IV isoform levels. Whereas COX IV-2 is up-regulated via HIF-1 interacting with a hypoxia-response element in the *cox4i2* gene, COX IV-1 is degraded by hypoxia- and HIF-1-mediated induction of mitochondrial protease LON expression (Fukuda et al. 2007; Fig. 13.2). Besides HIF, another hypoxia-responsive element in the human *cox4i2* promotor has been described (Hüttemann et al. 2007). However, a suggested role of HIF in neural regulation of COX IV isoforms remains to be elucidated.

HIF-1α is a transcription factor induced by low oxygen concentrations and found at high levels in malignant solid tumors. It is involved in the activation of numerous cellular processes including resistance against apoptosis, vascular remodeling, angiogenesis, as well as metastasis (Marín-Hernández et al. 2009). HIF-1α induces an overexpression and increased activity of several glycolytic protein isoforms and may also modulate mitochondrial function and oxygen consumption by inactivating pyruvate dehydrogenase complex and/or modulating COX subunit IV expression to increase oxidative phosphorylation in hypoxia-exposed cells, such as tumor cells or neural cells undergoing hypoxic/ischemic insults. The characterization of COX isoforms and other products of HIF activity could prove to be important for the development of new and more efficient strategies against tumor development (Marín-Hernández et al. 2009) and stroke (Arnold et al. unpublished). Inhibition or inactivation of COX IV-2 and/or Lon protease could be used therapeutically to inhibit tumor survival under hypoxic conditions by strategic modulation of tumor metabolism (Boutin and Johnson 2007) and to prevent deleterious consequences of stroke, such as oxidative stress, for energy metabolism and cell function in the brain (Arnold et al. unpublished).

13.3.4.2 COX Subunit IV Isoform Expression: A Role for Neural Cell–Cell Communication

We observed a cell-type specificity of COX IV-2 expression in the brain (Horvat et al. 2006). Comparing two major brain cell types, neurons and astrocytes, we demonstrated increased COX IV-2 expression levels in neurons, whereas astrocytes did show only marginal COX IV-2 levels. Based on the catalytic properties of the COX complex containing COX IV-2, we proposed that the neuronal COX IV-2 expression supports a constantly high neuronal activity (Horvat et al. 2006; Misiak et al. 2010a; Singh et al. 2010). Astroglia, however, rely on highly regulated aerobic and anaerobic energy production mechanisms as they represent the glia cell type which supports and modulates neuronal activity by metabolically communicating with neurons and providing them with energy intermediates, such as lactate, for a more efficient neuronal energy production (Arnold publication in preparation, Horvat et al. 2006; Pellerin and Magistretti 2004).

However, no functional relevance has been ascribed to the COX complex containing isoform IV-2 under physiological conditions until we demonstrated for the neuronal, but not for the astrocytic primary cell population, an abolishment of the allosteric COX inhibition by ATP (Horvat et al. 2006). This abolishment causes a suppression of the sensitivity of COX to detect the energy level and adjusting it to cellular energy needs (Horvat et al. 2006; Misiak et al. 2010a; Singh et al. 2010). With respect to a higher energy demand of neurons compared with astrocytes this mechanism could support the constantly high energy consumption by neurons.

A hypoxia-induced switch from COX isoform IV-1 to COX IV-2 in astrocytes could be relevant for a suppression of the metabolic communication between astrocytes and neurons during hypoxic, ischemic, and neurodegenerative processes in the CNS.

Under physiological conditions, astrocytes meet increased neuronal energy demand due to an enhanced neuronal activity by acting with a switch from oxidative phosphorylation to anaerobic lactate/ATP production to provide neurons with the essential energy intermediate lactate to enhance efficiency of neuronal oxidative phosphorylation (Arnold publication in preparation, Pellerin and Magistretti 2004). Under hypoxic and toxic conditions, astrocytes apparently favor oxidative phosphorylation due to increased COX IV-2 expression instead of the oxygen-preserving anaerobic ATP/lactate production. This would be accompanied by an enhanced oxygen consumption by astrocytes for aerobic ATP production under circumstances when energy substrates (oxygen under hypoxia) become limiting for both cell types, but with more severe consequences for the neuronal than the astrocytic survival (Arnold publication in preparation). The higher astrocytic survivability is based on the ability of astrocytes to cope with oxygen deprivation by preserving the cellular energy status by stimulation of glycolysis, whereas neurons rely on oxidative ATP production and will be even sooner limited by oxygen due to COX IV-2-mediated enhanced oxygen consumption by neighboring astrocytes. This could serve as an additional explanation for why neurons are more affected and are more vulnerable than astrocytes under hypoxic conditions.

Another aspect of diminished survivability of neurons due to increased COX IV-2 expression in neighboring astrocytes would be an enhancement of astrocytic mitochondrial peroxide production (Fig. 13.2, Singh et al. 2009). Thus, astrocytes would not only exhaust faster the limited energy substrates but also expose neurons to increased ROS, whereby neurons are more vulnerable to oxidative stress than astrocytes due to less pronounced antioxidant defense pathways in these cells (Bolanos et al. 1995).

13.4 The Role of Cytochrome *c* Oxidase in Neuropathophysiology

The brain is the highest consumer of energy substrates metabolizing approximately 60% of the glucose and 20% of oxygen, i.e., 20% of the total energy of the human organism at rest. This makes the brain extremely vulnerable toward energy deprivation. A 10-min cutoff from the energy substrate supply to the brain by blood stream causes permanent and irreversible brain damage. Thus, a sufficient energy supply by mitochondria is essential for proper functioning of brain cells.

The ATP produced by the respiratory chain is mainly used by neural cells to maintain the ion homeostasis, i.e., transmembrane ion gradients by ion pump activity of transporters and ATPases. Processes, such as axonal transport and synthesis of neurotransmitters and macromolecules, constitute only a minor fraction of neural energy expenditure (Lowry 1975). Mitochondrial oxidative energy metabolism is tightly coupled to neuronal activity and COX protein expression and/or activity is often used in brain studies as a marker of neural functional activity (Hevner et al. 1995; Wong-Riley 1989). COX activity has been demonstrated to differ among functionally different brain regions under physiological conditions and to change in response to altered neuronal activity in various brain regions to meet local metabolic demands (Hevner and Wong-Riley 1989; Horvat et al. 2006, reviewed in Wong-Riley 1989).

Neurons are susceptible to oxidative stress because of their high rate of oxygen consumption, high polyunsaturated fatty acid content, high transition metal ion content, and relatively limited antioxidant defense systems (Halliwell 1992). As COX consumes 85–90% of the cell's oxygen, mitochondria are the greatest source of neuronal ROS production (Shigenaga et al 1994). Respiratory chain defects caused by increased ROS levels, in turn, could facilitate further increased ROS production (Boveris and Chance 1973), leading to a vicious cycle that culminates in cell death. An impairment of the mitochondrial energetic competence to produce ATP and an accumulation of oxidative damage in neural cells undergoing aging, neurodegeneration or ischemic insults are the two underlying concepts explaining the involvement of mitochondria in diseases affecting the CNS. Besides the high energetic demand of the CNS, the brain shows a slow turnover of mitochondria and mitochondrial components in comparison with other organs of a mammalian organism, thereby accumulating dysfunctional mitochondria as a result of oxidative stress more than other organs and predisposing brain cells to a physiological deficit (Gould and McEwen 1993). A malfunction of COX has severe implications for cellular energy metabolism and causes increased production of ROS with a variety of deleterious consequences in humans (Wallace et al. 2010). Although the exact pathogenesis of neurodegenerative disorders is only poorly understood and the etiological role of mitochondria herein is still debated, these organelles together with COX play a critical role for disease etiology, pathogenesis, and progression.

13.4.1 Signaling and Oxidative Stress

Hypoxia, toxin exposure, and degenerative processes stimulate a variety of intracellular signaling pathways, such as intracellular calcium, protein phosphorylation/dephosphorylation, interaction with hormones, and cellular energy level, thereby affecting COX function which in turn influences the pathological processes (Fig. 13.2). An impairment of mitochondrial energy production and increased ROS production are two pathways that are widely accepted mechanisms involved in neurodegenerative diseases (Ayala et al. 2007; Beal 2005), whereby mitochondria represent the major source of cellular energy and ROS production. Interestingly, astrocytes after hypoxia and after toxic treatment did not show an increased cell death rate due to a reduction of the intracellular ATP level, but rather as a consequence of elevated mitochondrial peroxide production (Fig. 13.2, Boyalla et al. 2011; Horvat et al. 2006; Misiak et al. 2010b; Roemgens et al. 2010; Singh et al. 2009).

Although COX itself is not a source of ROS the enzyme exerts effects on the respiratory chain modulating the ROS production at complexes I and III. It is generally assumed that a decreased COX activity preserving $\Delta\Psi m$ at low levels diminishes the oxidative stress by decreasing ROS production (Cadenas et al. 1977; Lee et al. 2001). However, recent reports suggest a role for an elevation of COX activity in ROS production (Dröse and Brandt 2008). Thus, a higher COX electron transfer onto oxygen, such as caused by COX isoform IV-2 expression, leads to the ubiquinone pool to be oxidized to a higher degree, thereby enabling complex III to transfer electrons to oxygen and increase the ROS production.

Based on the "allosteric inhibition of COX by ATP" at high ATP/ADP ratios and with respect to an optimal ATP synthase activity at low $\Delta\Psi m$ values (100–120 mV; Kaim and Dimroth 1999) and increased basal proton leak of biological membranes at high $\Delta\Psi m$ values, the "second mechanism of respiratory control" is suggested to keep $\Delta\Psi m$ low, thereby protecting cells from ROS production (Kadenbach et al. 2010). In contrast, the well-known "first mechanism of respiratory control" leads to an inhibition of respiration only at high $\Delta\Psi m$ values (>140 mV) at which ROS is produced (Lee et al. 2001; Murphy 2009). Under circumstances, when COX isoform IV-2 is expressed, oxidative stress is increased and COX loses its ability to support cell function and survival.

The causal implication of COX isoform IV-2 in an increased COX activity accompanied by elevated mitochondrial peroxide production was verified by the application of an siRNA knockdown approach against COX IV-2. This knockdown approach attenuated the toxin-mediated increase of COX activity, ATP levels, and mitochondrial peroxide production (Misiak et al. 2010b; Singh et al. 2009). Our data highlight the crucial role for the COX isoform IV-2 in increased cell vulnerability under hypoxic and toxic conditions. In brain tissue, this in turn would not only influence on the astroglia function under pathological conditions, but would also indirectly impair the survival of neighboring neurons (Boyalla et al. 2011; Horvat et al. 2006; Misiak et al. 2010a, b; Roemgens et al. 2010; Singh et al. 2009, 2010).

With respect to increased oxidative stress leading to an impaired COX function and taking into account the small excess capacity of COX under physiological conditions, i.e., a tight regulation of mitochondrial respiratory rate by COX activity (Kunz et al. 2000), already small changes of COX amount/activity may cause alterations of mitochondrial respiration.

Nevertheless, ROS is not acting solely as a cellular oxidative stress factor, but is likely to function also as a second messenger in cell signaling including induction of autophagy (Forman et al. 2008; Scherz-Shouval et al. 2007; Valko et al. 2007) and mitohormesis (Ristow and Zarse 2010). Besides ROS, nitric oxide (NO) and hydrogen sulfur (H_2S) are similarly well recognized for playing a dual role as both deleterious and beneficial chemicals (Fig. 13.2, Collman et al. 2009; Valko et al. 2007). Deleterious effects are caused by NO and H_2S both binding to heme iron (NO) of COX and inhibiting COX activity (Collman et al. 2009; Cooper and Davies 2000; Taylor and Moncada 2010). Beneficial effects occur at low or moderate concentrations and involve physiological roles in cellular responses to noxia. This way, various ROS-mediated actions in fact protect cells against ROS-induced oxidative stress and re-establish or maintain the "redox homeostasis".

13.4.2 *Aging and Neurodegenerative Diseases*

13.4.2.1 Aging

Aging is a universal progressive phenomenon affecting all living organisms and organs. It is, however, characterized by a general decline of physiological performances which affects most of those functions that depend on the CNS. Harman was the first to propose a role of mitochondria in aging processes in 1972 (Harman 1972). Linnane and colleagues further expanded this theory by suggesting that the accumulation of somatic mtDNA mutations is a major cause of aging and age-related diseases (Linnane et al. 1989). This theory is controversially discussed until today (Lightowlers et al. 1997), which is mainly based on the fact that the level of individual mtDNA mutations rarely exceeds 1% thereby remaining well below the phenotypic expression threshold observed in mtDNA disorders (DiDonato et al. 1993; Ikebe et al. 1990; Lee et al. 1994; Simonetti et al. 1992). When a threshold of mutated mtDNA versus wild-type DNA of COX subunits I, II, or III is reached, a functionally intact COX complex cannot be assembled and COX activity becomes deficient. In patients with mtDNA defects, the CNS is the most severely affected tissue, because neurons represent a postmitotic and highly metabolically active cell population (Chinnery and Turnbull 1999). Defects of mtDNA often play a role in the hippocampus, an area involved in cognitive decline during aging (West 1993; West et al. 1994).

During aging, a reduction of mitochondrial content and phosphorylating capacity of the ATP synthase were ruled out in rodent brain (Navarro and Boveris 2004). However, an impairment of energy production due to decreased capacity to produce ATP, i.e., decreased rates of electron transfer, was identified (Benzi et al. 1992). This is reflected in a decreased respiratory chain protein content shown by Western Blot and immunohistochemical analysis and diminished rates of electron transfer by selectively reduced activities of complexes I and IV in senescent mice correlating well with neurological function and survival (Navarro and Boveris 2007). Individual COX complexes and COX being part of supercomplexes undergo a decrease by 20% and between 15 and 30%, respectively (Frenzel et al. 2010).

Age-related increase in COX-deficient cells has been reported in neurons of the substantia nigra (Itoh et al. 1996), dorsal lateral geniculate nucleus (Diaz et al. 1996), hippocampal pyramidal neurons and synapses (Bertoni-Freddari et al. 2004), choroid plexus epithelial cells correlating well to cell death and dysfunction in the aging CNS (Cottrell et al. 2001).The decreased activity of complexes I and IV by about one-third is close to the limit of a tolerable impairment of mitochondrial function with respect to basal ATP production of about 36% of the mitochondrial mass in metabolic state 3 of respiration (Boveris et al. 1999). Under conditions of increased ATP demand, brain mitochondria increase ATP synthesis by switching more mitochondria from the resting state 4 to the active state 3. However, neurons are long-living cells with a slow turnover of mitochondria and mitochondrial components with a half-life of 4–6 weeks instead of a few days as in other organs. And, neurons are spatially restricted and undergo quantitatively negligible neurogenesis

(Gould and McEwen 1993). Therefore, aged neurons with lower mitochondrial mass and enzyme activities as well as an increased number of dysfunctional mitochondria are close to be unable to respond to any increased ATP demand (Navarro and Boveris 2004). Thus, respiratory complexes I and IV are considered effective markers of aging (Navarro and Boveris 2007).

Strikingly, northern blot analyses revealed an increased transcription of mtDNA-encoded genes of complexes I, III, and V additionally to complex IV in 12- and 18-month-old mice compared with 2-month-old mice. This suggests a compensatory mechanism of overproduction of respiratory chain proteins. Apparently, the increased mRNA expression could not be sustained over a long time as it was decreased in 24-month-old mice (Manczak et al. 2005).

The increase in human lifespan in industrialized countries is accompanied by a marked prevalence of neurodegenerative diseases, such as Alzheimer's and Parkinson's diseases (AD and PD). Human neurodegenerative diseases are characterized by a progressive cellular damage which is encompassed by particular neurological deficits once neuronal loss reaches more than half of the neuronal population in a given CNS region. Ample evidence indicates that mitochondrial dysfunction and an impairment of respiratory chain complexes play a role in the neuronal loss, such as mainly the loss of complex IV activity in AD (Chagnon et al. 1995), decreased complex I activity in PD (Mizuno et al. 1989), and decreased complex I and II activities in Huntington's disease (HD, Parker et al. 1990a). A brain region-specific manifestation of different age-related and/or neurodegenerative diseases may result from the varying energetic roles and needs of the different brain areas.

13.4.2.2 Alzheimer's Disease

Alzheimer's disease (AD) is by far the most common neurodegenerative dementia in elderly affecting 5–15% of the population over the age of 65 years (Katzman 1986). Patients with AD initially show memory loss and develop impaired executive function, confusion, and personality change ending eventually with death as the disease progresses (Hauptmann et al. 2006). There are several pathological hallmarks of AD. One of them comprises hyperphosphorylation and aggregation of Tau, a microtubule-associated protein, appearing as neurofibrillary tangles, neuritic plaques, and neuronal threads within neural cells impairing their growth and function (Bandyopadhyay et al. 2007). A peptide containing residues 26–44 of Tau protein targeted COX and adenine nucleotide transporter causing impaired oxidative phosphorylation by the respiratory chain conferring deleterious effects on cellular availability of mitochondrial ATP (Atlante et al. 2008). Another hallmark of AD is the abnormal proteolytic processing of amyloid precursor protein which has been shown to contribute to Aβ deposition and formation of amyloid plaques in the cerebrovasculature of AD patients (Abraham et al. 1999; Chen et al. 2000; Davies et al. 1997). Aggregated Aβ, in turn, decreases redox activity, reduces mitochondrial membrane potential and ATP levels, and consequently the viability of neurons and astrocytes in cell culture and in the brain suggesting that Aβ impairs mitochondrial function and initiates a neurotoxic cascade (Eckert et al. 2008; Hauptmann et al. 2006; Kaneko et al. 1995; Kato et al. 1997; Shearman et al. 1995). However, the basis of Aβ toxicity is poorly understood.

Attention has been directed to the possible contribution of mitochondrial dysfunction and oxidative damage in late-onset familial and sporadic forms of AD (Beal 1995; Schapira 1996). It has been suggested that sporadic AD involves COX mtDNA mutations (Parker et al. 1990b; Swerdlow et al. 1997) and that AD is associated with deficient COX activity (Kish et al. 1992; Parker et al. 1994).

In age-related conditions, such as AD, higher levels of mtDNA mutations and oxidative damage (Bonilla et al. 1999) and lower levels of mtDNA-encoded mRNA and activity of COX have been reported (Chagnon et al. 1995; Simonian and Hyman 1994; Wong-Riley et al. 1997). In AD postmortem brain tissue from patients with AD, mRNA for *cox2* was determined by in situ hybridization to be down-regulated, especially in the entorhinal cortex and hippocampal formation (Chandrasekaran et al. 1998) and mitochondrial amyloid β decreases COX activity (Reddy et al. 2011). Paradoxically, in regions without significant neuronal loss there were higher percentages of COX-deficient neurons (Cottrell et al. 2001). The highest percentage of COX-deficient cells were found in CA2 region of the hippocampus, which is spared from neuronal loss and other pathological markers representing a neuronal population that is relatively tolerant to loss of COX activity and consequently diminished ATP levels (Cottrell et al. 2001). Pyramidal neurons of other hippocampal CA regions may be more susceptible to COX deficiency undergoing degeneration at an earlier stage.

However, how mtDNA mutations of COX would translate into a loss of specific neuronal populations, including cholinergic neurons in the forebrain, hippocampus, and neocortex, is unclear. In line with the oxidative stress hypothesis, increased generation of ROS causes oxidative damage of cell lipids, proteins, and DNA (Markesbery 1997). Deficient COX activity has been reported by several authors in different brain regions except motor cortex from AD patients (Bosetti et al. 2002; Kish et al. 1992; Maurer et al. 2000; Mutisya et al. 1994; Simonian and Hyman 1994; Wong-Riley et al. 1997), in platelets (Bosetti et al. 2002; Cardoso et al. 2004; Parker et al. 1990b), and fibroblasts (Curti et al. 1997). Moreover, COX activity and mRNA levels of COX subunits I and III have been shown to be decreased in association cortex from AD patients (Chandrasekaran et al. 1998). Controversially, COX at almost normal amount was found in mitochondria of the postmortem AD hippocampus (Schagger and Ohm 1995) and in AD platelets (Cardoso et al. 2004). The toxic Aβ fragment 25–35 selectively diminishes COX activity in rat brain mitochondria without affecting other components of the respiratory chain (Canevari et al. 1999). It was suggested that the reduced COX activity may render the tissue vulnerable to excitotoxicity or reduced oxygen viability.

Interestingly, complex IV together with complex III, but contrary to complex I, showed increased mRNA expressions in the brain specimens of both early and definite AD patients suggesting a great energy demand (Manczak et al. 2004). Based on their results, the authors proposed that an increased COX gene expression might be the result of functional compensation by the surviving neurons or an early mitochondrial alteration related to increased oxidative damage (Manczak et al. 2004).

Recent evidence suggests that mitochondria are significantly reduced in postmortem brains from AD patients and that amyloid precursor protein and amyloid β are targeted to mitochondria impairing COX activity. The proposed mechanisms of complex IV inhibition include (1) blockage of mitochondrial import channels to prevent the import of nuclear-encoded COX subunits (Devi et al. 2006) and, in consequence, disrupting electron transport and to provoke ROS production (Hirai et al. 2001; Reddy and Beal 2008), (2) sequestration and depletion of heme by Aβ forming a peroxidase (Atamna 2006, 2009; Atamna and Frey 2004), thereby interfering with heme metabolism which is essential for a catalytically active COX, and increasing oxidative stress, (3) interaction of Aβ and Aβ-binding alcohol dehydrogenase, thus promoting ROS production (Lustbader et al. 2004; Takuma et al. 2005).

Thus, an mtDNA-independent vicious cycle connecting Aβ, heme, COX, and ROS might participate in age-dependent exacerbation of AD pathology and complex IV defects (Fukui and Moraes 2008).

13.4.2.3 Parkinson's Disease

Parkinson's disease (PD), the second most common neurodegenerative disorder, shows an incidence that correlates well with sex (van den Eeden et al. 2003) occurring 1.5 times more frequently in men than in women (Dluzen and McDermott 2000; Fahn and Sulzer 2004; Haaxma et al. 2007; Wooten et al. 2004). PD is characterized by the degeneration of dopaminergic neurons in the *substantia nigra pars compacta* of the mesencephalon with a marked depletion of striatal dopamine. PD is suspected to evolve for years before typical motor signs appear, a moment when there is a loss of dopaminergic neurons for approximately 60%. The resulting striatal deficiency leads to the parkinsonian symptoms of bradykinesia, rigidity, tremor, and motor and postural instability (Bernheimer et al. 1973; Marsden 1990).

There are two forms of PD, the rare familial form and the more common sporadic form. Both forms are characterized by mitochondrial respiratory defects. Although the etiology of PD remains largely unknown, an accumulating body of evidence in experimental models of PD and in postmortem tissues of sporadic forms of PD suggests that mitochondrial dysfunction due to an impairment of mitochondrial energy production and oxidative stress may be involved. Exposure to environmental mitochondrial toxins leads to PD-like pathology and is the basis of various PD animal models. The identification of specific gene mutations and their influence on mitochondrial functions has further emphasized the relevance of mitochondrial structural and functional abnormalities in the disease pathogenesis (Banerjee et al. 2009). However, it is still under debate if respiratory chain deficiency is etiological in PD.

An impairment of complex I activity is central to the pathogenesis of the dopaminergic neuronal demise in PD (Dauer and Przedborski 2003). It predisposes to excitotoxicity by altering ATP levels and by impairing Ca^{2+} homeostasis. However, COX is apparently also involved in neurotoxicity in the substantia nigra. A large proportion of neurons in the substantia nigra have been shown by immunohistochemistry to lose COX with increasing age (Itoh et al. 1996). Interestingly, aged human substantia nigra contains very high levels of mtDNA deletions which are significantly higher in COX-deficient than in COX-positive neurons (Kraytsberg et al. 2006). Deletions of mtDNA are nonuniformly distributed among different brain areas with the substantia nigra, the primary site of neurodegeneration in PD, sustaining particularly high levels of mtDNA deletions compared with other brain areas (Soong et al. 1992). Interestingly, activities of the mitochondrial complexes I and IV were reduced in leukocytes from patients with idiopathic PD compared with age-matched controls, but no difference in COX activities was detected between controls and patients with parkin mutations (Muftuoglu et al. 2004).

Mitochondrial toxins are often used to trigger and mimic neurodegenerative processes. Thus, systemic application of 1-methyl-4-phenyl-1,2,3,6-tetrahydropyridine (MPTP) induces Parkinsonian symptoms indicating a causative or consequent involvement of mitochondria. The application of MPTP in vivo and of its toxic derivative 1-methyl-4-phenylpyridinium (MPP^+) in vitro represent a well-accepted experimental model of Parkinson's disease. Besides the known effects of MPP^+ on mitochondria and neural cell survival and with respect to the supportive role of astrocytes for neuronal function and survival (McGeer and McGeer 2008), we demonstrated the involvement of COX subunit IV isoform expression in energy and ROS production taking part in an impairment of astrocyte survival. Thus, MPP^+ caused a specific increase of COX IV-2 transcript and protein levels in male mesencephalic astrocytes accompanied by decreased ATP and increased ROS levels and elevated apoptotic cell death levels, which were more pronounced in mesencephalic than cortical astrocytes from male than female mice (Boyalla et al. 2011). The largest impairment of function and survival of astrocytes in the mesencephalon from male mice could exert negative effects on dopaminergic and nondopaminergic neurons, e.g., by releasing toxic factors such as ROS.

Although MPP^+ inhibits complex I, an electron entry still occurs via complex II. This could explain a maintained or diminished to a minor extent oxidative energy production by the respiratory chain during complex I inhibition. A recent study by Folbergrová and colleagues demonstrated that complex I inhibition was not accompanied by impaired ATP production, which was apparently due to excess capacity of complex I documented by energy thresholds (Folbergrová et al. 2010). Thus, elevated ROS production is apparently the major reason for cell death in mesencephalic astrocyte cultures. Our data suggest that MPP^+ acts on astrocytes in a sex- and brain region-specific manner involving COX isoform expression in an impairment of energy production and elevated oxidative stress levels. With respect to astrocytes influencing neuronal activity and survival, I suggest a negative bystander effect by astrocyte COX further impairing function and survival of neighboring neurons in PD.

13.4.2.4 Huntington's Disease

Huntington's disease (HD) is an autosomal-dominant inherited disorder caused by a CAG triplet codon repeat of variable length causing polyglutamine stretches in the widely expressed huntingtin, a protein of unknown function (Huntington's Disease Collaborative Research Group 1993). Abnormal energy metabolism in HD brain striatum is directly associated with severe deficiencies in the activities of respiratory chain complexes, most notably reduced activity of succinate-linked oxidation in postmortem HD caudate nucleus (Brennan et al. 1985). Further studies on postmortem HD brain tissue confirmed this and refined the defect in the caudate and putamen of striatum to a 30–65% and 35–60% reduced activity of respiratory complexes II/III and COX, respectively (Browne et al. 1997; Gu et al. 1996; Tabrizi et al. 1999). These abnormalities are confined to the striatum, the brain area that is most affected in HD (Browne et al. 1997). Additionally, in situ studies of HD brains revealed that the number of neurons expressing *cox1* mRNA tends to be lower in the striatum (Gourfinkel-An et al. 2002). A recent study of moderate-to-severe grade HD patients revealed a reduction of COX II protein levels that corresponded with disease severity and was accompanied by a significant down-regulation of Tfam, a regulator of mtDNA, and peroxisome proliferator-activated receptor-co-activator gamma-1 alpha (PGC-1α), a key transcriptional regulator of energy metabolism and mitochondrial biogenesis (Kim et al. 2010).

Toxin-mediated animal models may be induced by intrastriatal infusion of 3-nitropropionic acid (NPA) or malonate, both inhibitors of complex II of the respiratory chain, replicating features of HD (Borlongan et al. 1995, 1997; Greene et al. 1993). Applying a toxin-mediated in vitro HD model, we recently discovered a specific involvement of ncDNA-encoded COX genes. NPA mediated an up-regulation of COX isoform IV-2 mRNA transcription and protein expression in neural cells (Misiak et al. 2010b; Singh et al. 2009, 2010). The COX IV-2 increase was specific for striatal astrocytes and neurons (Misiak et al. 2010b). Furthermore, we demonstrated the effect of increased COX IV-2 expression on the kinetic properties of COX correlating it with intracellular ATP levels, mitochondrial peroxide production, and neural cell survival upon NPA treatment (Singh et al. 2009).

Spinobulbar muscular atrophy (SBMA) is a rare adult-onset neurodegenerative disease caused by the expansion of the polyglutamine stretches in the human androgen receptor polypeptide (LaSpada et al. 1991). One of the mechanisms of neurotoxicity is mediated by aberrant interactions with and possible sequestration of critical cellular proteins. One of the candidate proteins is COX subunit Vb. In a yeast tow hybrid system, COX Vb interacts with androgen receptor causing subsequently a sequestration of COX Vb. This may serve as a mechanism of mitochondrial dysfunction in SBMA thought to precede neuropathological symptoms in polyglutamine-expansion disorders (Beauchemin et al. 2001).

13.4.2.5 Amyotrophic Lateral Sclerosis

Amyotrophic lateral sclerosis (ALS) is a fatal and most common adult-onset motoneuron disease. More than 90% of ALS cases are sporadic, whereas the remaining ALS cases belong to familial ALS which is caused by mutations in mainly the Cu/Zn superoxide dismutase SOD1 (Jackson et al. 1997; Rosen et al. 1993). Investigations of the cellular changes that occur in motoneurons in ALS have shown that one of the earliest pathological changes in a mouse SOD1 transgenic model are abnormalities in mitochondrial structure and function. Besides dilated cristae and mitochondrial swelling, functional alterations occur before the onset of clinical symptoms. Mitochondria in different animal and cell culture models showed a reduction in respiratory chain activity, specifically complexes I and IV (Fukada et al. 2004, Mattiazzi et al. 2002; Menzies et al. 2002; Swerdlow et al. 1998; Wiedemann et al. 1998), reduced ATP generation (Mattiazzi et al. 2002), and mitochondrial membrane potential (Carri et al. 1997; Kruman et al. 1999). Similar changes have also been demonstrated in postmortem tissue from ALS patients (Fujita et al. 1996; Hirano et al. 1984; Sasaki and Iwata 1996, 2007; Siklos et al. 1996) suggesting that mitochondria including COX may represent an initial target for damage of neural cells in ALS.

13.5 The Role of Cytochrome *c* Oxidase in Neuroprotection

13.5.1 Steroid-Mediated Protection Pathways and Sex Specificity of Neuropathologies

Ovarian steroid hormones have well-established trophic and protective effects supporting both reproductive function and cognitive health. More recently, it has been recognized that these steroids also regulate metabolic functions sustaining the energetic demands of neuronal activities. Thus, brain mitochondria from hormone-treated rats displayed increased metabolic rates and enhanced respiratory efficiency (reviewed in Chen et al. 2009). There is indeed increasing evidence pointing to the mitochondrial respiratory chain as a putative and promising target for steroid action under physiological conditions and for steroid-mediated cell protection in the CNS under pathological conditions (Araujo et al. 2008; Arnold and Beyer 2009; Arnold et al. 2008, 2012; Brinton 2008; Klinge 2008; Simpkins and Dykens 2008).

Potential mechanisms of estrogen and glucocorticoid action involve estrogen (ERα and ERβ) and glucocorticoid receptors (GR) which have been identified in plasma membrane, nucleus, and mitochondria (Demonacos et al. 1996; Yang et al. 2004). A mitochondrial localization was primarily demonstrated for ERβ in primary neurons (Yang et al. 2004). Estrogens affect the mitochondrial energy and ROS production by two different pathways, i.e., the long-term classical estrogen receptor pathway and the short-term nonclassical pathway. The latter is characterized by direct interaction of estrogen with subunits of respiratory chain complexes and influencing intracellular signaling pathways, such as calcium and MAPK, which in turn affect mitochondrial proteins and activity. The classical long-term pathway requires the cooperation of both nuclear and mitochondrial genomes and ERs coordinating the transcriptional regulation of mtDNA- and ncDNA-encoded subunits of respiratory chain complexes (reviewed in Arnold et al. 2012).

Steroids, such as estrogen and glucocorticoids, up-regulate mitochondrial gene expression (Chen et al. 2005; Demonacos et al. 1995, 1996), specifically *cox1*, *cox2*, and *cox3* subunits in cortical astrocytes (Araujo et al. 2008), neurons (Nilsen et al. 2007), hippocampal neurons (Bettini and Maggi 1992), primary spinal cord neurons (Johann et al. 2010), and brain mitochondria from ovariectomized rats (Irwin et al. 2008). Although the exact mechanism is unknown, it has been speculated that the presence of sequences revealing strong similarity to glucocorticoid and estrogen response element consensus sequences within the mitochondrial genome, specifically *cox1* and *cox2*, may be responsible (Demonacos et al. 1995; Sekeris 1990). This supports a direct regulation of mitochondrial transcription by estrogen and glucocorticoids (Hatzoglou and Sekeris 1997). More recently, we have demonstrated that long-term estrogen exposure selectively increases the ratio of mtDNA versus ncDNA of COX indicative for an elevated mitochondrial mass and enzyme activity (Araujo et al. 2008). However, estrogen was also shown to increase mRNA expression of ncDNA-encoded COX genes, such as COX IV (Irwin et al. 2008), and COX VII-related protein (Watanabe et al. 1998).

The molecular mechanisms underlying the estrogen/ER-mediated effects on mtDNA and ncDNA expression are not completely understood. However, transcription factors, such as nuclear respiratory factors 1 and 2 (NRF-1 and NRF-2), PGC-1α, and mitochondrial transcription factor A (Tfam), appear to play a role. Tfam which controls mtDNA transcription is regulated by NRF-1 and NRF-2 indicating a link between the expression of nuclear and mitochondrial genes and a mechanism for coordinated regulation during mitochondrial biogenesis (Virbasius and Scarpulla 1994). One common feature of all ten nucleus-encoded COX subunits is the presence of an NRF-1 binding site in all ten promoters which is highly conserved among mammalian species and which was shown in in silico analysis, using EMSA and in vivo ChIP assays (Dhar et al. 2008). Silencing of NRF-1 expression by application of siRNA diminished all ten COX subunit transcripts indicating a significant role of NRF-1 in coordinating the transcription of all ten ncDNA- and mtDNA-encoded COX subunits in neural cells. Similar results were obtained for NRF-2 indicating that Nrf-2 is also an important mediator of a concerted COX gene expression (Ongwijitwat and Wong-Riley 2005), whereby NRF-2 has been demonstrated to sense the cellular energy demand in rat neurons (Ongwijitwat et al. 2006). Interestingly, the NRF-1 promotor contains an estrogen response element that specifically binds to ERα and ERβ. Estrogen induced an increase in NRF-1 followed by increased *Tfam* and Tfam-regulated *cox1* and *cox2* genes and increased mitochondrial biogenesis (Johann et al. 2010; Mattingly et al. 2008; Stirone et al. 2005). *Tfam* is also regulated directly by estrogen via ERs, maintaining mtDNA content in cells and increasing the gene expression of COX genes (Hsieh et al. 2006; Kanki et al. 2004).

The transcriptional changes exert profound effects on the activity of COX accompanied by elevated mitochondrial respiratory chain activity and efficiency together with a decreased rate of ROS production. Prolonged decrease of COX activity and mitochondrial ATP was observed in hippocampi of young- and middle-aged ovariectomized rats, whereas the treatment of rats with estradiol benzoate reversed this mitochondrial dysfunction induced by estrogen withdrawal (Shi and Xu 2008; Shi et al. 2008).

Studies on mouse cortical in comparison with mesencephalic astrocytes revealed structural and functional differences in gene expression and mitochondrial activity in a time- and brain region-dependent manner, but independent of nuclear ERs (Araújo et al. 2008).

Epidemiological studies have indicated that estrogen therapy reduced the risk of developing AD and PD in women. The incidence of AD is lower in young women than in men, but is increased among postmenopausal women. These observations indicate that estrogens have important protective effects against the development of AD. Treatment of cultured rat hippocampal neurons with estrogen prior to amyloid beta exposure significantly reduced the number of apoptotic neurons. These anti-apoptotic effects were attributed to (1) activation of antioxidant defense systems scavenging ROS (Nilsen 2008) and (2) improvement of respiratory chain activity, reducing DNA damage as well as improving COX activity, mitochondrial respiration, and ATP production under normal and stress conditions. In fact, estrogen treatment enhanced the expression of genes encoding respiratory chain proteins in rat hippocampus (Bettini and Maggi 1992) and in vivo treatment of rats with estrogen up-regulated brain mitochondrial proteins, COX activity, and respiratory chain function (Irwin et al. 2008; Nilsen et al. 2007) in AD animal models.

However, with respect to the potential carcinogenic effects of estrogen and the highly oxidative cellular environment in the brain favoring oxidative metabolism of estrogen, estrogen itself is a rather poor agent for treatment of existing AD (Nilsen 2008). In this regard, in cultured cells and animal models of AD, other estrogenic chemicals, some with less hormonal effects, have been shown to be neuroprotective, including 17α-estradiol (Dykens et al. 2005; Simpkins et al. 2004, 2005), selective agonists for ERα and ERβ (Zhao et al. 2004), phytoestrogens (Zeng et al. 2004; Zhao et al. 2009), and a hybrid structure of estrogen and vitamin E (Zhao et al. 2007).

PD is more frequent in men than in women and more prevalent in women with short reproductive life (Shulman 2007). Estrogens are considered neurotrophic for dopaminergic neurons and are neuroprotective against neurotoxic agents affecting dopaminergic neurons in vitro and in vivo (Rodriguez-Navarro et al. 2008). Neuroprotection by estrogen could be attributed to effects on respiratory chain biogenesis and function and on PD-related proteins that are involved in control of mitochondrial functions, such as PTEN-induced kinase 1 (PINK1), Parkin, DJ1, α-synuclein, and POLG (Clark et al. 2006; Davidzon et al. 2006; Dodson and Guo 2007; Henchcliffe and Beal 2008).

Steroid hormones, especially estrogen, induce mitochondrial alterations in the CNS supporting efficient and balanced bioenergetics and reducing oxidative stress.

13.5.2 Nutritional Antioxidant Protection Pathways

Nutritional interventions in the treatment of neurodegenerative disease mostly imply antioxidant therapies. The accumulation of ROS in neurons and subsequent oxidative stress are attenuated by free radical scavengers, which can be categorized as enzymatic or nonenzymatic antioxidants. Nonenzymatic antioxidants are represented by ascorbic acid (vitamin C), α-tocopherol (vitamin E), glutathione (GSH), and others.

Successful preliminary demonstration of α-tocopherol (vitamin E), a potent antioxidant, to limit oxidative damage and neurodegeneration led to clinical assessment of its potential in human HD (Peyser et al. 1995). Although α-tocopherol had no effect on neurologic or neuropsychiatric symptoms in the treatment group of patients with mild-to-moderate HD symptoms, post hoc analysis showed a significant effect on neurologic symptoms in HD patients early in the course of the disease concluding that α-tocopherol therapy may slow down the rate of motor decline early in the course of HD.

Conditions, such as vitamin E dietary supplementation in combination with caloric restriction, high spontaneous neurological activity, and moderate physical exercise, ameliorate mitochondrial dysfunction in aged and ALS brain (reviewed in Navarro and Boveris 2007).

In vivo studies revealed that systemically administered bilobalide, a constituent of Ginkgo biloba leaf extracts, can reduce cortical infarct volume in animal stroke models. One of the neuroprotective effects is associated with preservation of mitochondrial ATP synthesis by increasing the expression of mtDNA-encoded COX subunits (Defeudis 2002). The administration of paeonol (2′-hydroxy-4′-methoxyacetophenone; 1-(2-hydroxy-4-methoxyphenyl)ethan-1-one), a constituent of the bark of the Moutan Cortex and used as traditional Chinese medicine for its antioxidant properties, caused in an AD rat model an increase of COX protein levels paralleled by an improvement in behavioral indices of learning relative to the group receiving Aβ1-42 by intra-hippocampal injection alone (Zhou et al. 2011).

13.6 Conclusion

COX is the rate-limiting enzyme of the mitochondrial respiratory chain. This makes this enzyme predestined for studying its broad regulatory potential which in turn is crucial for an efficient energy transduction and cellular signaling under pathophysiological conditions in the brain as the organ with the highest energy demand (Fig. 13.2; reviewed in Arnold 2011). Indeed, enzyme electron transfer and proton-pumping efficiency of COX are subject to a broad range of regulatory factors and mechanisms, such as binding of various effector molecules, e.g., adenine nucleotides and hormones, and COX isoform expression. Mainly the latter represents a unique feature of COX distinct from the other mitochondrial respiratory chain complexes. These regulatory mechanisms are accomplished by the ten nucleus-encoded subunits of the COX complex. Five of them are expressed as isoforms in a developmental and tissue-specific way. Recently, we found that a switch between two isoforms of one particular COX subunit (COX IV-1 and IV-2) depends on the neural cell type, oxygen, and toxin concentrations. Apparently, COX subunit IV is a key regulatory subunit of the COX complex functioning as an energy and oxygen sensor in various cell types.

COX isoform IV-1 binds ATP causing an allosteric inhibition of the enzyme at high cellular energy levels. Based on this ΔΨm-independent feedback inhibition of COX activity by ATP described as the "second mechanism of respiratory control," COX IV-1 enables the enzyme to detect the cellular energy level and adjusts the ATP production to cellular energy demand. This has the advantage that the mitochondrial membrane potential ΔΨm cannot rise to levels which lead to increased ROS production, thereby avoiding oxidative stress which in turn could cause cell damage up to cell death. At high ATP demand, this mechanism is relieved due to an exchange of ATP by ADP at COX IV-1, but can also be modulated (switched off) by either the presence of COX isoform IV-2 instead of COX IV-1, by posttranslational modification/dephosphorylation of the enzyme and/or the presence of thyroid hormones. The *cox4i2* gene (COX isoform IV-2) apparently senses hypoxic and toxic conditions as its expression is induced under such. The functional consequences of COX IV-2 expression consists of triggering a desensitization of the enzyme toward the cellular energy level as COX is apparently unable to bind ATP and of an increased catalytic activity with increased ATP production, but at the expense of elevated ROS accompanied by impaired cell function and survival. Solving the question of the involvement of COX as a consequence and/or cause of pathological processes holds the promise for COX becoming not only a diagnostic marker, but also a potential target for the development of therapeutic strategies for many hitherto incurable diseases.

Acknowledgments I thank the Deutsche Forschungsgemeinschaft (AR 343/4-1) for continuous financial support and my former and current research team members and colleagues for sharing a common interest in mitochondria, for challenging discussions, and for successful experiments to gain a deeper understanding of these powerful organelles. I apologize to all scientists whose names were not mentioned in the acknowledgments and/or cited due to space limitations.

References

Abraham CR, Marshall DC, Tibbles HE, Otto K, Long HJ, Billingslea AM, Hastey R, Johnson R, Fine RE, Smith SJ, Simons ER, Davies TA (1999) Platelets and DAMI megakaryocytes possess beta-secretase-like activity. J Lab Clin Med 133:507–515

Allen LA, Zhao XJ, Caughey W, Poyton RO (1995) Isoforms of yeast cytochrome c oxidase subunit V affect the binuclear reaction center and alter the kinetics of interaction with the isoforms of cytochrome c. J Biol Chem 270:110–118

Antonini G, Malatesta F, Sarti P, Vallone B, Brunori M (1988) ATP-induced spectral changes in cytochrome c oxidase. A kinetic investigation. Biochem J 256:835–840

Araujo GW, Beyer C, Arnold S (2008) Oestrogen influences on mitochondrial gene expression and respiratory chain activity in cortical and mesencephalic astrocytes. J Neuroendocrinol 20:930–941

Arnold S (2012) The power of life – cytochrome c oxidase takes center stage in metabolic control, cell signalling and survival. Mitochondrion 12:46–56

Arnold S, Beyer C (2009) Neuroprotection by estrogen in the brain: the mitochondrial compartment as presumed therapeutic target. J Neurochem 110:1–11

Arnold S, Kadenbach B (1997) Cell respiration is controlled by ATP, an allosteric inhibitor of cytochrome c oxidase. Eur J Biochem 249:350–354

Arnold S, Kadenbach B (1999) The intramitochondrial ATP/ADP ratio controls cytochrome c oxidase activity allosterically. FEBS Lett 443:105–108

Arnold S, Goglia F, Kadenbach B (1998) 3,5-Diiodothyronine binds to subunit Va of cytochrome c oxidase and abolishes the allosteric inhibition of respiration by ATP. Eur J Biochem 252:325–330

Arnold S, de Araújo GW, Beyer C (2008) Gender-specific regulation of mitochondrial fusion and fission gene transcription and viability of cortical astrocytes by steroid hormones. J Mol Endocrinol 41:289–300

Arnold S, Victor MB, Beyer C (2012) Estrogen and the regulation of mitochondrial structure and function in the brain. J Steroid Biochem Mol Biol (in press)

Atamna H (2006) Heme binding to amyloid-β peptide: mechanistic role in Alzheimer's disease. J Alzheimers Dis 10:255–266

Atamna H (2009) Amino acids variations in amyloid-beta peptides, mitochondrial dysfunction, and new therapies for Alzheimer's disease. J Bioenerg Biomembr 41:457–464

Atamna H, Frey WH II (2004) A role for heme in Alzheimer's disease: heme binds amyloid β and has altered metabolism. Proc Natl Acad Sci USA 101:11153–11158

Atlante A, Amadoro G, Bobba A, de Bari L, Corsetti V, Pappalardo G, Marra E, Calissano P, Passarella S (2008) A peptide containing residues 26–44 of tau protein impairs mitochondrial oxidative phosphorylation acting at the level of the adenine nucleotide translocator. Biochim Biophys Acta 1777:1289–1300

Ayala A, Venero JL, Cano J, Machado A (2007) Mitochondrial toxins and neurodegenerative diseases. Front Biosci 12:986–1007

Babcock GT, Wikström M (1992) Oxygen activation and the conservation of energy in cell respiration. Nature 356:301–309

Bandyopadhyay B, Li G, Yin H, Kuret J (2007) Tau aggregation and toxicity in a cell culture model of tauopathy. J Biol Chem 282:16454–16464

Banerjee R, Starkov AA, Beal MF, Thomas B (2009) Mitochondrial dysfunction in the limelight of Parkinson's disease pathogenesis. Biochim Biophys Acta 1792:651–663

Beal MF (1995) Aging, energy, and oxidative stress in neurodegenerative diseases. Ann Neurol 38:357–366

Beal MF (2005) Mitochondria take center stage in aging and neurodegeneration. Ann Neurol 58:495–505

Beauchemin AMJ, Gottlieb B, Beitel LK, Elhaji YA, Pinsky L, Trifiro MA (2001) Cytochrome *c* oxidase subunit Vb interacts with human androgen receptor: a potential mechanism for neurotoxicity in spinobulbar muscular atrophy. Brain Res Bull 56:285–297

Beauvoit B, Rigoulet M (2001) Regulation of cytochrome c oxidase by adenylic nucleotides. Is oxidative phosphorylation feedback regulated by its end-products? IUBMB Life 52:143–152

Benzi G, Pastoris O, Marzatico F, Villa RF, Dagani F, Curti D (1992) The mitochondrial electron transfer alteration as a factor involved in the brain aging. Neurobiol Aging 13:361–368

Bernheimer H, Birkmayer W, Hornykiewicz O, Jellinger K, Seitelberger F (1973) Brain dopamine and the syndromes of Parkinson and Huntington. Clinical, morphological and neurochemical correlations. J Neurol Sci 20:415–455

Bertoni-Freddari C, Fattoretti P, Giorgetti B, Solazzi M, Balietti M, Casoli T, Di SG (2004) Cytochrome oxidase activity in hippocampal synaptic mitochondria during aging: a quantitative cytochemical investigation. Ann N Y Acad Sci 1019:33–36

Bettini E, Maggi A (1992) Estrogen induction of cytochrome c oxidase subunit III in rat hippocampus. J Neurochem 58:1923–1929

Bisson R, Schiavo G, Montecucco C (1987) ATP induces conformational changes in mitochondrial cytochrome c oxidase. Effect on the cytochrome c binding site. J Biol Chem 262:5992–5998

Bolanos JP, Heales SJR, Land JM, Clark JB (1995) Effect of peroxynitrite on the mitochondrial respiratory chain: differential susceptibility of neurones and astrocytes in primary culture. J Neurochem 64:1965–1972

Bonilla E, Tanji K, Hirano M, Vu TH, DiMauro S, Schon EA (1999) Mitochondrial involvement in Alzheimer's disease. Biochim Biophys Acta 1410:171–182

Borlongan CV, Koutouzis TK, Randall TS, Freeman TB, Cahill DW, Sanberg PR (1995) Systemic 3-nitropropionic acid: behavioural deficits and striatal damage in adult rats. Brain Res Bull 36:549–556

Borlongan CV, Koutouzis TK, Sanberg PR (1997) 3-Nitropropionic acid animal model and Huntington's disease. Neurosci Biobehav Rev 21:289–293

Bosetti F, Brizzi F, Barogi S, Mancuso M, Siciliano G, Tendi EA, Murri L, Rapoport SI, Solaini G (2002) Cytochrome c oxidase and mitochondrial F1F0-ATPase (ATP synthase) activities in platelets and brain from patients with Alzheimer's disease. Neurobiol Aging 23:371–376

Boutin AT, Johnson RS (2007) Waiting to inhale: HIF-1 modulates aerobic respiration. Cell 129:29–30

Boveris A, Chance B (1973) The mitochondrial generation of hydrogen peroxide. General properties and effect of hyperbaric oxygen. Biochem J 134:707–716

Boveris A, Costa LE, Cadenas E (1999) The mitochondrial production of oxygen radicals and cellular aging. In: Cadenas E, Packer L (eds) Understanding the process of aging: the roles of mitochondria, free radicals, and antioxidants, vol 8, Antioxidants in health and disease. Dekker, New York, pp 1–20

Boyalla SS, Victor MB, Roemgens A, Beyer C, Arnold S (2011) Gender- and brain region-specific role of cytochrome c oxidase in 1-methyl-4-phenylpyridinium-mediated astrocyte vulnerability. J Neurosci Res 89:2068–2082

Brennan WA, Bird ED, Aprille JR (1985) Regional mitochondrial respiratory activity in Huntington's disease brain. J Neurochem 44:1948–1950

Brinton RD (2008) The healthy cell bias of estrogen action: mitochondrial bioenergetics and neurological implications. Trends Neurosci 31:529–537

Browne SE, Bowling AC, MacGarvey U, Baik MJ, Berger SC, Muqit MM, Bird ED, Beal MF (1997) Oxidative damage and metabolic dysfunction in Huntington's disease: selective vulnerability of the basal ganglia. Ann Neurol 41:646–653

Burke PV, Poyton RO (1998) Structure/function of oxygen-regulated isoforms in cytochrome c oxidase. J Exp Biol 201:1177–1195

Cadenas E, Boveris A, Ragan CI, Stoppani AO (1977) Production of superoxide radicals and hydrogen peroxide by NADH-ubiquinone reductase and ubiquinol-cytochrome c reductase from beef-heart mitochondria. Arch Biochem Biophys 180:248–257

Canevari L, Clark JB, Bates TE (1999) Beta-Amyloid fragment 25–35 selectively decreases complex IV activity in isolated mitochondria. FEBS Lett 457:131–134

Capaldi RA (1990) Structure and function of cytochrome c oxidase. Annu Rev Biochem 59:569–596

Cardoso SM, Proença MT, Santos S, Santana I, Oliveira CR (2004) Cytochrome c oxidase is decreased in Alzheimer's disease platelets. Neurobiol Aging 25:105–110

Carri MT, Ferri A, Battistoni A, Famhy L, Gabbianelli R, Poccia F, Rotilio G (1997) Expression of a Cu, Zn superoxide dismutase typical of familial amyotrophic lateral sclerosis induces mitochondrial alteration and increase of cytosolic Ca^{2+} concentration in transfected neuroblastoma SH-SY5Y cells. FEBS Lett 414:365–368

Chagnon P, Betard C, Robitaille Y, Cholette A, Gauvreau D (1995) Distribution of brain cytochrome oxidase activity in various neurodegenerative diseases. Neuroreport 6:711–715

Chance B, Williams CR (1955) Respiratory enzymes in oxidative phosphorylation. III. The steady state. J Biol Chem 217:409–427

Chandrasekaran K, Hatanpaa K, Brady DR, Stoll J, Rapoport SI (1998) Downregulation of oxidative phosphorylation in Alzheimer disease: loss of cytochrome oxidase subunit mRNA in the hippocampus and entorhinal cortex. Brain Res 796:13–19

Chen M, Durr J, Fernandez HL (2000) Possible role of calpain in normal processing of beta-amyloid precursor protein in human platelets. Biochem Biophys Res Commun 273:170–175

Chen JQ, Yager JD, Russo J (2005) Regulation of mitochondrial respiratory chain structure and function by estrogens/estrogen receptors and potential physiological/pathophysiological implications. Biochim Biophys Acta 1746:1–17

Chen JQ, Cammarata PR, Baines CP, Yager JD (2009) Regulation of mitochondrial respiratory chain biogenesis by estrogens/estrogen receptors and physiological, pathological and pharmacological implications. Biochim Biophys Acta 1793:1540–1570

Chinnery PF, Turnbull DM (1999) Mitochondrial DNA and disease. Lancet 354:17–21

Clark IE, Dodson MW, Jiang C, Cao JH, Huh JR, Seol JH, Yoo SJ, Hay BA, Guo M (2006) Drosophila pink1 is required for mitochondrial function and interacts genetically with parkin. Nature 441:1162–1166

Collman JP, Ghosh S, Dey A, Decréau RA (2009) Using a functional enzyme model to understand the chemistry behind hydrogen sulfide induced hibernation. Proc Natl Acad Sci USA 106:22090–22095

Cooper CE, Davies NA (2000) Effects of nitric oxide and peroxynitrite on the cytochrome oxidase Km for oxygen: implications for mitochondrial pathology. Biochim Biophys Acta 1459:390–396

Cottrell DA, Blakely EL, Johnson MA, Ince PG, Borthwick GM, Turnbull DM (2001) Cytochrome *c* oxidase deficient cells accumulate in the hippocampus and choroid plexus with age. Neurobiol Aging 22:265–272

Curti D, Rognaoni F, Gasparini L, Cattaneo A, Paolillo M, Racchi M, Zani L, Trabucchi M, Bergamaschi S, Govoni S (1997) Oxidative metabolism in cultured fibroblasts derived from sporadic Alzheimer's disease (AD) patients. Neurosci Lett 236:13–16

Dalmonte ME, Forte E, Genova ML, Giuffre A, Sarti P, Lenaz G (2009) Control of respiration by cytochrome c oxidase in intact cells: role of the membrane potential. J Biol Chem 284:32331–32335

Dauer W, Przedborski S (2003) Parkinson's disease: mechanisms and models. Neuron 39:889–909

Davidzon G, Greene P, Mancuso M, Klos KJ, Ahlskog JE, Hirano M, DiMauro S (2006) Early-onset familial parkinsonism due to POLG mutations. Ann Neurol 59:859–862

Davies TA, Long HJ, Sgro K, Rathbun WH, McMenamin ME, Seetoo K, Tibbles H, Billingslea AM, Fine RE, Fishman JB, Levesque CA, Smith SJ, Wells JM, Simons ER (1997) Activated Alzheimer disease platelets retain more beta amyloid precursor protein. Neurobiol Aging 18:147–153

Defeudis FV (2002) Bilobalide and neuroprotection. Pharmacol Res 46:565–568

Demonacos C, Djordjevic-Markovic R, Tsawdaroglou N, Sekeris CE (1995) The mitochondrion as a primary site of action of glucocorticoids: the interaction of the glucocorticoid receptor with mitochondrial DNA sequences showing partial similarity to the nuclear glucocorticoid responsive elements. J Steroid Biochem Mol Biol 55:43–55

Demonacos CV, Karayanni N, Hatzoglou E, Tsiriyiotis C, Spandidos DA, Sekeris CE (1996) Mitochondrial genes as sites of primary action of steroid hormones. Steroids 61:226–232

Devi L, Prabhu BM, Galati DF, Avadhani NG, Anandatheerthavarada HK (2006) Accumulation of amyloid precursor protein in the mitochondrial import channels of human Alzheimer's disease brain is associated with mitochondrial dysfunction. J Neurosci 26:9057–9068

Dhar SS, Ongwijitwat S, Wong-Riley MT (2008) Nuclear respiratory factor 1 regulates all ten nuclear encoded subunits of cytochrome c oxidase in neurons. J Biol Chem 283:3120–3129

Diaz F, Villena A, Requena V, Gonzalez P, Peláez A, Pérez de Vargas I (1996) Quantitative histochemical study of cytochrome oxidase in the dLGN of aging rats. Mech Ageing Dev 91:47–54

DiDonato S, Zeviani M, Giovannini P, Savarese N, Rimoldi M, Mariotti C, Girotti F, Caraceni T (1993) Respiratory chain and mitochondrial DNA in muscle and brain in Parkinson's disease patients. Neurology 43:2262–2268

Dluzen DE, McDermott JL (2000) Gender differences in neurotoxicity of the nigrostriatal dopaminergic system: implications for Parkinson's disease. J Gend Specif Med 3:36–42

Dodson MW, Guo M (2007) Pink1, Parkin, DJ-1 and mitochondrial dysfunction in Parkinson's disease. Curr Opin Neurobiol 17:331–337

Dröse S, Brandt U (2008) The mechanism of mitochondrial superoxide production by cytochrome bc1 complex. J Biol Chem 283:21649–21654

Dykens JA, Moos WH, Howell N (2005) Development of 17alpha-estradiol as a neuroprotective therapeutic agent: rationale and results from a phase I clinical study. Ann N Y Acad Sci 1052:116–135

Eckert A, Hauptmann S, Scherping I, Rhein V, Muller-Spahn F, Gotz J, Muller WE (2008) Soluble betaamyloid leads to mitochondrial defects in amyloid precursor protein and tau transgenic mice. Neurodegener Dis 5:157–159

Fahn S, Sulzer D (2004) Neurodegeneration and neuroprotection in Parkinson disease. NeuroRx 1:139–154

Ferguson-Miller S, Babcock GT (1996) Heme/copper terminal oxidases. Chem Rev 96:2889–2908

Folbergrová J, Ješina P, Haugvicová R, Lisý V, Houšt k J (2010) Sustained deficiency of mitochondrial complex I activity during long periods of survival after seizures induced in immature rats by homocysteic acid. Neurochem Int 56:394–403

Forman HJ, Fukuto JM, Miller T, Zhang H, Rinna A, Levy S (2008) The chemistry of cell signaling by reactive oxygen and nitrogen species and 4-hydroxynonenal. Arch Biochem Biophys 477:183–195

Frank V, Kadenbach B (1996) Regulation of the H^+/e^--stoichiometry of cytochrome c oxidase from bovine heart by intraliposomal ATP/ADP ratios. FEBS Lett 382:121–124

Frenzel M, Rommelspacher H, Sugawa MD, Dencher NA (2010) Ageing alters the supramolecular architecture of OxPhos complexes in rat brain cortex. Exp Gerontol 45:563–572

Fujita K, Yamauchi M, Shibayama K, Ando M, Honda M, Nagata Y (1996) Decreased cytochrome c oxidase activity but unchanged superoxide dismutase and glutathione peroxidase activities in the spinal cords of patients with amyotrophic lateral sclerosis. J Neurosci Res 45:276–281

Fukada K, Zhang F, Vien A, Cashman NR, Zhu H (2004) Mitochondrial proteomic analysis of a cell line model of familial amyotrophic lateral sclerosis. Mol Cell Proteomics 3:1211–1223

Fukuda R, Zhang H, Kim JW, Shimoda L, Dang CV, Semenza GL (2007) HIF-1 regulates cytochrome c oxidase subunits to optimize efficiency of respiration in hypoxic cells. Cell 129:111–122

Fukui H, Moraes CT (2008) The mitochondrial impairment, oxidative stress and neurodegeneration connection: reality or just an attractive hypothesis? Trends Neurosci 31:251–256

Gould E, McEwen BS (1993) Neuronal birth and death. Curr Opin Neurobiol 3:676–682

Gourfinkel-An I, Vila M, Faucheux B, Duyckaerts C, Viallet F, Hauw JJ, Brice A, Agid Y, Hirsch EC (2002) Metabolic changes in the basal ganglia of patients with Huntington's disease: an in situ hybridization study of cytochrome oxidase subunit I mRNA. J Neurochem 80:466–476

Greene JG, Porter RH, Eller RV, Greenamyre JT (1993) Inhibition of succinate dehydrogenase by malonic acid produces an excitotoxic lesion in rat striatum. J Neurochem 61:1151–1154

Groen AK, Wanders RJA, Westerhoff HV, van der Meer R, Tager JM (1982) Quantification of the contribution of various steps to the control of mitochondrial respiration. J Biol Chem 257:2754–2757

Grossman LI, Lomax MI (1997) Nuclear genes for cytochrome c oxidase. Biochim Biophys Acta 1352:174–192

Gu M, Gash MT, Mann VM, Javoy-Agid F, Cooper JM, Schapira AH (1996) Mitochondrial defect in Huntington's disease caudate nucleus. Ann Neurol 39:385–389

Haaxma CA, Bloem BR, Borm GF, Oyen WJ, Leenders KL, Eshuis S, Booij J, Dluzen DE, Horstink MW (2007) Gender differences in Parkinson's disease. J Neurol Neurosurg Psychiatry 78:819–824

Halliwell B (1992) Reactive oxygen species and the central nervous system. J Neurochem 59:1609–1623

Harman D (1972) The biologic clock: the mitochondria? J Am Geriatr Soc 20:145–147

Hatzoglou E, Sekeris CE (1997) The detection of nucleotide sequences with strong similarity to hormone responsive elements in the genome of eubacteria and archaebacteria and their possible relation to similar sequences present in the mitochondrial genome. J Theor Biol 184:339–344

Hauptmann S, Keil U, Scherping I, Bonert A, Eckert A, Muller WE (2006) Mitochondrial dysfunction in sporadic and genetic Alzheimer's disease. Exp Gerontol 41:668–673

Helling S, Vogt S, Rhiel A, Ramzan R, Wen L, Marcus K, Kadenbach B (2008) Phosphorylation and kinetics of mammalian cytochrome c oxidase. Mol Cell Proteomics 7:1714–1724

Henchcliffe C, Beal MF (2008) Mitochondrial biology and oxidative stress in Parkinson disease pathogenesis. Nat Clin Pract Neurol 4:600–609

Hendler RW, Pardhasaradhi K, Reynafarje B, Ludwig B (1991) Comparison of energy-transducing capabilities of the two- and three-subunit cytochromes aa3 from *Paracoccus denitrificans* and the 13-subunit beef heart enzyme. Biophys J 60:415–423

Hevner RF, Wong-Riley MTT (1989) Brain cytochrome oxidase: purification, antibody production, and immunohistochemical/histochemical correlations in the CNS. J Neurosci 9:3884–3898

Hevner RF, Liu S, Wong-Riley MT (1995) A metabolic map of cytochrome oxidase in the rat brain: histochemical, densitometric and biochemical studies. Neuroscience 65:313–342

Hirai K, Aliev G, Nunomura A, Fujioka H, Russell RL, Atwood CS, Johnson AB, Kress Y, Vinters HV, Tabaton M, Shimohama S, Cash AD, Siedlak SL, Harris PL, Jones PK, Petersen RB, Perry G, Smith MA (2001) Mitochondrial abnormalities in Alzheimer's disease. J Neurosci 21:3017–3023

Hirano A, Nakano I, Kurland LT, Mulder DW, Holley PW, Saccomanno G (1984) Fine structural study of neurofibrillary changes in a family with amyotrophic lateral sclerosis. J Neuropathol Exp Neurol 43:471–480

Horvat S, Beyer C, Arnold S (2006) Effect of hypoxia on the transcription pattern of subunit isoforms and the kinetics of cytochrome c oxidase in cortical astrocytes and cerebellar neurons. J Neurochem 99:937–951

Hsieh YC, Yu HP, Suzuki T, Choudhry MA, Schwacha MG, Bland KI, Chaudry IH (2006) Upregulation of mitochondrial respiratory complex IV by estrogen receptor-beta is critical for inhibiting mitochondrial apoptotic signaling and restoring cardiac functions following trauma-hemorrhage. J Mol Cell Cardiol 41:511–521

Huntington's Disease Collaborative Research Group (1993) A novel gene containing a trinucleotide repeat that is expanded and unstable on Huntington's disease chromosomes. Cell 72:971–983

Hüttemann M, Kadenbach B, Grossman LI (2001) Mammalian subunit IV isoforms of cytochrome c oxidase. Gene 267:111–123

Hüttemann M, Jaradat S, Grossman LI (2003a) Cytochrome c oxidase of mammals contains a testes-specific isoform of subunit VIb – the counterpart to testes-specific cytochrome c? Mol Reprod Dev 66:8–16

Hüttemann M, Schmidt TR, Grossman LI (2003b) A third isoform of cytochrome c oxidase subunit VIII is present in mammals. Gene 312:95–102

Hüttemann M, Lee I, Liu J, Grossman LI (2007) Transcription of mammalian cytochrome c oxidase subunit IV-2 is controlled by a novel conserved oxygen responsive element. FEBS J 274:5737–5748

Ikebe S, Tanaka M, Ohno K, Sato W, Hattori K, Kondo T, Mizuno Y, Ozawa T (1990) Increase of deleted mitochondrial DNA in the striatum in Parkinson's disease and senescence. Biochem Biophys Res Commun 170:1044–1048

Irwin RW, Yao J, Hamilton RT, Cadenas E, Brinton RD, Nilsen J (2008) Progesterone and estrogen regulate oxidative metabolism in brain mitochondria. Endocrinology 149:3167–3175

Itoh K, Weis S, Mehraein P, Muller-Hocker J (1996) Cytochrome c oxidase defects of the human substantia nigra in normal aging. Neurobiol Aging 17:843–848

Jackson M, Al-Chalabi A, Enayat ZE, Chioza B, Leigh PN, Morrison KE (1997) Copper/zinc superoxide dismutase 1 and sporadic amyotrophic lateral sclerosis: analysis of 155 cases and identification of an ovelinsertion mutation. Ann Neurol 42:803–807

Johann S, Dahm M, Kipp M, Beyer C, Arnold S (2010) Oestrogen regulates mitochondrial respiratory chain enzyme transcription in the mouse spinal cord. J Neuroendocrinol 22:926–935

Kadenbach B (1986) Regulation of respiration and ATP synthesis in higher organisms: hypothesis. J Bioenerg Biomembr 18:39–54

Kadenbach B, Arnold S (1999) A second mechanism of respiratory control. FEBS Lett 447:131–134

Kadenbach B, Reimann A (1992) Cytochrome c oxidase: tissue-specific expression of isoforms and regulation of activity. In: Ernster L (ed) Molecular mechanisms in bioenergetics. Elsevier, Amsterdam, pp 241–263

Kadenbach B, Jarausch J, Hartmann R, Merle P (1983) Separation of mammalian cytochrome c oxidase into 13 polypeptides by a sodium dodecyl sulfate-gel electrophoretic procedure. Anal Biochem 129:517–521

Kadenbach B, Napiwotzki J, Frank V, Arnold S, Exner S, Hüttemann M (1998) Regulation of energy transduction and electron transfer in cytochrome c oxidase by adenine nucleotides. J Bioenerg Biomembr 30:25–33

Kadenbach B, Hüttemann M, Arnold S, Lee I, Bender E (2000) Mitochondrial energy metabolism is regulated via nuclear-coded subunits of cytochrome c oxidase. Free Radic Biol Med 29:211–221

Kadenbach B, Arnold S, Lee I, Hüttemann M (2004) The possible role of cytochrome c oxidase in stress- induced apoptosis and degenerative diseases. Biochim Biophys Acta 1655:400–408

Kadenbach B, Ramzan R, Vogt S (2009) Degenerative diseases, oxidative stress and cytochrome c oxidase function. Trends Mol Med 15:139–147

Kadenbach B, Ramzan R, Wen L, Vogt S (2010) New extension of the Mitchell Theory for oxidative phosphorylation in mitochondria of living organisms. Biochim Biophys Acta 1800:205–212

Kaim G, Dimroth P (1999) ATP synthesis by F-type ATP synthase is obligatorily dependent on the transmembrane voltage. EMBO J 18:4118–4127

Kaneko I, Yamada N, Sakuraba Y, Kamenosono M, Tutumi S (1995) Suppression of mitochondrial succinate dehydrogenase, a primary target of beta-amyloid, and its derivative racemized at Ser residue. J Neurochem 65:2585–2593

Kanki T, Ohgaki K, Gaspari M, Gustafsson CM, Fukuoh A, Sasaki N, Hamasaki N, Kang D (2004) Architectural role of mitochondrial transcription factor A in maintenance of human mitochondrial DNA. Mol Cell Biol 24:9823–9834

Kato M, Saito H, Abe K (1997) Nanomolar amyloid beta protein-induced inhibition of cellular redox activity in cultured astrocytes. J Neurochem 68:1889–1895

Katzman R (1986) Alzheimer's disease. N Engl J Med 314:964–973

Kim J, Moody JP, Edgerly CK, Bordiuk OL, Cormier K, Smith K, Beal MF, Ferrante RJ (2010) Mitochondrial loss, dysfunction and altered dynamics in Huntington's disease. Hum Mol Genet 19:3919–3935

Kish SJ, Bergeron C, Rajput A, Dozic S, Mastrogiacomo F, Chang LJ, Wilson JM, DiStefano LM, Nobrega JN (1992) Brain cytochrome oxidase in Alzheimer's disease. J Neurochem 59:776–779

Klinge CM (2008) Estrogenic control of mitochondrial function and biogenesis. J Cell Biochem 105:1342–1351

Kraytsberg Y, Kudryavtseva E, McKee AC, Geula C, Kowall NW, Khrapko K (2006) Mitochondrial DNA deletions are abundant and cause functional impairment in aged human substantia nigra neurons. Nat Genet 38:518–520

Kruman II, Pedersen WA, Springer JE, Mattson MP (1999) ALS-linked Cu/Zn-SOD mutation increases vulnerability of motor neurons to excitotoxicity by a mechanism involving increased oxidative stress and perturbed calcium homeostasis. Exp Neurol 160:28–39

Kunz WS, Kudin A, Vielhaber S, Elger CE, Attardi G, Villani G (2000) Flux control of cytochrome c oxidase in human skeletal muscle. J Biol Chem 275:27741–27745

Kwast KE, Burke PV, Poyton RO (1998) Oxygen sensing and transcriptional regulation of oxygen-responsive genes in yeast. J Exp Biol 201:1177–1195

Kwast KE, Burke PV, Staahl BT, Poyton RO (1999) Oxygen sensing in yeast: evidence for the involvement of the respiratory chain in regulating the transcription of a subset of hypoxic genes. Proc Natl Acad Sci USA 96:5446–5451

Lardy HA, Wellman H (1952) Oxidative phosphorylations; role of inorganic phosphate and acceptor systems in control of metabolic rates. J Biol Chem 195:215–224

LaSpada AR, Wilson EM, Lubahn DB, Harding AE, Fischbeck KH (1991) Androgen receptor gene mutations in X-linked spinal and bulbar muscular atrophy. Nature 352:77–79

Lee HC, Pang CY, Hsu HS, Wei YH (1994) Differential accumulations of 4,977 bp deletion in mitochondrial DNA of various tissues in human ageing. Biochim Biophys Acta 1226:37–43

Lee I, Bender E, Arnold S, Kadenbach B (2001) New control of mitochondrial membrane potential and ROS formation – a hypothesis. Biol Chem 382:1629–1636

Lee I, Salomon AR, Ficarro S, Mathes I, Lottspeich F, Grossman LI, Hüttemann M (2005) cAMP-dependent tyrosine phosphorylation of subunit I inhibits cytochrome c oxidase activity. J Biol Chem 280:6094–6100

Letellier T, Malgat M, Mazat JP (1993) Control of oxidative phosphorylation in rat muscle mitochondria: implications for mitochondrial myopathies. Biochim Biophys Acta 1141:58–64

Li Y, Park JS, Deng JH, Bai Y (2006) Cytochrome c oxidase subunit IV is essential for assembly and respiratory function of the enzyme complex. J Bioenerg Biomembr 38:283–291

Lightowlers RN, Chinnery PF, Turnbull DM, Howell N (1997) Mammalian mitochondrial genetics: heredity, heteroplasmy and disease. Trends Genet 13:450–455

Linnane AW, Marzuki S, Ozawa T, Tanaka M (1989) Mitochondrial DNA mutations as an important contributor to aging and degenerative diseases. Lancet 1:642–645

Lowry OH (1975) Energy metabolism in brain and its control. In: Ingvar DH, Lassen NA (eds) Brain work: the coupling of function, metabolism, and blood flow in the brain—proceedings of the Alfred Benzon symposium VII. Academic, New York, pp 48–64

Ludwig B, Bender E, Arnold S, Hüttemann M, Lee I, Kadenbach B (2001) Cytochrome c oxidase and the regulation of oxidative phosphorylation. Chembiochem 2:392–403

Lustbader JW, Cirilli M, Lin C, Xu HW, Takuma K, Wang N, Caspersen C, Chen X, Pollak S, Chaney M, Trinchese F, Liu S, Gunn-Moore F, Lue LF, Walker DG, Kuppusamy P, Zewier ZL, Arancio O, Stern D, Yan SS, Wu H (2004) ABAD directly links Aβ to mitochondrial toxicity in Alzheimer's disease. Science 304:448–452

Malatesta F, Antonini G, Sarti P, Brunori M (1987) Modulation of cytochrome oxidase activity by inorganic and organic phosphate. Biochem J 248:161–165

Manczak M, Park BS, Jung Y, Reddy PH (2004) Differential expression of oxidative phosphorylation genes in patients with Alzheimer's disease: implications for early mitochondrial dysfunction and oxidative damage. Neuromolecular Med 5:147–162

Manczak M, Jung Y, Park BS, Partovi D, Reddy PH (2005) Time-course of mitochondrial gene expressions in mice brains: implications for mitochondrial dysfunction, oxidative damage, and cytochrome c in aging. J Neurochem 92:494–504

Marín-Hernández A, Gallardo-Pérez JC, Ralph SJ, Rodríguez-Enríquez S, Moreno-Sánchez R (2009) HIF-1alpha modulates energy metabolism in cancer cells by inducing over-expression of specific glycolytic isoforms. Mini Rev Med Chem 9:1084–1101

Markesbery WR (1997) Oxidative stress hypothesis in Alzheimer's disease. Free Radic Biol Med 23:134–147

Marsden CD (1990) Parkinson's disease. Lancet 335:948–952

Mattiazzi M, D'Aurelio M, Gajewski CD, Martushova K, Kiaei M, Beal MF, Manfredi G (2002) Mutated human SOD1 causes dysfunction of oxidative phosphorylation in mitochondria of transgenic mice. J Biol Chem 277:29626–29633

Mattingly KA, Ivanova MM, Riggs KA, Wickramasinghe NS, Barch MJ, Klinge CM (2008) Estradiol stimulates transcription of nuclear respiratory factor-1 and increases mitochondrial biogenesis. Mol Endocrinol 22:609–622

Maurer I, Zierz S, Moller HJ (2000) A selective defect of cytochrome c oxidase is present in brain of Alzheimer disease patients. Neurobiol Aging 21:455–462

McGeer PL, McGeer EG (2008) Glial reactions in Parkinson's disease. Mov Disord 23:474–483

Menzies FM, Cookson MR, Taylor RW, Turnbull DM, Chrzanowska-Lightowlers ZMA, Dong L, Figlewicz DA, Shaw PJ (2002) Mitochondrial dysfunction in a cell culture model of familial amyotrophic lateral sclerosis. Brain 125:1522–1533

Misiak M, Beyer C, Arnold S (2010a) Gender-specific role of mitochondria in the vulnerability of 6-hydroxydopamine-treated mesencephalic neurons. Biochim Biophys Acta 1797:1178–1188

Misiak M, Singh S, Drewlo S, Beyer C, Arnold S (2010b) Brain region-specific vulnerability of astrocytes in response to 3-nitropropionic acid is mediated by cytochrome c oxidase isoform expression. Cell Tissue Res 341:83–93

Mitchell P (1961) Coupling of phosphorylation to electron and hydrogen transfer by a chemiosmotic type of mechanism. Nature 191:144–148

Mizuno Y, Ohta S, Tanaka M, Takamiya S, Suzuki K, Sato T, Oya H, Ozawa T, Kagawa Y (1989) Deficiencies in complex I subunits of the respiratory chain in Parkinson's disease. Biochem Biophys Res Commun 163:1450–1455

Müftüoglu M, Elibol B, Dalmizrak O, Ercan A, Kulaksiz G, Ogüs H, Dalkara T, Ozer N (2004) Mitochondrial complex I and IV activities in leukocytes from patients with parkin mutations. Mov Disord 19:544–548

Murphy MP (2009) How mitochondria produce reactive oxygen species. Biochem J 417:1–13

Murphy MP, Brand MD (1987) Variable stoichiometry of proton pumping by the mitochondrial respiratory chain. Nature 329:170–172

Mutisya EM, Bowling AC, Beal MF (1994) Cortical cytochrome oxidase activity is reduced in Alzheimer's disease. J Neurochem 63:2179–2184

Napiwotzki J, Kadenbach B (1998) Extramitochondrial ATP/ADP-ratios regulate cytochrome c oxidase activity via binding to the cytosolic domain of subunit IV. Biol Chem 379:335–339

Napiwotzki J, Shinzawa-Itoh K, Yoshikawa S, Kadenbach B (1997) ATP and ADP bind to cytochrome c oxidase and regulate its activity. Biol Chem 378:1013–1021

Navarro A, Boveris A (2004) Rat brain and liver mitochondria develop oxidative stress and lose enzymatic activities on aging. Am J Physiol Regul Integr Comp Physiol 287:R1244–R1249

Navarro A, Boveris A (2007) The mitochondrial energy transduction system and the aging process. Am J Physiol Cell Physiol 292:670–686

Nicholls DG, Ferguson SJ (2002) Bioenergetics, 3rd edn. Academic, London

Nilsen J (2008) Estradiol and neurodegenerative oxidative stress. Front Neuroendocrinol 29:463–475

Nilsen J, Irwin RW, Gallaher TK, Brinton RD (2007) Estradiol in vivo regulation of brain mitochondrial proteome. J Neurosci 27:14069–14077

Ongwijitwat S, Wong-Riley MT (2005) Is nuclear respiratory factor 2 a master transcriptional coordinator for all ten nuclear-encoded cytochrome c oxidase subunits in neurons? Gene 360:65–77

Ongwijitwat S, Liang HL, Graboyes EM, Wong-Riley MT (2006) Nuclear respiratory factor 2 senses changing cellular energy demands and its silencing down-regulates cytochrome oxidase and other target gene mRNAs. Gene 374:39–49

Ostermeier C, Harrenga A, Ermler U, Michel H (1997) Structure at 2.7 Å resolution of the *Paracoccus denitrificans* two-subunit cytochrome c oxidase complexed with an antibody FV fragment. Proc Natl Acad Sci USA 94:10547–10553

Papa S, Capitanio N, Capitanio G, De Nitto E, Minuto M (1991) The cytochrome chain of mitochondria exhibits variable $H+/e^-$ stoichiometry. FEBS Lett 288:183–186

Parker WD Jr, Boyson SJ, Luder AS, Parks JK (1990a) Evidence for a defect in NADH: ubiquinone oxidoreductase (complex I) in Huntington's disease. Neurology 40:1231–1234

Parker WD, Filley CM, Parks JK (1990b) Cytochrome oxidase deficiency in Alzheimer's disease. Neurology 40:1302–1303

Parker WD Jr, Parks J, Filley CM, Kleinschmidt-DeMasters BK (1994) Electron transport chain defects in Alzheimer's disease brain. Neurology 44:1090–1096

Pellerin L, Magistretti PJ (2004) Neuroenergetics: calling upon astrocytes to satisfy hungry neurons. Neuroscientist 10:53–62

Peyser CE, Folstein M, Chase GA, Starkstein S, Brandt J, Cockrell JR, Bylsma F, Coyle JT, McHugh PR, Folstein SE (1995) Trial of d-alpha-tocopherol in Huntington's disease. Am J Psychiatry 152:1771–1775

Poyton RO, Goehring B, Droste M, Sevarino KA, Allen LA, Zhao XJ (1995) Cytochrome c oxidase from *Saccharomyces cerevisiae*. Methods Enzymol 260:97–116

Ramzan R, Staniek K, Kadenbach B, Vogt S (2010) Mitochondrial respiration and membrane potential are regulated by the allosteric ATP-inhibition of cytochrome c oxidase. Biochim Biophys Acta 1797:1672–1680

Reddy PH, Beal MF (2008) Amyloid beta, mitochondrial dysfunction and synaptic damage: implication for cognitive decline in aging and Alzheimer's disease. Trends Mol Med 14:45–53

Reddy PH, Tripathy R, Troung Q, Thirumala K, Reddy TP, Anekonda V, Shirendeb UP, Calkins MJ, Reddy AP, Mao P, Manczak M (2011) Abnormal mitochondrial dynamics and synaptic degeneration as early events in Alzheimer's disease: implications to mitochondria-targeted antioxidant therapeutics. Biochim Biophys Acta. doi:10.1016/j.bbadis.2011.10.011

Reimann A, Huther FJ, Berden JA, Kadenbach B (1988) Anions induce conformational changes and influence the activity and photoaffinity labeling by 8-azido-ATP of isolated cytochrome c oxidase. Biochem J 254:835–840

Ristow M, Zarse K (2010) How increase oxidative stress promotes longevity and metabolic health: the concept of mitochondrial hormesis (mitohormesis). Exp Gerontol 45:410–418

Rodriguez-Navarro JA, Solano RM, Casarejos MJ, Gomez A, Perucho J, de Yebenes JG, Mena MA (2008) Gender differences and estrogen effects in parkin null mice. J Neurochem 106:2143–2157

Roemgens A, Singh S, Beyer C, Arnold S (2010) Inducers of chemical hypoxia act in a gender- and brain region-specific manner on primary astrocyte viability and cytochrome c oxidase. Neurotox Res 20:1–14

Rosen DR, Siddique T, Patterson D, Figlewicz DA, Sapp P, Hentati A, Donaldson D, Goto J, O'regan JP, Deng HX, Rahmani Z, Krizus A, McKenna-Yasek D, Cayabyab A, Gaston SM, Berger R, Tanzi RE, Halperin JJ, Herzfeldt B, Van den Bergh R, Hung WY, Bird T, Deng G, Mulder DW, Smyth C, Laing NG, Soriano E, Pericak–Vance MA, Haines J, Rouleau GA, Gusella JS, Horvitz HR, Brown RH Jr (1993) Mutations in Cu/Zn superoxide dismutase gene are associated with familial amyotrophic lateral sclerosis. Nature 362:59–62

Sasaki S, Iwata M (1996) Ultrastructural study of synapses in the anterior horn neurons of patients with amyotrophic lateral sclerosis. Neurosci Lett 204:53–56

Sasaki S, Iwata M (2007) Mitochondrial alterations in the spinal cord of patients with sporadic amyotrophic lateral sclerosis. J Neuropathol Exp Neurol 66:10–16

Schagger H, Ohm TG (1995) Human diseases with defects in oxidative phosphorylation. 2. F1F0-ATP synthase defects in Alzheimer disease revealed by blue native polyacrylamide gel electrophoresis (PAGE). Eur J Biochem 227:916–921

Schapira AH (1996) Oxidative stress and mitochondrial dysfunction in neurodegeneration. Curr Opin Neurol 9:260–264

Scherz-Shouval R, Shvets E, Fass E, Shorer H, Gil L, Elazar Z (2007) Reactive oxygen species are essential for autophagy and specifically regulate the activity of Atg4. EMBO J 26:1749–1760

Sekeris CE (1990) The mitochondrial genome: a possible primary site of action of steroid hormones. In Vivo 4:317–320

Shearman MS, Hawtin SR, Tailor VJ (1995) The intracellular component of cellular 3-(4,5-dimethylthiazol-2-yl)-2, 5-diphenyltetrazolium bromide (MTT) reduction is specifically inhibited by beta-amyloid peptides. J Neurochem 65:218–227

Shi C, Xu J (2008) Increased vulnerability of brain to estrogen withdrawal-induced mitochondrial dysfunction with aging. J Bioenerg Biomembr 40:625–630

Shi C, Xu XW, Forster EL, Tang LF, Ge Z, Yew DT, Xu J (2008) Possible role of mitochondrial dysfunction in central neurodegeneration of ovariectomized rats. Cell Biochem Funct 26:172–178

Shigenaga MK, Hagen TM, Ames BN (1994) Oxidative damage and mitochondrial decay in aging. Proc Natl Acad Sci USA 91:10771–10778

Shulman LM (2007) Gender differences in Parkinson's disease. Gend Med 4:8–18

Siklos L, Engelhardt J, Harati Y, Smith RG, Joo F, Appel SH (1996) Ultrastructural evidence for altered calcium in motor nerve terminals in amyotropic lateral sclerosis. Ann Neurol 39:203–216

Simonetti S, Chen X, DiMauro S, Schon EA (1992) Accumulation of deletions in human mitochondrial DNA during normal aging: analysis by quantitative PCR. Biochim Biophys Acta 1180:113–122

Simonian NA, Hyman BT (1994) Functional alterations in Alzheimer's disease: selective loss of mitochondrial-encoded cytochrome oxidase mRNA in the hippocampal formation. J Neuropathol Exp Neurol 53:508–512

Simpkins JW, Dykens JA (2008) Mitochondrial mechanisms of estrogen neuroprotection. Brain Res Rev 57:421–430

Simpkins JW, Yang SH, Liu R, Perez E, Cai ZY, Covey DF, Green PS (2004) Estrogen-like compounds for ischemic neuroprotection. Stroke 35:2648–2651

Simpkins JW, Wang J, Wang X, Perez E, Prokai L, Dykens JA (2005) Mitochondria play a central role in estrogen-induced neuroprotection. Curr Drug Targets CNS Neurol Disord 4:69–83
Singh S, Misiak M, Beyer C, Arnold S (2009) Cytochrome c oxidase isoform IV-2 is involved in 3-nitropropionic acid-induced toxicity in striatal astrocytes. Glia 57:1480–1491
Singh S, Misiak M, Beyer C, Arnold S (2010) Brain region specificity of 3-nitropropionic acid-induced vulnerability of neurons involves cytochrome c oxidase. Neurochem Int 57:297–305
Soong NW, Hinton DR, Cortopassi G, Arnheim N (1992) Mosaicism for a specific somatic mitochondrial DNA mutation in adult human brain. Nat Genet 2:318–323
Stirone C, Duckles SP, Krause DN, Procaccio V (2005) Estrogen increases mitochondrial efficiency and reduces oxidative stress in cerebral blood vessels. Mol Pharmacol 68:959–965
Swerdlow RH, Parks JK, Cassarino DS, Maguire DJ, Maguire JP, Bennett JP, Davis RE, Parker WD (1997) Cybrids in Alzheimer's disease: a cellular model of the disease? Neurology 49:918–925
Swerdlow RH, Parks JK, Cassarino DS, Trimmer PA, Miller SW, Maguire DJ, Sheehan JP, Maguire RS, Pattee G, Juel VC, Phillips LH, Tuttle JB, Bennett JP Jr, Davis RE, Parker WD Jr (1998) Mitochondria in sporadic amyotrophic lateral sclerosis. Exp Neurol 153:135–142
Tabrizi SJ, Cleeter M, Xuereb J, Taanman JW, Cooper JM, Schapira AVH (1999) Biochemical abnormalities and excitotoxicity in Huntington's disease brain. Ann Neurol 45:25–32
Takuma K, Yao J, Huang J, Xu H, Chen X, Luddy J, Trillat AC, Stern DM, Arancio O, Yan SS (2005) ABAD enhances Aβ-induced cell stress via mitochondrial dysfunction. FASEB J 19:597–598
Taylor CT, Moncada S (2010) Nitric oxide, cytochrome C oxidase, and the cellular response to hypoxia. Arterioscler Thromb Vasc Biol 30:643–647
Tsukihara T, Aoyama H, Yamashita E, Tomizaki T, Yamaguchi H, Shinzawa-Itoh K, Nakashima R, Yaono R, Yoshikawa S (1996) The whole structure of the 13-subunit oxidized cytochrome c oxidase at 2.8 Å. Science 272:1136–1144
Valko M, Leibfritz D, Moncol J, Cronin MT, Mazur M, Telser J (2007) Free radicals and antioxidants in normal physiological functions and human disease. Int J Biochem Cell Biol 39:44–84
van den Eeden SK, Tanner CM, Bernstein AL, Fross RD, Leimpeter A, Bloch DA, Nelson LM (2003) Incidence of Parkinson's disease: variation by age, gender, and race/ethnicity. Am J Epidemiol 157:1015–1022
Villani G, Attardi G (1997) In vivo control of respiration by cytochrome c oxidase in wild-type and mitochondrial DNA mutation-carrying human cells. Proc Natl Acad Sci USA 94:1166–1171
Villani G, Attardi G (2000) In vivo control of respiration by cytochrome c oxidase in human cells. Free Radic Biol Med 29:202–210
Villani G, Greco M, Papa S, Attardi G (1998) Low reserve of cytochrome c oxidase capacity in vivo in the respiratory chain of a variety of human cell types. J Biol Chem 273:31829–31836
Virbasius JV, Scarpulla RC (1994) Activation of the human mitochondrial transcription factor A gene by nuclear respiratory factors: a potential regulatory link between nuclear and mitochondrial gene expression in organelle biogenesis. Proc Natl Acad Sci USA 91:1309–1313
Wallace DC, Fan W, Procaccio V (2010) Mitochondrial energetics and therapeutics. Annu Rev Pathol 5:297–348
Watanabe T, Inoue S, Hiroi H, Orimo A, Kawashima H, Muramatsu M (1998) Isolation of estrogen responsive genes with a CpG island library. Mol Cell Biol 18:442–449
West MJ (1993) Regionally specific loss of neurons in the aging human hippocampus. Neurobiol Aging 14:287–293
West M, Coleman P, Flood D, Tronsco J (1994) Differences in the pattern of hippocampal neuronal loss in normal ageing and Alzheimer's disease. Lancet 344:769–772
Wiedemann FR, Winkler K, Kuznetsov AV, Bartels C, Vielhaber S, Feistner H, Kunz WS (1998) Impairment of mitochondrial function in skeletal muscle of patients with amyotrophic lateral sclerosis. J Neurol Sci 156:65–72
Wilson DF, Mokashi A, Chugh D, Vinogradov S, Osanai S, Lahiri S (1994) The primary oxygen sensor of the cat carotoid body is cytochrome a3 of the mitochondrial respiratory chain. FEBS Lett 351:370–374

Wong-Riley MTT (1989) Cytochrome oxidase: an endogenous metabolic marker for neuronal activity. Trends Neurosci 12:94–101

Wong-Riley M, Antuono P, Ho KC, Egan R, Hevner R, Liebl W, Huang Z, Rachel R, Jones J (1997) Cytochrome oxidase in Alzheimer's disease: biochemical, histochemical, and immunohistochemical analyses of the visual and other systems. Vision Res 37:3593–3608

Wooten GF, Currie LJ, Bovbjerg VE, Lee JK, Patrie J (2004) Are men at greater risk for Parkinson's disease than women? J Neurol Neurosurg Psychiatry 75:637–639

Yang SH, Liu R, Perez EJ, Wen Y, Stevens SM Jr, Valencia T, Brun-Zinkernagel AM, Prokai L, Will Y, Dykens J, Koulen P, Simpkins JW (2004) Mitochondrial localization of estrogen receptor beta. Proc Natl Acad Sci USA 101:4130–4135

Yoshikawa S, Shinzawa-Itoh K, Tsukihara T (1998) Crystal structure of bovine heart cytochrome c oxidase at 2.8 A resolution. J Bioenerg Biomembr 30:7–14

Zeng H, Chen Q, Zhao B (2004) Genistein ameliorates beta-amyloid peptide (25–35)-induced hippocampal neuronal apoptosis. Free Radic Biol Med 36:180–188

Zhao L, Wu TW, Brinton RD (2004) Estrogen receptor subtypes alpha and beta contribute to neuroprotection and increased Bcl-2 expression in primary hippocampal neurons. Brain Res 1010:22–34

Zhao L, Jin C, Mao Z, Gopinathan MB, Rehder K, Brinton RD (2007) Design, synthesis, and estrogenic activity of a novel estrogen receptor modulator–a hybrid structure of 17beta-estradiol and vitamin E in hippocampal neurons. J Med Chem 50:4471–4481

Zhao L, Mao Z, Brinton RD (2009) A select combination of clinically relevant phytoestrogens enhances estrogen receptor {beta}-binding selectivity and neuroprotective activities in vitro and in vivo. Endocrinology 150:770–783

Zhou J, Zhou L, Hou D, Tang J, Sun J, Bondy SC (2011) Paeonol increases levels of cortical cytochrome oxidase and vascular actin and improves behavior in a rat model of Alzheimer's disease. Brain Res 1388:141–147

Chapter 14
The Many Clinical Faces of Cytochrome *c* Oxidase Deficiency

Salvatore DiMauro, Kurenai Tanji, and Eric A. Schon

Abstract Cytochrome *c* oxidase (COX) catalyzes the last step in respiration, transferring electrons from cytochrome *c* to molecular oxygen and coupling electron transfer with proton translocation from the mitochondrial matrix to the intermembrane space. COX is composed of 13 subunits, three larger catalytic subunits encoded by mitochondrial DNA (mtDNA) and ten subunits encoded by nuclear DNA.

Clinically heterogeneous human diseases were attributed to COX deficiency since the 1970s, mostly based on histochemical or biochemical data in muscle biopsies. Here, we revisit the COX deficiencies described before the molecular era, assess the value of COX histochemistry in conjunction with succinate dehydrogenase (SDH) stain, and review the clinical presentations of primary COX deficiencies defined at the molecular level.

This chapter is dedicated to the memory of Eduardo Bonilla (1936–2010).

S. DiMauro (✉)
Department of Neurology, Columbia University Medical Center, College of Physicians & Surgeons, New York, NY 10032, USA
e-mail: sd12@columbia.edu

K. Tanji
Department of Neurology, and Department of Pathology and Cell Biology, Columbia University Medical Center, New York, NY, USA
e-mail: kt8@columbia.edu

E.A. Schon
Department of Neurology, and Department of Genetics and Development, Columbia University Medical Center, New York, NY, USA
e-mail: eas3@columbia.edu

B. Kadenbach (ed.), *Mitochondrial Oxidative Phosphorylation*, Advances in Experimental Medicine and Biology 748, DOI 10.1007/978-1-4614-3573-0_14,

In general, mutations in mtDNA COX genes are associated with milder and later onset clinical syndromes, probably due to heteroplasmy. Mutations affecting nuclear-encoded COX subunits ("direct hits") are extremely rare whereas mutations affecting assembly proteins ("indirect hits") account for most COX deficiencies and the list keeps growing. Onset is generally in infancy and survival into adolescence or adult life is infrequent. The most common neurological disorder is Leigh syndrome, either alone or associated with cardiopathy, hepatopathy, or nephropathy.

14.1 Introduction

The last review of cytochrome *c* oxidase (COX) deficiency by two of these authors (SDM and EAS) was published in 1990 (DiMauro et al. 1990), at the dawn of the molecular era. Although progress in our understanding of the molecular basis of COX deficiency has been astounding, we will keep the same general structure of that article and consider sequentially the enzyme, the premolecular era diseases, histochemical studies, and the postmolecular era diseases.

Historically, the first report of COX deficiency was published in 1977 from Nijmegen, The Netherlands, then and now a Mecca of mitochondrial disease research (Willems et al. 1977). The patient, a girl, was normal at birth and developed normally during the first year, except for poor weight gain and recurrent vomiting. At 2 years, she was admitted to the hospital because she could not walk and had lost the ability to stand without support. She was hypotonic, ataxic, and showed episodic respiratory abnormalities characterized by deep sighing and sobbing. She had vertical nystagmus, partial optic atrophy. Her neurological condition deteriorated steadily and at age 4 she could no longer sit unassisted and had lost the ability to speak in simple sentences. She died at age 5 of respiratory failure. Laboratory studies showed metabolic acidosis with slightly increased blood lactate and pyruvate. Neuropathology showed symmetrical cystic lesions in basal ganglia, thalamus, mesencephalon, pons, and dentate nucleus. Microscopically, there was neuronal loss, demyelination, astrocytosis, and proliferation of capillaries.

This is the typical presentation of what was then called subacute necrotizing encephalomyelopathy (SNE) and is now called Leigh syndrome (LS). This first report was received with some skepticism by clinical researchers, who doubted that the negligible residual COX activity in skeletal and cardiac muscle could be compatible with a 5-year survival.

However, that paper was, in fact, prophetic, because both clinical and biochemical features were confirmed over time. Van Coster et al. (1991) reviewed 34 patients and subdivided the course into three clinical stages that mirror the disease progression of the first reported child. The first stage is characterized by normal neurological development during the first 8–12 months, although somatic complaints often include decelerating body and head growth, diarrhea, and recurrent vomiting. The second stage, which occurs in late infancy or early childhood, is dramatic because it entails motor regression, altered breathing pattern, loss of vision, and pyramidal, extrapyramidal, and cerebellar signs. In the third stage, which may extend from 2 to 10 years, there is diffuse hypotonia and breathing problems requiring feeding and ventilatory assistance.

Although COX deficiency in LS is presumably generalized, it was soon noted that some patients also had cardiomyopathy and others had nephropathy (Van Coster et al. 1991). Accordingly, biochemical analysis of tissues from five children with LS showed variably severe COX deficiency in brain, skeletal muscle, kidney, and liver, but the activity was normal in the liver and fibroblasts from one patient (DiMauro et al. 1987). The selective involvement of individual nonneural tissues in COX-deficient LS heralded similar observations in the molecular era, when mutations in specific COX-assembly genes were associated with encephalo-cardiomyopathy, encephalo-hepatopathy, or encephalo-nephropathy (see below).

14.2 The Enzyme

Cytochrome *c* oxidase (complex IV of the mitochondrial respiratory chain) catalyzes the last step of respiration, the transfer of electrons from cytochrome *c* to molecular oxygen, while pumping protons across the inner mitochondrial membrane (IMM) to help establish a proton gradient for ATP synthesis. COX functions as a dimer. Each monomer contains 13 subunits, two heme groups found exclusively in COX (heme *a* and *a3*, both located in subunit I), three copper ions (two in the Cu_A site in subunit II and one in the Cu_B site in subunit I), a zinc ion, and a magnesium ion.

The three largest, highly hydrophobic and transmembrane subunits, COX I, COX II, and COX III, are encoded by mtDNA and form the catalytic core of the enzyme. They contain the prosthetic groups and perform the electron transfer and proton pumping functions.

The remaining 10 subunits surround the core of the enzyme and are encoded by nuclear DNA. They are required for the assembly and stability of the holoenzyme and for its dimerization. They also modulate the catalytic activity and protect the core from reactive oxygen species (ROS). In humans, five of them exist as tissue-specific isoforms capable of conferring different kinetic properties to the enzyme, thus probably optimizing the catalytic function to the metabolic requirements of different tissues: they are COX4, COX6A, COX6B, COX7A, and COX8-3 (Fontanesi et al. 2008; Huttemann et al. 2003). Another mechanism that contributes to the adaptability of COX is physiologically controlled phosphorylation.

COX assembly is a complex process requiring expression of the mtDNA-encoded subunits, expression and import of the nDNA-encoded subunits, insertion of the structural subunits into the protein milieu of the IMM, addition of prosthetic groups, assembly of the holoenzyme, dimerization, and additional assembly into supercomplexes (Fontanesi et al. 2006). Assembly occurs through the sequential and ordered addition of subunits and cofactors to an initial COX1 kernel in a multistep process through discrete short-term intermediates. In fact, analysis of human fibroblasts from patients with mutations in different COX-assembly genes revealed several subassemblies, all containing the core subunit COX1 (Williams et al. 2004).

Numerous nuclear-encoded factors are required for COX maturation and assembly, including those responsible for the synthesis of heme A (COX10 and COX15), and for the transport and insertion of copper ions (SCO1 and SCO2). In yeast, more than 30 chaperones assist in the assembly of COX and many of them have human homologues, including CMC1, COX10, COX11, COX15, COX17, COX18, COX19, COX20, LRPPRC, OXA1, PET191, SCO1, SCO2, and SURF1 (Brosel et al. 2010).

14.3 Diseases in the Premolecular Era

If the diseases are defined (as we think they should) by selective and severe deficiency of COX activity in one or more tissues, then identifying authentic COX deficiencies is not easy. The main difficulty is that in the past—and especially in the premolecular era—too many disorders were classified as COX deficiencies based on histochemical or biochemical criteria that did not emphasize the need for the selectivity of the enzyme defect. For example, it is instructive to look at the review on COX deficiency published by our group in 1986 (DiMauro et al. 1986). The table summarizing clinical phenotypes divides them in three groups, those affecting muscle exclusively or predominantly, those affecting predominantly the brain, and "other."

The only entity still acceptable nowadays as a bona fide COX deficiency is LS, which is arguably the most common and severe presentation. Other encephalopathic disorders included in the table were Alpers syndrome and Menkes disease. We now know that Alpers syndrome is due to mutations in the *POLG* gene and is associated with mtDNA depletion whereas Menkes disease is due to mutations in *ATP7A*, encoding an energy-dependent copper-transporting P-type membrane ATPase, also known as MNK. The general impairment of mitochondrial protein synthesis that accompanies mtDNA depletion easily explains the decreased COX activity of Alpers syndrome, which, however, is not isolated but is associated with defects in other respiratory chain complexes (Davidzon et al. 2005). In Menkes disease, the intracellular copper maldistribution results in defective copper incorporation into several essential enzymes, one of which is COX (Menkes 2008).

Among the myopathies, we listed a fatal infantile form associated with renal tubulopathy (DeToni–Fanconi–Debre syndrome) that was reported in seven children (Van Biervliet et al. 1977; DiMauro et al. 1980; Heiman Patterson et al. 1982; Minchom et al. 1983; Zeviani et al. 1985). Detailed studies of skeletal muscle in one of these children showed complete lack of COX histochemical stain, large aggregates of abnormal mitochondria by electron microscopy, almost undetectable COX activity, and an absent cytochrome aa_3 peak in the reduced-minus-oxidized spectrum (DiMauro et al. 1980). In a second child, the decrease of COX was documented both by immunohistochemistry and by enzyme-linked immunosorbent assay (ELISA), and the selective involvement of skeletal muscle was highlighted by the normal COX stain of intrafusal fibers of muscle spindles (Zeviani et al. 1985). The homogeneous clinical picture and the isolated lack of COX argued for a distinct entity: however, no additional cases have been reported in recent years and the molecular basis remains unknown. Immunological studies in an additional case of fatal infantile myopathy without renal involvement showed marked decrease of cross-reacting material and also documented that the subunit composition of the residual COX protein was normal (Bresolin et al. 1985), suggesting, in retrospect, a problem with COX assembly.

In contrast with these cases of "fatal infantile COX deficiency myopathy," in 1983 we reported an infant with what we called "benign mitochondrial myopathy due to reversible COX deficiency" (DiMauro et al. 1983). Despite neonatal onset of diffuse weakness and severe lactic acidosis, with appropriately vigorous support this child improved spontaneously and was only mildly weak at 33 months. His lactate declined steadily and was normal by 14 months of age and three muscle biopsies at 1, 7, and 36 months illustrated the reversibility of the COX deficiency. Fewer than 5% of fibers stained positively for COX in the first biopsy, 60% in the second, and all of them in the third. COX activity increased accordingly from 8 to 47% in the first two biopsies and was higher than normal in the third.

We postulated two scenarios to explain the reversibility of the enzyme defect. First, a mtDNA mutation could confer a selective disadvantage, such that fibers with high mutation load would be eliminated and replaced by fibers with lower abundance of the mutation. Second, a mutation in a nuclear gene encoding a muscle-specific and developmentally regulated COX subunit could be expressed only in the neonatal period and be "healed" by the expression of an adult isozyme. In support of the first hypothesis was the astute observation by Eduardo Bonilla that the muscle biopsy from the infant's mother showed sporadic COX-negative fibers, but it took 26 years to identify a homozygous mtDNA mutation (m.14674T>C) in $tRNA^{Glu}$ in most, though not all, patients with this disease (Horvath et al. 2009; Mimaki et al. 2010; Uusimaa et al. 2011). Even if this mtDNA mutation were sufficient to explain pathogenicity (which it is not, as we will see later), the reversible COX deficient myopathy would still not qualify as an authentic COX deficiency because other respiratory chain complexes are involved besides COX.

Yet another "mostly myopathic" syndrome attributed to COX deficiency in the table of our 1986 review (DiMauro et al. 1986) was "myopathy and cardiopathy," referring to a child with neonatal onset of generalized weakness and severe lactic acidosis, who at 4 months developed cardiomegaly and died at 8 months of cardiac arrest (Zeviani et al. 1986). COX activity was 7% in skeletal muscle and 12% in cardiac muscle, although normal amount of the protein was shown by immunotitration. This case is reminiscent of patients homozygous for the E140K mutation in the *SCO2* assembly gene of COX (see below).

The last item in the "mostly myopathic" group refers to two paternally related second cousins, one of whom had mitochondrial myopathy (including severe COX deficiency in muscle) and died at 18 weeks while the other had mitochondrial hepatopathy (including severe COX deficiency in liver) and died at 9 months (Boustany et al. 1983). Seven years later, these were the first patients in whom mtDNA depletion was documented in muscle and liver, respectively (Moraes et al. 1991). The molecular basis of the mtDNA depletion remains unknown in these infants, and, although COX deficiency was the dominant biochemical feature in the two affected tissues, this, too, is not a primary COX deficiency syndrome.

"Other" COX-deficiency syndromes in that table included "encephalomyopathy of adult onset" and "progressive external ophthalmoplegia." The patient with encephalomyopathy was a 52-year-old man with adult onset of slowly progressive limb weakness, sensorineural hearing loss, and complex partial seizures. Computed tomography of the brain showed cerebral atrophy and a muscle biopsy showed ragged-red fibers (RRF) and COX-negative fibers; COX activity in muscle was decreased to 30–40% of the normal mean (Servidei et al. 1987). In retrospect, this case has all the features of an mtDNA mutation, which once again excludes it from the primary COX deficiency syndromes.

Similarly, the cases of progressive external ophthalmoplegia (PEO) were probably associated with single or multiple mtDNA deletions affecting mitochondrial protein synthesis in toto, and causing multiple respiratory chain defects, not specifically COX deficiency. By the same token, the first case of MNGIE (myoneurogastrointestinal encephalopathy; the acronym has not changed but now means mitochondrial neurogastrointestinal encephalomyopathy) was attributed to COX deficiency (Bardosi et al. 1987). The enzyme defect was real but not causative, as we now know that MNGIE is a defect of intergenomic communication due to mutations in the *TYMP* gene and is associated with mtDNA depletion, multiple deletions, and point mutations (Hirano et al. 2005).

With hindsight, we have proven the point that the diagnosis of COX deficiency on purely clinical grounds is virtually impossible, although LS should raise this possibility. Primary COX deficiencies can only be defined with the help of molecular data, as we will show in detail below.

14.4 Histochemical Studies

Although the histochemical determination of COX deficiency may have limited value in the diagnosis of primary COX deficiencies, it has proven extremely useful for the diagnosis and initial classification of mitochondrial diseases in general. In fact, systematic analysis of frozen, cross-sectioned muscle with the histochemical reactions for COX, succinate dehydrogenase (SDH), or both stains superimposed has become a gold standard for diagnosis (Fig. 14.1).

It is important to remind the reader that the SDH stain reflects the activity of complex II of the respiratory chain (succinate–ubiquinone oxidoreductase), the only complex that is entirely encoded by nuclear DNA. This means that the SDH stain is unruffled by deleterious mutations of mtDNA while at the same time is an excellent marker of mitochondrial abundance. In contrast, the three catalytic subunits of COX are encoded by mtDNA and COX stain is a very good index of mtDNA function. Not only mutations in COX I, COX II, or COX III, but any mtDNA mutation impairing mitochondrial protein synthesis (i.e. large-scale deletions or point mutations in tRNA or rRNA genes) will result in decrease or lack of COX stain.

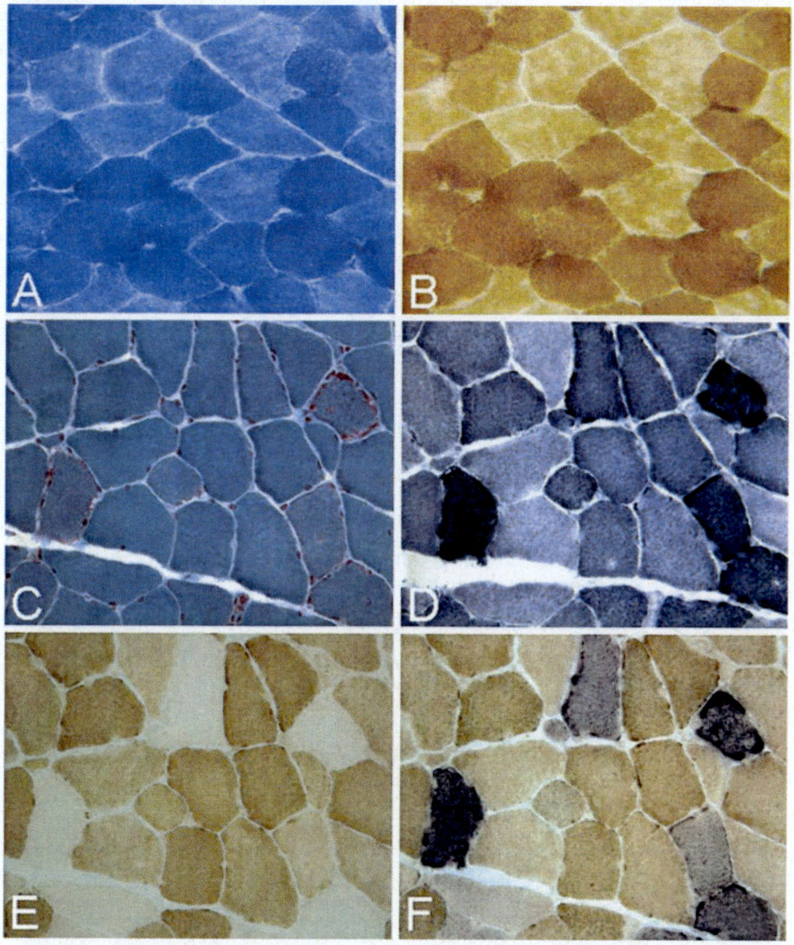

Fig. 14.1 In normal human muscle, all fibers stain with the succinate dehydrogenase (SDH) and the cytochrome *c* oxidase (COX) reactions (**a** and **b**, respectively). The different intensities correspond to type 1 (*darker stains*) and type 2 (*lighter stains*) fibers. In a patient with MERRF (myoclonus epilepsy and ragged-red fibers) due to the m.8344A>G mutation, two ragged-red fibers are evident with the modified Gomori trichrome stain (**c**); the same fibers appear "*ragged-blue*" with the SDH stain (**d**). In the same patient, two COX-negative fiber (**e**) appear *blue* with the combined COX/SDH stain (**f**)

Another important reminder is that pathogenic mtDNA mutations are usually heteroplasmic (i.e. mutated and wild-type mtDNAs coexist in a cell) and—in elongated syncytial muscle fibers—the mutation load is not uniformly distributed along the length of the fiber, such that adjacent segmental sections may have widely different amounts of mutant mtDNAs. It follows that cross sections of the muscle biopsy from patients harboring pathogenic mtDNA mutations will show a mosaic pattern of COX-negative and COX-positive fibers (Fig. 14.1e, f).

As mitochondria proliferate in response to energy failure (although the *primum movens* of this phenomenon remains obscure), COX deficiency is often accompanied by excess mitochondria, showing as hyperintense SDH stain (these fibers are dubbed "ragged-blue" because of their analogy with the more famous "ragged-red" fibers revealed with the modified Gomori trichrome stain) (Fig. 14.1c, d). Thus, finding a mosaic pattern of normally stained fibers admixed with "ragged-blue" COX-negative fibers in a muscle biopsy is a robust clue to the diagnosis of an mtDNA-related disease affecting mitochondrial protein synthesis (more rarely, affecting one of the three mtDNA-encoded COX subunits). A mosaic pattern of both ragged-blue and COX-positive fibers suggests a mutation in an mtDNA gene that encodes a respiratory chain subunit other than COX I, II, or III (for example, a complex I subunit or cytochrome *b*) (Andreu et al. 1999).

Heteroplasmy explains not only the mosaic distribution of COX-negative fibers but also the different degree of "COX-negativity," as some fibers have a deficiency rather than a lack of stain. To help reveal the COX-deficient fibers, Eduardo Bonilla introduced a modified stain that is as simple as it is effective. He noticed that when the COX and SDH stain are superimposed in normal fibers, the brown COX stain prevails and overshadows the blue SDH stain. However, even a small decrease in COX activity allows the SDH stain to shine through making the COX-deficient fibers appear blue. Nor is this technique limited to skeletal muscle, as it has been successfully applied to reveal COX-deficient motor neurons in patients with ALS (Borthwick et al. 1999), MELAS (Betts et al. 2006), other mtDNA-related diseases (Bortwick et al. 2006), COX-deficient smooth muscle cells in the GI of MELAS patients (Betts et al. 2008), and COX-deficient colonic crypt cells (Greaves et al. 2006).

As expected, in Mendelian disorders, for example COX-deficient LS, muscle histochemistry shows more or less severe but *diffuse* COX deficiency (Sue et al. 2000), which is an important diagnostic clue (Fig. 14.2). In fact, as routine muscle histochemistry is usually noninformative in LS, COX histochemistry is a useful way to rule in or out one important cause of the disease (Fig. 14.2). Children with the fatal infantile COX-deficient myopathy or the reversible COX-deficient myopathy described earlier have diffuse reduction of COX stain in muscle sections but normal COX stain in the intrafusal fibers of muscle spindles (Zeviani et al. 1985) and in the smooth muscle of blood vessels (DiMauro et al. 1986). The muscle specificity and the diffuse lack of COX stain suggest a nuclear DNA involvement, which has not yet been clarified.

In conclusion, COX histochemistry has proven invaluable both as a diagnostic test and as a research tool.

14.5 The Diseases in the Molecular Era

Numerous pathogenic mutations have been reported in all three mtDNA genes of COX and they have been associated with a multitude of symptoms and signs. By and large, however, these clinical pictures are less severe and have later onset than the

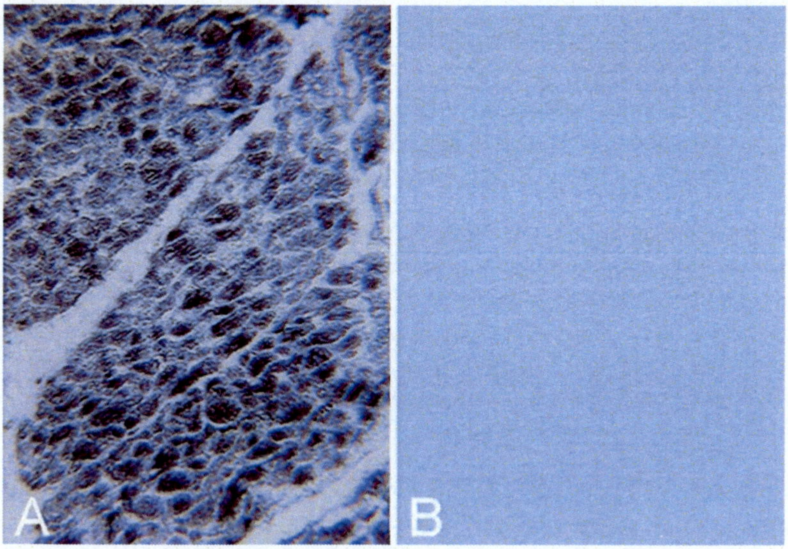

Fig. 14.2 In a patient with Leigh syndrome (LS) who was compound heterozygous for two mutations in *SCO2*, normal SDH activity (**a**) contrasts with diffusely negative COX activity (**b**)

syndromes (mostly LS) associated with Mendelian COX deficiencies (see below). A plausible explanation of this phenomenon is the "recessive" nature of heteroplasmic mtDNA mutations, requiring high mutation loads for clinical expression.

Aside from two infants with severe neonatal lactic acidosis, apnea, and bradycardia, who died soon after birth (Wong et al. 2001), all other cases affected older children or young adults. Interestingly, there were several examples of phenocopies: MELAS syndrome associated with mutations in COX II (Rossmanith et al. 2008) or COX III (Manfredi et al. 1995), LS associated with a mutation in COX III (Mkaouar-Rebai et al. 2011), Alpers syndrome associated with a mutation in COX II (Uusimaa et al. 2003), and even amyotrophic lateral sclerosis (ALS) associated with a mutation in COX I (Comi et al. 1998).

It is also notable that at least three sporadic patients with COX deficiency had isolated myopathy and myoglobinuria (Karadimas et al. 2000; McFarland et al. 2004; Kollberg et al. 2005).

Because of the frequent association of autosomal recessive LS and COX deficiency, we had come to the conclusion that mutations in one or more of the 10 nuclear encoded COX subunits would likely underlie most cases. Therefore, we and others cloned all the corresponding genes (for review, see DiMauro et al. 1990) and conducted Western and Northern analyses of different tissues from LS patients, to no avail. We therefore concluded rather disconsolately that "LS might be due to the mutation of a nuclear regulatory gene controlling the assembly or stability of the enzyme…" (DiMauro et al. 1990).

Although disconsolate, we were prophetic. Eight years later, two groups, using similarly elegant complementation analyses, documented that COX deficiency in human LS fibroblasts was rescued by addition of normal human chromosome 9 (Zhu et al. 1998; Tiranti et al. 1998). Candidate gene analysis led to the discovery of mutations in the *SURF1* gene, a homologue of the yeast assembly gene SHY1.

Although it soon became apparent that *SURF1* mutations explained most of the typical LS cases due to COX deficiency (Pequignot et al. 2001), the new notion that COX deficiency could be due to "indirect hits," together with the old knowledge that in yeast more than 30 chaperone proteins are needed to assemble COX (and many of these have human homologues), opened the traditional can of worms, as clinical scientists started fishing for new mutant COX-assembly genes.

Mutations in *SCO2*, which encodes a metallochaperone involved in mitochondrial copper delivery, cause a much more severe clinical phenotype than typical LS: a combination of neonatal hypertrophic cardiomyopathy with encephalopathy, which is fatal in the first weeks or months of life (Papadopoulou et al. 1999; Jaksch et al. 2000; Vesela et al. 2008; Knuf et al. 2007) and may be associated with early fetal lethality (Tay et al. 2004). The fetal infantile presentation is typical of compound heterozygous patients, who always harbor the common E140K mutation (Leary et al. 2006), whereas homozygosity for the same mutation is associated with delayed onset and longer survival (Vesela et al. 2008; Jaksch et al. 2001a).

Notably, *SCO2* mutations, both in homozygosity and in compound heterozygosity, can mimic spinal muscular atrophy (SMA), including the typical neurogenic histological pattern of the muscle biopsy (Tarnopolsky et al. 2004; Salviati et al. 2001; Pronicki et al. 2010). In fact, one postmortem study showed severe neuronal loss and astrocytosis in the anterior horns of the spinal cord (Salviati et al. 2001). Thus, in patients with SMA but without mutations in the *SMN* gene, it is important to exclude *SCO2* mutations.

Copper supplementation in the culture medium rescued COX activity in cultured fibroblasts and myoblasts from patients with SCO2 mutations (Jaksch et al. 2001b; Salviati et al. 2002), which led to the experimental treatment of a girl homozygous for the E140K mutation with subcutaneous and oral Cu-his (Freisinger et al. 2004). There was objective improvement of the cardiopathy but not of the encephalopathy and the girl died of pneumonia at 3½ years.

Mutations in *SCO1*, another metallochaperone needed for copper insertion into the COX holocomplex, were reported in a large family with several affected members (Valnot et al. 2000a). The child described in detail was hypotonic and lethargic at birth. Liver function tests were abnormal at day 4, and the infant developed hepatomegaly. He also had axial hypotonia, episodic apnea, bradycardia, and died at 2 months. A postmortem liver "biopsy" showed swollen hepatocytes with microvesicular steatosis, and a muscle biopsy showed lipid storage. A sibling presented with metabolic acidosis and "severe neurological distress" and died at 5 days.

It is not clear why mutations in the two COX metallochaperones, SCO2 and SCO1, should result in different clinical phenotypes, an encephalocardiopathy and an encephalohepatopathy. A comparative study of *SCO2* and *SCO1* transcription and protein expression in human and mouse tissues showed that both genes are expressed ubiquitously, but the expression of *SCO1* is especially robust in blood vessels and in liver, thus possibly explaining the encephalohepatopathy of patients with SCO1 deficiency (Brosel et al. 2010). The report of a single family suggests that mutations in *SCO1* are either very rare or incompatible with life. Several genetic screenings of patients with COX deficiency confirm the rarity of *SCO1* mutations (Coenen et al. 2006; Bohm et al. 2006; Horvath et al. 2000; Sacconi et al. 2003).

Mutations in *COX10*, which encodes a factor involved in the first step of heme-A biosynthesis, have been described in five patients from three families (Valnot et al. 2000b; Antonicka et al. 2003a). Three siblings from a consanguineous family died in childhood: the child described in more detail had severe encephalopathy (ataxia, weakness, ptosis, status epilepticus) and nephropathy (proximal renal tubulopathy) and his younger sister also had neurological involvement (Valnot et al. 2000b). One of the two unrelated children had transfusion-dependent macrocytic anemia, hypertrophic cardiomyopathy, and died at 5 months. The other child had a LS-like syndrome (both clinically and neuroradiologically), transfusion-dependent anemia, and died at 4 months (Antonicka et al. 2003a).

Mutations in COX15, which is also involved in heme-A biosynthesis, caused encephalocardiopathy in an infant girl who died at 24 days (Kennaway et al. 1990; Antonicka et al. 2003b), but typical LS was diagnosed in two patients, one of whom died at 3 years (Oquendo et al. 2004) whereas the other had an unusually protracted course and was still alive at 16 years of age (Bugiani et al. 2005).

A homozygous mutation in *C2orf64*, which encodes a factor involved in the early steps of COX assembly, caused cardiomyopathy in utero (reminiscent of *SCO2* mutations), fetal distress, and congenital biventricular hypertrophic cardiomyopathy in two siblings, who died 8 and 10 days after birth (Huigsloot et al. 2011).

A protein with the improbable name fas-activated serine-threonine kinase domain 2 (FASTKD2) has a role in mitochondrial apoptosis rather than in COX assembly. Yet, two siblings from a consanguineous family had COX deficiency associated with a homozygous mutation in the gene (*KIAA0971*) encoding FASTKD2 (Ghezzi et al. 2008). Both patients had progressive neurological deterioration and were severely incapacitated at 4 and 14 years of age, respectively.

Through integrated genomics, based on bioinformatics-generated intersection of DNA, mRNA, and protein data sets, Mootha et al. (2003) identified the gene *LRPPRC* responsible for the COX-deficient French Canadian type LS (LSFC) (Morin et al. 1993). As the LRPPRC (another protein with an improbable name, leucine-rich pentatricopeptide repeat cassette) controls the translation or stability of the mRNA of mtDNA-encoded COX subunits (Xu et al. 2004), LDFC does not really qualify as an "indirect hit" but rather as a defect of intergenomic communication and, more specifically, a defect of mtDNA translation (Chrzanowska-Lightowlers et al. 2011).

Another defect of mitochondrial translation affecting the COX I subunit is caused by mutations in the gene *TACO1* (translational activator of COX I) (Weraarpachai et al. 2009). Five children of a consanguineous Turkish family were affected with slowly progressive LS, more severe in girls than in boys: all patients were alive at the time of publication and three were in their 20s. The clinical picture was characterized by small stature, mental retardation, dystonia, dysarthria, spasticity, optic atrophy, and brain MRI showed symmetrical lesions of the basal ganglia (Seeger et al. 2010).

Because, as we had stated earlier, COX I is the "condensation nucleus" starting the biosynthesis of the COX holocomplex, it is hardly surprising that impairment of the initial assembly of other subunits with COX I would result in an unstable nascent enzyme complex and severe disease. This was, in fact, documented by a homozygous mutation in the *C12orf62* gene, which encodes a 6 kDa single-transmembrane protein that localizes to the mitochondria and is apparently involved in COX assembly initiation (Weraarpachai et al. 2012). Three of four siblings in a consanguineous family had rapidly fatal neonatal disease with lactic acidosis. One infant was dysmorphic and died at 24 h: autopsy showed involvement of multiple tissues, including the brain (hypertrophy, cavitation, dysmyelination), the heart (hypertrophic cardiomyopathy), the liver (hepatomegaly), the kidney (hypoplasia), and the adrenal glands (hyperplasia) (Weraarpachai et al. 2012).

Integrative genomics also made possible the identification of the *ETHE1* gene in children with ethylmalonic encephalopathy (EE), an early onset syndrome with microangiopathy, chronic diarrhea, and markedly increased levels of ethylmalonic acid and short-chain acylcarnitines in body fluids. On the basis of their studies in patients and in *Ethe 1*-null mice, Valeria Tiranti and Massimo Zeviani introduced a new pathogenic paradigm that could be dubbed "toxic indirect hit." They showed that ETHE1 is a mitochondrial matrix thioesterase and its dysfunction leads to accumulation of sulfide, a powerful COX inhibitor (Tiranti et al. 2009).

It is conceivable that most "direct hits" (i.e. mutations affecting directly nuclear COX subunits) are incompatible with life because years of research have revealed mutations in only one of the 10 subunits, COX6B1 (Massa et al. 2008). Two brothers had normal psychomotor development followed by muscle weakness, cognitive decline, visual problems, and lactic acidosis. Brain MRI showed cavitating leukodystrophy. One brother died at 10 years of age and the other was 8 year old and still alive at the time of publication. The severity of mutations in this gene may be better indicated by the fact that a sister of the patient had died immediately after birth and the mother had a second trimester miscarriage.

We now return to the conundrum of the reversible COX deficiency syndrome. As mentioned earlier, Horvath et al. (2009) identified a homoplasmic mutation (m.14674T>C) at the discriminator base of the $tRNA^{Glu}$ of mtDNA in 17 patients from 12 families. Clinically, eight patients had recovered completely, seven had residual mild myopathy, one had myopathy and seizures, and one had died. Clearly, the original term "benign reversible myopathy" is inappropriate because many patients are left with some weakness and the initial presentation is severe enough to cause death if vigorous supportive measures are not taken. The pathogenicity of the mutation was documented by Northern blots showing that steady-state levels of $tRNA^{Glu}$ were clearly decreased in the muscle biopsy of a 1-month-old patient, but were only mildly decreased in the same patient after recovery and in his asymptomatic mother. Accordingly, immunoblotting showed markedly decreased amounts of COX I, COX II, and of a complex I subunit at 1 month of age but not later in life (Horvath et al. 2009).

Either the same or a similar homoplasmic mutation (m.14674T>G) was also identified in eight Japanese patients with reversible COX deficiency (Mimaki et al.

2010). Two of them had LS-like bilateral basal ganglia lesions, confirming that in a few patients this disorder is not confined to muscle. In early muscle specimens, there were decreased levels of $tRNA^{Glu}$ and of mtDNA-encoded proteins, not just COX subunits.

The plot thickened when Uusimaa et al. (2011) reported eight new patients from six families: they all improved markedly from the infantile weakness but had persistent mild myopathy. Muscle biochemistry showed combined defects of respiratory chain enzymes rather than isolated COX deficiency. Importantly, while four families harbored the homoplasmic m.14674T>C mutation, two families had mutations in the nuclear *TRMU* gene, which encodes mtRNA 2-thiouridylase. Mutations in *TRMU* had been reported in infants with another reversible condition, a hepatopathy rather than a myopathy (Schara et al. 2011). The title of the paper by Uusimaa et al. redefines the "reversible COX deficiency myopathy" in the following way: "Reversible infantile respiratory chain deficiency is a unique, genetically heterogeneous mitochondrial disease" (Uusimaa et al. 2011).

We have come back full circle to a disease that we described in 1983 and we still cannot explain. This seems appropriate as defects of COX are often problematic: unraveling their causes not only helps the clinician in her bedside work but also provides clues to a better understanding of the complex control of our quintessential oxidative enzyme.

Acknowledgments This work has been supported by NICHD grant P01-H23062 and by the Marriott Mitochondrial Disorder Clinical Research Fund (MMDCRF).

References

Andreu AL, Hanna MG, Reichmann H et al (1999) Exercise intolerance due to mutations in the cytochrome b gene of mitochondrial DNA. N Engl J Med 341:1037–1044

Antonicka H, Leary SC, Guercin G-H et al (2003a) Mutations in COX10 result in a defect in mitochondrial heme A biosynthesis and account for multiple, early-onset clinical phenotypes associated with isolated COX deficiency. Hum Mol Genet 12:2693–2702

Antonicka H, Mattman A, Carlson CG et al (2003b) Mutations in COX15 produce a defect in the mitochondrial heme biosynthetic pathway, causing early-onset fatal hypertrophic cardiomyopathy. Am J Hum Genet 72:101–114

Bardosi A, Creutzfeldt W, DiMauro S et al (1987) Myo-, neuro-, gastrointestinal encephalopathy (MNGIE syndrome) due to partial deficiency of cytochrome-c-oxidase. A new mitochondrial multisystem disorder. Acta Neuropathol 74:248–258

Betts J, Jaros E, Perry RH et al (2006) Molecular neuropathology of MELAS: level of heteroplasmy in individual neurones and evidence of extensive vascular involvement. Neuropathol Appl Neurobiol 32:359–373

Betts J, Barron MJ, Schaefer AM, Taylor RW, Turnbull D (2008) Gastrointestinal tract involvement associated with the 3243A>G mitochondrial DNA mutation. Neurology 70:1290–1292

Bohm M, Pronicka E, Karczmarewicz E et al (2006) Retrospective, multicentric study of 180 children with cytochrome c oxidase deficiency. Pediatr Res 59:21–26

Borthwick GM, Johnson MA, Ince PG, Shaw PJ, Turnbull DM (1999) Mitochondrial enzyme activity in amyotrophic lateral sclerosis: implications for the role of mitochondria in neuronal cell death. Ann Neurol 46:787–790

Bortwick GM, Taylor RW, Walls TJ et al (2006) Motor neuron disease in a patient with a mitochondrial tRNAIle mutation. Ann Neurol 59:570–574

Boustany RN, Aprille JR, Halperin J, Levy H, DeLong GR (1983) Mitochondrial cytochrome deficiency presenting as a myopathy with hypotonia, external ophthalmoplegia, and lactic acidosis in an infant and as fatal hepatopathy in a second cousin. Ann Neurol 14:462–470

Bresolin N, Zeviani M, Bonilla E et al (1985) Fatal infantile cytochrome c oxidase deficiency: decrease of immunologically detectable enzyme in muscle. Neurology 35:802–812

Brosel S, Yang H, Tanji K, Bonilla E, Schon EA (2010) Unexpected vascular enrichment of SCO1 over SCO2 in mammsalian tissues. Am J Pathol 177:2541–2548

Bugiani M, Tiranti V, Farina L, Uziel G, Zeviani M (2005) Novel mutations in *COX15* in a long surviving Leigh syndrome patient with cytochrome *c* oxidase deficiency. J Med Genet 42:e28

Chrzanowska-Lightowlers ZMA, Horvath R, Lightowlers RN (2011) 175th ENMCinternational workshop: Mitochondrial protein synthesis in health and disease, 25-27th June, 2010, Naarden, The Netherlands. Neuromuscul Disord 21:142–147

Coenen MJH, Smeitink JAM, Pots JM et al (2006) Sequence analysis of the structural nuclear encoded subunits and assembly genes of cytochrome c oxidase in a cohort of 10 isolated complex IV-deficient patients revealed five mutations. J Child Neurol 21:508–511

Comi GP, Bordoni A, Salani S et al (1998) Cytochrome c oxidase subunit I microdeletion in a patient with motor neuron disease. Ann Neurol 43:110–116

Davidzon G, Mancuso M, Ferraris S et al (2005) *POLG* mutations and Alpers syndrome. Ann Neurol 57:921–924

DiMauro S, Mendell JR, Sahenk Z et al (1980) Fatal infantile mitochondrial myopathy and renal dysfunction due to cytochrome-c-oxidase deficiency. Neurology 30:795–804

DiMauro S, Nicholson JF, Hays AP et al (1983) Benign infantile mitochondrial myopathy due to reversible cytochrome c oxidase deficiency. Ann Neurol 14:226–234

DiMauro S, Zeviani M, Servidei S et al (1986) Cytochrome oxidase deficiency: clinical and biochemical heterogeneity. Ann N Y Acad Sci 488:19–32

DiMauro S, Servidei S, Zeviani M et al (1987) Cytochrome c oxidase deficiency in Leigh syndrome. Ann Neurol 22:498–506

DiMauro S, Lombes A, Nakase H et al (1990) Cytochrome c oxidase deficiency. Pediatr Res 28:536–541

Fontanesi F, Soto IC, Horn D, Barrientos A (2006) Assembly of mitochondrial cytochrome c-oxidase, a complicated and highly regulated cellular process. Am J Cell Physiol 291:C1129–C1147

Fontanesi F, Soto IC, Barrientos A (2008) Cytochrome c oxidase biogenesis: New levels of regulation. IUBMB Life 60:557–568

Freisinger P, Horvath R, Macmillan C, Peters J, Jaksch M (2004) Reversion of hypertrophic cardiomyopathy in a patient with deficiency of the mitochondrial copper binding protein Sco2: Is there a potential effect of copper? J Inherit Metab Dis 27:67–79

Ghezzi D, Saada A, D'Adamo P et al (2008) FASTK2 nonsense mutation in an infantile mitochondrial encephalomyopathy associated with cytochrome c oxidase deficiency. Am J Hum Genet 83:415–423

Greaves LC, Preston SL, Tadrous PJ et al (2006) Mitochondrial DNA mutations are established in human colonic stem cells, and mutated clones expand by cryptic fission. Proc Natl Acad Sci USA 103:714–719

Heiman Patterson TD, Bonilla E, DiMauro S, Foreman J, Schotland DL (1982) Cytochrome-c-oxidase deficiency in a floppy infant. Neurology 32:898–901

Hirano M, Lagier-Tourenne C, Valentino ML, Marti R, Nishigaki Y (2005) Thymidine phosphorylase mutations cause instability of mitochondrial DNA. Gene 354:152–156

Horvath R, Lochmuller H, Stucka R et al (2000) Characterization of human SCO1 and COX17 genes in mitochondrial cytochrome-c-oxidase deficiency. Biochem Biophys Res Comm 276:530–532

Horvath R, Kemp JP, Tuppen HAL et al (2009) Molecular basis of infantile reversible cytochrome *c* oxidase deficiency. Brain 132:3165–3174

Huigsloot M, Nijtmans LGJ, Szklarczyk R et al (2011) A mutation in *C2orf64* causes impaired cytochrome *c* oxidase assembly and mitochondrial cardiomyopathy. Am J Hum Genet 88:488–493

Huttemann M, Schmidt TR, Grossman L (2003) A third form of cytochrome *c* oxidase subunit VIII is present in mammals. Gene 312:95–102

Jaksch M, Ogilvie I, Yao J et al (2000) Mutations in SCO2 are associated with a distinct form of hypertrophic cardiomyopathy and cytochrome c oxidase deficiency. Hum Mol Genet 9:795–801

Jaksch M, Horvath R, Horn N et al (2001a) Homozygosity (E140K) in *SCO2* causes delayed infantile onset of cardiomyopathy and neuropathy. Neurology 57:1440–1446

Jaksch M, Paret C, Stucka R et al (2001b) Cytochrome *c* oxidase deficiency due to mutations in *SCO2*, encoding a mitochondrial copper-binding protein, is rescued by copper in human myoblasts. Hum Mol Genet 10:3025–3035

Karadimas CL, Greenstein P, Sue CM et al (2000) Recurrent myoglobinuria due to a nonsense mutation in the COX I gene of mtDNA. Neurology 55:644–649

Kennaway NG, Carrero-Valenzuela RD, Ewart G et al (1990) Isoforms of mammalian cytochrome *c* oxidase: correlation with human cytochrome *c* oxidase deficiency. Pediatr Res 28:529–535

Knuf M, Faber J, Huth RG, Freisinger P, Zepp F, Kampmann C (2007) Identification of a novel compound heterozygote SCO2 mutation cytochrome c oxidase deficient fatal infantile cardioencephalomyopathy. Acta Paediatr 96:128–134

Kollberg G, Moslemi A-R, Lindberg C, Holme E, Oldfors A (2005) Mitochondrial myopathy and rhabdomyolysis associated with a novel nonsense mutation in the gene encoding cytochrome c oxidase subunit I. J Neuropathol Exp Neurol 64:123–128

Leary SC, Mattman A, Wai T et al (2006) A hemizygous SCO2 mutation in an early onset rapidly progressive, fatal cardiomyopathy. Mol Genet Metab 89:129–133

Manfredi G, Schon EA, Moraes CT et al (1995) A new mutation associated with MELAS is located in a mitochondrial DNA polypeptide-coding gene. Neuromuscul Disord 5:391–398

Massa V, Fernandez-Vizarra E, Alshahwan S et al (2008) Severe infantile encephalomyopathy caused by a mutation in COX6B1, a nucleus-encoded subunit of cytochrome c oxidase. Am J Hum Genet 82:1281–1289

McFarland R, Taylor RW, Chinnery PF, Howell N, Turnbull DM (2004) A novel sporadic mutation in cytochrome *c* oxidase subunit II as a cause of rhabdomyolysis. Neuromuscul Disord 14:162–166

Menkes JH (2008) Disorders of copper metabolism: Wilson disease and Menkes disease. In: Rosenberg RN, DiMauro S, Paulson HL, Ptacek L, Nestler EJ (eds) The molecular and genetic basis of neurologic and psychiatric disease, 4th edn. Kluwer, Philadelphia, pp 721–726

Mimaki M, Hatakayama H, Komaki H et al (2010) Reversible infantile respiratory chain deficiency: A clinical and molecular study. Ann Neurol 68:845–854

Minchom PE, Dormer RL, Hughes IA et al (1983) Fatal infantile mitochondrial myopathy due to cytochrome c oxidase deficiency. J Neurol Sci 60:453–463

Mkaouar-Rebai E, Ellouze E, Chamkha I, Kammoun F, Triki C, Fakhfakh F (2011) Molecular-clinical correlation in a large family with a novel heteroplasmic Leigh syndrome missense mutation in the mitochondrial cytochrome *c* oxidase III gene. J Child Neurol 26:12–20

Mootha VK, Lepage P, Miller K et al (2003) Identification of a gene causing human cytochrome *c* oxidase deficiency by integrative genomics. Proc Natl Acad Sci USA 100:605–610

Moraes CT, Shanske S, Tritschler HJ et al (1991) MtDNA depletion with variable tissue expression: A novel genetic abnormality in mitochondrial diseases. Am J Hum Genet 48:492–501

Morin C, Mitchell G, Larochelle J et al (1993) Clinical, metabolic, and genetic aspects of cytochrome c oxidase deficiency in Saguenay-Lac-Saint-Jean. Am J Hum Genet 53:488–496

Oquendo CE, Antonicka H, Shoubridge EA, Reardon W, Brown GK (2004) Functional and genetic studies demonstrate that mutation in the COX15 gene can cause Leigh syndrome. J Med Genet 41:540–544

Papadopoulou LC, Sue CM, Davidson MM et al (1999) Fatal infantile cardioencephalomyopathy with COX deficiency and mutations in *SCO2*, a COX assembly gene. Nat Genet 23:333–337

Pequignot MO, Dey R, Zeviani M et al (2001) Mutations in the *SURF1* gene associated with Leigh syndrome and cytochrome *c* oxidase deficiency. Hum Mutat 17:374–381

Pronicki M, Kowalski P, Piekutowska-Abramczuk D et al (2010) A homozygous mutation in the SCO2 gene causes a spinal muscular atrophy like presentation with stridor and respiratory insufficiency. Eur J Paediatr Neurol 14:253–260

Rossmanith W, Freilinger M, Roka J et al (2008) Isolated cytochrome c oxidase deficiency as a cause of MELAS. J Med Genet 45:117–121

Sacconi S, Salviati L, Sue CM et al (2003) Mutation screening in patients with isolated cytochrome *c* oxidase deficiency. Pediatr Res 53:224–230

Salviati L, Sacconi S, Raslan MM et al (2001) A novel SCO2 mutation mimicking Werdig-Hoffman disease. Mitochondrion 1(Suppl 1):S82

Salviati L, Hernandez-Rosa E, Walker WF, Sacconi S, DiMauro S, Davidson MM (2002) Copper supplementation restores cytochrome *c* oxidase activity in cultured cells from patients with *SCO2* mutations. Biochem J 363:321–327

Schara U, von Kleist-Retzow J-C, Lainka E et al (2011) Acute liver failure with subsequent cirrhosis as the primary manifestation of *TRMU* mutations. J Inherit Metab Dis 34:197–201

Seeger J, Schrank B, Pyle A et al (2010) Clinical and neuropathological findings in patients with *TACO1* mutations. Neuromuscul Disord 20:720–724

Servidei S, Lazaro RP, Bonilla E, Barron KD, Zeviani M, DiMauro S (1987) Mitochondrial encephalomyopathy and partial cytochrome c oxidase deficiency. Neurology 37:58–63

Sue CM, Karadimas C, Checcarelli N et al (2000) Differential features of patients with mutations in two COX assembly genes, SURF-1 and SCO2. Ann Neurol 47:589–595

Tarnopolsky MA, Bourgeois JM, Fu M-H et al (2004) Novel SCO2 mutation (G1521A) presenting as a spinal muscular atrophy type I phenotype. Am J Med Genet 125A:310–314

Tay SKH, Shanske S, Kaplan P, DiMauro S (2004) Association of mutations in *SCO2*, a cytochrome *c* oxidase assembly gene, with early fetal lethality. Arch Neurol 61:950–952

Tiranti V, Hoertnagel K, Carrozzo R et al (1998) Mutations of SURF-1 in Leigh disease associated with cytochrome c oxidase deficiency. Am J Hum Genet 63:1609–1621

Tiranti V, Visconti C, Hildebrandt T et al (2009) Loss of ETHE1, a mitochondrial dioxygenase, causes fatal sulfide toxicity in ethylmalonic encephalopathy. Nat Med 15:200–205

Uusimaa J, Finnila S, Vainionpaa L et al (2003) A mutation in mitochondrial DNA-encoded cytochrome *c* oxidase II gene in a child with Alpers-Huttenlocher-like disease. Pediatrics 111:e262–e268

Uusimaa J, Jungbluth H, Fratter C et al (2011) Reversible infantile respiratory chain deficiency is a unique, genetically heterogeneous mitochondrial disease. J Med Genet 48:660–668

Valnot I, Osmond S, Gigarel N et al (2000a) Mutations of the SCO1 gene in mitochondrial cytochrome c oxidase deficiency with neonatal-onset hepatic failure and encephalopathy. Am J Hum Genet 67:1104–1109

Valnot I, von Kleist-Retzow J-C, Barrientos A et al (2000b) A mutation in the human heme-A:farnesyltransferase gene (COX 10) causes cytochrome c oxidase deficiency. Hum Mol Genet 9:1245–1249

Van Biervliet JPM, Bruinvis L, Ketting D et al (1977) Hereditary mitochondrial myopathy with lactic acidemia, a DeToni-Fanconi-Debré syndrome, and a defective respiratory chain in voluntary striated muscles. Pediatr Res 11:1088–1093

Van Coster R, Lombes A, DeVivo DC et al (1991) Cytochrome c oxidase-associated Leigh syndrome: phenotypic features and pathogenetic speculations. J Neurol Sci 104:97–111

Vesela K, Hulkova H, Hansikova H, Zeman J, Elleder M (2008) Structural analysis of tissues affected by cytochrome *c* oxidase deficiency due to mutations in the *SCO2* gene. Acta Pathol Microbiol Immunol Scand 116:41–49

Weraarpachai W, Antonicka H, Sasarman F et al (2009) Mutation in TACO1, encoding a translational activator of COX I, results in cytochrome *c* oxidase deficiency and late-onset Leigh syndrome. Nat Genet 41:833837

Weraarpachai W, Sasarman F, Nishimura T et al (2012) Mutations in C12orf62, a factor that couples COX I synthesis with cytochrome c oxidase assembly, cause fatal neonatal lactic acidosis. Am J Hum Genet 90:142–151

Willems JL, Monnens L, Trijbels J et al (1977) Leigh's encephalomyelopathy in a patient with cytochrome c oxidase deficiency in muscle tissue. Pediatrics 60:850–857

Williams SL, Valnot I, Rustin P, Taanman J-W (2004) Cytochrome c oxidase subassemblies in fibroblast cultures from patients carrying mutations in COX 10, SCO1, or SURF1. J Biol Chem 279:7462–7469

Wong LJ, Dai P, Tan D et al (2001) Severe lactic acidosis caused by a novel frame-shift mutation in mitochondrial-encoded cytochrome c oxidase subunit II. Am J Med Genet 102:95–99

Xu F, Morin C, Mitchell G, Ackerley C, Robinson BH (2004) The role of LRPPRC (leucine-rich pentatricopeptide repeat cassette) gene in cytochrome oxidase assembly: mutation causes lowered levels of COX (cytochrome c oxidase) I and COX III mRNA. Biochem J 382:331–336

Zeviani M, Nonaka I, Bonilla E et al (1985) Fatal infantile mitochondrial myopathy and renal dysfunction caused by cytochrome c oxidase deficiency: immunological studies in a new patient. Ann Neurol 17:414–417

Zeviani M, Van Dyke DH, Servidei S et al (1986) Myopathy and fatal cardiopathy due to cytochrome c oxidase deficiency. Arch Neurol 43:1198–1202

Zhu Z, Yao J, Johns T et al (1998) SURF1, encoding a factor involved in the biogenesis of cytochrome c oxidase, is mutated in Leigh syndrome. Nat Genet 20:337–343

Abbreviations

$\Delta\Psi_m$	Mitochondrial membrane potential
Δp_m, Δp	Proton motive force
A9-DA	Dopaminergic neurons A9-subtype of the substantia nigra pars compacta
AD	Alzheimer disease
ALS	Amyotrophic lateral sclerosis
AMPA	α-Amino-3-hydroxyl-5-methyl-4-isoxazolpropionic acid
Amplex Red/HRP	Amplex Red/horse raddish peroxidase
AOX	Alternative oxidase
ATP	Adenosine triphosphate
Aβ	Amyloidogenic peptide Aβ
BCS1L	Complex III assembly factor
BN-PAGE	Blue native polyacrylamide gel electrophoresis
cAMP	Cyclic adenosine monophosphate
CDNB	1-Chloro-2,4-dinitrobenzene
ChIP	Chromatin immunoprecipitation
CMC1	COX assembly factor
CMT2A	Charcot–Marie–Tooth disease type 2A
CNS	Central nervous system
Complex I = CI	NADH-CoQ oxidoreductase or NADH-ubiquinone oxidoreductase

B. Kadenbach (ed.), *Mitochondrial Oxidative Phosphorylation*, Advances in Experimental Medicine and Biology 748, DOI 10.1007/978-1-4614-3573-0,

Complex II = CII	Succinate-CoQ oxidoreductase or succinate-ubiquinone oxidoreductase
Complex III = CIII	CoQ-cytochrome c oxidoreductase or cytochrome bc_1 complex
Complex IV = CIV	Cytochrome c oxidase
Complex V = CV	ATP synthase (F_0F_1-ATP synthase)
CoQ	Coenzyme Q or ubiquinone
COS7	CV-1 in origin, and carrying the SV40 genetic material cell line
COX	Cytochrome *c* oxidase (complex IV)
COX10	Protein involved in biosynthesis of heme A
COX11,COX12, COX13, COX14	COX assembly factors
COX15	Protein involved in biosynthesis of heme A
COX16, COX17, COX18, COX19, COX20	COX assembly factors
Cytc	Cytochrome c
DBH	Decylubiquinol
DCU	Dicyclohexylurea
DIGE	Difference gel electrophoresis
DJ-1	Cytosolic chaperone protein that translocates to mitochondria in response to oxidative stress
DPI	Diphenyleneiodonium
DQA	2-*n*-Decyl-quinazolin-4-yl-amine (SAN 549)
Drp1	Dynamin related protein 1
EE	Ethylmalonic encephalopathy
EGFR	Epidermal growth factor receptor
ELISA	Enzyme-linked immunosorbent assay
EM	Electron microscopy
EMSA	Electrophoretic mobility shift
EPR	Electron paramagnetic resonance
ER	Estrogen receptor
ERRa	Estrogen-related receptor
ETF	Electron transferring flavoprotein
ETHE1	Mitochondrial matrix thioesterase
FAD	Flavin-adenine-dinucleotide
FAO	Fatty acid β-oxidation
FASTKD2	Fas-activated serine-threonine kinase domain 2
FMN	Flavin-mononucleotide
GFP	Green fluorescent protein

GPx	Glutathione peroxidase
GR	Glucocorticoid receptor
GSH	Glutathione
HD	Huntington's disease
HDAC6	Histone deacetylase 6
hFis1	Mitochondrial fission protein 1 (human homologue)
HIF-1α	Hypoxia inducible factor 1 alpha
IBMX	3-Isobutyl-1-methylxanthine
IMM	Inner mitochondrial membrane
IMS	Mitochondrial intermembrane space
INS1	Insulinoma cell line 1
ISCS	Cysteine desulfurase
ISCU	Scaffold proteins for the biosynthesis of Fe-S clusters
ITS	IMS targeting signal
K48	Lysine 48-linked ubiquitin
K63	Lysine 63-linked ubiquitin
Ka/Ks ratio	Ratio of non-synonymous over synonymous base replacements in DNA
KIF17	Kinesin superfamily protein
LRPPRC	Leucine-rich pentatricopeptide repeat cassette
LS	Leigh syndrome
LSFC	French Canadian type Leigh syndrome
MAPK	p38 mitogen-activated protein kinase
MELAS	Mitochondrial encephalmyopathy with lactic acidosis and stroke-like episodes
Mff	Mitochondrial fission factor
Mfn1, Mfn2	Mitofusin 1, mitofusin 2
Miro	Mitochondrial Rho GTPase
MISS	Mitochondrial intermembrane space sorting
mNFU1	Scaffold proteins for the biosynthesis of Fe–S clusters
MNGIE	Mitochondrial neurogastrointestinal encephalomyopathy
MNGIE	Myoneurogastrointestinal encephalopathy
MODS	Multiple organ dysfunction syndrome
MPP	Mitochondrial processing peptide

MPP^+	1-Methyl-4-phenylpyridinium
MPTP	1-methyl-4-phenyl-1,2,3,6-tetrahydropyridine
MRC	Mitochondrial respiratory chain
MRP	Mitochondrial RNA processing
mtDNA	Mitochondrial DNA
mTERF	Mitochondrial transcription termination factor
mtPA-GFP	Mitochondrial matrix-targeted photoactivatable GFP
mtPTP (PTP)	Mitochondrial permeability transition pore
mV	Millivolt
nDNA (ncDNA)	Nuclear DNA
NMDA	N-methyl-d-aspartic acid
NO	Nitric oxide
NOS	NO synthase, nNOS = neuronal, iNOS = in macrophages, eNOS = endothelial, mtNOS = mitochondrial
NPA	3-Nitropropionic acid
NRF-1, NRF-2	Nuclear respiratory factors 1 and 2
6-OHDA	6-Hydroxydopamine
OMA1	Metalloendopeptidase of the mitochondrial inner membrane
OPA1	Optic atrophy 1
OXA1, Oxa 1	Oxidase assembly mutant 1
OxPhos (OXPHOS)	Oxidative phosphorylation
PAGE	Polyacrylamide gel electrophoresis
PARL	Presenilin associated, rhomboid-like (mitochondrial intramembrane cleaving protease)
Pcp 1	Membrane-embedded rhomboid protease
PCR	Polymerase chain reaction
PD	Parkinson's Disease
PDGF	Platelet-derived growth factor
PDH	Pyruvate dehydrogenase
PDK	Pyruvate dehydrogenase kinase
PEG	Polyethylene glycol
PEO	Progressive external ophthalmoplegia
PET191	Protein involved in the assembly of COX

PGC-1or PGC-1α	Peroxisome proliferator-activated receptor gamma coactivator 1
PINK1	PTEN induced kinase 1
PKA	Protein kinase A
PKC	Protein kinase C
PPARα,	PPARγPeroxisome proliferator-activated receptor alpha and gamma
Prx3, Prx5	Mitochondrial peroxiredoxins 3 and 5
PTP (mtPTP)	Mitochondrial permeability transition pore
RET	Reverse electron transfer
RFP	Red fluorescent protein
RONS	Reactive oxygen and nitrogen species
ROS	Reactive oxygen species
RRF	Ragged-red fibers
SBMA	Spinobulbar muscular atrophy
SCO1, SCO2	Copper-binding COX assembly factors (proteins)
SDH	Succinate dehydrogenase
SDHAF1	SDH assembly factor 1
SDS-PAGE	Sodium dodecylsulfate-polyacrylamide gel electrophoresis
Shp-2	Protein tyrosine phosphatase
shRNA	Small hairpin interference RNA
SMA	Spinal muscular atrophy
SMP	Submitochondrial particles
SNE	Subacute necrotizing encephalomyelopathy = LS (Leigh syndrome)
SOD1 (Cu/Zn-SOD)	Superoxide dismutase 1
SOD2 (Mn-SOD)	Superoxide dismutase 2
SURF1	Surfeit locus protein 1, COX and complex III assembly protein
T2	3,5-Diiodo-L-thyronine
T3	3,3′,5-Triiodo-L-thyronine
TACO1	Translational activator of COX I (COX subunit I)
TCA cycle	Tricarboxylic acid cycle, also known as Krebs cycle
TFAM, TFB1M, TFB2M	Mitochondrial transcription factors A and B
TGF-β	Transforming growth factor β

TIM	Translocase of the inner mitochondrial membrane
TMRE	Tetramethylrhodamine ethyl ester perchlorate
TNFα	Tumor necrosis factor α
TOM	Translocase of the outer mitochondrial membrane
UQ	Ubiquinone or Coenzyme Q
UCP	Uncoupling protein
YFP	Yellow fluorescent protein
Yme1L	ATP-dependant metalloprotease YME1L1
YY1	Ying Yang 1

Glossary

Allosteric ATP-inhibition Sigmoidal inhibition kinetics of cytochrome c oxidase activity when oxygen consumption is measured at increasing cytochrome c concentrations

Apoptosis Programmed cell death

Autophagy Degradation of a cell's own components through the lysosomal machinery

Complementation Restoration of membrane potential and function in a depolarized mitochondrion after fusion with a more polarized unit

Fission Separation of one mitochondrion into two

Fragmentation Conversion of a highly branched network to solitary minimal units

Fusion Merging of membranes by two mitochondria

Metabolic control analysis Analysis of the overall flux of a metabolic pathway on the properties (activitiy, concentration) of individual components

Mitochondrial disorders Genetic defects of oxidative phosphorylation

Mitochondrial dynamics Balance of fusion and fission

Mitochondrial lifecycle Mitochondria existing within a cell in a constant state of flux between biogenesis, fused network, and solitary unit states, and eventual removal by autophagic degradation

Mitochondrial turnover Balance of mitochondrial biogenesis and degradation.

Mitophagy Autophagy of mitochondria

Motility Transport of mitochondria along the microtubule system of the cell

Network Refers to a collection of mitochondrial units with united membranes

Non-synonymous DNA codon substitution not resulting in replacement of the encoded amino acid

Quality control Quality control is upkeep of function by selective removal of damaged components

Respiratory control Control of mitochondrial respiration by the availability of ADP

Synonymous DNA codon substitution resulting in replacement of the encoded amino acid

Uncoupling of oxidative phosphorylation Dissipation of the mitochondrial membrane potential

B. Kadenbach (ed.), *Mitochondrial Oxidative Phosphorylation*, Advances in Experimental Medicine and Biology 748, DOI 10.1007/978-1-4614-3573-0,

Index

B. Kadenbach (ed.), *Mitochondrial Oxidative Phosphorylation*, Advances in Experimental Medicine and Biology 748, DOI 10.1007/978-1-4614-3573-0,